AF826943

Nanotechnology in Biology and Medicine

Research Advancements and Future Perspectives

Nanotechnology in Biology and Medicine

Research Advancements and Future Perspectives

Edited by
Pradipta Ranjan Rauta, Yugal Kishore Mohanta, and Debasis Nayak

CRC Press
Taylor & Francis Group
Boca Raton London New York

CRC Press is an imprint of the
Taylor & Francis Group, an **informa** business

CRC Press
Taylor & Francis Group
6000 Broken Sound Parkway NW, Suite 300
Boca Raton, FL 33487-2742

© 2020 by Taylor & Francis Group, LLC
CRC Press is an imprint of Taylor & Francis Group, an Informa business

No claim to original U.S. Government works

Printed on acid-free paper

International Standard Book Number-13: 978-0-367-20050-3 (Hardback)

This book contains information obtained from authentic and highly regarded sources. Reasonable efforts have been made to publish reliable data and information, but the author and publisher cannot assume responsibility for the validity of all materials or the consequences of their use. The authors and publishers have attempted to trace the copyright holders of all material reproduced in this publication and apologize to copyright holders if permission to publish in this form has not been obtained. If any copyright material has not been acknowledged please write and let us know so we may rectify in any future reprint.

Except as permitted under U.S. Copyright Law, no part of this book may be reprinted, reproduced, transmitted, or utilized in any form by any electronic, mechanical, or other means, now known or hereafter invented, including photocopying, microfilming, and recording, or in any information storage or retrieval system, without written permission from the publishers.

For permission to photocopy or use material electronically from this work, please access www.copyright.com (http://www.copyright.com/) or contact the Copyright Clearance Center, Inc. (CCC), 222 Rosewood Drive, Danvers, MA 01923, 978-750-8400. CCC is a not-for-profit organization that provides licenses and registration for a variety of users. For organizations that have been granted a photocopy license by the CCC, a separate system of payment has been arranged.

Trademark Notice: Product or corporate names may be trademarks or registered trademarks, and are used only for identification and explanation without intent to infringe.

Visit the Taylor & Francis Web site at
http://www.taylorandfrancis.com

and the CRC Press Web site at
http://www.crcpress.com

Contents

Preface ix
Acknowledgments xi
Editors xiii
Contributors xv

SECTION 1 Nanoscale Materials: Design, Manufacture, and Physicochemical Properties

Chapter 1 Polymeric Nanostructures: Their Synthesis Approaches and Morphological and Physicochemical Characterizations 3

Debasis Nayak and Pradipta Ranjan Rauta

Chapter 2 Synthesis, Characterization, and Applications of Silica Nanomaterials from a Nanobiotechnological Perspective 11

Chityal Ganesh Kumar, Kanugala Sirisha, and Puvvada Naga Prasad

Chapter 3 Assessing Interactions of Microalgae in Synthesizing Nanoparticles 29

Manoj Kumar Enamala, Azzuliani Binti Supangat, Chandrasekhar Kuppam, Sudhakar Reddy Pamanji, Murthy Chavali, and Maria P. Nikolova

Chapter 4 Liposome Nanocarrier: Synthesis, Characterization, and Applications 45

Prabhu Sukumaran and V. Sandhya

SECTION 2 Nano Biotechnology Applications on Food and Agriculture

Chapter 5 New Insights into the Pharmaceutical and Food-Based Applications of Starch Nanocrystals 63

Neethu Hari, Supriya Radhakrishnan, Athira S. Dev, and Ananthakrishnan Jayakumaran Nair

Chapter 6 Recent Developments on Nanotechnology in Agriculture: Challenges and Prospects 79

Yugal Kishore Mohanta, Tapan Kumar Mohanta, and Debasis Nayak

Chapter 7 An Overview of Mycogenic Nanoparticles: Synthesis and Their Bio-Prospective Applications 87

Sagarika Mohanta, Sameer Kumar Singdevsachan, Yugal Kishore Mohanta, Sujogya Kumar Panda, and Tapan Kumar Mohanta

Chapter 8 Nanonutraceutical Delivery Systems: Application in Food and Feed Industry 99

Visha Pasuvalingam

SECTION 3 Nano Biotechnology in Health Care/Therapeutic Applications

Chapter 9 Nanomedicine for Wound Healing: Challenges and Future Perspectives 115

Sourav Das, Papia Basu Thakur, Aparna Harri, Manali Bagade, and Chitta Ranjan Patra

Chapter 10 Recent Advancements on Nano-Based Drug Delivery System for Targeted Cancer Therapy 133

Arindam Pramanik

Chapter 11 Recent Nanotechnological Advancements in Delivery of Peptide and Protein Macromolecules 143

Sunita Dahiya and Rajiv Dahiya

Chapter 12 Nanotechnology: The Hope for New Generation and Its Therapeutic Applications 159

Dnyaneshwar Rathod, Avinash Kharat, Varsha Wankhade, and Mahendra Rai

Chapter 13 Origami-Engineered Structures and Their Biomedical Applications 167

Dipti Mohanta, Snehaprava Mohapatra, Sonali Khuntia, Suchilagna Sahoo, and Pradipta Ranjan Rauta

SECTION 4 Biosensors/Nanochip/Nanodevice/Nanocomputing

Chapter 14 Nanobiosensors and Their Biomedical Applications 177

Pradipta Ranjan Rauta, Debasis Nayak, and Pankaj Kumar Singh

Chapter 15 Theranostic Advances of Graphene-Based Nanomaterial in Diabetes 185

Kunal Biswas, Siddharth Shaw, Sudhanshu Mishra, Debashis De, and Jaya Bandyopadhyay

Chapter 16 Role of Gold Nanoparticles (GNPs) in Cancer Diagnosis and Treatment 193

Kalpana Panati, Parasuraman Aiya Subramani, Dharaneeswara Reddy Devireddy, and Venkata Ramireddy Narala

Chapter 17 Graphene: A Revolutionary 2D Nanomaterial Heralding a New Horizon for Diagnostics and Therapeutic Applications 205

Kunal Biswas, Debashis De, and Jaya Bandyopadhyay

Chapter 18 Applications of Nanosensor System in the Detection of Heavy Metals 215

Arun Kumar Pradhan and Soudamini Acharya

SECTION 5 Nanobiotecnology: Health Safety Toxicity and Environmental Impact

Chapter 19 Epigenetic Landscape–Modifying Nanoparticles 227

Dipta Sengupta and Debasis Nayak

Chapter 20 Nanotoxicology: Evaluating Toxic Potential of Drug Nanoparticles 235

Pavani Gonnabathula, Nageswararao Mekala, and Praveen Kumar Relangi

Chapter 21 Nanobioremediation: A Novel and Sustainable Biological Advancement for Ecological Cleanup 245

Manoj Kumar Enamala, P. Divya Sruthi, Silpi Sarkar, Murthy Chavali, Induri Vasavi, and Chandrashekar Kuppam

Chapter 22 Nanotoxicity: Physicochemical and Biological (PK-PD) Perspectives 259

Mithun Rudrapal, Dipak Chetia, Khomendra Kumar Sarwa, and Soumya Bhattacharya

Index 269

Preface

> I seem to have been only like a boy playing on the seashore, and diverting myself in now and then finding a smoother pebble or a prettier shell than ordinary, whilst the great ocean of truth lay all undiscovered before me.
>
> **–Isaac Newton**

This quote by the famous scientist and philosopher Isaac Newton always inspires us to acquire more and more knowledge in the ocean of science. It is in this pursuit of knowledge that this work was conceptualized.

Nanobiotechnology, also termed bionanotechnology, is an interface between two emerging fields of the 21st century, i.e., nanotechnology and biotechnology. It communicates scientific and technological advances in the field of biotechnology, with an emphasis in their interface with nanosciences. This deals with designing, manufacture, investigation, and exploitation of systems at the nanoscale, whereas biotechnology gives us a way to understand biological systems and to utilize our knowledge for industrial build-up. Nanotechnology is a multidisciplinary field that currently recruits the strategies, technologies, and facilities available in conventional and advanced avenues of engineering, physics, chemistry, and biology.

Nanotechnology is foreseen to be next big revolution in medicine and biology. For the past 20 years, much research has been carried out for the development of new and innovative ways of exploring/using novel nanomaterials for biotechnological applications. Nanomaterials are also becoming progressively more important in medical and healthcare applications, with the development of new drugs and diagnostic tools based on nanotechnology. Every year (particularly in the last 2 decades), hundreds of innovative ideas using nanomaterials are applied in biotechnological applications. Moreover, increasing numbers of new enterprises are also searching for market opportunities using these technologies. This book serves as an interdisciplinary platform presenting state-of-the-art, multidisciplinary research and technological advances in theory, instrumentations, and methods as well as applications in a variety of areas of nanotechnology related to food, agriculture, and medicine. It presents the basics, fundamental results/current applications, latest achievements of nanobiotechnological research, and future prospective of nanobiotechnology. Just as the progress of the last two decades has been vast, so a comprehensive update is highly required to bridge the gap and put forward the pros and cons to establish the methods and principles for real use by human beings. Based on broad surveys of this field, this book has been designed from the combined contributions from bio-organic and bio-inorganic chemistry, molecular biology, materials science, and bioanalytics to reflect the full scope of current and future developments. This is the most updated version with wide-ranging references to the last 20 years of nano research.

Overall, the book is focused on providing an overview of the developments made in nanotechnology to date, along with ongoing trends and future prospects. This book presents the basics, fundamental results, and latest achievements of nanobiotechnological research in the worldwide scientific arena. Also, this book extends greater promising applications of novel nanomaterials with newly discovered nanoeffects. Besides these, the fundamentals of nanotechnology are covered along with several specific bionanotechnology applications including nanobioimaging, drug delivery, food, and agriculture which have now rapidly grown into a more than US$100 billion industry. Additionally, a large variety of nano-bio-analytical methods are presented as a core asset to the young budding nanotechnologist as well as to entrepreneurs. Besides these, one of the major goals of this book is to highlight the multifaceted issues surrounding of nanotechnology on the basis of case studies, academic and theoretical articles, technology transfer (patents and copyrights), innovation, economics, and policy management. More than fifty experts with their in-depth understanding, perspectives, and resources have provided a very broad audience with an overview of the latest developments in the field of nanobiotechnology. This book has been designed for scientists, academicians, students, and entrepreneurs engaged in nanotechnology research and development. Nevertheless, it should be of interest to those in a variety of scientific disciplines including food and agricultural science, genetics, pharmacology, medicine, and consumer products.

This book is classified into 5 sections consisting of 22 chapters that are dedicated to the new developments and prospects of nanotechnology in biology and medicine. **Section 1** provides an inclusive overview of nanoscale materials: design, manufacture, and physicochemical properties. **Section 2** throws light onto the prospects of nanobiotechnology, which have certainly impacted on the food and agriculture sector. **Section 3** is a synopsis of the recent advances in health care and therapeutic applications with the help of nanobiotechnology and also highlights future perspectives with pros and cons. **Section 4** gives an overview of the most promising nanomaterials, biosensors, nanodevices, and nanotheranostics along with their applications in medical diagnostics and therapy. **Section 5** is very important in the current scenario, as it discuss the toxicity and environmental risks of nanomaterials along with various challenges and future requirements for safe environmental nanotechnology.

Pradipta Ranjan Rauta

Yugal Kishore Mohanta

Debasis Nayak

Acknowledgments

We take this opportunity to express our indebtedness, deep gratitude, and sincere thanks to all our contributing authors who have worked diligently on their tasks and produced chapters of high excellence.

On a very personal note, we are very much thankful to Ms Renu Upadhyay (Commissioning Editor, CRC Press) and Ms Shikha Garg (Editorial Assistant, CRC Press) for their continuous editorial support and suggestions in shaping the book in the current form.

We would also like to extend special thanks to all our teachers, mentors, family members, colleagues, friends, and well-wishers for their constant help, motivation, and encouragement in completing this endeavor.

Pradipta Ranjan Rauta

Yugal Kishore Mohanta

Debasis Nayak

Editors

Pradipta Ranjan Rauta is a postdoctoral research fellow at the Center for Radiation Oncology Research, The University of Texas MD Anderson Cancer Center, Houston, Texas. He earned his PhD in Life Science from the National Institute of Technology, Rourkela, Odisha, India; MSc. in Microbiology from the Orissa University of Agriculture and Technology, Bhubaneswar, Odisha, India; and BSc in Zoology from the Utkal University, Bhubaneswar, Orissa, India. He has also worked as a PhD mobility fellow at Instituto Superior Técnico, Lisbon, Portugal during his PhD through the Erasmus Mundus fellowship program. Dr Rauta has more than 10 years of research experience in various fields, e.g., microbiology, immunology, molecular biology, nanotechnology, and cell biology. He has published 23 peer-reviewed journal articles and 2 popular articles, and has presented at many international conferences. He has expertise on nanobiomedical research in the following areas: nanotechnology, cell biology, proteomics, molecular biology, immunology, in vivo study, bio-techniques (instrumentation), etc. His research areas are mainly involved in developing smart delivery systems for anticancer drugs. Specifically, he has developed multifunctional nanoplatforms comprising a biocompatible polymer, an anticancer drug, and cancer-specific ligands that specifically and efficiently target only cancer cells, thereby significantly reducing systemic toxicity in response to chemotherapy. This involves, but is not limited to, fabrication of polymeric nanoparticles, effective drug encapsulation, controlled drug release, functionalization with cancer cell-specific ligands for specific targeting, and achieving enhanced therapeutic efficacy of drugs without systemic toxicity. His current research is focused on designing novel multifunctional gold-based nanotheranostics for advanced cancer. He has actively contributed to all phases of the research project such as, concept development and design, wet-lab experiments, data analysis, and manuscript (and grant) preparation. He is associated with many scientific societies as a fellow member, such as the Indian Science Congress Association, the American Society for Microbiology, etc. He has reviewed many research articles for international journals from Elsevier, Springer, etc. He also serves as review editor for international journals such as *Frontiers in Bioengineering and Biotechnology* and *Frontiers in Molecular Biosciences*.

Yugal Kishore Mohanta works as a visiting researcher at Natural and Medical Sciences Research Center (NMSRC), University of Nizwa, Oman. He has received his PhD in Life Sciences from North Orissa University, Baripada, India. He has also successfully completed the BCIL industrial training program on plant tissue culture sponsored by the Department of Biotechnology, Government of India, New Delhi. Among other academic/research achievements, he was awarded as visiting research scholar fellowship at the Department of Biology, KU Leuven, Belgium sponsored by KU Leuven. Dr Mohanta has more than 9 years of post-MSc research experience on natural products chemistry, nanotechnology, microbiology, and plant biotechnology. He has published 21 research articles, 8 book chapters, and 5 popular science articles, and has presented at many national and international conferences. He is currently working on the application of nanotechnology on different aspects such as food safety, biomedicine, and conservation of natural resources. Besides such scientific activities, Dr Mohanta is passionately involved in wildlife photography and nature conservation. Currently, he works as a volunteer for movements like **"WeR4River"** and **"Youth4River"** for the conservation of rivers and water resources. He also serves as a reviewer for many international journals from Elsevier, Springer, etc.

Debasis Nayak is an assistant professor and Head of the Department of Zoology, Seemanta Mahavidyalaya, Odisha, India. Dr Nayak has earned his PhD in Life Science, National Institute of Technology Rourkela, Odisha, India, MSc in Life Science from the National Institute of Technology Rourkela, Odisha, India, and BSc with Zoology (Honors) from Government Autonomous College, Rourkela, Odisha, India. Dr Nayak has 7 years research experience in nanotechnology, microbiology, immunology, and molecular biology. He has published 15 research articles in peer-reviewed international journals and has attended and presented at many international conferences. Dr Nayak has expertise in synthesis of metallic nanoparticles through the green synthesis route, and preparation of novel drug delivery carriers using gelatin, carrageenan, chitosan, and various natural and synthetic polymers for oral, transdermal, and parenteral drug delivery. He has wide expertise on handling high-end instruments such as scanning electron microscope, atomic force microscope, XRD, FTIR, flow cytometer, and fluorescence microscope. His research is mainly focused on the development of cost-effective and novel therapeutics using indigenous materials as anticancer agents by inducing oxidative stress and mitochondrial damage inside the specific cancer cell without affecting the neighboring normal cells or producing any additional side effects. Dr Nayak has reviewed many research articles for international journals, and his current research interests are on developing topical gels and ointments containing novel indigenous phytochemicals and organic and inorganic nanoparticles against various forms of cutaneous and malignant skin carcinomas.

Contributors

Soudamini Acharya
Centre for Biotechnology
Siksha 'O' Anusandhan University
Odisha, India

Manali Bagade
Department of Applied Biology
CSIR-Indian Institute of Chemical Technology
Hyderabad, India

Jaya Bandyopadhyay
Department of Biotechnology
Maulana Abul Kalam Azad University of Technology
West Bengal, India

Soumya Bhattacharya
Guru Nanak Institute of Pharmaceutical Science and Technology, Kolkata
West Bengal, India

Kunal Biswas
Department of Biotechnology
Maulana Abul Kalam Azad University of Technology
West Bengal, India

Murthy Chavali
MCETRC
Tenali, Guntur
and
Shree Velagapudi Ramakrishna Memorial College (SVRMC-PG Studies)
Andhra Pradesh, India

Dipak Chetia
Department of Pharmaceutical Sciences
Dibrugarh University
Dibrugarh, India

Rajiv Dahiya
School of Pharmacy, Faculty of Medical Sciences
The University of the West Indies
Trinidad and Tobago, West Indies

Sunita Dahiya
School of Pharmacy, Medical Sciences Campus
University of Puerto Rico
San Juan, Puerto Rico

Sourav Das
Department of Applied Biology
CSIR-Indian Institute of Chemical Technology
Hyderabad, India
and
Academy of Scientific and Innovative Research (AcSIR)
Uttar Pradesh, India

Debashis De
Department of Computer Science and Engineering
Maulana Abul Kalam Azad University of Technology
West Bengal, India

Athira S. Dev
Department of Biotechnology
University of Kerala
Thiruvananthapuram, India

Dharaneeswara Reddy Devireddy
Micco Laboratories Private Limited
Tirupati, India

Manoj Kumar Enamala
Bioserve Biotechnologies Pvt. Ltd. Hyderabad
Telangana, India

Pavani Gonnabathula
College of Science, Engineering, and Technology
Texas Southern University
Houston, Texas

Neethu Hari
Department of Biotechnology
University of Kerala
Thiruvananthapuram, India

Aparna Harri
Department of Applied Biology
CSIR-Indian Institute of Chemical Technology
Hyderabad, India

Ananthakrishnan Jayakumaran Nair
Department of Biotechnology
University of Kerala
Thiruvananthapuram, India

Avinash Kharat
Regenerative Medicine Laboratory
Dr. D. Y. Patil Dental College and Hospital
Pune, India

Sonali Khuntia
District Institute of Education and Training, Puri
Government of Odisha
Odisha, India

Chityal Ganesh Kumar
Medicinal Chemistry and Biotechnology Division
CSIR-Indian Institute of Chemical Technology
Hyderabad, India

Chandrashekar Kuppam
School of Applied Biosciences
Kyungpook National University
Bukgu, South Korea

Nageswararao Mekala
Department of Engineering Chemistry
Andhra University
Visakhapatnam, India

Shudhanshu Mishra
Department of Biotechnology
Maulana Abul Kalam Azad University of Technology
West Bengal, India

Dipti Mohanta
District Institute of Education and Training, Puri
Government of Odisha
Odisha, India

Sagarika Mohanta
Department of Botany
Berhampur University
Odisha, India

Tapan Kumar Mohanta
Natural Products and Medical Sciences Research Center
University of Nizwa
Nizwa, Oman

Yugal Kishore Mohanta
Department of Botany
North Orissa University
Odisha, India

Snehaprava Mohapatra
District Institute of Education and Training, Puri
Government of Odisha
Odisha, India

Venkata Ramireddy Narala
Department of Zoology
Yogi Vemana University
Kadapa, India

Debasis Nayak
Department of Zoology
Seemanta Mahavidyalaya
Odisha, India

Maria P. Nikolova
Department of Material Science and Technology
University of Ruse
Ruse, Bulgaria

Sudhakar Reddy Pamanji
Department of Zoology
Vikrama Simhapuri University
Andhra Pradesh, India

Kalpana Panati
Department of Biotechnology
Government College for Men
Kadapa, India

Sujogya Kumar Panda
Department of Biology
KU Leuven
Leuven, Belgium

Visha Pasuvalingam
Department of Veterinary Physiology and Biochemistry
Veterinary College and Research Institute
Tamil Nadu, India

Chitta Ranjan Patra
Department of Applied Biology
CSIR-Indian Institute of Chemical Technology
Hyderabad, India
and
Academy of Scientific and Innovative Research (AcSIR)
Uttar Pradesh, India

Arun Kumar Pradhan
Centre for Biotechnology
Siksha 'O' Anusandhan University
Odisha, India

Arindam Pramanik
School of Biomedical Science
University of Leeds
Leeds, United Kingdom

Puvvada Naga Prasad
Applied Biology Division
CSIR-Indian Institute of Chemical Technology
Hyderabad, India

Supriya Radhakrishnan
Department of Biotechnology
University of Kerala
Thiruvananthapuram, India

Mahendra Rai
Department of Biotechnology
Sant Gadge Baba Amravati University
Maharashtra, India

Dnyaneshwar Rathod
Department of Zoology
Savitribai Phule Pune University
Maharashtra, India

Pradipta Ranjan Rauta
Radiation Oncology Research
The University of Texas M. D. Anderson Cancer Center
Houston, Texas

Praveen Kumar Relangi
College of Science, Engineering, and Technology
Texas Southern University
Houston, Texas

Mithun Rudrapal
Department of Pharmaceutical Sciences
Dibrugarh University
Dibrugarh, India

Suchilagna Sahoo
District Institute of Education and Training, Puri
Government of Odisha
Odisha, India

V. Sandhya
Department of Biotechnology
Sri Venkateswara College of Engineering
Chennai, India

Silpi Sarkar
Department of Biotechnology
VFSTR University
Andhra Pradesh, India

Khomendra Kumar Sarwa
Government Girls' Polytechnic
Raipur, India

Dipta Sengupta
Institute of Genetics and Molecular Medicine
University of Edinburgh
Edinburgh, United Kingdom

Siddharth Shaw
Department of Biotechnology
Maulana Abul Kalam Azad University of Technology
West Bengal, India

Sameer Kumar Singdevsachan
Process and Analytical Science Department
Biocon Research Limited-SEZ Unit
Bangalore, India

Pankaj Kumar Singh
Center for Radiation Oncology Research
The University of Texas M.D. Anderson Cancer Center
Houston, Texas

Kanugala Sirisha
Medicinal Chemistry and Biotechnology Division
CSIR-Indian Institute of Chemical Technology
Hyderabad, India

P. Divya Sruthi
ITC Spices (ITC Agri Business Division)
Guntur, India

Parasuraman Aiya Subramani
School of Life Sciences
Vels Institute of Science, Technology and Advanced Studies
Chennai, India

Prabhu Sukumaran
Department of Biotechnology
Sri Venkateswara College of Engineering
Chennai, India

Azzuliani Binti Supangat
Department of Physics
University of Malaya
Kuala Lumpur, Malaysia

Papia Basu Thakur
Department of Applied Biology
CSIR-Indian Institute of Chemical Technology
Hyderabad, India

Induri Vasavi
Department of Clinical Genomics and Bioinformatics
Positive Bioscience
Mumbai, India

Varsha Wankhade
Department of Zoology
Savitribai Phule Pune University
Pune, India

Section 1

Nanoscale Materials: Design, Manufacture, and Physicochemical Properties

1 Polymeric Nanostructures
Their Synthesis Approaches and Morphological and Physicochemical Characterizations

Debasis Nayak and Pradipta Ranjan Rauta

CONTENTS

1.1 Introduction 3
1.2 Types of Polymeric Nanoparticles 4
1.2.1 Natural Polymeric Nanoparticles 4
1.2.2 Synthetic Polymeric Nanoparticles 4
1.3 Methods of Synthesis of Polymeric Nanoparticles 5
1.3.1 Solvent Evaporation 5
1.3.2 Nanoprecipitation 5
1.3.3 Emulsification–Solvent Diffusion 5
1.3.4 Salting Out 5
1.3.5 Dialysis 5
1.3.6 Supercritical Fluid Technology 6
1.4 Characterization of Polymeric Nanoparticles 6
1.4.1 Size and Morphology 6
1.4.2 Surface Properties and Stability 6
1.4.2.1 Zeta Potential 6
1.4.2.2 Energy-Dispersive Spectroscopy 7
1.4.2.3 Fourier Transform Infrared Spectroscopy 7
1.4.3 X-Ray Diffraction 7
1.5 Drug Loading Efficiency Studies 7
1.6 Application of Polymeric Nanoparticles 8
References 8

1.1 INTRODUCTION

Nanobiotechnology is an amalgamation of material, chemical, and biological science with a common goal of manufacturing nanoparticles (NPs) that possess unique optical, magnetic, and biological properties. Basically, nanoparticles exhibit totally different characteristic properties from their bulk material owing to quantum confinement, which increases their surface-to-volume ratio, and their spectacular optical properties are due to the surface plasmon resonance (SPR) phenomenon (Fu & Yao, 2001). According to pharmaceutical science, nanoparticles are defined as submicronic solid therapeutics-carrying particles that are below 100 nm and/or can be biodegradable in nature; hence, in pharmaceutical terms they are referred as nanocapsules or nanospheres (Soppimath, Aminabhavi, Kulkarni, & Rudzinski, 2001). Further, based on their dimensions, nanomaterials can be categorized as follows. **Zero-dimensional nanomaterials**, such as nanoparticles, can be amorphous or crystalline and have applications in textiles and health care products. **One-dimensional nanomaterials**, such as thin film, have applications in various repository systems, biochemical sensors, laser systems, and magnetic and optical devices. **Two-dimensional nanomaterials**, such as carbon nanotubes (single and multi-walled), have applications in electrical and conductive devices. And **three-dimensional nanomaterials**, such as fullerenes, dendrimers, and quantum dots, have applications in electronics, diagnostics, and targeted drug delivery (Castagnola et al., 2017; Glezer, 2011).

Based on their composition, nanomaterials are a classified into metallic, ceramic, and polymeric nanoparticles. **Metallic nanoparticles** are basically composed from their metallic precursor materials, showing vibrant optical, electrical, magnetic, and biological properties owing to their large surface-area-to-volume ratio, plasmon excitation, quantum confinement, and large surface energies (Buzea & Pacheco, 2017; Mody, Siwale, Singh, & Mody, 2010). **Ceramic nanoparticles** are mostly oxides, phosphates, nitrides, and carbides of various metallic and non-metallic elements. As ceramic nanoparticles are typically insulative to harsh chemical environments, electricity, and heat, they are mostly preferred in major orthopedic, dental, and other medical applications (Ajayan, Schadler, & Braun, 2006; Sternitzke, 1997).

Polymeric nanoparticles have the benefit of permitting the entrapment of bioactive molecules and defending them against hydrolytic and enzymatic degradation. Polymeric nanoparticles, the most versatile nanoparticles among the metallic and ceramic nanoparticles, are basically prepared from natural or synthetic polymers, which allows them to be utilized as drug carrier molecules for various drug delivery approaches (oral, parenteral, transdermal, and nasal). The polymeric nanoparticles provide a wide range of advantages such as the availability of a wide array of polymers which can be chosen according to their application, biodegradability, hemocompatibility, and targeted delivery to specified tissues (Farokhzad & Langer, 2009; Liu, Jiao, Wang, Zhou, & Zhang, 2008; Owens III & Peppas, 2006; R. Singh & Lillard Jr, 2009).

Polymers provide a complete diverse platform for drug delivery applications due to the presence of various natural polymers such as chitosan, gelatin, sodium alginate, albumin, and synthetic polymers such as polylactic acid (PLA), polyglycolic acid (PGA), poly(lactide co-glycolic acid) (PLGA), polyanhydrides, polyorthoesters, polycyanoacrylates, polycaprolactone, polyglutamic acid, poly (malic acid), poly(N-vinyl pyrrolidone), poly(methyl methacrylate), poly(vinyl alcohol) (PVA), poly(acrylic acid), polyacrylamide, poly(ethylene glycol) (PEG), and poly(methacrylic acid), which are biocompatible and biodegradable (Wang et al., 2007). Further, based on their preparatory routes, polymeric nanoparticles can be manufactured with various combinations that provide additional functional groups to the synthesized polymeric nanoparticle, which enables higher cell recognition and encapsulation efficiency for the targeted delivery of the encapsulated drug (Elsabahy & Wooley, 2012; Guterres, Alves, & Pohlmann, 2007). The modulation of polymers with nanotechnology offers multifaceted opportunities for designing polymeric nanoparticles with preferred shape, size, biocompatibility, and controlled drug release profile, which are exceptionally significant for biomedical applications. Supplementary to this, they can also be explored for their applicability in the biomedical field as nano-carriers for targeted delivery and as novel therapeutics with combinations of diverse natural and synthetic drug molecules (Shenhar, Norsten, & Rotello, 2005).

Pharmaceutically, polymeric nanoparticles are the most preferred nanoparticles owing to their considerably versatile physiochemistry, hemocompatibility, biocompatibility, biodegradability, and ease of functionalization properties (Petros & DeSimone, 2010). The various preparation methodologies, characterization techniques, bio-functionalization, and drug delivery applicability are discussed in this chapter.

1.2 TYPES OF POLYMERIC NANOPARTICLES

Based on their composition, polymeric nanoparticles are either of two types: natural or synthetic.

1.2.1 Natural Polymeric Nanoparticles

Naturally occurring renewable resources have been the primary source for preparation of polymers for ages. Among various naturally occurring substances, the most abundantly used and ideal materials for polymeric nanoparticle synthesis are the polysaccharides. Polysaccharides are the most preferred base materials because of their easy availability, low cost, biocompatibility, and the presence of various functional nucleophilic reactive groups (Kumari, Yadav, & Yadav, 2010). Further, naturally occurring polysaccharides provide a wide range of molecular weights with numerous structural and chemical compositions that affect their overall physical and chemical properties in their nanometer size range. The polysaccharide polymeric material can be modulated during the synthesis procedures, which assists in modifying their reactive functional groups, such as hydroxyl, carboxyl, and amino groups, that can form non-covalent bonds with the mucosal and epithelial tissues of animals (Barron & Zuckerman, 1999; Park, 1979). Other polymers, such as chitosan, gelatin, lignin, starch, and albumin, are most commonly used for the synthesis of polymeric nanoparticles. Traditionally, gelatin, starch, and chitosan have been employed as binders and diluents, but now they have emerged as the most popular polymers for targeted drug delivery carriers. Easy modulation of particle size, charge, and molecular weight, drug-loading efficacy supported by non-toxicity, biodegradability, and hemocompatibility during their synthesis steps have all strongly backed their potential as targeted drug delivery vehicles (Elzoghby et al., 2016).

1.2.2 Synthetic Polymeric Nanoparticles

Synthetic polymeric nanoparticles are basically synthesized chemically using various chemicals and physical and chemical methodologies. Hence, these synthetic polymers can be tailored according to their function as drug carrier molecules with a wide array of possibilities. Further, the synthetic polymers do not require additional purification steps to remove additional functional groups which are present in natural polymers as contaminants (Sabir, Xu, & Li, 2009). Various examples of synthetic polymers used for the preparation of nanoparticles are PLA, PLGA, polyurethane (PU), poly(methyl methacrylate), polyester, polyvinylpyrrolidone (PVP), silicone rubber, and polyvinyl alcohol. Most of the synthetic polymers employed in drug delivery are FDA approved owing to their biocompatibility and biodegradability. These polymers generally prove a wide array of properties such as higher mechanical strength, fatigue and wear resistance, high elastic modulus, high elasticity, and thermal formability. These properties provide additional weightage for the utilization of synthetic polymers in bone tissue engineering, film preparation, and the synthesis of micro- and nanospheres, nanocapsules, and nanoparticles as vehicles for the controlled release of encapsulated drug (Baldwin & Kiick, 2010; Hornig, Heinze, Becer, & Schubert, 2009). However, synthetic polymers have the drawback of being less biocompatible compared to natural polymers, and they are less easily degraded to other bioproducts.

1.3 METHODS OF SYNTHESIS OF POLYMERIC NANOPARTICLES

The preparation methodology applied for the synthesis of polymeric nanoparticle is a very important criterion which can be modulated according to requirements. The synthesis methods employed have a direct relationship with the physiochemical characteristics of the entrapped drug. Depending upon the application of the carrier, its composition (natural/synthetic), and the drug to be encapsulated, the preparatory methods are selected. Based upon the application, polymeric nanoparticles can be manufactured by polymerization of selected polymers, by dispersal of drug in selected polymeric matrix, or by ionic gelation or emulsification of hydrophilic polymers (Vauthier & Bouchemal, 2009).

1.3.1 Solvent Evaporation

The solvent evaporation technique is one of the leading approaches developed to prepare polymeric nanoparticles from a previously formed polymeric layer. This method generally employs two liquid phases: one phase is the solvent and the other phase is partially or totally immiscible. Other active molecules known as surfactants are employed to reduce the surface tension between the two phases. This technique was first successfully applied by Gurny et al. (1981) for the preparation of biocompatible drug carrier polymers (Gurny, Peppas, Harrington, & Banker, 1981). The first step in this preparatory method is the dissolution of polymer matrix in a volatile solvent such as ethyl acetate. Then, the formed emulsion is allowed to evaporate by continuous overnight stirring at a specified rate and temperature condition. Conventionally, the emulsion is prepared through oil in water (o/w), i.e., single emulsion, or by water in oil in water (w/o/w), i.e., double emulsion (Murakami, Kobayashi, Takeuchi, & Kawashima, 1999; Reis, Neufeld, Ribeiro, & Veiga, 2006). The emulsion phases can be mixed thoroughly by using homogenizer or sonication (probe/water). Finally, the formed nanoparticles are collected though centrifugation, and additional purification steps can be employed to finally get the nanoparticle through lyophilization.

1.3.2 Nanoprecipitation

The nanoprecipitation technique, which is also known as the solvent displacement method, was first demonstrated by Fessi et al. (1989). This method employs an organic liquid phase where the polymer is dissolved followed by subsequent dissemination of the organic solvent in the aqueous medium in the presence or absence of a surfactant. To simplify, we can say that the nanoparticles are formed when the polymer diffuses into the solution of a water-miscible solvent, which is continuously stirred in a dropwise or controlled manner to another organic solvent (Rao & Geckeler, 2011). Due to the sudden diffusion of the polymer solution into the aqueous phase, the nanoparticles are formed instantly in an endeavor to escape the water molecules. The efficacy of this technique is confined to the array of water-miscible solvents in which the distribution frequency is adequate to produce spontaneous emulsification. Acetone and dichloromethane are most commonly used to dissolve and modulate the mean size of the nanoparticles formed through precipitation. This technique is principally relevant for hydrophobic and lipid-based drugs constructed on the basis of the miscibility of the solvent with the aqueous phase, and this method isn't applicable for the encapsulation of water-soluble drugs (Hornig et al., 2009).

1.3.3 Emulsification–Solvent Diffusion

This technique of emulsification–solvent diffusion was first developed by Leroux et al. (1995). This method employs two solvents: a partially water-miscible solvent which contains the polymer and drug dissolved in it and an aqueous solution which contains the surfactant. Hence, basically it is a conventional oil-in-water (o/w) emulsion where one phase contains the drug and the other provides shape and size to the overall polymeric matrix. For the accomplishment of this technique, the partially water-miscible solvent where the polymer is dissolved and the aqueous water phase which contains the surfactant are mutually saturated at room temperature to maintain a thermodynamic equilibrium between the two liquid phases. In comparison to the solvent evaporation technique, the solvent diffusion technique provides relatively uniform and smaller particle size within a fraction of a millisecond (Crucho & Barros, 2017; Hans & Lowman, 2002).

1.3.4 Salting Out

The salting out technique is a modified version of the emulsification/solvent diffusion method employed for the preparation of polymeric nanoparticles. In this synthesis method, the nanoparticles are formed by separation of a water-miscible solvent from an aqueous solution (Ibrahim, Bindschaedler, Doelker, Buri, & Gurny, 1992). Then, the polymer and drug of interest are dissolved in a water-miscible organic solvent such as acetone, which is then sequentially emulsified into an aqueous gel by employing salting-out agents such as magnesium chloride, calcium chloride, and magnesium acetate. The most prominently water-soluble stabilizer is PVP. The advantage of this technique is that it increases the encapsulation efficiency of protein-based drugs and does not require additional steps of temperature and pressure. However, the greatest disadvantage associated with this technique is that it requires extensive washing and purification steps, which can hamper the encapsulation efficiency of the associated drug (Allémann, Gurny, & Doelker, 1992).

1.3.5 Dialysis

Dialysis, one of the simplest methods, is relevant for the manufacture of very small polymeric nanoparticles with uniform size distribution. This method employs a semipermeable membranous tube with a required molecular weight cutoff as a physical barrier for the polymers (Chronopoulou,

Fratoddi, Palocci, Venditti, & Russo, 2009). In this technique, the applicable polymer is dissolved in an organic solvent and positioned inside the dialysis membranous sheath, which is dialyzed, contrary to a non-solvent solution. Successive and gradual vibrations by a magnetic stirrer at a particular rpm and temperature help increase the interfacial tension of the polymer in the dialysis tube, which progresses towards the development of colloidal suspension of nano-sized polymeric materials (Rane, Kanny, Abitha, & Thomas, 2018). The morphology and particle size distribution of the nanoparticles can be controlled by altering the reaction factors such as solvent/non-solvent pair, molecular weight cutoff of the dialysis membrane, temperature of the overall reaction condition, pH of the solvent/non-solvent pair, rpm of the magnetic stirrer, and concentrations of the polymer and solvent used (Eftekhari & Saito, 2017; El-Say & El-Sawy, 2017). The most commonly used organic solvents are DMSO (dimethylsulfoxide), DMF (dimethylformamide), DMAc (dimethylacetamide), and NMPy (N-methyl-2-pyrrolidinone).

1.3.6 Supercritical Fluid Technology

All of the above-mentioned processes generally use various concentrations of organic solvents and chemicals which are not only toxic but also environmentally hazardous. Further, most of the techniques require additional purification or washing steps, which consequently increase the bio-burden to the environment. Hence, to overcome such drawbacks, the supercritical fluid technology was developed to produce polymeric nanoparticles which do not release any additional biohazardous elements into the environment (Soppimath et al., 2001). The supercritical fluid technique utilizes a supercritical fluid, which is a fluid that is compressed and heated above its critical temperature and pressure. The most extensively used supercritical fluid is supercritical carbon dioxide ($scCO_2$) because it has mild critical conditions and is abundant, inexpensive, non-flammable, non-toxic, and environmentally benign (Crucho & Barros, 2017).

1.4 CHARACTERIZATION OF POLYMERIC NANOPARTICLES

After the synthesis of nanoparticles, one of the most essential steps is to study their characteristic features in order to confirm the synthesis of the desired nanoparticles. Most of the polymeric nanoparticles synthesized are basically prepared with applicability towards the pharmaceutical and biomedical fields. Hence, they require a specific set of tests to be performed, which are prerequisite for various preclinical in vitro and in vivo studies (Nimesh, 2013b). It is well known that physiological interactions mostly depend on the physiochemical properties of the nanoparticles, and thus their overall credibility can be screened in a sequential manner for size and morphology through dynamic light scattering (DLS) and scanning electron microscopy (SEM), for atomic arrangement and chemical bond distribution through X-ray diffraction (XRD) studies and Fourier-transmission infrared (FT-IR) spectroscopy; and for drug loading studies through high-performance liquid chromatography (HPLC), nuclear magnetic resonance (NMR), and UV spectroscopy (Kumar & Dixit, 2017).

1.4.1 Size and Morphology

In drug delivery systems, some of the important criteria for efficient delivery of therapeutics are the shape and particle size distribution profiles. Some important size-based parameters are their ability to target specific cells, their in vivo controlled drug distribution rate, and their biological activity such as degradability and biocompatibility. The encapsulation efficiency of the drug as well as its uniform performance distribution has a direct relationship with the size and morphology of the synthesized nanoparticle. Compared with microparticles nanoparticles exhibit various advantageous properties of higher intracellular uptake basically due to their efficient mobility, which is attributed primarily to their nano size range. It is a well-known property of nanoparticles that with decrease in size, their surface-to-volume ratio increases, which potentially influences their drug-carrying capacity. In contrast to nanoparticles, microparticles have the ability to carry considerably huge quantities of drug/therapeutic components, but with abrupt distribution. Hence, nanoparticles are generally considered as most efficient drug carrier entities in targeted delivery systems. The size of the nanoparticles can be efficiently analyzed using DLS spectroscopy. DLS uses a monochromatic light source, which can provide a prominent size distribution profile by volume as well as determine the number of particles. The morphology of the synthesized particles are prominently visualized through electron microscopy such as SEM), transmission electron microscopy (TEM), and atomic force microscopy (AFM). Electron microscopy uses electron beam energy, which, when it falls on the gold/platinum-coated samples, scatter the energy, which is then captured by the detectors.

1.4.2 Surface Properties and Stability

The surface properties of the polymeric nanoparticles are most significant, keeping in view that upon internalization by any route (oral/transdermal or parenteral), the polymeric nanoparticle will come directly under the scanner of bodily fluids. Hence, various techniques such as zeta potential, energy-dispersive spectroscopy (EDS), X-ray photoelectron spectroscopy (XPS), and FT-IR techniques are utilized, which provides a comprehensive assessment of their chemical composition and additional functionalization.

1.4.2.1 Zeta Potential

The surface zeta potential of a nanoparticle is commonly used to characterize the surface charge property of nanoparticles. Zeta potential, otherwise known as the electrokinetic potential, is measured at the surface of the colloidal particles and has a great significance in relating the overall stability of colloidal dispersions. The zeta potential indicates the degree of repulsion between adjacent particles and correspondingly charged particles in the dispersed liquid medium. For molecules and particles that are small enough, a high zeta potential confers stability (dispersion will resist aggregation). Nanoparticles with a zeta potential (surface charge) above (+/–) 30 mV have been shown to be stable in suspension, as the surface charge prevents aggregation of the particles. The zeta potential can also be used to determine whether a charged active material is encapsulated within the core of the nanocapsule or adsorbed onto the surface (Nayak, Minz, et al., 2016b).

1.4.2.2 Energy-Dispersive Spectroscopy

EDX is a technique which analyzes the chemical composition of the synthesized nanoparticle. The EDX technique, also commonly known as EDS, uses the X-rays emitted from an electron source generally attached with a scanning electron microscope that provides an X-ray spectrum. The EDS spectrum is basically the energy frequency of the emitted X-rays which bombard the sample and that can correctly identify an element (Crucho & Barros, 2017). X-ray photoelectron spectroscopy (XPS) is another technique that uses X-ray excitation of electrons to disturb the energy levels of the elemental composition of the samples to quantify chemical composition directly through the scanning mode of the X-ray spectra.

1.4.2.3 Fourier Transform Infrared Spectroscopy

The FT-IR spectrum represents the molecular absorption and transmission peaks which correspond to the frequencies of vibrations between the bonds of the atoms making up the material. Thus, it provides a unique molecular fingerprint which is totally unique for the sample (Nayak, Ashe, Rauta, Kumari, & Nayak, 2016a). The strong vibrations produced in the spectra are characteristic of the particle, which can be prepared from the basic polymeric materials into the final product. This technique can also confirm the presence of various impurities and of the characteristic functional groups associated with the solute, solvent, and chemicals used during the preparatory steps (Nimesh, 2013a).

1.4.3 X-Ray Diffraction

Apart from size and surface properties, the structural integrity and composition based on their atomic grain size or arrangement of lattice are very prominent criteria for the characterization of nanoparticles. In this regard, XRD provides the grain width and shape of the nanoparticles by direct bombardment of X-ray on the samples where the data is collected as diffraction beams against an angle (2θ) with respect to the incident X-ray beam. A powder diffraction pattern supplies information from the nanomaterial on phases present, phase concentrations, and structure, as well as the degree of crystallinity or amorphous content followed by details on crystallite size/strain. The peak widths in a given phase pattern provide information on the average crystallite size where large crystallites give rise to sharp peaks and any increase in the peak width indicates a reduced crystallite size. The measure of the crystallite size does not equal to the size of the particle, as a single nanoparticle can be an aggregate of several crystallites (Chirayil, Abraham, Mishra, George, & Thomas, 2017).

1.5 DRUG LOADING EFFICIENCY STUDIES

The ideal nanoparticulate system should have a high drug-loading capacity, thereby reducing the drug quantity for the administration. Generally, drug loading can be achieved by two methods: the incorporation method (adding at the time of nanoparticle formulation) and the adsorption/absorption technique (absorbing the drug after formation of nanoparticles). Encapsulation of protein antigens depends on a series of factors like the nature of the polymer, the method of formulations, molecular interaction between polymers, etc. The amount of drug bound to NPs depends on the chemical structure of the drug/polymer ratio and the conditions of drug loading. Hydrophilic polymers show higher encapsulation levels than hydrophobic polymers due to enhanced molecular interactions between the polymer and the drug. Entrapment efficiency is very much dependent on the solid-state drug solubility in the matrix material or polymer (solid dispersion or dissolution), which is related to the molecular weight, polymer composition, drug–polymer interaction, and the presence of functional end groups. In the case of macromolecules or proteins, higher loading efficiency is achieved when it is loaded at or near its isoelectric point (with minimum solubility and maximum adsorption). For smaller molecules, the use of ionic interaction between the drug and matrix materials can be a very effective approach to increase the drug encapsulation. The loading efficiencies of NPs can be calculated using spectroscopy methods by utilizing standard curve techniques prepared by varying concentrations of encapsulated drug.

Drug release and polymer degradation are important factors to develop an ideal nanoparticulate system. Generally, the rate of drug release from nanoparticles depends on many factors: (i) solubility of the drug; (ii) drug diffusion through the nanoparticle matrix; (iii) desorption of the surface-attached/adsorbed drug; (iv) nanoparticle matrix erosion/degradation. Thus, properties like solubility, diffusion, and biodegradation of the matrix materials govern the release process. The rapid initial burst release is particularly attributed to weak bonding or adsorption of the drug to the surface of nanoparticles. The solubility and diffusivity of the drug in polymer matrix also play measured roles in controlling drug release. When the drug is encapsulated by the incorporation method, the system offers a relatively small burst effect and better-sustained release characteristics. The in vitro drug release can be studied by various methods using high-end techniques such as HPLC, which is generally used for confirming the identity of the drug against a standard sample as well as providing quantitative data regarding the entrapped drug in the carrier polymeric matrix (A. K. Singh, 2016). In HPLC, the UV detector is mostly preferred, as it offers excellent linearity and rapid analysis of multiple drugs against standard drug concentrations quantitatively.

1.6 APPLICATION OF POLYMERIC NANOPARTICLES

Rapid and remarkable research into polymeric nanoparticles has revolutionized the overall aspect of natural and synthetic polymer-based nanoparticles in the field of medicine and in the diagnosis industry. However, polymer-based nanoparticle therapeutic, diagnosis, and treatment practices have not achieved much success, as most of the results are confined to the laboratory scale and a negligible proportion has reached the clinical trial stage (Wilczewska, Niemirowicz, Markiewicz, & Car, 2012). In the area of drug delivery, polymeric nanoparticles propose high dividends in cases of targeted drug therapy, where they can be modulated chemically for reaching specific targeted sites, thereby having minimal side effects on the surrounding cellular environment (Girija, 2019). The various drawbacks of the conventional medicinal system are non-specific drug distribution with high doses of drug concentration, and non-specific interactions with surrounding normal and healthy tissues, with uncontrolled rates of drug release profile that can ultimately result in additional drug-based side effects (Ladj et al., 2013). These drawbacks can be mediated through particular polymer-based nanosystems. The small size with potentially larger surface-to-volume ratio aids in the entrapping the drugs with both hydrophilic and hydrophobic characteristics (Dudefoi et al., 2018). Further, polymer-based nanoparticles provide additional properties of surface charge modulations that can help in better uptake efficacy to the targeted cells with effective, uniform, and controlled release profiles (Saini, Bagri, Bajpai, & Mishra, 2018). The research in advanced nano drug delivery strategies has opened new avenues for precise and smart drug delivery to targeted tissues/organs. Owing to their exciting and intrinsic nano-dimensional properties, various polymeric nanomaterials are being explored for smart drug delivery, which can sense the presence of particular molecular groups associated within the damaged tissue or cancerous cells (Kumari et al., 2010; Sharma, Chatterjee, Singh, & Satapathi, 2019).

Apart from drug delivery, polymeric nanoparticles are also highly capable of having their basic structural chemical properties reorganized in order to exploit them for various contrasting agents in imaging purposes. Due to their high stability, structural flexibility, and incorporation of various reactive functional groups, polymeric nanoparticles can be modulated easily for any activity without hampering their biochemical properties. The polymeric nanoparticles assist themselves as imaging materials based on their extremely small nanometer size range, which directly helps in their enhanced uptake rate in targeted cells, accumulation of contrast agents at the targeted site, multimodal imaging potential, multiplexing capability, and theragnostic potential, which incorporates both therapeutic and diagnostic capabilities (Kumar, Jalodia, Kumar, & Gautam, 2017; Margineanu, 2019; Saeedi, Eslamifar, Khezri, & Dizaj, 2019; Yi, Son, Yoo, Park, & Koo, 2019). Thus, polymeric nanoparticles have a wide range of applications as antimicrobial agents, drug delivery carrier molecules, and contrasting agents in various imaging systems such as in MRI; moreover, various research activities can help them further to exhibit themselves as theranostatic, wound healing, regenerative, and nutraceutical innovations.

REFERENCES

Ajayan, P. M., Schadler, L. S., & Braun, P. V. (2006). *Nanocomposite science and technology*. John Wiley & Sons.

Allémann, E., Gurny, R., & Doelker, E. (1992). Preparation of aqueous polymeric nanodispersions by a reversible salting-out process: Influence of process parameters on particle size. *International Journal of Pharmaceutics*, *87*(1), 247–253. doi: 10.1016/0378-5173(92)90249-2

Baldwin, A. D., & Kiick, K. L. (2010). Polysaccharide-modified synthetic polymeric biomaterials. *Peptide Science: Original Research on Biomolecules*, *94*(1), 128–140.

Barron, A. E., & Zuckerman, R. N. (1999). Bioinspired polymeric materials: In-between proteins and plastics. *Current Opinion in Chemical Biology*, *3*(6), 681–687.

Buzea, C., & Pacheco, I. (2017). Nanomaterials and their classification. A. K. Shukla (Ed.), *Nanomaterials and their classification BT - EMR/ESR/EPR spectroscopy for characterization of nanomaterials* (3–45). New Delhi: Springer India. 10.1007/978-81-322-3655-9_1

Castagnola, V., Cookman, J., De Araujo, J. M., Polo, E., Cai, Q., Silveira, C. P., Krpetić, Ž., Yan, Y., Boselli, L., & Dawson, K. A. (2017). Towards a classification strategy for complex nanostructures. *Nanoscale Horizons*, *2*(4), 187–198.

Chirayil, C. J., Abraham, J., Mishra, R. K., George, S. C., & Thomas, S. (2017). Chapter 1 – Instrumental techniques for the characterization of nanoparticles. In S. Thomas, R. Thomas, A. K. Zachariah, & R. K. B. T.-T. and R. M. T. for N. C. Mishra (Eds.), *Micro and nano technologies* (pp. 1–36). Elsevier. doi: 10.1016/B978-0-323-46139-9.00001-3

Chronopoulou, L., Fratoddi, I., Palocci, C., Venditti, I., & Russo, M. V. (2009). Osmosis based method drives the self-assembly of polymeric chains into micro- and nanostructures. *Langmuir*, *25*(19), 11940–11946. doi: 10.1021/la9016382

Crucho, C. I. C., & Barros, M. T. (2017). Polymeric nanoparticles: A study on the preparation variables and characterization methods. *Materials Science and Engineering: C*, *80*, 771–784. doi: 10.1016/j.msec.2017.06.004

Dudefoi, W., Villares, A., Peyron, S., Moreau, C., Ropers, M.-H., Gontard, N., & Cathala, B. (2018). Nanoscience and nanotechnologies for biobased materials, packaging and food applications: New opportunities and concerns. *Innovative Food Science & Emerging Technologies*, *46*, 107–121. doi: 10.1016/j.ifset.2017.09.007

Eftekhari, A., & Saito, T. (2017). Synthesis and properties of polymerized ionic liquids. *European Polymer Journal*, *90*, 245–272. doi: 10.1016/j.eurpolymj.2017.03.033

Elsabahy, M., & Wooley, K. L. (2012). Design of polymeric nanoparticles for biomedical delivery applications. *Chemical Society Reviews*, *41*(7), 2545–2561.

El-Say, K. M., & El-Sawy, H. S. (2017). Polymeric nanoparticles: Promising platform for drug delivery. *International Journal of Pharmaceutics*, *528*(1), 675–691. doi: 10.1016/j.ijpharm.2017.06.052

Elzoghby, A., Abd-Elwakil, M. M., Abd-Elsalam, K., Elsayed, M. T., Hashem, Y., & Mohamed, O. (2016). Natural polymeric nanoparticles for brain-targeting: Implications on drug and gene delivery. *Current Pharmaceutical Design*, *22*(22), 3305–3323.

Farokhzad, O. C., & Langer, R. (2009). Impact of nanotechnology on drug delivery. *ACS Nano*, *3*(1), 16–20.

Fessi, H., Puisieux, F., Devissaguet, J. P., Ammoury, N., & Benita, S. (1989). Nanocapsule formation by interfacial polymer deposition following solvent displacement. *International Journal of Pharmaceutics*, *55*(1), R1–R4. doi: 10.1016/0378-5173(89)90281-0

Fu, H.-B., & Yao, J.-N. (2001). Size effects on the optical properties of organic nanoparticles. *Journal of the American Chemical Society*, *123*(7), 1434–1439.

Girija, A. R. (2019). 12 – Medical applications of polymer/functionalized nanoparticle systems. In K. Pielichowski & T. M. B. T.-P. C. with F. N. Majka (Eds.), *Micro and nano technologies*, (pp. 381–404). Elsevier. doi: 10.1016/B978-0-12-814064-2.00012-3

Glezer, A. M. (2011). Structural classification of nanomaterials. *Russian Metallurgy (Metally)*, *2011*(4), 263.

Gurny, R., Peppas, N. A., Harrington, D. D., & Banker, G. S. (1981). Development of biodegradable and injectable latices for controlled release of potent drugs. *Drug Development and Industrial Pharmacy*, *7*(1), 1–25. doi: 10.3109/03639048109055684

Guterres, S. S., Alves, M. P., & Pohlmann, A. R. (2007). Polymeric nanoparticles, nanospheres and nanocapsules, for cutaneous applications. *Drug Target Insights*, *2*, 147-157. doi: 10.1177/117739280700200002.

Hans, M. L., & Lowman, A. M. (2002). Biodegradable nanoparticles for drug delivery and targeting. *Current Opinion in Solid State and Materials Science*, *6*(4), 319–327.

Hornig, S., Heinze, T., Becer, C. R., & Schubert, U. S. (2009). Synthetic polymeric nanoparticles by nanoprecipitation. *Journal of Materials Chemistry*, *19*(23), 3838–3840.

Ibrahim, H., Bindschaedler, C., Doelker, E., Buri, P., & Gurny, R. (1992). Aqueous nanodispersions prepared by a salting-out process. *International Journal of Pharmaceutics*, *87*(1), 239–246. doi: 10.1016/0378-5173(92)90248-Z

Kumar, A., & Dixit, C. K. (2017). Methods for characterization of nanoparticles. S. Nimesh, R. Chandra, & N. Gupta (Eds.), *Advances in nanomedicine for the delivery of therapeutic nucleic acids* (pp. 43–58). Woodhead Publishing. 10.1016/B978-0-08-100557-6.00003-1

Kumar, B., Jalodia, K., Kumar, P., & Gautam, H. K. (2017). Recent advances in nanoparticle-mediated drug delivery. *Journal of Drug Delivery Science and Technology*, *41*, 260–268. doi: 10.1016/j.jddst.2017.07.019

Kumari, A., Yadav, S. K., & Yadav, S. C. (2010). Biodegradable polymeric nanoparticles based drug delivery systems. *Colloids and Surfaces B: Biointerfaces*, *75*(1), 1–18.

Ladj, R., Bitar, A., Eissa, M. M., Fessi, H., Mugnier, Y., Le Dantec, R., & Elaissari, A. (2013). Polymer encapsulation of inorganic nanoparticles for biomedical applications. *International Journal of Pharmaceutics*, *458*(1), 230–241. doi: 10.1016/j.ijpharm.2013.09.001

Leroux, J.-C., Alléman, E., Doelker, E., & Gurny, R. (1995). New approach for the preparation of nanoparticles for the preparation of nanoparticles by an emulsification diffusion method. *European Journal of Pharmaceutics and Biopharmaceutics*, 41, 14–18.

Liu, Z., Jiao, Y., Wang, Y., Zhou, C., & Zhang, Z. (2008). Polysaccharides-based nanoparticles as drug delivery systems. *Advanced Drug Delivery Reviews*, *60*(15), 1650–1662.

Margineanu, A. (2019). Chapter 14 – Biological applications of nanoparticles in optical microscopy. In C. B. T.-P. N. in N. Vasile (Ed.), *Micro and nano technologies* (pp. 469–495). Elsevier. doi: 10.1016/B978-0-12-813932-5.00014-5

Mody, V. V, Siwale, R., Singh, A., & Mody, H. R. (2010). Introduction to metallic nanoparticles. *Journal of Pharmacy and Bioallied Sciences*, *2*(4), 282.

Murakami, H., Kobayashi, M., Takeuchi, H., & Kawashima, Y. (1999). Preparation of poly (DL-lactide-co-glycolide) nanoparticles by modified spontaneous emulsification solvent diffusion method. *International Journal of Pharmaceutics*, *187*(2), 143–152.

Nayak, D., Ashe, S., Rauta, P. R., Kumari, M., & Nayak, B. (2016a). Bark extract mediated green synthesis of silver nanoparticles: Evaluation of antimicrobial activity and antiproliferative response against osteosarcoma. *Materials Science and Engineering: C*, *58*, 44–52. doi: 10.1016/j.msec.2015.08.022

Nayak, D., Minz, A. P., Ashe, S., Rauta, P. R., Kumari, M., Chopra, P., & Nayak, B. (2016b). Synergistic combination of antioxidants, silver nanoparticles and chitosan in a nanoparticle based formulation: Characterization and cytotoxic effect on MCF-7 breast cancer cell lines. *Journal of Colloid and Interface Science*, *470*, 142–152.

Nimesh, S. (2013a). Characterization of nanoparticles: In vitro and in vivo. S. B. T.-G. T. Nimesh (Ed.), *Gene therapy: Potential applications of nanotechnology* (pp. 65–88). Woodhead Publishing. 10.1533/9781908818645.65

Nimesh, S. (2013b). Tools and techniques for physico-chemical characterization of nanoparticles. S. B. T.-G. T. Nimesh (Ed.), *Gene therapy: Potential applications of nanotechnology* (pp. 43–63). Woodhead Publishing. 10.1533/9781908818645.43

Owens III, D. E., & Peppas, N. A. (2006). Opsonization, biodistribution, and pharmacokinetics of polymeric nanoparticles. *International Journal of Pharmaceutics*, *307*(1), 93–102.

Park, J. B. (1979). Polymeric materials. In *Biomaterials* (pp. 73–96). Springer.

Petros, R. A., & DeSimone, J. M. (2010). Strategies in the design of nanoparticles for therapeutic applications. *Nature Reviews Drug Discovery*, *9*(8), 615.

Rane, A. V., Kanny, K., Abitha, V. K., & Thomas, S. (2018). Methods for Synthesis of Nanoparticles and Fabrication of Nanocomposites. S. Mohan Bhagyaraj, O. S. Oluwafemi, N. Kalarikkal, & S. Thomas (Eds.), *Synthesis of inorganic nanomaterials* (pp. 121–139). Woodhead Publishing. 10.1016/B978-0-08-101975-7.00005-1

Rao, J. P., & Geckeler, K. E. (2011). Polymer nanoparticles: Preparation techniques and size-control parameters. *Progress in Polymer Science*, *36*(7), 887–913.

Reis, C. P., Neufeld, R. J., Ribeiro, A. J., & Veiga, F. (2006). Nanoencapsulation I. Methods for preparation of drug-loaded polymeric nanoparticles. *Nanomedicine: Nanotechnology, Biology and Medicine*, *2*(1), 8–21.

Sabir, M. I., Xu, X., & Li, L. (2009). A review on biodegradable polymeric materials for bone tissue engineering applications. *Journal of Materials Science*, *44*(21), 5713–5724.

Saeedi, M., Eslamifar, M., Khezri, K., & Dizaj, S. M. (2019). Applications of nanotechnology in drug delivery to the central nervous system. *Biomedicine & Pharmacotherapy*, *111*, 666–675. doi: 10.1016/j.biopha.2018.12.133

Saini, R. K., Bagri, L. P., Bajpai, A. K., & Mishra, A. (2018). Responsive polymer nanoparticles for drug delivery applications. A. S. H. Makhlouf & N. Y. Abu-Thabit (Eds.), *Stimuli responsive polymeric nanocarriers for drug delivery applications* (pp. 289–320). Woodhead Publishing. 10.1016/B978-0-08-101997-9.00015-1

Sharma, M., Chatterjee, M., Singh, K., & Satapathi, S. (2019). Design of a multimodal colloidal polymeric drug delivery vesicle: A detailed pharmaceutical study. *Nano-Structures & Nano-Objects*, *18*, 100245. doi: 10.1016/j.nanoso.2019.01.004

Shenhar, R., Norsten, T. B., & Rotello, V. M. (2005). Polymer-mediated nanoparticle assembly: Structural control and applications. *Advanced Materials*, *17*(6), 657–669.

Singh, A. K. (2016). Chapter 4 – Experimental methodologies for the characterization of nanoparticles. In A. K. B. T.-E. N. Singh (Ed.) (pp. 125–170). Boston: Academic Press. doi: 10.1016/B978-0-12-801406-6.00004-2

Singh, R., & Lillard Jr, J. W. (2009). Nanoparticle-based targeted drug delivery. *Experimental and Molecular Pathology*, *86*(3), 215–223.

Soppimath, K. S., Aminabhavi, T. M., Kulkarni, A. R., & Rudzinski, W. E. (2001). Biodegradable polymeric nanoparticles as drug delivery devices. *Journal of Controlled Release*, *70*(1–2), 1–20.

Sternitzke, M. (1997). Structural ceramic nanocomposites. *Journal of the European Ceramic Society*, *17*(9), 1061–1082.

Vauthier, C., & Bouchemal, K. (2009). Methods for the preparation and manufacture of polymeric nanoparticles. *Pharmaceutical Research*, *26*(5), 1025–1058.

Wang, X., Hall, J. E., Warren, S., Krom, J., Magistrelli, J. M., Rackaitis, M., & Bohm, G. G. A. (2007). Synthesis, characterization, and application of novel polymeric nanoparticles. *Macromolecules*, *40*(3), 499–508.

Wilczewska, A. Z., Niemirowicz, K., Markiewicz, K. H., & Car, H. (2012). Nanoparticles as drug delivery systems. *Pharmacological Reports*, *64*(5), 1020–1037. doi: 10.1016/S1734-1140(12)70901-5

Yi, G., Son, J., Yoo, J., Park, C., & Koo, H. (2019). Emulsan-based nanoparticles for in vivo drug delivery to tumors. *Biochemical and Biophysical Research Communications*, *508*(1), 326–331. doi: 10.1016/j.bbrc.2018.11.106

2 Synthesis, Characterization, and Applications of Silica Nanomaterials from a Nanobiotechnological Perspective

Chityal Ganesh Kumar, Kanugala Sirisha, and Puvvada Naga Prasad

CONTENTS

2.1 Introduction 11
2.2 Methods Used for the Synthesis of Silica-Based Nanomaterials 11
2.2.1 Physical Approaches 12
2.2.2 Chemical Approaches 12
2.2.3 Biological Approaches 13
2.3 Characterization of Silica-Based Nanomaterials 13
2.4 Applications of Silica-Based Nanomaterials 17
2.4.1 Drug Delivery (Theranostics) 17
2.4.2 Biosensors 18
2.4.3 Antimicrobial Coatings 21
2.4.4 Food Industry 21
2.4.5 Agriculture 21
2.4.6 Cosmetics 21
2.5 Conclusions and Future Perspectives 24
References 24

2.1 INTRODUCTION

The term "Nano" generally refers to one billionth of a metric unit depicting any physical entity in the range of 0.000000001/10^{-9}. As the physical dimensionality goes deep down into nanometer and sub-nanometer scales, the potential applications for the corresponding material properties become meteoric, leaving a nano-footprint in every known discipline, technology, and industry. Today, nanomaterials and nanotechnology play a key role in human lives ranging from food, medicine, agriculture, and drinking water to energy, high-end digital systems, electronics, security, and spacecraft. The historical origin of nanomaterials dates back to Mesopotamian civilizations, while modern science's focus on nanomaterials dates from the experiments of Faraday on gold colloids in the nineteenth century. Later, in early 1959, Richard Feynman highlighted the initial ideas and concepts behind the fascinating physical world at nanoscale dimensions in a talk entitled "There's Plenty of Room at the Bottom," at the American Physical Society meeting at the California Institute of Technology. In 1974, Prof. Norio Taniguchi coined the term "nanotechnology."

Nanosized silica was initially used as a supplement in rubber products during the 1940s in the United States and the first synthesis of nanosized silica or silicon dioxide was reported in a US patent (Page et al. 1970). Mesoporous silica nanoparticles were later synthesized by Japanese researchers (Yanagisawa et al. 1990), as mesoporous molecular sieves at Mobil Corporation Laboratories, Texas and named Mobil Crystalline Materials (MCM) (Beck et al. 1992; Trewyn et al. 2007). Subsequently, silica nanoparticles of larger pores (4.6–30 nm) were prepared at University of California, Santa Barbara and were named as Santa Barbara Amorphous type material or SBA-15 (Zhao et al. 1998). Currently, silica nanomaterials find wide range of applications in diverse niche areas including drug delivery (theranostics), diagnostics, biosensors, agriculture, food, cosmetics, bioseparations, nanoimmunotherapy, etc. Considering the aforementioned opportunities for different silica nanomaterials, this chapter focuses on various synthetic routes, characterization aspects, and their application in diverse biotechnological niches.

2.2 METHODS USED FOR THE SYNTHESIS OF SILICA-BASED NANOMATERIALS

Conventionally, silica-based nanomaterials were synthesized employing physical, chemical, and biological methods. The synthetic methods can be followed by two approaches: "bottom-up," where the nanostructured building blocks (nanoparticles) were synthesized followed by self-assembly into the final structures, and "top-down," where the suitable starting material is imparted the functionality from the material.

2.2.1 Physical Approaches

Pyrolysis is a bottom-up, physical decomposition process of organic or inorganic materials at higher temperatures with liquid or vapor precursors. Precursors are supersaturated followed by nucleation, gas–solid surface reaction, and finally nanoparticle formation. These precursors are subjected to gas phase reaction or supersaturation to form reaction intermediates. This step would be generally followed by nucleation. In some cases, reaction intermediates participate in gas–solid surface reaction. Both nucleation and gas–solid surface reactions lead to the formation of desired primary particles that either undergo chain elongation to form nanoparticles or agglomerate resulting in nanoclusters. Flame spray pyrolysis is a bottom-up method wherein silica precursor solution drops are sprayed followed by their concentration-dependent evaporation into microparticles and nucleation. Spray pyrolysis allows monitoring and varying the reaction conditions, precursor quantities, and droplet size, which leads to the formation of desired sizes of nanoparticles through particle mobility scanning systems (Mädler et al. 2002). Laser pyrolysis is a CO_2 laser-dependent bottom-up process that decomposes silica precursors in gas or liquid phase, in a laser and silica absorption resonance – and photon absorption – dependent manner. Major types of silica precursors used in laser pyrolysis are those that produce desired products immediately after laser pyrolysis, precursors that depend on energy or other sensitizers, and those that assist in the decomposition of moieties (Borsella et al. 2011). Ultrasonic spray pyrolysis is a bottom-up process involving ultrasonic spraying of the silica precursors followed by their atomization and deposition, the so-called aerosol technique. The parameters used in this approach comprise tunable flow rates, different amplitudes of ultrasonic rays, choice of silica precursors, adjustable temperature for depositions as nanofilms, and temperature control on substrates for the fabrication of coatings and films using nanomaterials (Patra et al. 1999). Chemical vapor deposition is a bottom-up approach consisting of deposition of solid silica precursors under hot conditions. This approach allows the achievement of materials of robust physicochemical and topological properties by offering tunable parameters like reaction conditions, precursors, temperature, mixture of gases used in reactions, pressure, flow rate, etc. This is a better choice for the fabrication of uniform coatings and the selective fabrication of coatings upon substrates (Katsui and Goto 2018). Physical vapor deposition is a bottom-up approach involving the conversion of silica source materials from solid or liquid phase to a gas phase by evaporation, supersaturation, and/or nucleation. This is followed by their transportation and deposition into a starting phase, as a film or coating at higher temperatures or temperature gradients. Physical vapor deposition is a collective term that refers to many nanoparticle deposition strategies. Generally, the reactions are carried out with the help of furnaces, through which the carrier gases are passed. These carrier streams reach the substrates to be coated with nanomaterials. Temperature gradients are attained with water circulation (Cakir-Omur et al. 2017). Pulsed laser deposition is a bottom-up approach where the precursors are exposed to a high-pulse laser under vacuum, resulting in thermal or electronic excitement. This phenomenon results in solvent evaporation and surface deposition. In between, the plasma state of the materials is created, after ablation of the precursors over to the substrates occurs. The ablated reactants are deposited over the target surfaces, which is followed by the seeding and growth elongation of nanomaterials as films (Gontad et al. 2017). Atomic layer deposition is a bottom-up approach where the silica precursors are deposited by surface reaction in a controlled pattern. The major strategy of this approach is the sequential passage of the precursors in separate pulses, one after other instead of mixing all the precursors together. This strategy allows homogenous and uniform occupying of the substrate sites by the precursors (Huang et al. 2017). Molecular beam epitaxy, a thin film deposition method, is a bottom-up approach that involves the layering of nanocrystalline materials over crystalline surfaces carried out under vacuum under tightly controlled reaction rates, leading to highly pure films. The substrates are initially heated to very high temperatures of ~1000°F. The precursors are focused from effusion cells onto the substrate. These precursor molecules settle and form ultrathin layers on the substrate (Lazarenko et al. 2017).

2.2.2 Chemical Approaches

Sol-gel synthesis is a bottom-up approach that involves the formation of solid silica materials, like silica oxides, into monomers; these formed silicate monomers are built into colloidal solutions that act as a scaffold for the solid aggregates or polymers. The colloidal stage of the silica is followed by a hybrid phase of solid and liquid material. Solid silica particle formation is then achieved by the elimination of liquid or solvent by various methods like drying, firing, and centrifugation, or by a combination of these methods. The silica sol-gel may transform into particles of various dimensions or may be casted as films depending on the presence or absence of solid support by acid or base catalysis. The silica sol-gel process involves the catalyst-dependent hydrolysis of an alkoxide silica precursor, leading to siloxane intermediates, which polymerize into aggregates to give particles or polymers or coatings. Sonication, or the Pechini process, may improve the efficiency of sol-gel synthesis in terms of enhancing the rate of polymerization and avoiding the formation of multiphase and heterogenous particles. The well-known Stober synthesis follows the sol-gel approach (Dixit et al. 2016). The microemulsion technique is a bottom-up approach involving biphasic liquid dispersions made of water, oil, and a co-surfactant. The co-surfactant follows a phase diagram, formed by the water/oil/surfactant triad in a triangular form, with each vertice representing the water or oil or surfactant and the region towards the center of the triangle representing the possible combinations of the water/oil/surfactant triad.

The mean sizes of emulsions formed in this system may vary approximately between 1 and 100 nm (Luwang et al. 2011). Hydrothermal synthesis is a bottom-up synthesis in which silica precursors are maintained in hot aqueous conditions under maintained pressure in a specialized autoclave. In one strategy, the temperature is reduced during the supersaturation of silica particles, which can be termed as the temperature-difference method. In another strategy, the solubility differences between the stable and metastable phases play a vital role in silica nanoparticle formation; this can be termed as the metastable phase technique. The silica precursor chemistry, temperature, pressure, solvents used, surfactants, and duration of aging control are various properties that control silica nanoparticle synthesis (Liang et al. 2014). Etching is a chemical top-down manufacturing process which involves the bathing of materials in chemicals under regulated physical parameters to prepare products of desired dimensions, which could be achieved using acids or alkalis and various solvents and co-solvents. Etching is a simple and scalable process that can yield silica nanoparticles of various physical and chemical features (Chen et al. 2009). Some examples of silica nanomaterials synthesized using various physical and/or chemical methods are summarized in Table 2.1.

2.2.3 Biological Approaches

The biogenic synthetic approach generally occurs through reduction or an enzyme activity mediation. This silica nanoparticle generation by natural process has gained a renewed interest among researchers in recent years. The exploitation of natural sources or mimicking the natural phenomenon for the nature-inspired fabrication of silica nanoparticles was found to be very superior as compared to the conventional physical or chemical routes, since they are very expensive and potentially harmful. The chemicals and reagents applied might be toxic at systemic and local levels to non-target species and the environment. Further, biogenic synthetic approaches are superior in that they offer simplicity, low cost, cleanliness, safety, biocompatibility, and sustainability. In this regard, various agro-wastes like rice husk, rice straw, *Sorghum vulgare* seed heads, sugar beet bagasse, corncobs, agricultural waste ash, sugarcane bagasse, and wheat straw were investigated for biogenic synthesis of silica nanomaterials, which is a **S**imple, **S**afe, and **S**ustainable (SSS) mode of silica nanofabrication. A few other approaches use silicate salts as starting material for the biological synthesis mediated by bacteria like *Actinobacter* sp. (Singh et al. 2008) and thermophilic bacterium BKH1 (Show et al. 2015). Fungi like *Fusarium oxysporum* were also explored for bioleaching efficacy for the generation of nanosilica (Bansal et al. 2005a, b). Other biological sources like viruses could act as templates for silica nanoparticle synthesis (Royston et al. 2009). In addition, *Cynodon dactylon* leaf extracts (Babu et al. 2018) and extracts from mushrooms like *Lentinus edodes* and *Ganoderma lucidum* (Vetchinkina et al. 2018) were explored for fabrication of silica nanoparticles. Some typical silica nanomaterials synthesized using various biomaterials are summarized in Table 2.2.

2.3 CHARACTERIZATION OF SILICA-BASED NANOMATERIALS

The confirmation of synthesized silica nanomaterials relies on their unique and specific physical, chemical, physiochemical, and analytical properties, which together is called characterization. Scanning electron microscopy (SEM) and transmission electron microscopy (TEM) are used for visualization of shape, structure, diameter, aggregation, dispersion, distribution, pore size, morphology, and uniformity of the synthesized silica nanoparticles. Moreover, these microscopic techniques enable us to observe and confirm loading, doping, or surface modifications of the silica nanoparticles with external chemical moieties like drugs, biomolecules, enzymes, dyes, etc. (Figure 2.1).

Energy-dispersive X-ray spectroscopy is interfaced to SEM or TEM for better understanding the elemental composition of silica nanoparticles with the help of characteristic patterns of X-rays. Thermogravimetric analysis enables assessment of the silica nanoparticle purity and the modification of silane surface chemistry, and it quantifies weight of silica nanoparticles and weight changes in them when subjected to a temperature-wide gradient ranging from –196°C to ~1000°C. Brunauer–Emmett–Teller (BET) analyses are used for probing the surface area and porosity of silica nanoparticles using nitrogen adsorption and desorption by plotting silica nanoparticles (cc/g) against the relative pressure changes upon the adsorption/desorption (P/P_0). Depending on the porosity of the silica nanoparticles, there are six types of BET isotherms, which are depicted in Figure 2.2. Microporous materials with pore diameter <2 nm exhibit Type-I isotherm, as the partial pressure will be less than 1 and the BET constant will be greater than 1. Type-II isotherms are formed when a microporous material is used leading to monolayer formation initially followed by multilayer formation. In this case, the BET constant will be greater than 1. Type-III isotherms exhibit an asymptote curve with a BET constant of less than 1, with only multilayer formation but no monolayer of adsorption. Type-IV and Type-V BET isotherms are applicable to mesoporous materials with monolayer formation followed by multilayer formation. The N_2 gas undergoes capillary condensation in the mesopores of silica nanoparticles. Type-IV exhibits a characteristic hysteresis loop with an analogy to Type-II. For example, the case of Type-IV isotherms depicts the mesoporous nature of the silica nanoparticles with pore diameters ranging between 2 and 50 nm. Whereas Type-V shows an analogy to Type-III isotherms. Type-VI isotherms are exhibited by the non-porous materials with adsorption of N_2 gas in multiple steps, as represented in the peaks.

Fourier Transform Infrared (FT-IR) spectroscopy allows probing of the surface silane chemistry of the synthesized silica nanoparticles by analyzing the surface functional groups and the associated chemical bond vibrations, such as the Si-O bond vibrations (1000–1200 cm^{-1}), Si-OH bond vibrations (~940–960 cm^{-1}), and Si-O bond vibrations (790–810 cm^{-1}). Modification of the silica nanoparticle surface with amines, oxides, esters, and other functional groups leads to

TABLE 2.1
Silica Nanomaterials Synthesized Using Different Physical and/or Chemical Methods

Type of silica nanomaterials	Synthetic approach	Reagent(s) used	Condition	Reference
		Physical methods		
Silica nanoparticles	Ball milling	Graphite and SiO_2 powders	Solid phase reaction	Lam et al. (2000)
Native	Flame pyrolysis	Hexamethyldisiloxane, ethanol	Air and oxygen	Mädler et al. (2002)
Au coated	Ultrasonic irradiation	TEOS	Alkaline	Pol et al. (2003)
Native	Ball milling	Rice husk	Heating rate (5 K/min), activation energy (166±10 kJ/mol)	Liou (2004)
Mesoporous	Microfluidics	TEOS, hexadecane, C_2H_5OH	Acidic, room temperature	Carroll et al. (2008)
PEEK/nano-SiO_2 composites	Ball milling	Poly(etherether ketone)	Room temperature for NPs, later compression molding, melt mixing	Zhang et al. (2008)
PEEK silica nanoparticles	Ball milling	Poly(etherether ketone), 3-glycidoxypropyltrimethoxy silane	Vial volume (600 ml); ball to powder weight ratio (26:1); 280 rpm; ball and vial material (hardened Cr steel); room temperature; Ar atmosphere	Hedayati et al. (2011)
Native	Laser pyrolysis	SiO_2, Si, SiC, Si_3N_3	Gas phase inert atmosphere	Borsella et al. (2011)
		Physicochemical methods		
ZnS NPs on Si nanospheres	Sonochemical	EOS, zinc acetate, thioacetamide	Ultrasonic irradiation (20 kHz, 100 Wcm^{-2})	Arul Dhas et al. (1999)
Eu_2O_3, Tb_2O_3 doped in SiO_2 NPs; Eu_2O_3, Tb_2O3-coated	Stober, sonochemical	TEOS, Europium oxide (Eu_2O_3), terbium oxide (Tb_2O_3)	Alkaline	Patra et al. (1999)
Si-coated ZnO nanoparticles	Sonochemical	TEOS, aqueous ethanolic ammonia	Alkaline	Siddiquey et al. (2012)
Iron oxide nanoparticles coated with silica	Iron oxide NPs (co-precipitation); Si coating (sonochemical synthesis)	TEVS, $FeCl_3.6H_2O$, $FeCl_2.4H_2O$, NaCl	Acidic ferrofluid and silica mixture, pH >3; sonicated (20 kHz, 6 W cm^{-2}) for 1 h; NH_4OH addition	Sodipo and Aziz (2013)
Si-coated terbium fluorescent nanoparticles	Water-in-oil microemulsion	TEOS, Triton X-100, N,N,N1,N1-[2,6-bis(3'-aminomethyl-1'-pyrazolyl)-phenylpyridine] tetrakis(acetate)-Tb^{3+}	Room temperature, surface modification (4°C)	Ye et al. (2004)
		Chemical methods		
Native	Diffusion reactor	TEOS	Gradients in TEOS, gas flow, and composition	Jang (2001)
Native	Electrospray-assisted chemical vapor deposition	TEOS	Aging at 80°C	Nakaso et al. (2003)

(Continued)

TABLE 2.1 (CONTINUED)
Silica Nanomaterials Synthesized Using Different Physical and/or Chemical Methods

Type of silica nanomaterials	Synthetic approach	Reagent(s) used	Condition	Reference
γ-Fe_2O_3@SiO_2 core/shell	Multi-step continuous flow microreactor synthesis	APTES, ammonic TEOS, EDC, NHS, and RITC	Optimized coaxial flow	Abou-Hassan et al. (2009)
YVO_4:Ln^{3+} coated	Microemulsion	CTAB, n-butanol, hexane, water	Room temperature for NPs, coatings at 500°C and 900°C	Luwang et al. (2011)
Gadolinium-DOTAGA-based	Top-down synthesis	$GdCl_3$, NaOH, TEOS, APTES, DOTAGA	Room temperature	Mignot et al. (2013)
Large-cavity mesoporous silica nanoparticles	Chemical etching and reverse microemulsion, doxorubicin loading	Cetyltrimethylammonium chloride	Room temperature	Chen et al. (2014)
Silica–nylon 6 nanocomposites	Gelation of polymer nylon	SNPs, estradiol. estriol, estrone, methanol, formic acid, nylon 6	Acidic, room temperature	Reyes-Gallardo et al. (2017)
ε-Polylysine-capped MSNPs	TEOS hydrolysis	TEOS, CTAB, safranine O, rhodamine B, Hoechst 33342	Room temperature	De la Torre et al. (2018)
Polyethylenimine-coated Fe_3O_4 core shell-siRNA and daunorubicin loaded	TEOS hydrolysis	TEOS, Anti-TWIST siRNA, daunorubicin, NH_4NO_3	Alkaline, room temperature	Sanchez-Salcedo et al. (2018)

TABLE 2.2
Silica Nanomaterials Synthesized Using Various Biomaterials

Type of silica nanomaterials	Method	Approach/technique	Reagent(s) used	Condition	Reference
Hexagonal, ladder-like, petal-like silica structures	Biomimetic	Poly-L-lysine reaction with TMOS	Poly-L-lysine, TMOS	Stirring of reaction mixtures	Patwardhan et al. (2003)
Quasispherical, mono-dispersed, hollow silica nanoparticles	Biogenic	Bioleaching by *Fusarium oxysporum*	*Fusarium oxysporum* biomass, sand, phenol/chloroform	Room temperature	Bansal et al. (2005a)
Quasispherical silica nanoparticles	Biogenic	Hydrolysis of aqueous anionic complex of SiF_6^{2-} by *Fusarium oxysporum*	Fungal biomass, K_2SiF_6	Aqueous K_2SiF_6 (pH 3.1), reactants mixed (200 rpm at 27°C)	Bansal et al. (2005b)
Silica nanoparticles	Biogenic	Biofabrication using *Actinobacter* sp.	*Actinobacter* biomass, K_2SiF_6	27°C; 200 rpm	Singh et al. (2008)
Silica nanoparticles	Biogenic	Alkali dissolution-precipitation	Semi-burned rice straw ash, NaOH, HCl	Box-Behnken design, 2 N NaOH, 100°C, 4 h, 700 rpm	Zaky et al. (2008)
Nanosized silica layers (1 nm and ~20 nm) by thick shell silica-modified TMV templates, metal-coated thick shell silica-modified TMV templates	Biogenic	Layering of TEOS onto aniline-coated tobacco mosaic virus	TEOS, aniline, TMV, $AgClO_4{\cdot}H_2O$, $HAuCl_4{\cdot}3H_2O$, $K_2PtCl_4{\cdot}H_2O$, DMAB, $C_6H_{16}O_3SSi$	Different conditions for various metal-coated TMV templates	Royston et al. (2009)
Silica nanotubes	Biomimetic	Deposition of Si on I_3K peptide nanotubes through lysine mediated sol-gel process	I_3K, TEOS	Ambient temperature; varied concentrations of peptide and silica precursor, solution pH, and reaction time	Wang et al. (2011)
Cobalt rice husk silica nanoparticles	Biogenic	Acid digestion	Rice husk, HNO_3, 20 wt.% Co^{2+}	pH 3, Room temperature	Adam et al. (2011)
Nano silica powder (25 nm)	Biogenic	Alkali extraction followed by acid precipitation	Rice husk, NaOH, HCl	2.5 N NaOH, HCl, pH 2	Yuvakkumar et al. (2012)
Amorphous silica nanoparticles	Biogenic	Laser ablation and NaOH treatment of sugar beet bagasse	Sugarbeet bagasse, NaOH, HCl, HNO_3	(Nd:YLF laser 527 nm, pulse duration 100 ns, output power 16 W, pulse repetition rate 1 kHz, pulse energy 16 mJ); acidic pH followed by alkaline conditions with NaOH	San et al. (2014)
Amorphous silica nanoparticles (10–15 nm)	Biogenic	Alkali dissolution-precipitation	Rice straw ash, NaOH, H_2SO_4	Triple factorial design; 2 N NaOH, 20% H_2SO_4, 100°C, 4 h, 700 rpm	Khorsand et al. (2013)
Human mesenchymal stem cell–compatible silica nanoparticles	Biogenic	Acid digestion and calcination	Rice husk, HCl	Rice husk pretreated with HCl followed by incineration at 120°C under pressure	Athinarayanan et al. (2014)
Silica nanoparticles (anti-corrosion coating)	Green synthesis	Alkali treatment followed by acid digestion	Rice husk, 6 M H_2SO_4,	Progression of reaction from alkaline to acidic condition	Vijayalakshmi et al. (2015)
Silica nanoparticles	Biogenic	Acid digestion	Rice husk, HCl	120°C under pressure followed by calcination process	Alshatwi et al. (2015)
Si nanowires (~25 nm)	Biogenic	Adam procedure	HNO_3, green beans, NaOH	100°C for 24 h, later pH 9 maintained by NaOH addition	Ghorbani (2018)
Amorphous silica nanoparticles (antimicrobial)	Biogenic	Modified Adam procedure	HNO_3, NaOH, *Cynodon dactylon* leaf extract	Progression of reaction from acidic to alkaline and heating final product to 80°C for 12 h	Babu et al. (2018)
Silica nanoparticles	Biogenic	Biofabrication with basidiomycetes	Na_2SiO_3, extracellular extracts of *Lentinus edodes*, *Ganoderma lucidum*, *Pleurotus ostreatus*	Bioreduction	Vetchinkina et al. (2018)
AgNP-coated mesoporous silica nanoparticles	Green synthesis	Tyrosine reduction	Tyrosine, ICPTES, TEOS, DMF, CTAB, NaOH	Room temperature	Ratirotjanakul et al. (2018)

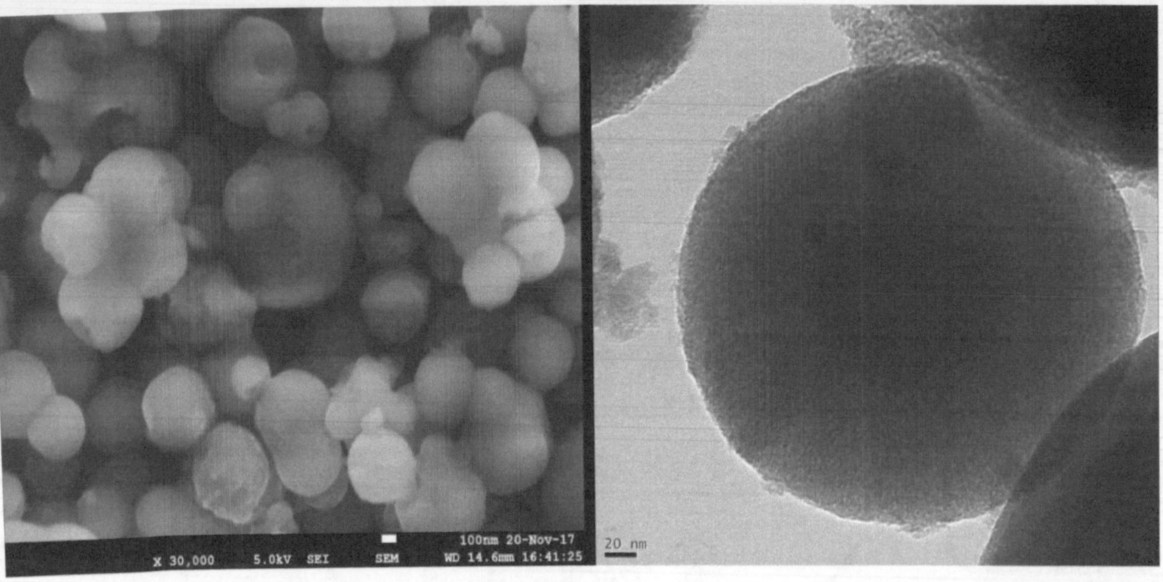

FIGURE 2.1 Micrographs of (A) FE-SEM, and (B) TEM representing mesoporous silica nanoparticles (Unpublished data).

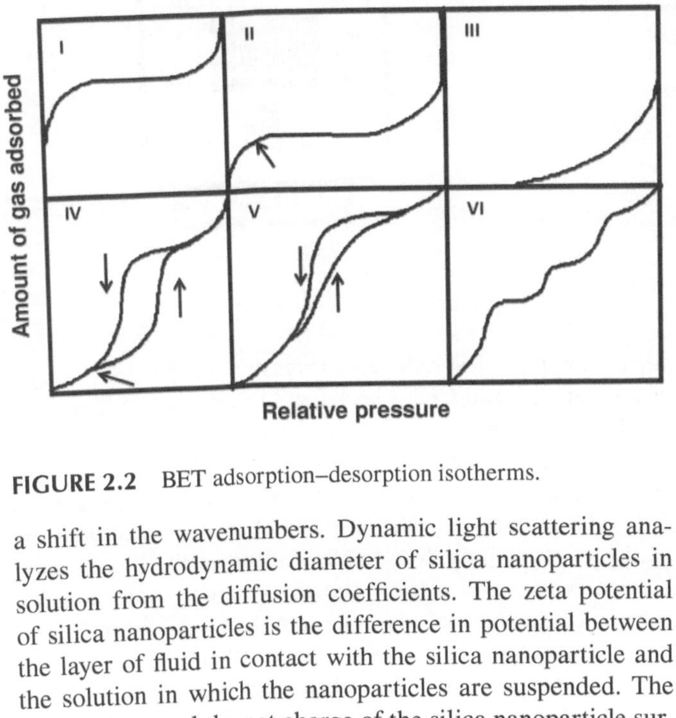

FIGURE 2.2 BET adsorption–desorption isotherms.

a shift in the wavenumbers. Dynamic light scattering analyzes the hydrodynamic diameter of silica nanoparticles in solution from the diffusion coefficients. The zeta potential of silica nanoparticles is the difference in potential between the layer of fluid in contact with the silica nanoparticle and the solution in which the nanoparticles are suspended. The ionic groups and the net charge of the silica nanoparticle surface play a significant role in determining the zeta potential of the dispersed nanoparticles. ^{29}Si solid-state NMR (Nuclear Magnetic Resonance) spectroscopy enables the silica nanoforms to be subjected to magnetic fields and the magnetic field lines are passed. The nuclei get excited by the magnetic field lines, and a resonance is generated. Each intramolecular magnetic field inside a particular molecule exerts a specific resonance pattern. The XRD technique enables assessment of the purity and characterizes whether the nanomaterial has a crystalline nature. When X-rays hit the silica nanoparticles, they get diffracted with interference by following Bragg's rule: $2d \sin \theta = n\lambda$. X-ray photoelectron spectroscopy measures the electron binding energies of the silica nanoparticles. As the energy of incident X-rays are known, the measured kinetic energies of the sample electrons allow the measurement of the electron-binding energy of the sample. Time-of-flight secondary ion mass spectrometry (ToF-SIMS) is a highly sensitive method for probing the surface composition of silica nanoparticles. The silica nanoparticles are hit by primary ion beams that sputter the nanoparticle surface to remove secondary ions from the surface, which are received by a time-of-flight detector, and the mass/charge ratio is represented in terms of *m/z* versus intensity.

2.4 APPLICATIONS OF SILICA-BASED NANOMATERIALS

Researchers have prepared diverse silica nanoforms of various shapes, including spheres, vesicles, core-shell structures, tubes, rods, wires, cubes, hexagonal, honeycomb-shaped, etc. Their salient topological, chemical, and physical properties make them attractive candidates for use in diverse niche applications (Figure 2.3).

2.4.1 Drug Delivery (Theranostics)

The distinct and dynamic tailorable features of silica nanoforms, including size, shape, volume, surface area, surface chemistry, diameter, pore volume, pore diameter, and drug loading ability have tuned them as a nanocargo for drug loading, specific targeting, slow release, and monitoring and imaging drug/gene delivery. Single fluorescein molecules were trapped in 3-(propyldisulfanyl) propionic acid–modified sponge-like silica nanoparticles. These modified groups enabled the controlled release of fluorescein in response to intracellular agents like dihydrolipoic acid and dithiothreitol in a dose-dependent pattern in human cervical cancer cell lines (Giri et al. 2005). Novel silica nanorod-based versatile nanoprobes with highly paramagnetic Gd^{3+} and fluorescein

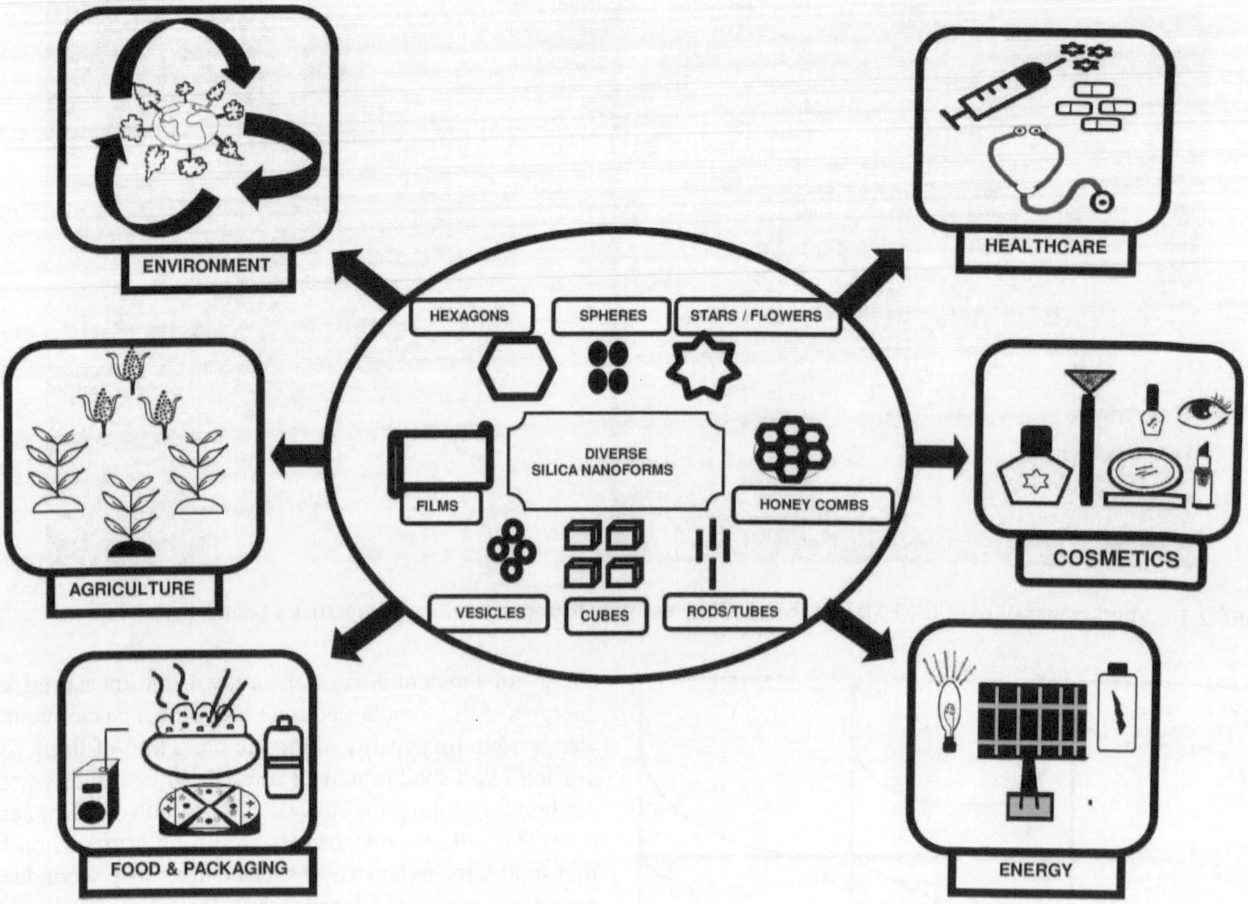

FIGURE 2.3 Diverse silica nanoforms employed in various niche application areas.

isothiocyanate were demonstrated for cell imaging, which could find application as MRI contrast agents and cell tracking owing to their T2-enhancing ability (Tsai et al. 2008). Amine-functionalized hollow mesoporous silica nanoparticles (MSNPs) were potent against *Mycobacterium smegmatis* strain mc2 651, and also effective against A549 cancer cells with several-fold lower IC_{50} values (Hao et al. 2015). Moghaddam et al. (2018) synthesized hollow MSNPs with disulfide bridges, loaded with doxorubicin, that were effective against MCF-7 cells *in vitro*, while being non-toxic to the normal RAW 264.7 macrophages and NIH 3T3 fibroblasts. The diverse functionality of Si nanotubes that act as dual functional theranostic tools for drug delivery vehicles and imaging was demonstrated by Namgung et al. (2011). Huang et al. (2015) demonstrated new superparamagnetic iron oxide nanoparticles (SPION)-incorporated magnetic silica nanotubes impregnated with Mn_3O_4 that were effective against HeLa cell lines with outstanding MRI-T1 scanning outputs. The ideal cargo capacity of silica nanosystems was also extended to nucleic acids and RNA interference. The sensitivity of nucleic acids, especially RNA, demands a stable, inert, and specific delivery vehicle. Finlay et al. (2015) reported the application of silica nanoparticles as a nanocargo transportation vehicle for TWIST1-siRNA in MDA-MB-435S melanoma cells and female NOD.Cg-Prkdc^{scid}Il2rg^{tm1Wjl}/SzJ (NSG) mice. These silica nanoparticles reduced expression of TWIST1 target genes, tumors, and CCL2 levels and prevented angiogenesis. Gold-nanoparticle capped on chlorine-doped MSNPs were fabricated and explored for application in photothermal therapy or photodynamic therapy against tumor cells (Sun et al. 2017). In another step ahead, researchers have explored the ability of nanosilica drug carriers to cross the blood–brain barrier. Kong et al. (2012) showed that Fe_3O_4-silica nanospheres could successfully penetrate the blood–brain barrier under a magnetic field in mice. In another study, arginine-glycine-aspartate–modified silica nanoparticles promoted the blood–brain barrier penetration of anti-glioblastoma selenocysteine derivative in mice (You et al. 2016). Tamba et al. (2018) showed that glucose and glucose-poly(ethylene glycol) methyl ether amine–modified silica nanoparticles could penetrate the blood–brain barrier in mice. Some examples of MSNPs with different functionalization agents and their target-based applications are shown in Table 2.3.

2.4.2 Biosensors

Nano-immunosensors find application in rapid veterinary diagnostics for detection of retroviral leukemia in cows and healthy cows by exploring the advantage of the optical property of silica nanotubes in their hydrogel form (Viter et al. 2011).

TABLE 2.3
Silica Nanomaterials with Different Functionalization Agents and Their Target-Based Applications

Type of silica nanomaterials	Functionalization agent	Drug loaded	Target site	Application	Reference
Mesoporous silica nanoparticles capped with superparamagnetic Fe_3O_4 NPs	3-(propyldisulfanyl)propionic acid	Fluorescein	Cervical carcinoma cell line (HeLa)	Stimulus-responsive controlled release delivery	Giri et al. (2005)
Mesoporous silica nanoparticles	Fluorescein isothiocyanate (FITC)	Paclitaxel	Human pancreatic cancer cells (PANC-1)	Anticancer therapy	Lu et al. (2007a)
Mesoporous silica nanoparticles	FITC	Camptothecin	Human pancreatic cancer cells (PANC-1)	Anticancer therapy	Lu et al. (2007b)
Mesoporous silica nanoparticles	Lactobionic acid as cell-specific targeting moiety, collagen end-capping agent by disulfide linker	FITC	Intracellular	Redox-based intracellular drug delivery	Luo et al. (2010)
Mesoporous silica nanoparticles	Amine and folic acid	Methotrexate	HeLa cell line	Endocytosis of drug-loaded MSNPs	Rosenholm et al. (2010)
Mesoporous silica nanoparticles	Gelatin/chitosan (E2-MSN@PEM) via layer-by-layer assembly technique on titanium substrates	β-Estradiol	Osteoblasts/osteoclasts	Bone homeostasis	Hu et al. (2010)
Mesoporous silica nanoparticles	FITC or Alexafluor 750 labeling	γ-secretase inhibitors	Notch signal	*In vitro* and *in vivo* inhibition of Notch signaling	Mamaeva et al. (2011)
Mesoporous silica nanoparticles	TAT peptide	Doxorubicin	Nucleus	Anticancer therapy	Pan et al. (2012)
Mesoporous silica nanoparticles	Folic acid	Camptothecin	Pancreatic cancer xenografts in mice	*In vivo* pancreatic tumor suppression	Lu et al. (2012)
Mesoporous silica nanoparticles	Photothermal-responsive [2] rotaxane-modified	Curcumin	Zebra fish larvae	Photothermal-controlled drug delivery	Yan et al. (2012)
Mesoporous silica nanoparticles	Fluorescein and galactose	Camptothecin	Human colorectal, pancreatic, and breast cancer cell lines	Combination of photodynamic therapy, specific targeting, and drug delivery	Gary-Bobo et al. (2012)
Mesoporous silica nanoparticles	Thiol, PEG, TRC105 Ig 64, Cu labeling	Doxorubicin	4T1 tumor-bearing mice	Image-guided drug delivery	Chen et al. (2013)
Mesoporous silica nanoparticles	Functionalization with COOH groups and Cy-5 dye	Doxorubicin	MCF-7 cell line	Anticancer therapy	Xie et al. (2013)
Mesoporous silica nanoparticles	Hyaluronic acid conjugation	Hyaluronic acid (HA)	Human colon cancer cell line	Targeted delivery of CD44 over-expressing tumors via HA-mediated endocytosis	Yu et al. (2013)
Mesoporous silica nanoparticles	RGDFFFFC peptide and benzoic imine modification, PEGylation, disulfide linkage, click chemistry	Doxorubicin	Human glioblastoma cell line (U-87)	Anticancer therapy	Xiao et al. (2013)

(Continued)

TABLE 2.3 (CONTINUED)
Silica Nanomaterials with Different Functionalization Agents and Their Target-Based Applications

Type of silica nanomaterials	Functionalization agent	Drug loaded	Target site	Application	Reference
Mesoporous silica nanoparticles	β-Cyclodextrin anchoring, Arg-Gly-Asp, Leu-Gly-Val-Arg	Doxorubicin	Squamous cell carcinoma cells, human colon cancer cells	Glutathione induced easy uptake	Zhang et al. (2013)
Mesoporous silica nanoparticles	Piperazine-based methoxymethyl (MOM)-protected N-diazeniumdiolate	n.s.	*Streptococcus mutans* adhered dental restorative materials	Nitric oxide (NO) delivery for plaque dental restorative treatment	Carpenter et al. (2013)
Mesoporous silica nanoparticles	15-mer single-stranded DNA (ssDNA)	Doxorubicin	Cervical carcinoma cell line (HeLa)	Anticancer therapy	Ma et al. (2013)
Mesoporous silica nanoparticles	L-cysteine-derivatized gold nanoparticles (AuNPs), bound to amino-functionalized MSNs using Cu^{2+} as a bridging ion	Doxorubicin	Cervical carcinoma cell line (HeLa)	Stimulus-responsive controlled drug release delivery vehicle	Chen et al. (2014)
Mesoporous silica nanoparticles	Poly(ethylene glycol) modification and amino-β-cyclodextrin functionalization	Doxorubicin	MDA-MB-231 xenograft model	Anticancer therapy	Zhang et al. (2014)
Mesoporous, dual-compartment Janus particles	Upconversion nanoparticle@ SiO_2@ mesoporous nanoSiO_2 @periodic mesoporous organosilica	Paclitaxel and Doxorubicin	Cervical carcinoma cell line (HeLa)	Anticancer therapy	Li et al. (2014)
Mesoporous silica nanoparticles	Acetal immobilization, lanthanide gate keeping	Doxorubicin	Cervical carcinoma cell line (HeLa) and murine hepatocarcinoma cell lines	Anticancer therapy; enhanced permeation-retention, MRI imaging	Chen et al. (2015)
Mesoporous silica nanoparticles	poly(2-(diethylamino)ethyl methacrylate) modification, disulfide cleavage, O-nitrobenzyl ester cleavage	Doxorubicin	Cervical carcinoma cell line (HeLa)	Anticancer therapy	Zhang et al. (2015)
Mesoporous silica nanoparticles	Pores supported by rotaxanes, azido-GFLGR7RGDS peptide modification	Doxorubicin	αvβ3-positive HeLa cell line	Anticancer therapy	Cheng et al. (2015)
Mesoporous silica nanoparticles	Folic acid-Chitosan-NR-MSP	Anisomelic acid	Folate receptor (FR)-positive SiHa cell line	Cervical cancer treatment	Senthilkumar et al. (2015)
No-releasing mesoporous silica nanoparticles	Esterase- and pH-responsive poly(β-amino ester)	Doxorubicin	Breast carcinoma cell line (MDA-MB-231)	Anticancer therapy	Fernando et al. (2015)
Mesoporous silica nanoparticles	Folic acid conjugated, FITC	Cisplatin	Cervical carcinoma cell line (HeLa)	Cisplatin(IV)-prodrug Si NP-based delivery	Alvarez-Berríos and Vivero-Escoto (2016)
Mesoporous silica nanoparticles	Acetylcholinesterase	Rhodamine	n.s.	Diisopropyl fluorophosphate (DFP) detection in aqueous environments	Pascual et al. (2016)
Mesoporous silica nanocomposites	Folic acid	Polydopamine	Hepatocellular carcinoma cell line (HepG2)	Chemo-photothermal therapy	Zhang et al. (2018)
Mesoporous silica nanoparticles	Folic acid conjugated	Topotecan	Retinoblastoma cancer cells (Y79)	Retinoblastoma cancer treatment	Qu et al. (2018)

n.s.: not specified.

Silica nanotubes were used for the fabrication of quantum dot barcoded probes as DNA sensors for specific detection of *E. coli*. These silica nanotubes were initially synthesized using anodic aluminum oxide templates, followed by plasma treatment (Kim et al. 2011). With regard to food systems, Long et al. (2016) fabricated nanosilica-based molecular imprinting for sensitive detection of rhodamine B, an allergen and cytotoxic agent present in red wine. These rhodamine-based nanosilica materials were used for the detection of rhodamine B as observed from HPLC combined with an extraction step. Further, Luo et al. (2018) demonstrated a sensitive detection method for ethyl carbamate in red wine based on fluorescence-based immunoassay. The aqueous o-phenylenediamine chromogenic substrate was combined with the fluorescent silicon nanoparticles for supporting the ELISA system based on ratiometric fluorescence, which were referenced against 2,3-diaminophenazine standard quenching-response signals, which showed 33-fold more sensitivity.

2.4.3 Antimicrobial Coatings

Nanosilica offers a good choice for fabrication of coatings owing to its minimal toxicity, long shelf life, tailorable silane chemistry, and thermal stability. Wang et al. (2017) reported gentamicin-based silica nanocoating on titanium implants that promisingly showed controlled drug release and inhibited *Staphylococcus aureus*. Mustafa (2018) reported organic and inorganic hybrid silica nanocoatings that inhibited *Salmonella gallinarum*, *Pseudomonas fluorescens*, *Xanthomonas campestris*, and *Burkholderia pseudomallei*. Verma and Bhattacharya (2018) reported a silica-titania core-shell nanoparticle-based coating that efficiently killed up to 70% of the pathogenic *Fusarium solani* and significantly diminished the growth up to ~90% of the pathogenic *Acremonium kiliense* and *A. strictum*. Further, silica nanocoatings applied over glass exhibited promising antibacterial activity apart from their anti-reflective properties (Wang et al. 2018). A recent study from our laboratory demonstrated that mesoporous silica nanoparticles functionalized with phenazine-1-carboxamide derived from *Pseudomonas aeruginosa* strain CGK-KS-1 exhibited promising role as antibiofilm coatings on silicone urethral catheters, which showed sustained release and targeted delivery of the bioactive compound at specific pH and inhibited mixed populations of uropathogenic *Candida albicans* and *Staphylococcus aureus* biofilms (Kanugala et al. 2019).

2.4.4 Food Industry

Food packaging is one of the key components in the food industry since it prevents food from degradation by various physical, chemical, and biological factors. Food-packing materials play a key role in terms of their safety and biocompatibility. Silica has been explored as a food-packaging material owing to its non-toxic, non-irritant nature and its biocompatibility with food. Xiong et al. (2008) reported that combination of silica nanoparticles with starch scaffolds increased the tensile strength apart from reducing the hydrophilicity of starch. Vladimirov et al. (2006) reported that combination of silica nanoparticles with isotactic polypropylene increased the storage profiles by offering stiffness and enhanced O_2 barrier capacity. Jia et al. (2007) fabricated nanocomposites made of polyvinyl alcohol and SiO_2 nanoparticles using vinyl acetate copolymerization. These hybrid silica-based nanocomposites were more thermostable and mechanically flexible as compared with polyvinyl alcohol, which was attributed to the covalent interactions between the silica nanoparticles and polyvinyl groups.

2.4.5 Agriculture

Several studies demonstrated that silica nanoparticles exhibit an important role in the seed germination process and/or in plant growth promotion. Shah and Belozerova (2009) reported aminopropyl-functionalized silica nanoparticles as an effective germination-promoting agent in lettuce, which enhanced the shoot/root proportions. Yuvakkumar et al. (2011) reported the effect of nanosilica particles on the growth and survival of maize crops *in vitro* and observed an increase in stem height and width, average number of foliage, seed germination percentage, efficacy of water utility, net chlorophyll quantity, node length, mean count of nodes, weights of dry and fresh seeds, cob weights and lengths, and number of granules. Haghighi et al. (2012) studied the effect of silica nanoparticles on the germination of tomato seeds to circumvent the adverse effects of salinity. The germination percentage, germination index, lengths of roots and shoots, and weights of fresh and dry seeds were enhanced in the presence of 1 mM silica nanoparticles. Suriyaprabha et al. (2012) reported biogenic silica nanoparticles synthesized from rice husk and their uptake effects on hydroponics of *Zea mays* seeds. The germination percentage, mean germination duration, number of roots formed, root lengths, and number of shoots were observed to be very high in nano-SiO_2-treated seeds as compared to other experimental groups (Micro-SiO_2, Na_2SiO_3, and H_4SiO_4). Some examples of different silica-based nanomaterials using different functionalization agents for application in agriculture sector are shown in Table 2.4.

2.4.6 Cosmetics

Liposomes were first introduced in cosmetics way back in 1986 by the Christian Dior Company. Subsequently, many cosmetic companies have applied nanotechnology in their formulations. Silica in its amorphous or nanoform(s) offers many promising advantages in cosmetic formulations (Wu 2012). Nanosilica or fumed silica has unique properties including versatile surface chemistry, minimal bulk density, large surface area, emulsification, dispersion, pigment retention, aggregate prevention, smooth flow, anti-grease, adsorption of debris, etc., which makes them an attractive choice for use in various cosmetic products for skin, hair, face, lips, and nails. All these features provide anti-caking, anti-sag, anti-setting, thickening, rheotropic, and thixotropic properties to silica.

TABLE 2.4
Silica Nanomaterials with Different Functionalization Agents for Application in Agriculture Sector

Type of Silica Nanomaterials	Modification/Functionalization	Target	Highlight	Reference
Mesoporous silica nanoparticles	Amine cross-linking, fluorescein isothiocyanate labeling	Wheat, lupin, *Arabidopsis*	Localization of nanoparticles to roots and leaves	Hussain et al. (2013)
Mesoporous silica nanoparticles	Fluorescein or rhodamine doping, *N*-trimethoxysilylpropyl-*N,N,N*-trimethylammonium chloride and 3-trihydroxysilyl)propylmethyl phosphonate functionalization, plasmid DNA loading	Tobacco BY-2 protoplasts, *Arabidopsis* roots	Enzyme free, biocompatible and efficient gene delivery into plants	Chang et al. (2013)
Amorphous nanosilica	Native	Maize seeds	Promotion of seed germination, positive effect on growth of plant beneficial bacteria	Karunakaran et al. (2013)
Colloidal luminescent SiO_2	$Cs_2[Mo_6Br_{14}]@SiO_2$	*Brassica napus*	No phytotoxicity	Aubert et al. (2013)
Nanosized SiO_2	Native silica nanoparticles	*Agropyron elongatum* L.	Promoted seed germination, dry weights of roots, shoots, and seedlings	Azimi et al. (2014)
SiO_2 nanoparticles	Native	*Crataegus* sp.	Enhanced resistance to drought stress	Ashkavand et al. (2015)
Silicon nanoparticles	Cr(VI)	*Pisum sativum* (L.) seedlings	Protection from oxidative stress	Tripathi et al. (2015)
Nano and micro SiO_2 particles	Native	Maize seeds and roots	Nanoparticles enhanced seed germination, microparticles enhanced seed germination and root elongation	Karunakaran et al. (2016)
Mesoporous silica nanoparticles	Native	Wheat and lupin	Enhanced total leaf protein and chlorophyll, MSNP absorption by roots, and transport to stem and leaves	Sun et al. (2016)
SiO_2 nanoparticles	Native	*Astragalus squarrosus*	Enhanced seed germination	Azimi et al. (2016)
Nanosized silica	Native	*Triticum aestivum*	Protection of seedlings from UV-B stress	Tripathi et al. (2017)
Nanosized silica	Native	*Phaseolus vulgaris*	Enhanced germination, dry mass of root and shoot, length of root and shoot	Alsaeedi et al. (2017)
SiO_2 nanoparticles	Native	*Nigella sativa* L.	Enhanced gene expression of geranyl diphosphate synthase	Kahila et al. (2017)
Mesoporous Au/SiO_2	Auxin loaded	Flax	Enhanced ploidy numbers, embryogenesis, induction, and methylation level	Kokina et al. (2017)
Nanosilicon	Native	Rice plants in cadmium-contaminated land	Reduced cadmium accumulation	Chen et al. (2018)

(*Continued*)

TABLE 2.4 (CONTINUED)
Silica Nanomaterials with Different Functionalization Agents for Application in Agriculture Sector

Type of Silica Nanomaterials	Modification/Functionalization	Target	Highlight	Reference
Silica nanoparticles	Native	*Pimpinella anisum* L.	Enhanced number of seeds, umbels, root diameter, thickness of epidermis and cortex	Menesy et al. (2018)
Mesoporous silica nanoparticles	Cinnamaldehyde encapsulation, aldehyde modification	*Pseudomonas syringae* pv. *pisi* caused blight in pea	143.58% disease free plants	Cadena et al. (2018)
SiO_2 nanoparticles	Native	*Solaum tubersosum*	Enhanced average tube mass, root length, sprout length, chlorophyll content, and carotenoid content	Mushinskiy et al. (2018)
Nanosilica	Native SiNP and *Phoma* sp. GS8-3	Cucumber	Reduced cucumber mosaic virus titers	Elsharkawy et al. (2018)
Nanosilica	NPK fertilizer with nanosilica (Nanochisil)	*Zea mays* L.	Enhanced plant height, number of leaves, wet weight, and dry weight	Pertaminingsih et al. (2018)
Nanosilica	NPK fertilizer with nanosilica	*Canna edulis* Ker.	Enhanced plant height, dry weight, and wet weight of the plant	Barita et al. (2018)
Nanosilica	Native	Cucumber	Improved seed germination by maintaining K^+/Na^+ ratio	Alsaeedi et al. (2018)
Nanosilica	Oligochitosan modified	*Neoscytalidium dimidiatum* infection in dragon fruit plants	Promoted chitinase levels and reduced infection	Tuan et al. (2018)

Cosmetics with various silica salts like hydrated silica, sodium potassium aluminum silicate, alumina magnesium metasilicate, and other silica products were included in primers and foundation creams, conditioners, shampoos, moisturizers, sunscreen lotions, face powders, lipsticks, eye makeup, gels for face and hair, deodorants, etc. (SCCS and Hoet 2016). The Evonik Company from Germany markets various speciality silica-based formulations for use in cosmetics and personal care products. Recent studies showed that encapsulation of organic sunscreens into silica particles enable the reduction in its phototoxicity and degradation (Zhang et al. 2014).

2.5 CONCLUSIONS AND FUTURE PERSPECTIVES

Diverse silica nanoforms which find application in a wide array of sectors like drug delivery, food, energy, agriculture, medicine, environment, etc., are known. There is no doubt that the exploration for the limitless uses of silica nanoparticles in various domains of science and technology is still in its budding stage. A better understanding of the interaction of these silica nanoforms of different sizes and compositions with human biology will enable the use of these smart materials as suitable drug delivery nanocarriers. Size and surface modifications have been critical aspects which need to be considered for assessing toxicity of these materials. Although there are pros and cons in terms of human and ecological safety regarding the use of silica nanoparticles, a better understanding of basic silica nanoscience, silica nanomaterials, and the balanced combination of silica nanotechnology with other areas of science and technology would be necessary to bring fascinating and boundless transformations into our day-to-day lives.

REFERENCES

Abou-Hassan, A., R. Bazzi, and V. Cabuil. 2009. Multistep continuous-flow microsynthesis of magnetic and fluorescent γ-Fe_2O_3@SiO_2 core/shell nanoparticles. *Angewandte Chemie* 121:7316–9.

Adam, F., J. Andas, and I. A. Rahman. 2011. The synthesis and characterization of cobalt-rice husk silica nanoparticles. *The Open Colloid Science Journal* 4:12–8.

Alshatwi, A. A., J. Athinarayanan, and V. S. Periasamy. 2015. Biocompatibility assessment of rice husk-derived biogenic silica nanoparticles for biomedical applications. *Materials Science and Engineering: C* 47:8–16.

Alsaeedi, A., H. El-Ramady, T. A. Alshaal, et al. 2017. Enhancing seed germination and seedlings development of common bean (*Phaseolus vulgaris*) by SiO_2 nanoparticles. *Egyptian Journal of Soil Science* 57:407–15.

Alsaeedi, A., H. El-Ramady, T. A. Alshaal, et al. 2018. Exogenous nanosilica improves germination and growth of cucumber by maintaining K^+/Na^+ ratio under elevated Na^+ stress. *Plant Physiology and Biochemistry* 125:164–71.

Alvarez-Berríos, M. P., and J. L. Vivero-Escoto. 2016. *In vitro* evaluation of folic acid-conjugated redox-responsive mesoporous silica nanoparticles for the delivery of cisplatin. *International Journal of Nanomedicine* 11:6251–65.

Arul Dhas, N., A. Zaban, and A. Gedanken. 1999. Surface synthesis of zinc sulfide nanoparticles on silica microspheres: Sonochemical preparation, characterization, and optical properties. *Chemistry of Materials* 11:806–13.

Ashkavand, P., M. Tabari, M. Zarafshar, et al. 2015. Effect of SiO_2 nanoparticles on drought resistance in hawthorn seedlings. *Forest Research Papers* 76:350–9.

Athinarayanan, J., V. S. Periasamy, M. Alhazmi, et al. 2014. Synthesis of biogenic silica nanoparticles from rice husks for biomedical applications. *Ceramics International* 41:275–81.

Aubert, T., F. Cabello-Hurtado, M. -A. Esnault, et al. 2013. Extended investigations on luminescent $Cs_2[Mo_6Br_{14}]$@SiO_2 nanoparticles: Physico-structural characterizations and toxicity studies. *The Journal of Physical Chemistry C* 117:20154–63.

Azimi, R., M. J. Borzelabad, H. Feizi, et al. 2014. Interaction of SiO_2 nanoparticles with seed prechilling on germination and early seedling growth of tall wheatgrass (*Agropyron elongatum* L.). *Polish Journal of Chemical Technology* 16:25–9.

Azimi, R., G. Heshmati, and R. K. Habib. 2016. Evaluation of SiO_2 nanoparticle effects on seed germination in *Astragalus squarrosus*. *Journal of Rangeland Science* 6:135–43.

Babu, R. H., P. Yugandhar, and N. Savithramma. 2018. Synthesis, characterization and antimicrobial studies of biosilica nanoparticles prepared from *Cynodon dactylon* L.: A green approach. *Bulletin of Materials Science* 41:65. doi: 10.1007/s12034-018-1584-4

Bansal, V., A. Sanyal, D. Rautaray, et al. 2005a. Bioleaching of sand by the fungus *Fusarium oxysporum* as a means of producing extracellular silica nanoparticles. *Advanced Materials* 17:889–92.

Bansal, V., D. Rautaray, A. Bharde, et al. 2005b. Fungus-mediated biosynthesis of silica and titania particles. *Journal of Materials Chemistry* 15:2583–9.

Barita, Y., E. Prihastanti, S. Haryanti, et al. 2018. The influence of granting NPK fertilizer and nanosilica fertilizers on the growth of Ganyong plant (*Canna edulis* Ker.). *Journal of Physics: Conference Series* 1025:012054. doi: 10.1088/1742-6596/1025/1/012054

Beck, J. S., J. C. Vartuli, W. J. Roth, et al. 1992. A new family of mesoporous molecular sieves prepared with liquid crystal templates. *Journal of the American Chemical Society* 114:10834–43.

Borsella, E., R. D'Amato, G. Terranova, et al. 2011. Synthesis of nanoparticles by laser pyrolysis: from research to applications. *Energie, Ambiente e Innovazione* 4–5:54–64.

Cadena, M. B., G. M. Preston, R. A. Van der Hoorn, et al. 2018. Species-specific antimicrobial activity of essential oils and enhancement by encapsulation in mesoporous silica nanoparticles. *Industrial Crops and Products* 122:582–90.

Cakir-Omur, T., R. Gozneli, and Y. Ozkan. 2019. Effects of silica coating by physical vapor deposition and repeated firing on the low-temperature degradation and flexural strength of a zirconia ceramic. *Journal of Prosthodontics* 28:e186–94. doi: 10.1111/jopr.12618

Carpenter, A. W., K. P. Reighard, J. E. Saavedra, et al. 2013. O^2-Protected diazeniumdiolate-modified silica nanoparticles for extended nitric oxide release from dental composites. *Biomaterials Science* 1:456–9.

Carroll, N. J., S. B. Rathod, E. Derbins, et al. 2008. Droplet-based microfluidics for emulsion and solvent evaporation synthesis of monodisperse mesoporous silica microspheres. *Langmuir* 24:658–61.

Chang, F., L. Kuang, C. Huang, et al. 2013. A simple plant gene delivery system using mesoporous silica nanoparticles as carriers. *Journal of Materials Chemistry B* 2013:5279–87.

Chen, Y., H. Chen, L. Guo, et al. 2009. Hollow/rattle-type mesoporous nanostructures by a structural difference-based selective etching strategy. *ACS Nano* 4:529–39.

Chen, W., X. Wang, Q. Tao, et al. 2013. Lotus-like paper/paperboard packaging prepared with nano-modified overprint varnish. *Applied Surface Science* 266:319–25.

Chen, X., X. Cheng, A. H. Soeriyadi, et al. 2014. Stimuli-responsive functionalized mesoporous silica nanoparticles for drug release in response to various biological stimuli. *Biomaterials Science* 2:121–30.

Chen, Y., K. Ai, J. Liu, et al. 2015. Multifunctional envelope-type mesoporous silica nanoparticles for pH-responsive drug delivery and magnetic resonance imaging. *Biomaterials* 60:111–20.

Chen, R., C. Zhang, Y. Zhao, et al. 2018. Foliar application with nano-silicon reduced cadmium accumulation in grains by inhibiting cadmium translocation in rice plants. *Environmental Science and Pollution Research International* 25:2361–8.

Cheng, Y. -J., G. -F. Luo, J. -Y. Zhu, et al. 2015. Enzyme-induced and tumor-targeted drug delivery system based on multifunctional mesoporous silica nanoparticles. *ACS Applied Materials & Interfaces* 7:9078–87.

De la Torre, C., L. Domínguez-Berrocal, J. R. Murguía, et al. 2018. ε-Polylysine-capped mesoporous silica nanoparticles as carrier of the C9h peptide to induce apoptosis in cancer cells. *Chemistry – A European Journal* 24:1890–7.

Dixit, C. K., S. Bhakta, A. Kumar, et al. 2016. Fast nucleation for silica nanoparticle synthesis using a sol–gel method. *Nanoscale* 8:19662–7.

Elsharkawy, M. M., H. Suga, and M. Shimizu. 2018. Systemic resistance induced by *Phoma* sp. GS8-3 and nanosilica against cucumber mosaic virus. *Environmental Science and Pollution Research International*. doi: 10.1007/s11356-018-3321-3

Fernando, I. R., D. P. Ferris, M. Frasconi, et al. 2015. Esterase- and pH-responsive poly(β-amino ester)-capped mesoporous silica nanoparticles for drug delivery. *Nanoscale* 7:7178–83.

Finlay, J., C. M. Roberts, J. Dong, et al. 2015. Mesoporous silica nanoparticle delivery of chemically modified siRNA against TWIST1 leads to reduced tumor burden. *Nanomedicine: Nanotechnology, Biology and Medicine* 11:1657–66.

Gary-Bobo, M., O. Hocine, D. Brevet, et al. 2012. Cancer therapy improvement with mesoporous silica nanoparticles combining targeting, drug delivery and PDT. *International Journal of Pharmaceutics* 423:509–15.

Ghorbani, H. R. 2018. The formation of silica nanowires using green beans. *Revue Roumaine de Chimie* 63:55–8.

Giri, S., B. G. Trewyn, M. P. Stellmaker, et al. 2005. Stimuli-responsive controlled-release delivery system based on mesoporous silica nanorods capped with magnetic nanoparticles. *Angewandte Chemie* 117:5166–72.

Gontad, F., A. Caricato, M. Cesaria, et al. 2017. Decoration of silica nanowires with gold nanoparticles through ultra-short pulsed laser deposition. *Applied Surface Science* 418:430–6.

Haghighi, M., Z. Afifipour, and M. Mozafarian. 2012. The effect of N-Si on tomato seed germination under salinity levels. *Journal of Biological and Environmental Science* 6:87–90.

Hao, N., K. W. Jayawardana, X. Chen, et al. 2015. One-step synthesis of amine-functionalized hollow mesoporous silica nanoparticles as efficient antibacterial and anticancer materials. *ACS Applied Materials & Interfaces* 7:1040–5.

Hedayati, M., M. Salehi, R. Bagheri, et al. 2011. Ball milling preparation and characterization of poly (ether ether ketone)/surface modified silica nanocomposite. *Powder Technology* 207:296–303.

Hu, Y., K. Cai, Z. Luo, et al. 2010. Layer-by-layer assembly of β-estradiol loaded mesoporous silica nanoparticles on titanium substrates and its implication for bone homeostasis. *Advanced Materials* 22:4146–50. doi: 10.1002/adma.201000854

Huang, L., L. Ao, W. Wang, et al. 2015. Multifunctional magnetic silica nanotubes for MR imaging and targeted drug delivery. *Chemical Communications* 51:3923–6.

Huang, L., B. Han, M. Fan, et al. 2017. Design of efficient monoaminosilane precursors for atomic layer deposition of SiO_2 thin films. *RSC Advances* 7:22672–8.

Hussain, H. I., Z. Yi, J. E. Rookes, et al. 2013. Mesoporous silica nanoparticles as a biomolecule delivery vehicle in plants. *Journal of Nanoparticle Research* 15:1676. doi: 10.1007/s11051-013-1676-4

Jang, H. D. 2001. Experimental study of synthesis of silica nanoparticles by a bench-scale diffusion flame reactor. *Powder Technology* 119:102–8.

Jia, X., Y. Li, Q. Cheng, et al. 2007. Preparation and properties of poly(vinyl alcohol)/silica nanocomposites derived from copolymerization of vinyl silica nanoparticles and vinyl acetate. *European Polymer Journal* 43:1123–31.

Kahila, M. M., A. M. Najy, M. Rahaie, et al. 2017. Effect of nanoparticle treatment on expression of a key gene involved in thymoquinone biosynthetic pathway in *Nigella sativa* L. *Natural Product Research* 32:1858–62.

Kanugala, S., S. Jinka, N. Puvvada, et al. 2019. Phenazine-1-carboxamide functionalized mesoporous silica nanoparticles as antimicrobial coatings on silicone urethral catheters. *Scientific Reports* 9:6198. doi: 10.1038/s41598-019-42722-9

Karunakaran, G., P. Manivasakan, R. Yuvakkumar, et al. 2013. Effect of nanosilica and silicon sources on plant growth promoting rhizobacteria, soil nutrients and maize seed germination. *IET Nanobiotechnology* 7:70–7.

Karunakaran, G., V. Rajendran, R. Suriyaprabha, et al. 2016. Influence of ZrO_2, SiO_2, Al_2O_3 and TiO_2 nanoparticles on maize seed germination under different growth conditions. *IET Nanobiotechnology* 10:171–7.

Katsui, H., and T. Goto. 2018. Coatings on ceramic powders by rotary chemical vapor deposition and sintering of the coated powders. *Journal of the Ceramic Society of Japan* 126:413–20.

Khorsand, H., N. Kiayee, and A. H. Masoomparast. 2013. Optimization of amorphous silica nanoparticles synthesis from rice straw ash using design of experiments technique. *Particulate Science and Technology* 31:366–71.

Kim, S. Y., S. J. Son, and J. Min. 2011. A DNA sensor using gold-coated barcode silica nanotubes. In *Nanotech Conference and Expo Technical Proceedings*, Vol. 3, Nanotechnology 2011: Bio Sensors, Instruments, Medical, Environment and Energy, ISBN 978-1-4398-7138-6.

Kokina, I., I. Jahundoviča, I. Mickeviča, et al. 2017. Target transportation of auxin on mesoporous Au/SiO_2 nanoparticles as a method for somaclonal variation increasing in flax (*L. usitatissimum* L.). *Journal of Nanomaterials* 2017:7143269. doi: 10.1155/2017/7143269

Kong, S. D., J. Lee, S. Ramachandran, et al. 2012. Magnetic targeting of nanoparticles across the intact blood–brain barrier. *Journal of Controlled Release* 164:49–57.

Lam, C., Y. Zhang, Y. Tang, et al. 2000. Large-scale synthesis of ultrafine Si nanoparticles by ball milling. *Journal of Crystal Growth* 220:466–70.

Lazarenko, A. A., T. N. Berezovskaya, sD. V. Denisov, et al. 2017. Preparation of a silicon surface for subsequent growth of dilute nitride alloys by molecular-beam epitaxy. *Journal of Physics: Conference Series* 917:032003. doi: 10.1088/1742-6596/917/3/032003

Li, X., L. Zhou, L. Wei, et al. 2014. Anisotropic growth-induced synthesis of dual-compartment janus mesoporous silica nanoparticles for bimodal triggered drugs delivery. *Journal of the American Chemical Society* 136:15086–92.

Liang, J., X. Li, Y. Zhu, et al. 2014. Hydrothermal synthesis of nano-silicon from a silica sol and its use in lithium ion batteries. *Nano Research* 8:1497–504.

Liou, T. 2004. Preparation and characterization of nano-structured silica from rice husk. *Materials Science and Engineering: A* 364:313–23.

Long, Z., W. Xu, Y. Lu, et al. 2016. Nanosilica-based molecularly imprinted polymer nanoshell for specific recognition and determination of rhodamine B in red wine and beverages. *Journal of Chromatography B* 1029–1030:230–8.

Lu, J., M. Liong, S. Sherman, et al. 2007a. Mesoporous silica nanoparticles for cancer therapy: Energy-dependent cellular uptake and delivery of paclitaxel to cancer cells. *Nanobiotechnology* 3:89–95.

Lu, J., M. Liong, J. I. Zink, et al. 2007b. Mesoporous silica nanoparticles as a delivery system for hydrophobic anticancer drugs. *Small* 3:1341–6.

Lu, J., Z. Li, J. I. Zink, et al. 2012. *In vivo* tumor suppression efficacy of mesoporous silica nanoparticle-based drug delivery system: Enhanced efficacy by folate modification. *Nanomedicine in Cancer* 8:215–34.

Luo, Z., K. Cai, Y. Hu, et al. 2010. Mesoporous silica nanoparticles end-capped with collagen: redox-responsive nanoreservoirs for targeted drug delivery. *Angewandte Chemie* 123:666–9.

Luo, L., Y. Song, C. Zhu, et al. 2018. Fluorescent silicon nanoparticles-based ratiometric fluorescence immunoassay for sensitive detection of ethyl carbamate in red wine. *Sensors and Actuators B: Chemical* 255:2742–9.

Luwang, M. N., R. S. Ningthoujam, S. K. Srivastava, et al. 2011. Preparation of white light emitting YVO_4:Ln^{3+} and silica-coated YVO_4: Ln^{3+} (Ln^{3+} = Eu^{3+}, Dy^{3+}, Tm^{3+}) nanoparticles by CTAB/*n*-butanol/hexane/water microemulsion route: Energy transfer and site symmetry studies. *Journal of Materials Chemistry* 21:5326–37.

Ma, X., O. S. Ong, and Y. Zhao. 2013. Dual-responsive drug release from oligonucleotide capped mesoporous silica nanoparticles. *Biomaterials Science* 1:912–7.

Mädler, L., H. Kammler, R. Mueller, et al. 2002. Controlled synthesis of nanostructured particles by flame spray pyrolysis. *Journal of Aerosol Science* 33:369–89.

Mamaeva, V., J. M. Rosenholm, L. T. Bate-Eya, et al. 2011. Mesoporous silica nanoparticles as drug delivery systems for targeted inhibition of notch signaling in cancer. *Molecular Therapy* 19:1538–46.

Menesy, F., M. Khalfallah, N. Rashed, et al. 2018. Response of *Pimpinella anisum* L plant for graphite and silica nanoparticles. *Journal of Sustainable Agricultural Sciences* 44:1–10.

Mignot, A., C. Truillet, F. Lux, et al. 2013. A top-down synthesis route to ultrasmall multifunctional Gd-based silica nanoparticles for theranostic applications. *Chemistry* 19:6122–36.

Moghaddam, S. P. H., M. Yazdimamaghani, and H. Ghandehari. 2018. Glutathione-sensitive hollow mesoporous silica nanoparticles for controlled drug delivery. *Journal of Controlled Release* 282:62–75.

Mushinskiy, A. A., E. V. Aminova, and A. M. Korotkova. 2018. Evaluation of tolerance of tubers *Solanum tuberosum* to silica nanoparticles. *Environmental Science and Pollution Research* 25:34559–69.

Mustafa, B. 2018. Preparation, characterization and antimicrobial properties of silica based nanocoatings. *Journal of Advanced Chemical Engineering* 8:184. doi: 10.4172/2090-4568.1000184

Nakaso, K., B. Han, K. Ahn, et al. 2003. Synthesis of non-agglomerated nanoparticles by an electrospray assisted chemical vapor deposition (ES-CVD) method. *Journal of Aerosol Science* 34:869–81.

Namgung, R., Y. Zhang, Q. Fang, et al. 2011. Multifunctional silica nanotubes for dual-modality gene delivery and MR imaging. *Biomaterials* 32:3042–52.

Page, M. L., R. Beau, and J. Duchene. 1970. Porous silica particles containing a crystallized phase and method. *US Patent* US 3493341A (Applicant: Produits Chimiques Pechiney Saint Gobain, France).

Pan, L., Q. He, J. Liu, et al. 2012. Nuclear-targeted drug delivery of TAT peptide-conjugated monodisperse mesoporous silica nanoparticles. *Journal of the American Chemical Society* 134:5722–5.

Pascual, L., S. El Sayed, R. Martínez-Máñez, et al. 2016. Acetylcholinesterase-capped mesoporous silica nanoparticles that open in the presence of diisopropylfluorophosphate (a sarin or soman simulant). *Organic Letters* 18:5548–51.

Patra, A., E. Sominska, S. Ramesh, et al. 1999. Sonochemical preparation and characterization of Eu_2O_3 and Tb_2O_3 doped in and coated on silica and alumina nanoparticles. *Journal of Physical Chemistry B* 103:3361–5.

Patwardhan, S. V., N. Mukherjee, M. Steinitz-Kannan, et al. 2003. Bioinspired synthesis of new silica structures. *Chemical Communications* 10:1122–3.

Pertaminingsih, L. D., E. Prihastanti, S. Parman, et al. 2018. Application of inorganic fertilizer with nanochisil and nanosilica on black corn plant growth (*Zea mays* L.). *Journal of Physics: Conference Series* 1025:012128. doi: 10.1088/1742-6596/1025/1/012128

Pol, V. G., A. Gedanken, and J. Calderon-Moreno. 2003. Deposition of gold nanoparticles on silica spheres: A sonochemical approach. *Chemistry of Materials* 15:1111–8.

Qu, W., B. Meng, Y. Yu, et al. 2018. Folic acid-conjugated mesoporous silica nanoparticles for enhanced therapeutic efficacy of topotecan in retina cancers. *International Journal of Nanomedicine* 13:4379–89.

Ratirotjanakul, W., T. Sioloetwong, T. Suteewong, et al. 2018. Green synthesis of AgNPs coated mesoporous silica nanoparticles using tyrosine as reducing/ stabilising agent. *Materials Science Forum* 928:89–93.

Reyes-Gallardo, E. M., R. Lucena, and S. Cárdenas. 2017. Silica nanoparticles–nylon 6 composites: Synthesis, characterization and potential use as sorbent. *RSC Advances* 7:2308–14.

Rosenholm, J. M., E. Peuhu, L. T. Bate-Eya, et al. 2010. Cancer-cell-specific induction of apoptosis using mesoporous silica nanoparticles as drug-delivery vectors. *Small* 6:1234–41.

Royston, E. S., A. D. Brown, M. T. Harris, et al. 2009. Preparation of silica stabilized tobacco mosaic virus templates for the production of metal and layered nanoparticles. *Journal of Colloid and Interface Science* 332:402–7.

San, N. O., C. Kurşungöz, Y. Tümtaş, et al. 2014. Novel one-step synthesis of silica nanoparticles from sugarbeet bagasse by laser ablation and their effects on the growth of freshwater algae culture. *Particuology* 17:29–35.

Sanchez-Salcedo, S., M. Vallet-Regí, S. A. Shahin, et al. 2018. Mesoporous core-shell silica nanoparticles with anti-fouling properties for ovarian cancer therapy. *Chemical Engineering Journal* 340:114–24.

SCCS, and P. H. M. Hoet. 2016. Opinion of the Scientific Committee on Consumer Safety (SCCS) – Revision of the opinion on the safety of the use of silica, hydrated silica, and silica surface modified with alkyl silylates (nano form) in cosmetic products. *Regulatory Toxicology and Pharmacology* 74:79–80.

Senthilkumar, R., D. S. Karaman, P. Paul, et al. 2015. Targeted delivery of a novel anticancer compound anisomelic acid using chitosan-coated porous silica nanorods for enhancing the apoptotic effect. *Biomaterials Science* 3:103–11.

Shah, V., and I. Belozerova. 2009. Influence of metal nanoparticles on the soil microbial community and germination of lettuce seeds. *Water, Air, and Soil Pollution* 197:143–8.

Show, S., A. Tamang, T. Chowdhury, et al. 2015. Bacterial (BKH1) assisted silica nanoparticles from silica rich substrates: A facile and green approach for biotechnological applications. *Colloids and Surfaces B: Biointerfaces* 126:245–50.

Siddiquey, I. A., T. Furusawa, M. Sato, et al. 2012. Sonochemical synthesis, photocatalytic activity and optical properties of silica coated ZnO nanoparticles. *Ultrasonics Sonochemistry* 19:750–5.

Singh, S., U. M. Bhatta, P. V. Satyam, et al. 2008. Bacterial synthesis of silicon/silica nanocomposites. *Journal of Materials Chemistry* 18:2601–6.

Sodipo, B. K., and A. A. Aziz. 2013. Sonochemical synthesis of silica coated super paramagnetic iron oxide nanoparticles. *Materials Science Forum* 756:74–9.

Sun, D., H. I. Hussain, Z. Yi, et al. 2016. Mesoporous silica nanoparticles enhance seedling growth and photosynthesis in wheat and lupin. *Chemosphere* 152:81–91.

Sun, Q., Q. You, X. Pang, et al. 2017. A photoresponsive and rod-shape nanocarrier: Single wavelength of light triggered photothermal and photodynamic therapy based on AuNRs-capped & Ce6-doped mesoporous silica nanorods. *Biomaterials* 122:188–200.

Suriyaprabha, R., G. Karunakaran, R. Yuvakkumar, et al. 2012. Silica nanoparticles for increased silica availability in maize (*Zea mays* L.) seeds under hydroponic conditions. *Current Nanoscience* 8:902–8.

Tamba, B., Streinu, V., Foltea, G., et al. 2018. Tailored surface silica nanoparticles for blood-brain barrier penetration: Preparation and *in vivo* investigation. *Arabian Journal of Chemistry* 11:981–90.

Trewyn, B. G., I. I. Slowing, S. Giri, et al. 2007. Synthesis and functionalization of a mesoporous silica nanoparticle based on the sol–gel process and applications in controlled release. *Accounts of Chemical Research* 40:846–53.

Tripathi, D. K., V. P. Singh, S. M. Prasad, et al. 2015. Silicon nanoparticles (SiNP) alleviate chromium (VI) phytotoxicity in *Pisum sativum* (L.) seedlings. *Plant Physiology and Biochemistry* 96:189–98.

Tripathi, D. K., S. Singh, V. P. Singh, et al. 2017. Silicon nanoparticles more effectively alleviated UV-B stress than silicon in wheat (*Triticum aestivum*) seedlings. *Plant Physiology and Biochemistry* 110:70–81.

Tsai, C. -H, M. Xu, P. Kunal, et al. 2018. Aerobic oxidative esterification of primary alcohols over Pd-Au bimetallic catalysts supported on mesoporous silica nanoparticles. *Catalysis Today* 306:81–8.

Tuan, L. N., B. D. Du, L. D. Ha, et al. 2019. Induction of chitinase and brown spot disease resistance by oligochitosan and nanosilica–oligochitosan in dragon fruit plants. *Agricultural Research* 8:184–90. doi: 10.1007/s40003-018-0384-9

Verma, J., and A. Bhattacharya. 2018. Development of coating formulation with silica–titania core–shell nanoparticles against pathogenic fungus. *Royal Society Open Science* 5:180633. doi: 10.1098/rsos.180633

Vetchinkina, E., E. Loshchinina, M. Kupryashina, et al. 2018. Green synthesis of nanoparticles with extracellular and intracellular extracts of basidiomycetes. *PeerJ* 6:e5237. doi: 10.7717/peerj.5237

Vijayalakshmi, U., V. Vaibhav, M. Chellappa, et al. 2015. Green synthesis of silica nanoparticles and its corrosion resistance behavior on mild steel. *International Conference on Emerging Trends in Chemical Sciences* 92:621–3.

Viter, R., N. Starodub, V. Smyntyna, et al. 2011. Immune biosensor based on silica nanotube hydrogels for rapid biochemical diagnostics of bovine retroviral leukemia. *Procedia Engineering* 25:948–51.

Vladimirov, V., C. Betchev, A. Vassiliou, et al. 2006. Dynamic mechanical and morphological studies of isotactic polypropylene/fumed silica nanocomposites with enhanced gas barrier properties. *Composites Science and Technology* 66:2935–44.

Wang, S., X. Ge, J. Xue, et al. 2011. Mechanistic processes underlying biomimetic synthesis of silica nanotubes from self-assembled ultrashort peptide templates. *Chemistry of Materials* 23:2466–74.

Wang, J., G. Wu, X. Liu, et al. 2017. A decomposable silica-based antibacterial coating for percutaneous titanium implant. *International Journal of Nanomedicine* 12:371–9.

Wang, Y., Y. Wang, X. Li, et al. 2018. Dendritic silica particles with well-dispersed Ag nanoparticles for robust anti-reflective and antibacterial nanocoatings on polymeric glass. *ACS Sustainable Chemistry & Engineering* 6:14071–81.

Wu, X. 2012. Nanotechnology in cosmetics: A review. *Cosmetics and Toiletries* 6:266–72.

Xiao, D., H. Jia, J. Zhang, et al. 2013. A dual-responsive mesoporous silica nanoparticle for tumor-triggered targeting drug delivery. *Small* 10:591–8.

Xie, M., H. Shi, K. Ma, et al. 2013. Hybrid nanoparticles for drug delivery and bioimaging: Mesoporous silica nanoparticles functionalized with carboxyl groups and a near-infrared fluorescent dye. *Journal of Colloid and Interface Science* 395:306–14.

Xiong, H., S. Tang, H. Tang, et al. 2008. The structure and properties of a starch-based biodegradable film. *Carbohydrate Polymers* 71:263–8. doi: 10.1016/j.carbpol.2007.05.035

Yan, H., C. Teh, S. Sreejith, et al. 2012. Functional mesoporous silica nanoparticles for photothermal-controlled drug delivery *in vivo*. *Angewandte Chemie* 51:8373–7.

Yanagisawa, T., T. Shimizu, K. Kuroda, et al. 1990. The preparation of alkyltrimethylammonium-kanemite complexes and their conversion to microporous materials. *Bulletin of the Chemical Society of Japan* 63:988–92.

Ye, Z., M. Tan, G. Wang, et al. 2004. Preparation, characterization, and time-resolved fluorometric application of silica-coated terbium(III) fluorescent nanoparticles. *Analytical Chemistry* 76:513–8.

You, Y., L. Yang, L. He, et al. 2016. Tailored mesoporous silica nanosystem with enhanced permeability of the blood–brain barrier to antagonize glioblastoma. *Journal of Materials Chemistry B* 4:5980–90.

Yu, M., S. Jambhrunkar, P. Thorn, et al. 2013. Hyaluronic acid modified mesoporous silica nanoparticles for targeted drug delivery to CD44-overexpressing cancer cells. *Nanoscale* 5:178–83.

Yuvakkumar, R., V. Elango, V. Rajendran, et al. 2011. Influence of nanosilica powder on the growth of maize crop (*Zea mays* L.). *International Journal of Green Nanotechnology* 3:180–90.

Yuvakkumar, R., V. Elango, V. Rajendran, et al. 2012. High-purity nano silica powder from rice husk using a simple chemical method. *Journal of Experimental Nanoscience* 9:272–81.

Zaky, R. R., M. M. Hessien, A. A. El-Midany, et al. 2008. Preparation of silica nanoparticles from semi-burned rice straw ash. *Powder Technology* 185:31–5.

Zhang, G., A. Schlarb, S. Tria, et al. 2008. Tensile and tribological behaviors of PEEK/nano-SiO_2 composites compounded using a ball milling technique. *Composites Science and Technology* 68:3073–80.

Zhang, J., Z. -F. Yuan, Y. Wang, et al. 2013. Multifunctional envelope-type mesoporous silica nanoparticles for tumor-triggered targeting drug delivery. *Journal of the American Chemical Society* 135:5068–73.

Zhang, J., A. P. Raphael, Y. N. Yang, et al. 2014. Nanodispersed UV blockers in skin-friendly silica vesicles with superior UV-attenuating efficiency. *Journal of Materials Chemistry B* 2:7673–8.

Zhang, Y., C. Y. Ang, M. Li, et al. 2015. Polymer coated hollow mesoporous silica nanoparticles for triple-responsive drug delivery. *ACS Applied Materials & Interfaces* 7:18179–87.

Zhang, H., X. Wang, P. Wang, et al. 2018. One-pot synthesis of biodegradable polydopamine-doped mesoporous silica nanocomposites (PMSNs) as pH-sensitive targeting drug nanocarriers for synergistic chemo-photothermal therapy. *RSC Advances* 8:37433–40.

Zhao, D., J. Feng, Q. Huo, et al. 1998. Triblock copolymer syntheses of mesoporous silica with periodic 50 to 300 angstrom pores. *Science* 279:548–52.

3 Assessing Interactions of Microalgae in Synthesizing Nanoparticles

Manoj Kumar Enamala, Azzuliani Binti Supangat, Chandrasekhar Kuppam, Sudhakar Reddy Pamanji, Murthy Chavali, and Maria P. Nikolova

CONTENTS

3.1 Introduction 29
3.2 The Chemical Composition of Algae 31
3.3 Algal Interactions with Nanoparticles 31
3.3.1 Properties of Gold Nanoparticles 33
3.3.2 Properties of Silver Nanoparticles 33
3.4 Characteristics and Biosynthesis of Nanoparticles 33
3.4.1 Biosynthesis of Gold Nanoparticles 33
3.4.2 Biosynthesis of Silver Nanoparticles 34
3.4.3 Biosynthesis of Iron Particles 35
3.5 Nanoparticles in the Treatment of Cancer 36
3.6 Mechanism of Nanoparticle Synthesis 36
3.7 The Function of Polysaccharides in Releasing Nanoparticles 36
3.8 Effect of Metals on the Growth of Microalgae 38
3.9 The Relation between Nanoparticle Synthesis and Bioenergy 38
3.10 Short Note on Coccoliths 39
3.11 Applications of Algae in Nanotechnology 39
3.11.1 Application of Nanospheres in the Transesterification Reaction 39
3.11.2 Application of Diatoms in Waste Degradation and Metal Toxicity 40
3.11.3 Application of Nano-Cellulose from Algae in Electrical Insulators 40
3.11.4 Applications of Diatoms in the Electronics Industry 40
3.11.5 Applications of Coccoliths in Nanotechnology 40
3.12 Conclusion 41
References 42

3.1 INTRODUCTION

In today's world, the engineering of molecules has been scaled down from huge sizes to nano sizes. Nanotechnology is playing a vital role in the field of science and engineering, and is also an emerging field of science and technology that includes the synthesis and development of countless nanostructures and fabricating nanodevices. These nanotechnology materials when combined with certain research and development methodologies will definitely play serious roles in the field of science and engineering (Figure 3.1). Nanotechnology research is integrating different fields like separation, differentiation, and processing of various materials (Sharma et al. 2016). A nanoparticle (NP) can be defined as an object in the nanoscale that performs as a complete unit as far as its properties are concerned. Basically, nanomaterials are defined as the particles which lie in the range of 10^{-9} and have dimensions of between 1 and 100 nm in size (Sharma et al. 2016). Some of the various properties of the nanoparticles are tabulated in Table 3.1. In the current generation, the study of nanoparticles will definitely give a boost to the research because of their various chemical, optical, and mechanical properties. Various generations of nanotechnologies evolved during the process have been depicted in Figure 3.2. Apart from their other properties, some nanoparticles have been found to be health hazards to various organisms or even to human beings through the process of toxicity. Developing a process by which these nanoparticles can be made non-toxic to the environment. Many researchers have begun exploring "green technology," in which bacteria, fungi, algae, yeast, and many other organisms play their own roles by reducing or even eliminating their production of toxic substances which pose a threat to human beings, and also by producing nanoparticles to protect the environment. Various stable nanoparticles like silver, gold, platinum, selenium, etc., have been successfully synthesized by bacteria, yeast, fungi, algae, and plants. Comparing among these various organisms, the nanoparticle synthesis from green algae (Figure 3.1) is found to be more efficient because

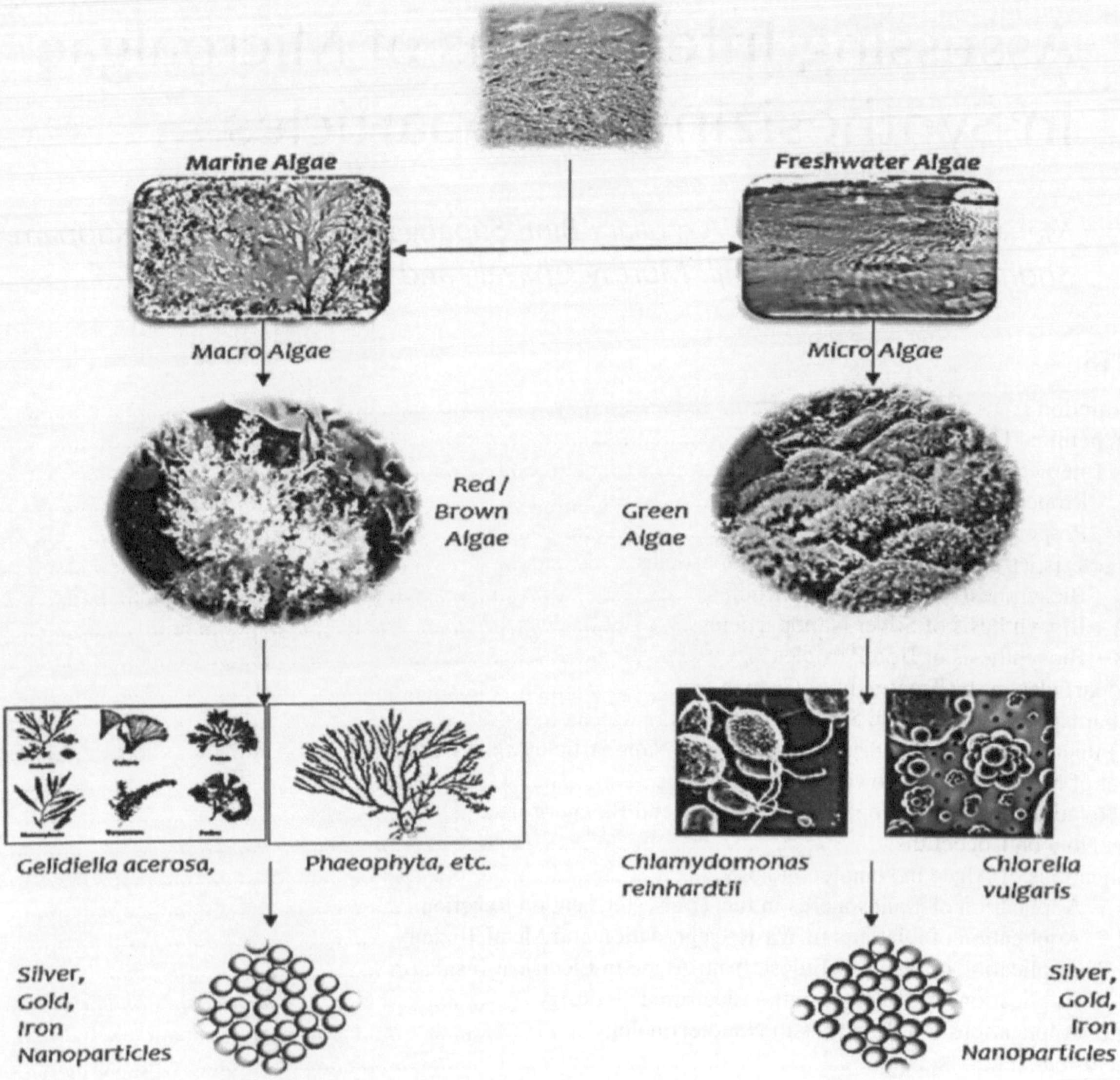

FIGURE 3.1 Major metallic nanoparticles synthesized by algae.

of their abundant sources of phytochemicals, which can be exploited for a good number of applications (Figure 3.3) in the various fields of life science (Karthikeyan et al., 2015). Nanoparticles have also found much effectiveness in work on bacterial pathogenicity and various other biological activities. Much of the research is currently being focused to studying the various antibacterial properties isolated from different bacterial species (Rajeshkumar et al., 2012b). Over the last five years, work has focused on algal species, which are very diverse and which can grow in various habitats like marine water, freshwater lakes, and even the modified mediums which we prepare in the laboratories.

Currently, there are many nanoparticles available in silver, gold, copper, platinum etc., among these, silver nanoparticles have been the most studied because of their diverse properties like magnetism, polarizability, thermal conductivity, etc., which make this particular chemical prominent in the field of nanotechnology (Karthikeyan et al., 2015). The application of this algal nanotechnology has been found in various fields like nutraceuticals, the drug industry, cosmetics, biofuels, and various other applications in the field. *Strauroneis* sp. was used in the preparation of silicon–germanium nanobio composite, which will have a great amount of uses in the near future in the field of electronics because of its simple and efficient structure (Asmathunisha and Kathiresan, 2013). Various kinds of algae have been thoroughly studied, and scientists have identified numerous of them as potentially useful in the field of nanotechnology (Sharma et al. 2016). Nanotechnology also has an impact on various other biotechnological and pharmaceutical applications, but its use in the field of bioenergy research is less (Figure 3.3). Until now, many researchers have produced biofuels through activated sludge systems and even by use of nanosilver, which reduces the adverse effect on microalgal growth. Researchers should also focus on the production of biofuels through nanotechnology (Rahman et al., 2016). In this article, we will discuss the interactions of microalgae in synthesizing the various nanoparticles which are currently being researched and their role in the field of nanotechnology.

TABLE 3.1
Various Groups of Algae and Their Corresponding Nanoparticles Synthesized

Algal group	Division	Major component(s) of cell wall	Chlorophylls and carotenoids	Nanoparticles synthesized
Chlorophyta	Chlorophyceae Micromonadophyceae Charophyceae Pleurastrophyceae Ulvophyceae	Cellulose	Chlorophyll a, b, β carotene	Gold, silver, iron
Phaeophyta	Phaeophyceae	Cellulose & alginic acid	Chlorophyll a, b, β carotene, & fucoxanthin	Gold, silver, iron, copper, zinc
Pyrrophyta	Dinophyceae Adinophyceae	Cellulose plates covering the cells	Chlorophyll a, c,	Gold, silver, iron
Chrysophyta	Cryptophyceae	NA	Chlorophyll a, c, fucoxanthinlutin	Gold, silver, iron
Rhodophyta	Faucheaceae Champiaceae Lomentariaceae Rhodymeniaceae	Cellulose	Chlorophyll a, d, β carotene, zeaxanthin	Gold, silver, iron
Euglenophyta	Euglenophyceae	NA	Chlorophyll a, b, Xanthophyceae	Gold, silver, iron

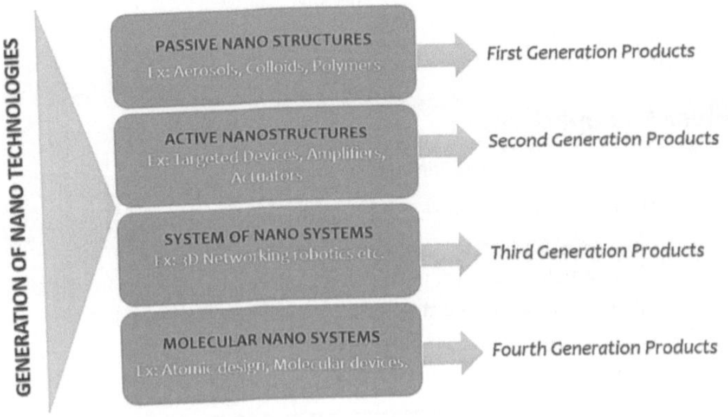

FIGURE 3.2 Generation of nanotechnologies.

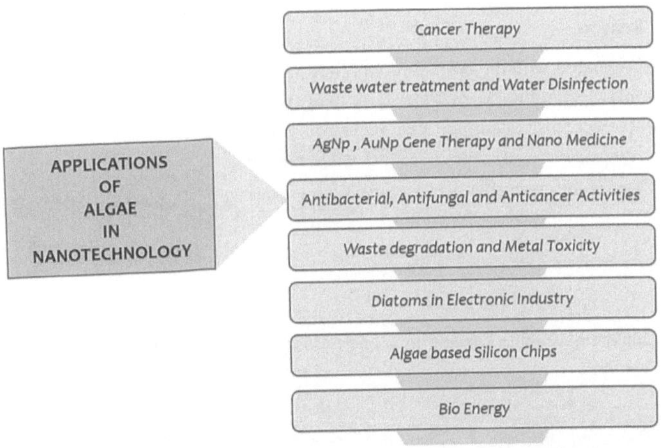

FIGURE 3.3 Applications of algae in the field of nanotechnology.

3.2 THE CHEMICAL COMPOSITION OF ALGAE

Algae have been used as traditional food sources for many years. The high protein content in many seaweeds makes them useful as sources for feeding in many aquaculture industries. Many trials have been conducted to evaluate algae as fish feed, and whole and dried algae have been used as a partial or complete replacement for other fish feed. The cell walls of microalgae contain various proportions of carbohydrates, proteins, amino acids, and even lipids, the latter of which are considered as the key component as the future the food source (Abo-Shady et al. 1993). Rodrigues et al. (2015) characterized various species of algae and analyzed the composition of molecules, and found that they contain various molecular components like lipids, proteins, sugar contents, and phenolic contents. These components resembled the molecules present in various plant sources.

3.3 ALGAL INTERACTIONS WITH NANOPARTICLES

The abundant availability of algae makes them a good source of biomaterial for the massive production of metallic NPs (Kannan et al., 2013). Nanoparticles are synthesized using algae in three significant steps:

a. the extract of algae is prepared in either aqueous or organic media,
b. concentrations of metallic compounds are prepared in ionic state, and
c. the algal extract and ionic metallic concentrations are allowed to concentrate for a certain time.

This is followed either with or without stirring under measured conditions (Rauwel et al., 2015). The synthesis of nanoparticles is completely dosage dependent and also associated with the variety of algae used. A number of biomolecules, including peptides, polysaccharides, and pigments, are accountable for the reduction of metals. Stabilizing and capping the metal NPs in aqueous solutions is usually done by proteins through $-NH_2$ groups or residues of cysteine and sulphated polysaccharides (Singaravelu et al., 2007). Synthesis of NPs using algae takes moderately less time than other biosynthesizing methods (Thakkar et al., 2010). To date, several seaweeds like *Sargassum wightii* and *Fucus vesiculosus* have been used for synthesizing silver NPs of diverse sizes and shapes (Singaravelu et al., 2007). Marine algae have been meagerly explored for the synthesis of nanoparticles. Tetrachloroaurate ions have strong binding ability towards *Chlorella vulgaris* to form algal-bound gold, finally reducing into $Au^{(0)}$.

Some advantages of using algae as a source for the synthesis of nanoparticles over other biomaterials are as follows:

- Easily available
- Cost-effective
- Safe to handle
- Environmentally friendly
- Easy to handle
- Contain many metabolites which help in reduction

Synthesized nanoparticles (Ashokkumar and Vijayaraghavan, 2016) using algae are explored in Section 3.11 (applications):

a. Synthesis of palladium nanocrystals (spherical shape) via aqueous Na_2 $[PdCl_4]$ solution (photosynthetic reaction within *C. vulgaris*) that can be used as a material for recycling as a catalyst in the Mizoroki–Heck cross-coupling reaction.
b. The antioxidant potentials of Ag nanoparticles synthesized using *G. corticata* were also obtained.
c. Au nanoparticles were produced using dried biomass of *Lemanea fluviatilis* (L.) (an edible freshwater epilithic red alga) as both reductant and stabilizer; further, its antioxidant property was estimated using DPPH assay.

The advantages and disadvantages (Praveen et al., 2016) of gold and silver (see Figure 3.3) and the applications of various nanoparticles in the production of biofuel are given in Tables 3.2 and 3.3 respectively.

TABLE 3.2
Advantages and Disadvantages of Gold and Silver Nanoparticles

Advantages	Disadvantages
• Nanoparticles are small, making penetration much easier • Several areas of industry benefit from the nanotechnology • Materials made out of the nanoparticles are durable, strong, and light • Nanoparticles increase therapeutic efficiency	• Because of their small size, nanosized weapons are now easily accessible • Nanoparticles are a risk to health, including inhalation problems because of their small size • Products from nanotechnology are more expensive • Loss of jobs in the industrial sector

TABLE 3.3
Applications of Various Nanoparticles in the Production of Biofuel

Nanotechnological product	Application	Reason	Reference
Immobilized enzymes	Biomass transformation and pre-treatment in biofuel production	Compared with the normal immobilized enzymes, nanotechnological enzymes have many properties like increase in surface area, increased production rates, and easier catalyst recovery and recycling	(Biofuels, 2012)
Nano-catalysts for transesterification	Act as better catalysts during transesterification	Reusable, separation is easy, time-saving process	(Biofuels, 2012)
Metal nanocatalysts for biogasification	Conversion of algal biomass into hydrogen, methane, and other synthetic fuel	NA	(Biofuels, 2012)
Zeolites	Removal of moisture, which results in the production of glycerine, a by-product of biofuel production	Inexpensive, high surface area, and porosity	(Biofuels, 2012)
Nanohybrid catalysts	Separation of bio-oil, which is partially soluble in water and organic solvents	NA	(Biofuels, 2012)

3.3.1 Properties of Gold Nanoparticles

The basic color of the gold nanoparticles is wine red. Usually, gold nanoparticles vary in size between 1 nm and 8 μm and are in various shapes like spherical, octahedral, icosahedral, tetrahedral, etc. (Alaqad and Saleh, 2016). Gold nanoparticles are non-toxic to nature, have a large surface area, and can bind with most biological molecules. Gold nanoparticles are one of the best therapeutic agents because of their ease of transport in the diseased cell. Because of their small size, colloidal gold nanoparticles can be easily introduced into any biological molecules, including DNA, RNA, protein, etc. Au nanoparticles have been extensively used in the field of biomedicine due to their unique optical properties. In accordance with the theory of surface plasmon resonance (SPR), when light scatters on these gold nanoparticles, they excite at a particular wavelength. This particular wavelength is dependent upon the size of the gold particles. Moreover, certain features like surface charge, agglomeration, coating, and dissolution rate are very important for the biological interactions. Some of the frequently used nanostructures in the field of medicine are nanorods, nanoplates, nanocubes, etc. (Sharma and Kothiyal, 2015c).

3.3.2 Properties of Silver Nanoparticles

Silver nanoparticles have special properties like electrical conductivity, thermal insulation, Raman scattering, optical properties, catalytic activity, etc. These properties are usually used in the electronic industries. Silver nanoparticles also possess antibacterial properties and are used in many everyday commercial products. Apart from these properties, these silver nanoparticles also exhibit several shapes like spherical, hexagonal, octagonal, triangular, etc. These shapes are important for the determination of the toxicity to cells. Smaller particles cause high toxicity to cells, and it is the reverse for the large particles; moreover, the surface charge of these silver nanoparticles also determines the toxicity effect in cells. If the surface charge is positive, then the toxic cells stay on the cells for a longer time when compared with the negative, by which the anticancer agents can be easily administered into the bloodstream (Zhang et al., 2016). Properties of various nanoparticles are given in Table 3.4.

3.4 CHARACTERISTICS AND BIOSYNTHESIS OF NANOPARTICLES

3.4.1 Biosynthesis of Gold Nanoparticles

Metallic gold is expensive and highly stable; its recovery from various industrial effluents is found to be expensive when done with various conventional technologies like electrochemical treatments, precipitation, cyanide leaching, etc. Therefore, there is the need for a substitute technology: a cleaner and simpler substitute known as biosorption is being utilized to recover the metal at very low concentrations (Mata et al., 2009). Gold nanoparticles have found applications in many fields, including the fabrication of electronic devices, cosmetics, food materials, etc. When these gold nanoparticles are added to the mixture, surface plasmon resonance plays an important role in determining the color change in the medium. The color visible to our eyes is due to the reflection of the light scattered at 520-nm radiation in the UV-Vis spectrophotometer, and the electrons in the medium tend to oscillate in different directions (Journal et al., 2015). The brown seaweed *Sargassum wightii* has found to synthesize gold particles in the size range of 8–10 nm, which was found to be more stable. Gold nanoparticles have a lot of bactericidal effects which depend upon their size and shape. These nanoparticles attach to the surface of the microorganisms, causing changes in the structure and damage to the vital cell functions, which leads finally to the death of an organism (Naveena and Prakash, 2013 and Vincy et al., 2017). Some of the species used in the

TABLE 3.4
Properties of Various Nanoparticles

Parameter	0-D	1-D	2-D	3-D	Reference
Size	These are measured in nanoscale (no dimensions)	These are outside the nanoscale	Not confined to nanoscale	Not confined to any dimensions	(Tiwari et al., 2012)
Appearance	Exhibits various shapes and forms	Needle shape	Plate shapes	Shapes not available	(Tiwari et al., 2012)
Existence	Exists individually	Embedded within another medium	Deposited on substrate	NA	(Tiwari et al., 2012)
Direction of confinement	x, y, z	x, y	NA	NA	(Tiwari et al., 2012)
Relation among other nanomaterials	All dimensions at the nanoscale	Two dimensions at the nanoscale, one dimension at the macro scale	One dimension at the nanoscale, Two dimensions at the macro scale	No dimensions at nanoscale, All dimensions are at the macroscale	(Tiwari et al., 2012)
Examples	NA	Nanotubes, nanoribbons, nanorods, and nano-belts	Nanofilms, nanolayers, and nanocoatings	A nanocomposite, thick films	(Tiwari et al., 2012)

study synthesizing gold nanoparticles are listed as follows: *Fucus vesiculosus*, *Tetra selmiskochinensis*, *Padina gymnospora*, *Sargassum muticum*, and certain cyanobacterial species like *Lyngbya majuscula*, *Spirulina subsalsa*, etc. Alkaline pH and room temperature were the ideal conditions for synthesizing gold nanoparticles, whereas the size and shape of the nanoparticles were controlled by the pH conditions in the solution. Several other species of brown algae like *Navicula atomus* and *Diadesmis gallica* have been found to synthesize these gold nanoparticles and silica–gold nanocomposites (Asmathunisha and Kathiresan, 2013). The synthesis of the nanogold particles from various species of algae has found very interesting diverse applications like antibacterial, antiviral, anti-inflammatory, antifungal, antioxidant, antiprotozoal, cytotoxic, contraceptive, hypoglycemic, gastrointestinal, anti-enzymatic, cardiovascular, spasmolytic, and allelopathic effects (Vincy et al., 2017). The chemicals which are used in the synthesis of these gold nanoparticles are citrate, sodium borohydride, and some organic compounds (Soisuwan and Warisnoicharoen, 2010). In an experiment with interesting results, Abdel-raouf (2017) made two different samples: one with only algal powder and the other the algal powder sample mixed with ethanol. The colors of the samples were green and dark green, respectively. After 3 h of incubation, a color change was observed visually, confirming the synthesis of gold nanoparticles. The color change is due to the excitation of surface plasmon vibrations and also due to the transition of Au^+ to Au. There were various shapes of nanoparticles reported, like spherical, truncated triangular, and hexagonal, out of which the majority of the particles were spherical. In further investigation, fatty acid components were isolated from the selected species of algae, and andrographolide and alloaromadendrene oxide were reported at the highest concentrations (37–86% and 21.94%); glutamic acid, terpenoids, aliphatic amines, etc., acted to reduce the gold particles and also prevented the gold from aggregation, which showed how important these different fatty acids, phytochemicals, amino acids, etc., were in preventing the gold particles from forming aggregates. Some researchers have found that when they initially set a wavelength band at 325 nm, it decreased with time, and then a band at 540 nm was shown, which indicated that the macroalga *Osmudaria obtusiloba* was acting as both a reducing and a capping agent. The faster synthesis of these nanoparticles in the brown algae is due to the rich source of polysaccharides known as fucoidans, which have good reducing properties, and the sefucoidans contain L-fucose and sulfate ester groups (Ramakrishna et al., 2016 and Asmathunisha and Kathiresan, 2013). Fucoidan is a rich source of polysaccharide material which is extracted from the cell walls of brown algal species like *Chordarilaes*, *Desmarestiales*, *Dictyotales*, etc. The structure of the fucoidan is the linear backbone of α1-3-linked fucopyranose with a sulfate group (Soisuwan and Warisnoicharoen, 2010). Mata et al. (2009) have found that pH also influenced the synthesis of these nanoparticles. The optimum pH stood in the range of 4–9, and during this process -OH groups from the algae were involved in the process of reducing gold as well as the formation of nanoparticles. Aarthi et al. (2012) have determined that hydroxyl groups in brown seaweeds were responsible for the reduction of gold nanoparticles. Ramakrishna et al. (2016) studied the catalytic efficiency of gold nanoparticles for the reduction of p-nitroaniline in the presence of sodium borohydride and found that both 4-AP and p-PDA are very useful as key players in the intermediates during organic synthesis. Researchers have found that the synthesis of these organic compounds is a very costly and tiresome process; an alternative method available for synthesis using algal nanotechnology was found to be efficient, and henceforth drug manufacturing companies have found this method as an efficient production process to manufacture drugs at the required scales. Parial et al. (2012) had used various cyanobacterial and green algal species like *Phormidium valderianum*, *P. tenue*, *Rhizocolonium fontinale*, *Microcoleus chthonoplastes*, *Ulva intestinalis*, *Charazeylanica*, and *Pithophorao deogoniana*. Among these, *Rhizocolonium fontinale* and *Ulva intestinalis* synthesized gold particles of varying shapes and sizes. *Phormidium valderianum* synthesized almost spherical nanoparticles but with hexagonal and triangular shapes at basic as well as neutral pH. *Chara zeylanica* and *Pithophoraode ogoniana* could synthesize these nanoparticles. Singaravelu et al. (2007) studied the synthesis of gold nanoparticles by *Sargassum wightii* and found that when this species was exposed to gold ions, it exhibited ruby color and an absorption band was visible at 510–540 nm, which indicated that the gold nanoparticles were well dispersed in the solution by not forming clumps and that the size of these NPs was in the range of 8–12 nm. The extracellular polysaccharides present in the seaweed *Sargassum wightii* was found to be 35%, which is responsible for the stabilization of the nanoparticles.

3.4.2 Biosynthesis of Silver Nanoparticles

Silver has been considered one of safest and least toxic antibacterials/antifungals for many centuries. Silver nanoparticles are of high interest because they exhibit various optical, magnetic, and electrical properties which can be incorporated into various applications like biosensors, antimicrobial applications, composite fibers, cosmetic products, and various other electronic items. The application of these silver nanoparticles is vast and therefore cannot be covered in this single review paper. Among the various chemical methods used for synthesizing these silver nanoparticles, the Turkevich method and the Brust method have been the most popular methods to date. The Turkevich method was found to be less hazardous. Jena et al. (2013) have conducted an experiment in synthesizing the silver nanoparticles, finding that the *Chlorococcum* species which they used in their experiment had various substances present inside them like carbohydrates, proteins, lipophilic substances, and water-soluble phenolic substances. Among all these substances, various cellular proteins played an important role in synthesizing silver nanoparticles, which was revealed by Fourier transform infrared spectroscopy (FTIR), and they have shown that the algal species *Chlorococcum humicola* was an efficient species in terms of the synthesis of intracellular as well as extracellular silver nanoparticles. Vivek et al. (2011) have conducted an experiment in synthesizing the silver nanoparticles from the red seaweed known as *Gelidiella acerosa*, in which the synthesis of silver nanoparticles was confirmed by the color change through surface plasmon resonance activity; they noticed that the presence of certain aromatic compounds or amines or alkanes was responsible for the synthesizing of these silver nanoparticles. Finally, they concluded that this method has great potential for studying the toxic effects of various bacterial and fungal species and that new varieties of antibacterial as well as antifungal compounds can be synthesized. During their research on algal polysaccharides involving silver NPs forming on fabrics, El-Rafie et al. 2013 found by FT-IR measurements that several bands, which were identified as various groups like OH, the -C-O-O group, the S-O of the sulphated polysaccharide group, the C-N group of aromatic amine, -C-O groups of polyols, and -C-O-SO_4, were involved in reducing silver nanoparticles which were synthesized by the microalgal species. The various algal species used in this study were *U. faciata*, *P. capillacea*, *J. rubins*, and *C. sinusa*. Mahdieh et al. (2012) conducted an experiment in which they found that metal ions are trapped inside the cells of algae through electrostatic force of interaction between the ions and the negatively charged carboxyl groups present inside the cell surface. *Spirulinaplantesis* released the cellular reductive, which could be a possibility for the reduction of the silver nanoparticles in the solution. Another possibility for reducing these nanoparticles was by the electron transport chain mechanism of the cyanobacteria through the NADH cofactor, which played an important role. Patel et al. (2015) conducted recent research on the role of the proteinaceous pigment from cyanobacteria known as C-phycocyanin, which was used for biosynthesis of silver nanoparticles. In their work, they found that only one cyanobacterial and one green algal strain were not able to synthesize the Ag-NP. The formation of nanoparticles was due to an extracellular polysaccharide and was found to be dependent on the light source. C-phycocyanin is a blue-colored accessory pigment consisting of two polypeptide chains which are covalently attached to the linear tetrapyrrole-phycocyanobilin, and this C-phycocyanin binds only to the heavy metals. Rajeshkumar et al. (2012b) conducted an experiment in which they found that the brown algae could synthesize the silver nanoparticles due to the presence of amide and polyphenol groups which were identified by the FT-IR method. Yousef et al. (2014) reported the first synthesis of Ag-NP from green algae which were collected from the Persian Gulf, and they found that the reduction of silver nanoparticles was because of various amines and peptides as well the presence of secondary metabolites (Yousefzadi et al. 2014). Azizi et al. (2013) were able to synthesize the Ag-NP because of the sulfated polysaccharides and hydroxyl moieties which were present in brown marine alga, *Sargassum muticum*. During the solution's color change, the initial pH of the solution was about 7.5, and the pH had decreased to 5.6 at the end of the reaction. The marine alga *Porphyra vietnamensis* contained a sulfated polysaccharide, which had a strong capacity to synthesize these Ag-NP. Razack et al. (2016) synthesized nanoparticles using *C. vulgaris* and found that the silver nitrate nanoparticles were responsible for the synthesis and release of the lipids and carbohydrates from the cell wall for the production of biofuel. The shape of these nanoparticles was spherical and crystalline. These silver nanoparticles were bound to and distorted the composition of the cell wall, thus rupturing the cell wall to extract the carbohydrates as well as the lipids for the production of biofuel. The Kannan et al. (2013) research group conducted research on silver nanoparticles using microalgae and found an interesting point: the peptides present inside the cells of the marine microalga *Chaetomorpha linum* synthesized these silver nanoparticles. The synthesis of these nanoparticles is due to the strong absorption band observed at 1535 cm^{-1} and 1610 cm^{-1}, which corresponds to the (NH)C=O. These peptide bonds are also responsible for stabilizing these nanoparticles, which makes them occupy a very small area, in turn making them unmovable from their positions and hence stronger and more rigid. These researchers also found a few other absorption peaks at 1725 cm^{-1}, which corresponded to free amines, which were also used as stabilizing silver nanoparticles. Castro et al. (2013) conducted their research on dead biomass of microalgae for the synthesis of metallic nanoparticles and concluded that the dead biomass consisted largely of pectins, which are found on the extracellular cell walls. Due to this, the hydroxyl group present in the cell walls might act as a good reducing agent for the formation of Ag salt.

3.4.3 Biosynthesis of Iron Particles

In recent years, many researchers have been prompted by the physiochemical properties of iron to explore Iron in various fields like magnetic media storage, biosensors, catalysts, separation processes, etc. Mahdavi et al. (2013) used a brown seaweed, *Sargassum muticum*, and they synthesized cubic nanoparticles of 18 $\pm$ 4 nm (average) in size. The synthesis of these nanoparticles is because of the various functional groups present like hydroxyl, sulphate, and aldehyde groups. The rapid reaction was completed in one step by a color change from yellow to dark brown. This compound exhibits electrical and magnetic properties due to the cubic inverse structure present, which makes it unique and accounts for its many applications. These nanoparticles also possess various physical, chemical, thermal, and magnetic properties. Subramaniyam et al. (2015) have done their experiments using the soil alga *Chlorococcum* sp. and they have synthesized spherical iron nanoparticles of 20–50 nm. Biomolecules like carbonyl, glycoproteins, and amine were the reason for the synthesis of iron nanoparticles. These results were confirmed by TEM. El-Kassas et al. (2016) reported that sulfated polysaccharides play a dual role in reducing the $FeCl_3$ and stabilizing the obtained nanoparticles. This experiment shed light on the purification of wastewater from harmful industries. In their experimental procedure, after synthesizing these nanoparticles from *Padina pavonica* and *Sargassum cinarium*, they were incorporated into calcium alginate beads, which made them much stronger such that their retention capacity was increased up to 78%.

3.5 NANOPARTICLES IN THE TREATMENT OF CANCER

Cancer is the world's most dreaded disease next to cardiovascular diseases and has taken the lives of many. Though there are many treatments available, they are highly expensive and thus out of reach for the poor. The current treatments for cancer include radiotherapy, surgery, chemotherapy, and photodynamic therapy (Kuddus, 2017); all have their own limitations. These limitations account for the deaths of many people suffering from this disease. These researchers used cerium oxide, also known as nanoceria.

The major disadvantage of this element is the stability of the cerium element. Much of the research going on currently is to remove this drawback without affecting its original activity (Kuddus, 2017). To overcome this, scientists and researchers have come up with a novel technology, namely, drug delivery through nanomachines. Researchers from Australia have studied and successfully developed a method to deliver the drug through genetically engineered microalgae in the year 2015.

According to a report published in *Nature*, Delalat et al. (2015) demonstrated delivering the drug through diatoms. Diatoms are tiny, unicellular, photosynthesizing algae and they are covered with silica, which can be used as a therapeutic nanoparticle in delivering the drug to the specific site in the human body. Delalat et al. (2015) have genetically engineered the diatoms to produce antibody binding proteins on their surface. These binding proteins bind particularly only to specific molecules found on the cancer cells, making delivery more specific. They have successfully killed 90% of the cells in cultured human cells and were also able to kill tumor cells which were developed in mice.

Krishnamoorthy et al. (2015) studied how cell death takes place after exposing cells to nanoparticles synthesized from microalgae. For their experimental procedure, they incubated the cells (HL60 and HeLa) with silver nanoparticles for five days and have measured cell viability using the MTT assay. During their experiment, they have shown key factors which are involved in cell death, like DNA fragmentation, caspase activation, and annexin V binding. The apoptosis of the cells was inhibited by the silver nanoparticles.

3.6 MECHANISM OF NANOPARTICLE SYNTHESIS

Microalgae are considered as one of the most valuable species. These microalgae are eukaryotic organisms which are found in most bodies of water, like freshwater, marine water, and also wastewater. Shape and size variations in nanoparticles synthesized by microalgae are given in Table 3.5. Microalgae function through the process of photosynthesis by producing O_2. Microalgal species are also considered as the primary producers in various aquatic environments. In one review, the authors mentioned that among the various algae, *Chlorella* sp. was found to accumulate various heavy metals like cadmium, uranium, copper, and nickel. Algal powder (dried) was found to have a strong affinity towards tetrachloroaurate, which reduced Au to $Au^{(0)}$. After the synthesis of these particles, they were reduced to various structures like tetrahedra, decahedra, and icosahedra and these particles were synthesized with well-controlled shape and size (Velusamy et al., 2016). To date, only *C. vulgaris* which contain thc gold shape-directing protein (GSP) was used in bioreduction and in shape/size synthesis and produced hexagonal as well as triangular shapes. An additional study has shown that brown algae are better at biosorption of heavy metals.

Not many researchers have proposed any mechanism for the synthesis of nanoparticles through enzyme mediation reaction using algae. The mechanism was studied through two different cyanobacteria, *Calothrix* and *Anabaena*. These two species of cyanobacteria possess two classes of reducing enzymes, nitrogenase and hydrogenase (Brayner et al., 2007).

The hydrogenase enzyme reduces in the presence of electron acceptors through the presence of thylakoids and ferredoxins. These work as a link between hydrogenase enzyme and various other electron donors and acceptors. Two different species, *Leptolyngbya* and *Anabaena*, were selected for their studies since the former does not contain cysts and the latter contains many heterocysts. *Leptolyngbya* fixes nitrogen only in the dark and *Anabaena* produces high concentrations of nitrogenases. It was found that these metallic nanoparticles like gold, silver, etc., could not penetrate within the cells and were formed near the cell walls, which indicated that nitrogenase is involved in the reduction process. It has been observed huge sizes of these metallic nanoparticles when an experiment was conducted in vivo and concluded that these enzymes play a major role in particle size control and that nitrogenase is present only on the cell membrane.

3.7 THE FUNCTION OF POLYSACCHARIDES IN RELEASING NANOPARTICLES

Many metallic nanoparticles like silver (Ag), gold (Au), iron (Fe), etc., are released into the medium through the intercellular channel and embedded in the polysaccharide network. These polysaccharides are rich in uronic acids and hence are used in the treatment of toxic wastewaters. These polysaccharides are also used in natural chelators binding cation etc. Brayner et al., (2007) compared the UV absorption spectral values from various cyanobacteria species and found that red-colored spherical colloidal gold particles had an absorption spectrum of 530 nm whereas *Anabaena* species had a pink–purple color at an absorption spectrum at 537 nm. The genera *Colothrix* and *Leptdyngbya* presented a purple–blue color at 577 nm and 562 nm, while the silver nanoparticles for genera *Colothrix* and *Anabaena* exhibited red color at 410 nm. This color is due to the polysaccharide being absorbed on the surfaces of the metal nanoparticles. The polysaccharides attached are released into the medium surrounded by the stabilized polysaccharide and could be therefore used as synthesized homogeneous catalysts.

TABLE 3.5
Shapes and Sizes of Various Nanoparticles Synthesized by Microalgae

Species	M & MOx NPs	Shape and size	Reference
Anabaena sp. L31	Zinc oxide	Spherical and 80 nm	(Sharma et al. 2016)
Lyngbyama juscula	Silver	Spherical and >80 nm	(Sharma et al. 2016)
Microcoleus sp.	Silver	Spherical and 44–79 nm	(Sharma et al. 2016)
Nostocellipsosporum	Gold	Nanorods and 137–209 nm	(Sharma et al. 2016)
Phormidium willei	Silver	Spherical and 100–200 nm	(Sharma et al. 2016)
Leptolyngbya tenius	Cadmium sulfide	Spherical (5 nm)	(Sharma et al. 2016)
Leptolyngbya boryana	Gold, platinum, silver	Octahedral, spherical and octahedral	(Sharma et al. 2016)
Chlamydomonas reinhardtii	Silver	Round and rectangular and 5–15, 5–35 nm	(Sharma et al. 2016)
Auxenochlorella pyrenoidosa	Gold	Spherical and 25–35 nm	(Sharma et al. 2016)
Chlorella vulgaris	Silver, palladium	Nanoplates, spherical, and 28, >15, 7	(Sharma et al. 2016)
Chlorococcum sp.	Iron	Spherical and 20–50 nm	(Sharma et al. 2016)
Klebsormidium flaccidum	Silver	Spherical and 9 ± 3.4	(Sharma et al. 2016)
Tetraselmisko chinensis	Silver	Spherical and 15 nm	(Sharma et al. 2016)
Tetraselmis suecica	Silver	Spherical and 79 nm	(Sharma et al. 2016)
Caulerpa racemosa	Silver	Spherical and triangular and 5–25 nm	(Sharma et al. 2016)
Prasiola crispa	Gold	Spherical and 5–25 nm	(Sharma et al. 2016)
Ulva intestinalis	Gold	Spherical to irregular and 40–50 nm	(Sharma et al. 2016)
Ulva reticulata	Silver	Spherical and 40–50 nm	(Sharma et al. 2016)
Padina gymnospora	Silver	Spherical and 25–40 nm	(Sharma et al. 2016)
Padina pavonica	Silver	Spherical and 45–64 nm	(Sharma et al. 2016)
Padina tetrastromatica	Silver	Spherical and 14 nm	(Sharma et al. 2016)
Sargassum plagiophyllum	Silver	Spherical and 20–50 nm	(Sharma et al. 2016)
Stoechospermum polypodioides	Gold	Spherical and 62.5 nm	(Sharma et al. 2016)
Gelidiella acerosa	Silver	Spherical and 22 nm	(Sharma et al. 2016)
Gracilaria dura	Silver	Spherical and 6 nm	(Sharma et al. 2016)
Kappaphycus alvarezii	Gold	Spherical and 6nm	(Sharma et al. 2016)
Sargassum wightii	Gold	Thin planar and 8 to 12 nm	(Singaravelu et al., 2007)
Phormidium valderianum, P. tenue, Microcoleus chthonoplastes,	Gold	Spherical along with triangular and hexagonal	(Parial et al., 2012)
Rhizoclonium fontinale, Ulva intestinales	Gold	Spherical	(Parial et al., 2012)
Tetraselmis kochinensis	Gold	Spherical, triangular, and 15–35 nm	(Senapati et al., 2012)
Padina pavonica	Gold	Spherical and 30–100 nm	(Isaac and Renitta, 2015)
T. conoides, S. tenerrimum	Gold	Mostly spherical and 27–35 nm	(Ramakrishna et al., 2016)
Stoechospermum marginatum	Gold	Spherical, hexagonal, and triangular and 18.7 to 93.7 nm	(Aarthi et al., 2012)
Galaxaura elongata	Gold	Spherical, few rods; Triangular, truncated, Triangular, hexagonal, and 3.85–77.13 nm	(Abdel-raouf, 2017)
Kjellmaniella crassifolia	Gold	Spherical and 8–10 nm	(Soisuwan and Warisnoicharoen, 2010)
Turbinaria ornata	Gold	Spherical and 7–11 nm	(Ashokkumar and Vijayaraghavan, 2016)
Sargassum muticum	Iron	Cubic and 18 ± 4	(Mahdavi et al., 2013)
Chlorococcum	Iron chloride	Spherical and 20–50 nm	(Saif et al., 2016)
Nanochloropsis, Isochrysis	α-Iron oxide & γ-iron oxide	Spherical and 10–90 nm	(Reeder et al., 2016)
Sargassum muticum	Gold	Spherical and 5.42 ± 1.18 nm	(Sharma et al., 2015)
Tetraselmis kochinensis	Gold	Spherical, triangular, and 5–34 nm	(Sharma et al., 2015a)
Ecklonia cava	Gold	Spherical, triangular, and 30 ± 0.25 nm	(Sharma et al., 2015a)
Caulerpa racemosa	Silver	Spherical, triangular, and 5–25 nm	(Sharma et al., 2015a)
Cystophora moniliformis	Silver	Spherical and 50–100 nm	(Sharma et al., 2015a)
Chlamydomonas reinhardtii	Silver	Round/rectangular and 5–35 nm	(Sharma et al., 2015a)
Chlorella vulgaris	Gold	Spatial array of self-assembled structures, 2–10 nm	(Sharma et al., 2015a)

(*Continued*)

TABLE 3.5 (CONTINUED)
Shapes and Sizes of Various Nanoparticles Synthesized by Microalgae

Species	M & MOx NPs	Shape and size	Reference
Padina gymnospora	Gold	Spherical and 53–67 nm	(Sharma et al., 2015a)
Euglena gracilis	Ferrihydrite	Spherical and 0.6–1.0 nm	(Sharma et al., 2015a)
Sargassum muticum	Silver	Spherical and 5–15 nm	(Azizi et al., 2013)
Chlorella vulgaris	Silver	Spherical and 50–70 nm	(Karthikeyan et al., 2015)
Chaetoceros calcitrans	Silver	Spherical and 30–55 nm	(Karthikeyan et al., 2015)
Spirulina plantesis	Silver	Spherical and 30–50 nm	(Sharma et al. 2015)
Acanthophora spicifera	Silver	Cubic and 33–81 nm	(Ibm et al., 2016)
Sargassum plagiophyllum	Silver	Spherical and 15–24 nm	(Pharm and Sci, 2015)
Spirulina plantesis	Silver	Cubic, octahedral, and 11.6 nm	(Mahdieh et al., 2012)
Sargassum wightii	Gold	Well dispersed and 8–12 nm	(Manivasagan and Kim, 2015)
Gelidiella acerosa	Silver	Spherical and 22 nm	(Manivasagan and Kim, 2015)
Ulva fasciata	Silver	Spherical and 28–41 nm	(Manivasagan and Kim, 2015)
Chlorella vulgaris	Silver	Triangular and 44 nm	(Manivasagan and Kim, 2015)
Chlorella vulgaris	Gold	Triangular and spherical	(Manivasagan and Kim, 2015)
Stoechospermum marginatum	Gold	Spherical and 18.7–93.7 nm	(Manivasagan and Kim, 2015)
Turbinaria conoides	Silver	Spherical and 96 nm	(Rajeshkumar et al., 2012a)
Kappaphycus alvarezii	Gold	Spherical and 10–40 nm	(Rajasulochana et al., 2010)
Sargassum muticum	Gold	Spherical and 5.42 ± 1.18	(Siddiqi and Husen, 2016)
Stoechospermum marginatum	Gold	Spherical, hexagonal, and 18.7–93.7 nm	(Siddiqi and Husen, 2016)
Tetraselmis kochinensis	Gold	Spherical, triangular, and 5–35 nm	(Siddiqi and Husen, 2016)
Ecklonia cava	Gold	Spherical, triangular, and 30 ± 0.25	(Siddiqi and Husen, 2016)
Padina gymnospora	Gold	Spherical and 53–67 nm	(Siddiqi and Husen, 2016)
Turbinaria conoides	Gold	Triangular and 2–19 nm	(Siddiqi and Husen, 2016)
Cystophora moniliformis	Silver	Spherical and 50–100 nm	(Siddiqi and Husen, 2016)
Chlamydomonas reinhardtii	Silver	Round, rectangular, and 5–35 nm	(Siddiqi and Husen, 2016)
Turbinaria conoides	Silver	Spherical and 2–17 nm	(Siddiqi and Husen, 2016)
Pithophora oedogonia	Silver	Cubical, hexagonal, and 25–44 nm	(Siddiqi and Husen, 2016)
Bifurcaria bifurcata	Copper oxide	Spherical and 5–45 nm	(Siddiqi and Husen, 2016)

3.8 EFFECT OF METALS ON THE GROWTH OF MICROALGAE

Metals are considered as one of the important sources which determine the growth of every living organism. The positive and negative effects of metals on the growth of algae are given in Tables 3.6 and 3.7, respectively. These metals are helpful in performing cellular functions whenever needed, especially for the electron transport chain, other metabolic cycles, DNA transcription, etc., which take place in our cells (Mahdavi et al., 2013). However, these metals when accumulated in high concentrations cause severe negative effects like cell division blockage and photosynthetic activity loss within the cells. This metal accumulation in the algal cells happens through two types of mechanisms:

- Metal absorption on the cell wall surface that contains various functional groups like -COOH, -OH, -NH_2, sulfhydryl, etc.
- Metal transport systems assist in the absorption of metals inside the cells.

The reason for the inhibitory effects of these nanoparticles on these microalgal species is primarily due not only to the reactive oxygen species and the mechanical damage triggered by these nanoparticles but also to the interactions with growth media components, light shading effects, age of suspensions, and size.

3.9 THE RELATION BETWEEN NANOPARTICLE SYNTHESIS AND BIOENERGY

The algal biomass can be converted into biofuel in three different ways: thermal, chemical, and biological process. During this process, the size of the molecules and the various other contaminants impact the conversion process. These nanoparticles can be from both natural and artificial sources. Several different kinds of nanotechnological products are available in terms of the efficient transformation of algal biomass into a valuable product known as algal biofuel (Kumar et al., 2014). These nanotechnological products have shown an impact in obtaining an efficient volume of biofuel.

Currently, many researchers, as well as industries, are facing a huge problem in the commercialization of algal biofuels in various areas of research, which are addressed in the following points (Kumar et al., 2014).

TABLE 3.6
Positive Effects of Metals on the Growth of Algae

Algal species	Metal	Effect	Reference
Dunaliella tertiolecta	Lead, aluminum, and cobalt (low conc.)	Stimulatory effect on their growth	(Mahdavi et al., 2013)
Monoraphidium minutum	Lead, aluminum, and cobalt (low conc.)	Stimulatory effect on their growth	(Mahdavi et al., 2013)
Nostoc minutum	Arsenic	Improve growth of *Nostoc minutum*	(Mahdavi et al., 2013)
Desmodesmus	Cerium	Increased cell number in the culture	(Mahdavi et al., 2013)
Thalassiosiraweiss flogii	Cadmium (low conc.)	Stimulates growth and maintains the activity of carbonic anhydrase	(Mahdavi et al., 2013)

TABLE 3.7
Negative Effects of Metals on the Growth of Algae

Algal species	Metal	Effect	Reference
Chlamydomonas acidophila	Cadmium (Cd)	Increase in cell size, decomposition of polyphosphate bodies	(Mahdavi et al., 2013)
Chlorella sorokiniana	Lead (Pb)	Formation of colonies of *Chlorella* and chloroplasts lost their shape	(Mahdavi et al., 2013)
Synechocystis sp.	Thallium (Tl)	Fragmentation of thylakoid membranes	(Mahdavi et al., 2013)
Desmidium swartzii	Zinc (Zn)	Mitochondria enlarged and bloated	(Mahdavi et al., 2013)
Dunaliella tertiolecta	Aluminum (Al) and lead (Pb)	Cell membrane lysis	(Mahdavi et al., 2013)
Chlamydomonas reinhardtii	Lithium (Li)	Alters the length of flagella	(Mahdavi et al., 2013)
Synedraacus	Germanium (Ge), titanium (Ti), zirconium (Zr), or tin (Sn)	Caused alterations in shape, size, and mechanical strength of silica valves	(Mahdavi et al., 2013)
Phaeodactylum tricornutum, *Cyclotella cryptica*, *Thalassiosiraweiss flogii*	Nickel	Inhibitory for the growth of these diatoms	(Mahdavi et al., 2013)

- Several algal strains are available which need to be genetically engineered for more accumulation of oils into the cells.
- Harvesting techniques need to be improved for cost-effectiveness, and the algal biomass should be recovered during this process. Currently, the techniques available are not efficient enough, which makes the cost equivalent to the biofuel production.
- Methods need to be developed for efficient recovery of the biodiesel. During the extraction process, glycerin needs to be eliminated; this is primarily responsible for obtaining a considerable amount of biodiesel during the process.
- An efficient downstream process must be developed in order to recover more of the product and decrease the cost of production.

3.10 SHORT NOTE ON COCCOLITHS

Coccoliths are usually found in marine ocean systems, where they are represented by the phytoplankton. *Emiliania huxleyi* is one of the most prevalent species which produces the calcite; it is categorized under the family of red algae. The structure of these coccoliths consists of two elliptical discs like grooved shields which are connected by a central tube. More than 10–15 coccoliths merge together to form a cover which protects them from external sources. *Emiliania huxleyi* can be grown at the laboratory scale, which yields a high amount of coccolith production, but the danger is from the cultivation vessel where it grows becomes malformed or partially destroyed (Jakob et al., 2017). But as of the latest research, the highest concentration they could achieve was 5g/L.

3.11 APPLICATIONS OF ALGAE IN NANOTECHNOLOGY

3.11.1 Application of Nanospheres in the Transesterification Reaction

In the normal transesterification of fatty acids from algal oil, sodium methoxide or an acidic or basic compound is used as a catalyst for increasing the speed of the reaction. Catalysts are now replaced with the nanospheres, where these nanospheres are loaded with the acidic or basic catalysts to react with the free fatty acids. The advantages of using nanospheres as a replacement to the normal chemical catalysts are that it eliminates various steps like water washes, separations, acid neutralization, etc. Normal catalysts cannot be used again and again, but nanospheres which are loaded with catalysts can be. Overall, this economical, recyclable method produces cleaner biodiesel, reduces water consumption, and has a great impact on the various environmental factors (Sap and Demmers, 2016).

3.11.2 Application of Diatoms in Waste Degradation and Metal Toxicity

In recent times, the disposal of waste has become one of the major problems for people all around the world. There are many different kinds of wastes being disposed of like industrial waste, sewage waste, kitchen waste, etc., and the list of these types of waste goes on. The quantity of waste depends upon the quantity of the raw materials used and their processing in the particular industry. To eliminate all these kinds of wastes, many physical as well as chemical process have to come into existence. Apart from these chemical and physical processes, nowadays much biological decomposition of the wastes is taking place. The advantage of using bacteria, viruses, and even algae for decomposition of the waste is to remove the toxic compounds and components otherwise released into the environment. In searching for these various components, diatoms have been found to be more capable of degrading wastes compared to bacteria and viruses. This is because of their unique physiological structure, which can be used for material degradation. Certain freshwater species of algae like *Chlorella vulgaris*, *Scendesmus platy discus*, *Scendesmus quadricauda*, etc., have found to be effective in degrading aromatic hydrocarbons. The degradation of these toxic substances was carried out by the metabolism of phenanthrene and pyrene, which takes place by enzyme oxidation, converting the toxic forms into less toxic forms. Apart from this waste degradation, metal toxicity is also gaining a lot of concern. This metal toxicity is happening because of the heavy metals being discharged into reservoirs and lakes nearby. This toxic metal pollution is found mainly in the context of water pollution. However, there are various methods available for the physical/mechanical detection of these metal pollutants. These days, diatoms are one of the best indicators to monitor the concentrations of metal toxicity near the site. This particular approach is taken by comparing the sensitivity of microalgal species towards certain ions. The most toxic metals identified to date are Zn, Cd, and Cu. However, the concentration of Cu is useful at low concentrations since it is helpful for growth and metabolic processes; if the concentration is high, then Cu is dangerous (Jamali et al., 2012).

3.11.3 Application of Nano-Cellulose from Algae in Electrical Insulators

A research group from Uppsala University performed an experiment to characterize the dielectric properties from two different sources, namely, wood and an algal species *Cladophora*. From their research work, this group concluded that the algal nano-cellulose sample was more crystalline, had a lower moisture capacity, was more porous, which made it lower in strength, and had a higher dielectric loss, which had considerable impact on the dielectric properties. After comparing the quality of the nano-cellulose material from both of the samples, they concluded that the conductivity increased with relative humidity and at low frequencies; *Cladophora* resulted in frequency-independent conductivity, which makes it best product for electric insulators (Le Bras et al., 2015).

3.11.4 Applications of Diatoms in the Electronics Industry

Diatoms are considered one of the best sources for various applications in the different fields of the electronics industry. These diatoms are single-celled photosynthetic algae which come under the class of Bacillariophyceae. These diatoms were found useful when researchers (Jeffryes et al., 2011) who studied the metabolic pathways of these diatoms found that silicic acid is taken up from the environment and then precipitated in amorphous SiO_2 through the silicon deposition vesicle (SDV), which is found within the frustule. According to them (Jeffryes et al., 2011), this explains the formation of silicon oxide, which will be then converted into nanospheres. The silicon hydroxide is catalyzed by specialized proteins known as silafins, which link up with the monomers to form solid amorphous silica through the (Si-O-Si) bond formation. By the aqueous phase separation reaction, the silicon molecules are embedded inside these silafins. These specialized proteins, which help to inbound the amorphous silica, will then be used as a product to design/fabricate the nanostructured metal oxides which are to be used in the electronic devices.

According to the researchers (Jeffryes et al., 2011), they have identified that none of the products of diatom nano-based technology devices are available but many of the prototypes have been reported in the literature. Many of these devices are also being patented. They have given a detailed description of diatom dye-sensitized solar cell applications. Since this paper focuses on only various applications of algae in the field of nanotechnology, we are not able to describe these functions in detail.

Apart from their normal applications, these diatoms also have a large variety of applications in various fields like biogas sensing systems, drug delivery systems, batteries, solar cells, etc. Since the diatom surfaces can be highly modified because of their specific porous structures and highly organized frustule architecture, a wide variety of numerous applications have come into existence. Diatom applications are not very far away in time and soon will be widespread (Medarevic et al., 2015).

3.11.5 Applications of Coccoliths in Nanotechnology

Generally, coccoliths are defined as individual plates of $CaCO_3$ formed by coccolithophores. The most famous algal species in this category is *Emiliania huxleyi*. This particular species is mostly found in ocean ecosystems. It produces a group of characteristic chemicals known as alkenones which are used by many scientists to estimate about the past sea surface (Boeckel and Baumann, 2008; McIntyre and Bé, 1967; Okada and Honjo, 1973). Applications from the coccoliths are widespread nowadays in the fields of medical diagnostics, drug delivery vehicles, filtration equipment, manipulating the properties of light like the light-scattering effect, etc. Iwasaka and Mizukawa (2013) conducted an experiment with the light-scattering effect with the coccolithophores produced from *Emiliania huxleyi* and they found that when the cells of coccolithophores produced colors, they had undergone structural changes, and when a magnetic field was applied in the direction of observation from 0T to 5T, the colors underwent dynamic transitions (Iwasaka and Mizukawa, 2014). When the magnetic field was more than 4T, the measurement of these coccoliths decreased gradually. The formed coccolith discs were acting as photonic crystals because the $CaCO_3$ was arranged on them by the force of the magnetic field, which had a power of up to 5T. When the magnetic power was more than 1T they could observe the structural colors of coccoliths.

A large number of species like *Umbilicosphaera sibogae*, *Emiliania huxleyi*, *M. elegans*, *Scyphosphaera porosa*, *Pleris japonica*, etc., have been used in the field of nanotechnology. In the recent past, many researchers have sequenced the genomes of these coccoliths, which prompted them to explore more about the structures of coccoliths and also helped them understand the pathways and mechanism of their synthesis process. Until then it was hard to imagine the structures of these coccoliths. Currently, this branch of study is also gaining prominence by modifying the structures and understanding the mechanisms of these species. The *Umbilicosphaera sibogae* eyelets of 1–2-μm in diameter were used in the filtration, whereas *Calcidiscus leptoporous* was used for studying the optical properties of the light by the light-scattering technique. These coccoliths were also used in the development of nanoactuators, which help in the movement of objects at nanoscales. Many coccoliths like *Discosphaera tubifera* and *Rhabdosphaera clavigera* are used in the applications of nanoactuators because of their hollow funnel and tubular shape and also because they have nanopores or slits which are helpful in designing these devices. Since these coccoliths have a wide variety of applications in the field of nanotechnology, there are quite a lot of challenges being faced with the growth of these coccoliths: for instance, there must be proper cultivation systems, the cultivation systems should be improved for many species, and genetic manipulations of coccolithophores should be developed and studied on a large-scale basis so that the calcium formation pathways can be exploited further for the efficient development of products (Boeckel and Baumann, 2008).

3.12 CONCLUSION

Nanotechnology is considered a modern advancement in various fields of applications like electronics, medical, pharmaceutical, etc. Although nanotechnology is advancing in various fields, nanoparticles have been found to be hazardous to human health in terms of the size of the particles being synthesized during the process. The bigger the size, the more dangerous it is for human cells to absorb these particles. Research should focus on the precise and specific movements of nanoparticles inside the body and study the toxicity of these particles, the effect of nanoparticles on cellular level, and also how human organs respond to various doses and sizes of the nanoparticles. Future research should focus on these points and issues to craft nanotechnology into a successful project for the generations to come (Ray et al., 2009).

Microalgae are considered one of the most valuable sources for various applications like biofuel, phyconanotechnology, manufacturing of drugs, manufacturing of food products, etc. The role of nanotechnology in many products is evident and clearly stated. The production of nanotechnological products from various biological sources is being widely studied and applied in each and every field. When it comes to the field of phycology, algae have been considered as the lowest priority in terms of developing nanotechnological products compared with other biological sources like plants, bacteria, fungus, etc. Moreover, the oil content of these microalgae is very high compared with these other biological sources. From the above literature studies, it has already been proven that microalgae are an excellent source not only for the production of biofuels but also for the synthesis of metallic nanoparticles, with diverse applications in various fields like clinical diagnostics and agriculture and other important areas like paints, electronics, coatings, and packing. Since the algal growth rate is very high when compared with other sources, alga-mediated approaches for the synthesis of metallic nanoparticles is better than chemical synthesis. During our experimentation process, it is also important to select correct species for the synthesis of the selected metal. Researchers should evaluate which particular species of algae is able to reduce the particular metal with higher yield. There are lots of papers, as discussed above, and many have reported on various species of algae and the different shapes and sizes of the nanoparticles being synthesized using various microalgal species. Currently, there are many species of macro- as well as microalgae which can synthesize metallic as well as nonmetallic nanoparticles, and these have been found to have a wide variety of applications in many fields of science. However, researchers should also study the synthesis process not only with these metallic nanoparticles but also with the other available materials like more oxide- and more chalcogenide-based ones. In synthesizing these chalcogenides with the algae, there would more opportunities for various other applications related to these chalcogenides. Currently, the synthesizing of the chalcogenides with micro- and macroalgae is very much limited. As evident from the above material, there are many species of macro- and microalgae reported in the literature but the studies with red algae are very limited. Algal species also have the ability to fight against the toxicity of any nanoparticles and make them into a useful form. It was clearly stated that algae hold a promising role in the field of nanotechnology due to their unique properties exhibited during the synthesis of various metallic nanoparticles.

REFERENCES

Aarthi, F. A., Parthiban, C., Kumar, V. G., and Anantharaman, P. (2012). Biosynthesis of antibacterial gold nanoparticles using brown alga, *Stoechospermum marginatum* (kützing). *Spectrochim. Acta Part A Mol. Biomol. Spectrosc.* 99, 166–173. doi: 10.1016/j.saa.2012.08.081.

Abdel-raouf, N. (2017). Green biosynthesis of gold nanoparticles using *Galaxaura elongata* and characterization of their antibacterial activity. *Arab. J. Chem.* 10, S3029-S3039. doi: 10.1016/j.arabjc.2013.11.044.

Abdul Razack, S., Duraiarasan, S., and Mani, V. (2016). Biosynthesis of silver nanoparticle and its application in cell wall disruption to release carbohydrate and lipid from *C. vulgaris* for biofuel production. *Biotechnol. Reports* 11, 70–76. doi: 10.1016/j.btre.2016.07.001.

Abo-Shady, A. M., Mohamed, Y. A., and Lasheen, T. (1993). Chemical composition of the cell wall in some green algae species. *Biol. Plant* 35, 629–632. doi: 10.1007/BF02928041.

Alaqad, K., and Saleh, T. A. (2016). Gold and silver nanoparticles: Synthesis methods, characterization routes and applications towards drugs. *J. Environ. Anal. Toxicol.* 6, 384. doi: 10.4172/2161-0525.1000384.

Ashokkumar, T., and Vijayaraghavan, K. (2016). Brown seaweed-mediated biosynthesis of gold nanoparticles. *J. Environ. Biotechnol. Res.* 2, 45–50.

Asmathunisha, N., and Kathiresan, K. (2013). A review on biosynthesis of nanoparticles by marine organisms. *Colloids Surf. B Biointerfaces* 103, 283–287. doi: 10.1016/j.colsurfb.2012.10.030.

Azizi, S., Namvar, F., Mahdavi, M., Ahmad, M. B., and Mohamad, R. (2013). Biosynthesis of silver nanoparticles using brown marine macroalga, Sargassum muticum aqueous extract. *Materials (Basel)* 6, 5942–5950. doi: 10.3390/ma6125942.

Boeckel, B., and Baumann, K.-H. (2008). Vertical and lateral variations in coccolithophore community structure across the subtropical frontal zone in the South Atlantic Ocean. *Mar. Micropaleontol.* 67, 255–273. doi: 10.1016/j.marmicro.2008.01.014.

Brayner, R., Barberousse, H., Hemadi, M., Djedjat, C., Yéprémian, C., Coradin, T., et al. (2007). Cyanobacteria as bioreactors for the synthesis of Au, Ag, Pd, and Pt nanoparticles via an enzyme-mediated route. *J. Nanosci. Nanotechnol.* 7, 2696–2708. doi: 10.1166/jnn.2007.600.

Castro, L., Blázquez, M. L., Muñoz, J. A., González, F., and Ballester, A. (2013). Biological synthesis of metallic nanoparticles using algae. *IET Nanobiotechnol.* 7, 109–116. doi: 10.1049/iet-nbt.2012.0041.

Delalat, B., Sheppard, V. C., Rasi Ghaemi, S., Rao, S., Prestidge, C. A., McPhee, G., et al. (2015). Targeted drug delivery using genetically engineered diatom biosilica. *Nat. Commun.* 6, 8791. doi: 10.1038/ncomms9791.

El-Kassas, H. Y., Aly-Eldeen, M. A., and Gharib, S. M. (2016). Green synthesis of iron oxide (Fe3O4) nanoparticles using two selected brown seaweeds: Characterization and application for lead bioremediation. *Acta Oceanol. Sin.* 35, 89–98. doi: 10.1007/s13131-016-0880-3.

El-Rafie, H. M., El-Rafie, M. H., and Zahran, M. K. (2013). Green synthesis of silver nanoparticles using polysaccharides extracted from marine macro algae. *Carbohydr. Polym.* 96, 403–410. doi: 10.1016/j.carbpol.2013.03.071.

Ibm, I., Bee, A. E., Wf, S., and Wa, F. (2016). Green biosynthesis of silver nanoparticles using marine red algae Acanthophora specifera and its antimicrobial activity. *J Nanomed. & Nanotechnol.* 7, 7–10. doi: 10.4172/2157-7439.1000409.

Isaac, G., and Renitta, R. E. (2015). Brown algae mediated synthesis, characterization of gold nano particles using *Padina pavonica* and their antibacterial activity against human pathogens. *Int. J. PharmTech. Res.* 8, 31–39.

Iwasaka, M., and Mizukawa, Y. (2014). Magneto-optical properties of biogenic photonic crystals in algae. *J. Appl. Phys.* 115, 17B501. doi: 10.1063/1.4861556.

Jakob, I., Chairopoulou, M. A., Posten, C., Teipel, U., Georg, N., and Ohm, S. (2017). Biogenic calcite particles from microalgae — Coccoliths as a potential raw material. *Eng. Life Sci.* 17, 605–612. doi: 10.1002/elsc.201600183.

Jamali, A. A., Akbari, F., Ghorakhlu, M. M., de la Guardia, M., and Khosroushahi, A. Y. (2012). Applications of diatoms as potential microalgae in nanobiotechnology. *BioImpacts* 2, 83–89. doi: 10.5681/bi.2012.012.

Jeffryes, C., Campbell, J., Li, H. Y., Jiao, J., and Rorrer, G. (2011). The potential of diatom nanobiotechnology for applications in solar cells, batteries, and electroluminescent devices. *Energy Environ. Sci.* 4, 3930–3941. doi: 10.1039/c0ee00306a.

Jena, J., Pradhan, N., Dash, B. P., Sukla, L. B., and Panda, P.K. (2013). Biosynthesis and characterization of silver nanoparticles using microalga *Chlorococcum humicola* and its antibacterial activity. *Int. J. Nanomater. Bios.* 3, 1–8.

Journal, A., Adorno, L., Cordero, M., Vélez, C., Feliciano, I., and Cabrera, C. (2015). Biosynthesis of gold nanoparticles using *Osmudaria obtusiloba* extract and their potential use in optical sensing applications. *Austin J. Biosens. & Bioelectr.* 1, 1–9.

Kannan, R. R. R., Arumugam, R., Ramya, D., Manivannan, K., and Anantharaman, P. (2013). Green synthesis of silver nanoparticles using marine macroalga *Chaetomorpha linum*. *Appl. Nanosci.* 3, 229–233. doi: 10.1007/s13204-012-0125-5.

Karthikeyan, P., Mohan, D., Abishek, G., and Priya, R. (2015). Synthesis of silver nanoparticles using phytoplankton and its characteristics. *Int. J. Fish. Aquat. Stud.* 2, 398–401.

Krishnamoorthy, K., Alsagaby, S. A., Singaravelu, G., Govindaraju, K., and Premanathan, M. (2015). Green synthesis of silver nanoparticles for selective toxicity towards cancer cells. *IET Nanobiotechnol.* 9, 325–330. doi: 10.1049/iet-nbt.2015.0001.

Kuddus, S. A. (2017). Nanoceria and its perspective in cancer treatment. *J. Cancer Sci. Ther.* 09, 368–373. doi: 10.4172/1948-5956.1000444.

Kumar, E. M., Jagadish, D., and Kumar, R. B. (2014). A note on algae as potential source for alternate fuels – Biodiesel. *Int. J. PharmTech Res.* 6, 1783–1793.

Le Bras, D., Strømme, M., and Mihranyan, A. (2015). Characterization of dielectric properties of nanocellulose from wood and algae for electrical insulator applications. *J. Phys. Chem. B* 119, 5911–5917. doi: 10.1021/acs.jpcb.5b00715.

Mahdavi, M., Namvar, F., Ahmad, M. B., and Mohamad, R. (2013). Green biosynthesis and characterization of magnetic iron oxide (Fe3O4) nanoparticles using seaweed (Sargassum muticum) aqueous extract. *Molecules* 18, 5954–5964. doi: 10.3390/molecules18055954.

Mahdieh, M., Zolanvari, A., Azimee, A. S., and Mahdieh, M. (2012). Green biosynthesis of silver nanoparticles by *Spirulina platensis*. *Sci. Iran* 19, 926–929. doi: 10.1016/j.scient.2012.01.010.

Manivasagan, P., and Kim, S.-K. (2015). Biosynthesis of nanoparticles using marine algae: A review. *Mar. Algae Extr.* Chapter 17. 295–304. doi: 10.1002/9783527679577.ch17.

Mata, Y. N., Torres, E., Blázquez, M. L., Ballester, A., González, F., and Muñoz, J. A. (2009). Gold(III) biosorption and bioreduction with the brown alga Fucus vesiculosus. *J. Hazard. Mater.* 166, 612–618. doi: 10.1016/j.jhazmat.2008.11.064.

McIntyre, A., and Bé, A. W. H. (1967). Modern coccolithophoridae of the atlantic ocean—I. Placoliths and cyrtoliths. *Deep Sea Res. Oceanogr. Abstr.* 14, 561–597. doi: 10.1016/0011-7471(67)90065-4.

Medarevic, D., Losic, D., and Ibric, S. (2015). Diatoms – Nature materials with great potential for bioapplications. *Hem. Ind.* 70. 613–627. doi: 10.2298/HEMIND150708069M.

Naveena, B., and Prakash, E. (2013). Biological synthesis of gold nanoparticles using marine algae *Gracilaria corticata* and its aplplication as a potent antimicrobial and antioxidant agent. *Asian J. Pharm. Clin. Res.* 6, 4–7.

Okada, H., and Honjo, S. (1973). The distribution of oceanic coccolithophores in the {P}acific. *Deep Sea Res.* 20, 355–374.

Parial, D., Patra, H. K., Dasgupta, A. K. R., and Pal, R. (2012). Screening of different algae for green synthesis of gold nanoparticles. *Eur. J. Phycol.* 47, 37–4110.1080/09670262.2011.653406.

Parveen, K., Banse, V., and Ledwani, L. (2016). Green synthesis of nanoparticles : Their advantages and disadvantages. *AIP Conf. Proc.* 1724, 020048. doi: 10.1063/1.4945168.

Patel, V., Berthold, D., Puranik, P., and Gantar, M. (2015). Screening of cyanobacteria and microalgae for their ability to synthesize silver nanoparticles with antibacterial activity. *Biotechnol. Reports* 5, 112–119. doi: 10.1016/j.btre.2014.12.001.

Pharm, C. A. J., and Sci, B. (2015). Biosynthesis and characterization of silver nanoparticles using marine macro algae sargassum plagiophyllum biosynthesis and characterization of silver nanoparticles using marine macro algae *Sargassum plagiophyllum* C. Agardh. *J. Pharm. Biomed. Sci.* 05, 705–712.

Rahman, K. M., Melville, L., Huq, S. M. I., and Khoda, S. K. (2016). Understanding bioenergy production and optimisation at the nanoscale – A review. *J. Exp. Nanosci.* 11, 762–775. doi: 10.1080/17458080.2016.1157905.

Rajasulochana, P., Dhamotharan, R., Murugakoothan, P., Murugesan, S., Krishnamoorthy, P., and Academic, D. (2010). Biosynthesis and characterization of gold nanoparticles using the alga Kappaphycus alvarezii. *Int. J. Nanosci.* 9, 511–516. doi: 10.1142/S0219581X10007149.

Rajeshkumar, G., Kannan, S., and Chellapandian, A. (2012). Green synthesis of silver nanoparticles using marine brown algae *Turbinaria conoides* and its antibacterial activity. *Int. J. Pharma Bio. Sci.* 3, 502–510.

Ramakrishna, M., Dandamudi, R., Gengan, R. M., Chandra, S., and Rao, G. N. (2016). Green synthesis of gold nanoparticles using marine algae and evaluation of their catalytic activity. *J. Nanostructure Chem.* 6, 1–13. doi: 10.1007/s40097-015-0173-y.

Ray, P. C., Yu, H., and Fu, P. P. (2009). Toxicity and environmental risks of nanomaterials: challenges and future needs. *J. Environ. Sci. Health. C. Environ. Carcinog. Ecotoxicol. Rev.* 27, 1–35. doi: 10.1080/10590500802708267.

Reeder, S. B., Hu, H. H., Sirlin, C. B., Group, L. I., and Diego, S. (2016). HHS public access. *J. Magn. Reson. Imaging* 36, 1011–1014. doi: 10.1002/jmri.23741.Proton.

Rodrigues, D., Freitas, A. C., Pereira, L., Rocha-Santos, T. A. P., Vasconcelos, M. W., Roriz, M., et al. (2015). Chemical composition of red, brown and green macroalgae from Buarcos bay in Central West Coast of Portugal. *Food Chem.* 183, 197–207. doi: 10.1016/j.foodchem.2015.03.057.

Saif, S., Tahir, A., and Chen, Y. (2016). Green synthesis of iron nanoparticles and their environmental applications and implications. *Nanomaterials* 6, 209. doi: 10.3390/nano6110209.

Sap, K. A., and Demmers, J. A. A. (2016). World's largest Science, Technology & Medicine Open Access book publisher c. *Intech* 6, 111–133. doi: 10.5772/711.

Senapati, S., Syed, A., Moeez, S., Kumar, A., and Ahmad, A. (2012). Intracellular synthesis of gold nanoparticles using alga Tetraselmis kochinensis. *Mater. Lett.* 79, 116–118. doi: 10.1016/j.matlet.2012.04.009.

Sharma, D., Kanchi, S., and Bisetty, K. (2015a). Biogenic synthesis of nanoparticles: A review. *Arab. J. Chem.* ARABJC 1795, Open Access, 1-25. doi: 10.1016/j.arabjc.2015.11.002.

Sharma, G., Jasuja, N. D., Kumar, M., and Ali, M. I. (2015b). Biological synthesis of silver nanoparticles by cell-free extract of spirulina platensis. *J. Nanotechnol.* 2015, 6. doi: 10.1155/2015/132675.

Sharma, N., Bhatt, G., and Kothiyal, P. (2015c). Gold nanoparticles synthesis, properties, and forthcoming applications – A review. *Indian J. Pharm. Biol. Res.* 3, 13–27.

Sharma, A., Sharma, S., Sharma, K., Chetri, S. P. K., Vashishtha, A., Singh, P., et al. (2016). Algae as crucial organisms in advancing nanotechnology : A systematic review. *J. Appl. Phycol.* 28, 1759–1774. doi: 10.1007/s10811-015-0715-1.

Siddiqi, K. S., and Husen, A. (2016). Fabrication of metal and metal oxide nanoparticles by algae and their toxic effects. *Nanoscale Res. Lett.* 11, 363. doi: 10.1186/s11671-016-1580-9.

Singaravelu, G., Arockiamary, J. S., Kumar, V. G., and Govindaraju, K. (2007). A novel extracellular synthesis of monodisperse gold nanoparticles using marine alga, Sargassum wightii Greville. *Colloids Surf. B Biointerfaces* 57, 97–101. doi: 10.1016/j.colsurfb.2007.01.010.

Soisuwan, S., and Warisnoicharoen, W. (2010). Eco-friendly synthesis of fucoidan-stabilized gold nanoparticles kriengsak lirdprapamongkol and 3 Jisnuson Svasti Pharmaceutical Technology (International Program), Department of Food and Pharmaceutical Chemistry, Faculty of Pharmaceutical Sciences, V. 7, 1038–1042.

Subramaniyam, V., Subashchandrabose, S. R., Thavamani, P., Megharaj, M., Chen, Z., and Naidu, R. (2015). Chlorococcum sp. MM11—A novel phyco-nanofactory for the synthesis of iron nanoparticles. *J. Appl. Phycol.* 27, 1861–1869. doi: 10.1007/s10811-014-0492-2.

Tiwari, J. N., Tiwari, R. N., and Kim, K. S. (2012). Zero-dimensional, one-dimensional, two-dimensional and three-dimensional nanostructured materials for advanced electrochemical energy devices. *Prog. Mater. Sci.* 57, 724–803. doi: 10.1016/j.pmatsci.2011.08.003.

Velusamy, P., Kumar, G. V., Jeyanthi, V., Das, J., and Pachaiappan, R. (2016). Bio-inspired green nanoparticles : Synthesis, mechanism, and antibacterial application. *Toxicol. Res.* 32, 95–102. doi: 10.5487/TR.2016.32.2.095.

Vincy, W., Mahathalana, T. J., and Sukumaran, S. (2017). Algae as source of nanoparticles: A review. *Int. J. Lat. Trends Engg. Tech.*, Special issue - ICON - The Fruition of Science 5–9.

Vivek, M., Kumar, P. S., Steffi, S., and Sudha, S. (2011). Biogenic silver nanoparticles by gelidiella acerosa extract and their antifungal effects. *Avicenna J. Med. Biotechnol.* 3, 143–148.

Yousefzadi, M., Rahimi, Z., and Ghafori, V. (2014). The green synthesis, characterization and antimicrobial activities of silver nanoparticles synthesized from green alga Enteromorpha flexuosa (wulfen) J. Agardh. *Mater. Lett.* 137, 1–4. doi: 10.1016/j.matlet.2014.08.110.

Zhang, X.-F., Liu, Z.-G., Shen, W., and Gurunathan, S. (2016). Silver nanoparticles: Synthesis, characterization, properties, applications, and therapeutic approaches. *Int. J. Mol. Sci.* 17, 1534. doi: 10.3390/ijms17091534.

4 Liposome Nanocarrier
Synthesis, Characterization, and Applications

Prabhu Sukumaran and V. Sandhya

CONTENTS

4.1 Introduction 46
4.2 Structure of Liposomes 46
4.2.1 Phospholipids 46
4.2.2 Cholesterol 47
4.3 Classification of Liposomes 47
4.3.1 Conventional Liposomes 47
4.3.2 pH-Sensitive Liposomes 47
4.3.3 Cationic Liposomes 47
4.3.4 Long-Circulating Liposomes 48
4.3.5 Immune Liposomes 48
4.3.6 UV 48
4.3.7 MUV 48
4.4 Types of Drug Delivery by Liposomes 48
4.4.1 Passive Drug Delivery 48
4.4.2 Active Drug Delivery 48
4.5 Mode of Action 49
4.5.1 Adsorption 49
4.5.2 Fusion 49
4.5.3 Endocytosis or Phagocytosis 49
4.6 Parameters to Be Considered for the Formulation of Liposome 49
4.6.1 Bilayer Fluidity 49
4.6.2 Surface Charge 50
4.6.3 Surface Hydration 50
4.6.4 Size of the Liposome 50
4.6.5 Method of Preparation 50
4.6.6 Stability of Liposomes 50
4.6.6.1 Physical Stability 50
4.6.6.2 Chemical Stability 50
4.6.6.3 Advantages 50
4.6.6.4 Disadvantages 51
4.7 Synthesis of Liposomes 51
4.7.1 Liposome Formation 51
4.7.1.1 Mechanical Dispersion Methods 51
4.7.1.2 Solvent Dispersion Methods 52
4.7.1.3 Detergent Solubilization 52
4.7.2 Sizing of Liposomes 52
4.7.2.1 Without Sizing Step 53
4.7.2.2 With a Sizing Step 53
4.7.3 Drug Loading 53
4.7.4 Removal of Non-Encapsulated Material 53
4.8 Ligand Coupling Strategies 53
4.8.1 Binding of Ligands to the Surface of PEG-Containing Liposomes 53
4.8.2 Addition of PEG after Coupling of Ligand 53
4.8.3 Incorporation of Ligand–Anchor Conjugates Prior to Liposome Formation 53
4.8.4 Non-Covalent Coupling Method 54

4.9 Lyophilization ... 54
4.10 Characterization of Liposomes ... 54
4.10.1 Physical Parameters ... 54
4.10.1.1 Vesicle Shape and Lamellarity Analysis ... 54
4.10.1.2 Size Analysis ... 54
4.10.1.3 Zeta Potential ... 55
4.10.1.4 Stability of Liposomes ... 55
4.10.1.5 Leakage Analysis ... 55
4.10.1.6 Encapsulation Efficiency ... 55
4.10.1.7 In Vitro Drug Release ... 55
4.10.2 Chemical Parameters ... 56
4.10.2.1 Analysis of Lipid ... 56
4.10.3 Biological Parameters ... 56
4.11 Applications ... 56
4.11.1 Clinical Applications ... 56
4.11.1.1 Cancer Therapy ... 56
4.11.1.2 Gene Therapy ... 56
4.11.1.3 As Vaccine Adjuvant ... 57
4.11.1.4 Against Parasitic and Fungal Infections ... 57
4.11.1.5 Antimicrobial Therapy ... 57
4.11.2 Therapeutic Applications ... 58
4.11.2.1 As Brain Targeting ... 58
4.11.2.2 Ocular Applications ... 58
4.11.2.3 Pulmonary Applications ... 58
4.11.3 Applications in Food Industry ... 58
4.11.4 Cosmetic Applications ... 58
4.12 Conclusion ... 59
References ... 59

4.1 INTRODUCTION

Liposomes are micro-vesicles which are made up of natural or synthetic phospholipids. They were described in 1964 by Dr. Alec D. Bangham, initially called "smectic mesophases" and renamed as "liposomes" later. The name is derived from the Greek words **lipos** (meaning fats), and **soma** (meaning body) (Daraee et al., 2014). They are said to be amphipathic and lipophilic molecules which are made up of phospholipid bilayer, consisting of a hydrophilic head group and a hydrophobic tail group, and are usually in the range of 0.5–5.0 μm in diameter. As these have some attractive biological properties such as biodegradability, biocompatibility, non-toxicity, and non-immunogenicity, they can be potent drug delivery systems (Sharma and Sharma, 1997).

Liposomes can encapsulate drugs with hydrophilic and hydrophobic properties, which can be entrapped in the liposomes' lipid bilayer interface and provide various therapeutic applications. They can combat the disadvantages of other delivery systems by protecting the drugs from several types of resistance such as enzymes from the body (from mouth to stomach), stomach acids, bile salts, and free radicals (Akbarzadeh et al., 2013). The main aim of a liposomal delivery system is to provide a high therapeutic index and to increase the half-life of the entrapped drug, since they are stable in both in vitro and in vivo and have a high systemic residence time and improved sustained release of therapeutics (Banerjee et al., 2015).

Liposome encapsulation technology has developed novel strategies to design liposome carriers based on their stability and reactivity with our system. Two types of liposomes have been designed based on the interaction with our system: they are non-interactive sterically stabilized (long-circulating) liposomes (LCL) and highly interactive cationic liposomes. Both differ only by their formulation of drugs and affinity towards their uptake by the cells. LCL can entrap hydrophilic long-chain polymers; due to the reduction in recognition of liposomes by opsonins to the reticuloendothelial system (RES), they possess a greater half-life in systemic circulation and exhibit target-specific delivery. Cationic liposomes, however, are said to be non-viral drug delivery mechanism which can transfect cells by exogenous materials such as RNA or plasmid and which shows high affinity towards cell membranes (Sharma and Sharma, 1997).

4.2 STRUCTURE OF LIPOSOMES

Liposomes are made up of two components, phospholipids and cholesterol, as shown in Figure 4.1.

4.2.1 PHOSPHOLIPIDS

Phospholipids consist of glycerol or sphingosine, fatty acids, alcohol, and a phosphate group (Figure 4.1). Phospholipids are classified as glycerophospholipids and sphingophospholipids. Phospholipids which contain glycerol are said to be

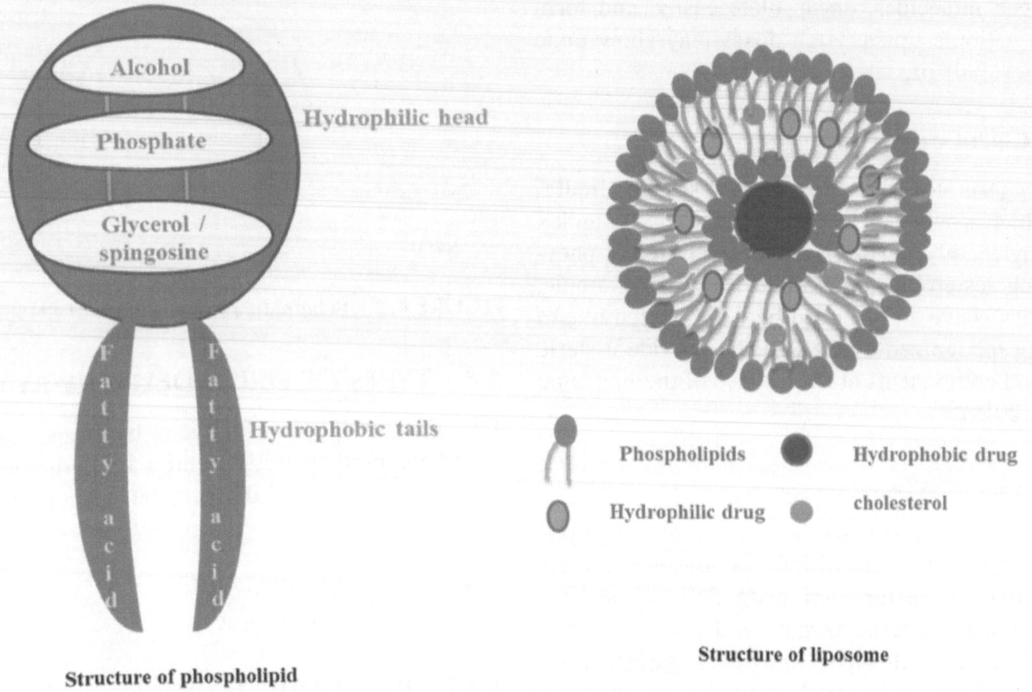

FIGURE 4.1 Schematic representation of structure of phospholipid with hydrophilic heads and hydrophobic tails and liposome structure shows that it can entrap hydrophilic and hydrophobic drugs.

glycerophospholipids, and they are classified as phosphatidyl choline (lecithin; PC), phosphatidyl ethanolamine (cephalin; PE), phosphatidyl serine (PS), phosphatidyl inositol (PI), DOPE (dioleoylphosphatidylethanolamine), and phosphatidyl glycerol (PG). Phospholipids which contain amino alcohol instead of glycerol are known as sphingophospholipids. The most common phopholipid used is phosphatidyl choline. They are not soluble in water or any aqueous media, and they exist in closely planar sheets to minimize unwanted interaction between the aqueous phase and the fatty acid chain (Daraee et al., 2014; Dua et al., 2012).

4.2.2 Cholesterol

Cholesterol cannot form a bilayer by itselfbut can be added in molar ratios of 1:1 or 2:1 of cholesterol to phosphatidyl choline and it orients its hydroxyl group towards aqueous phase and aligns its aliphatic chain with acyl chain, which presents in the center of bilayer parallely. It increases lipid bilayer stability and fluidity and reduces permeability in aqueous media (Laouini et al., 2012; Dua et al., 2012).

4.3 CLASSIFICATION OF LIPOSOMES

Based on the composition and mode of drug delivery, liposomes are classified into five types:

- Conventional liposomes
- pH-sensitive liposomes
- Cationic liposomes
- Immune liposomes
- Long-circulating liposomes

4.3.1 Conventional Liposomes

These are made up of negatively charged or neutral natural phospholipids or cholesterols such as sphingomyelin, 1,2-Distearoyl-sn-glycero-3-phosphocholine (DSPC) (DSPC), and egg phosphatidylcholine. They target the RES and have a short circulation half-life (Sharma and Sharma, 1997).

4.3.2 pH-Sensitive Liposomes

These membranes are comprised of PE, DOPE, oleic acid (OA), or cholesterol hemisuccinate (CHEMS). These are designed based on the observation of the mechanism by which viruses interact with endosomes and deliver genetic material to the cytoplasm before reaching lysosomes. They deliver the entrapped cargo to the target, where they undergo phase transition due to change in pH (acidic) when they fuse with the endosome (pH 6.50) (Mellman et al., 1986).

4.3.3 Cationic Liposomes

They are made up of cationic lipids such as dioctadecyldimethyl ammonium chloride (DOGS), 2,3-dioleyloxy-N-[2(sperminecarboxamido)-ethyl]-N,N-dimethyl-l-propanaminium trifluoroacetate (DOSPA), 1,2-dioleoyl-3-dimethyl-hydroxyethyl ammonium bromide (DORIE), 1,2-dioleoyloxy-3-(trimethylammonio)-propane (DOTAP), 1,2-dimyristyloxypropyl-3-dimethyl-hydroxylethyl ammonium bromide (DMRIE), and dimethyldioctadecyl ammonium bromide (DDAB) combined with DOPE, which are used for the delivery of DNA and RNA since they are negatively charged molecules, encapsulate easily, and form a complex with cationic lipids. High doses may show toxic effects with a short half-life.

4.3.4 Long-Circulating Liposomes

As the name suggests, they have extended systemic circulation in blood, which is achieved by the formation of a complex between polyethylene glycol (PEG) and neutral high-T_c phospholipids (with cholesterol at about 5% to 10%). For example, "stealth liposomes" have an extended half-life, which masks them from being recognized by RES. They provide a steric barrier with blood components at the surface of the liposome (Banerjee et al., 2015).

4.3.5 Immune Liposomes

They are conventional liposomes or long-circulating liposomes with antibody or recognition sequences on their surfaces, and they are a targeted drug delivery system where they bind with specific targets and provide therapeutic effects. They are otherwise known as "guided missiles" or "immunological adjuvants"; they elicit an immune response, when an antigen or any other recognition sequences are present on the liposome's surface (Daraee et al., 2014).

Based on their size, they are further classified as follows:

- UV (unilamellar vesicles)
- MLV (multilamellar vesicles)

4.3.6 UV

These are made up of single phospholipid bilayer encapsulated in an aqueous solution (one lamella) and they are classified as small unilamellar vesicles (SUV) and large unilamellar vesicles (LUV). SUV are less than 100 nm and LUV are in the range of 100 nm to 1000 nm.

4.3.7 MUV

They are made up of more than one lamella, have an onion-like structure, and range between 500 nm and 10,000 nm (Dua et al., 2012; Popovska et al., 2013) (Figure 4.2).

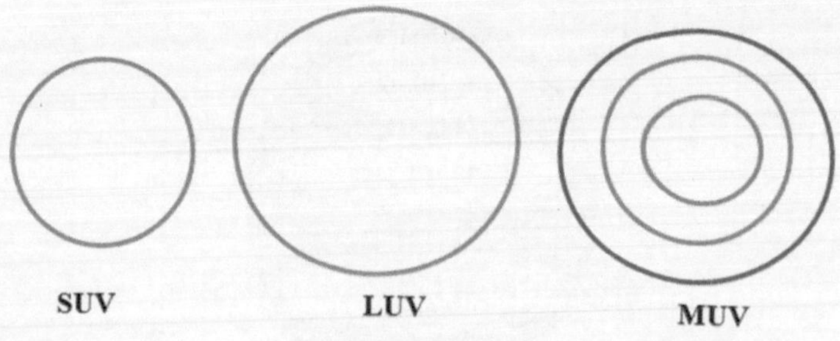

FIGURE 4.2 Schematic representation of SUV, LUV, and MUV.

4.4 TYPES OF DRUG DELIVERY BY LIPOSOMES

The aim of the drug delivery by liposomes is to provide improved biodistribution and a high therapeutic index. To do so, two modes of delivery have been utilized. They are as follows:

- Passive drug delivery
- Active drug delivery

4.4.1 Passive Drug Delivery

This mode of delivery tends to utilize the natural defense mechanism of the cells such as macrophages to internalize them. It has been utilized to treat tumors, since their leaky vasculature and poor lymphatic drainage makes the liposome enter the target site after extravasation, as shown in Figure 4.3. The tumor vessels filter the nanoparticles as a sieve and retain them in the tumor tissue (Turánek, 2009). The liposomes used in this delivery are coated by PEG, which resists the interaction of blood components with them, which causes them to escape from RES. This phenomenon is called the enhanced permeation and retention (EPR) effect. However, it shows increased toxicity and decreased drug retention, which becomes a barrier in using this mode of delivery.

4.4.2 Active Drug Delivery

Damaged cells possess certain biomarkers such as receptors, where the cells can be targeted via certain ligands which will have specificity towards those markers; this is called active drug delivery or ligand-mediated drug delivery. Ligands such as antibodies, proteins, and carbohydrates which can attach

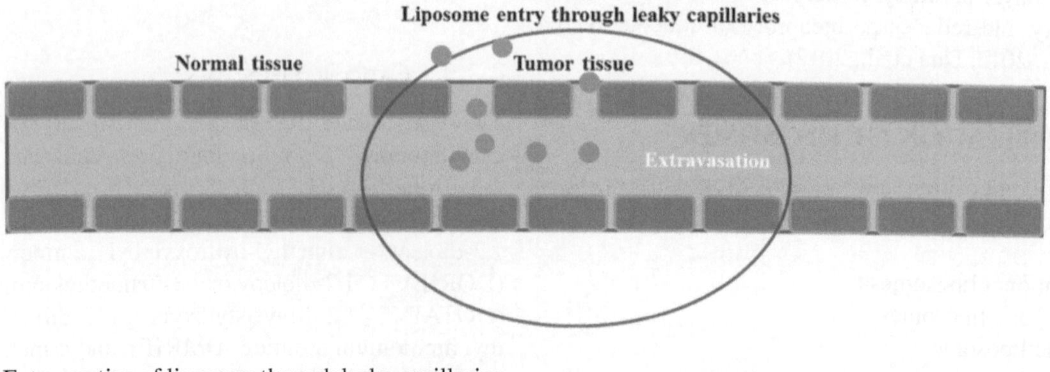

FIGURE 4.3 Extravasation of liposome through leaky capillaries.

to the surface of the liposome have been involved in this type of delivery. Examples are folate- and transferrin-mediated drug delivery. This kind of delivery provides low toxicity and increased therapeutic effect (Figure 4.4).

4.5 MODE OF ACTION

The drug which is encapsulated in liposome is released to the target site by interacting with the cells of the system. This mode of action happens in a step-by-step process (Figure 4.5). The steps are as follows:

4.5.1 Adsorption

The liposome makes contact with the cell membrane, where it is adsorbed onto the surface. Ligands or receptors are involved in this kind of interaction.

4.5.2 Fusion

This kind of interaction is mediated by fusogens such as surfactants or detergents, and the liposome is fused with the cell membrane and enters into the cytosol. This mediation does not usually occur; the rapidness of phagocytosis makes them skip this step.

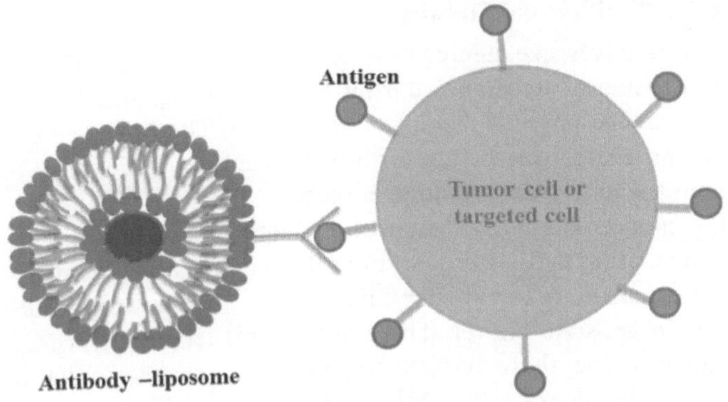

FIGURE 4.4 Antibody-mediated targeted drug delivery by liposome.

4.5.3 Endocytosis or Phagocytosis

In a natural phenomenon of biological systems, phagocytic cells such as macrophages can engulf the liposome and, due to the action of lysosomes, the lipid bilayer of liposome breaks and releases the encapsulated drug to the outer environment. The access of liposome to other cell organelles can be mediated by endocytosis, and cell surface receptors are required to elicit this effect.

In the case of tumor or cancerous cells, the liposomes can be easily taken up by the cells as they recognize liposomes as a nutritional supplement. The internalization of liposomes by cancer cells causes them to release anti-cancer drugs which can destroy them easily (Yadav et al., 2017).

4.6 PARAMETERS TO BE CONSIDERED FOR THE FORMULATION OF LIPOSOME

There are certain factors to be considered during the formulation of liposome that will influence in vivo characteristics such as stability, drug characteristics, bilayer fluidity, bio-distribution patterns, size, and surface hydration. These parameters are explained below.

4.6.1 Bilayer Fluidity

The phase transition temperature (T_c) characterizes the bilayer fluidity. The bilayer exists in different physiological states below and above T_c. Below T_c, it shows a rigid, gel-like phase and above T_c it exists in a crystalline phase. The bilayer fluidity can be altered by using phospholipids of different T_cs, which depend upon the nature and length of fatty acids. High-T_c lipids make the bilayer less fluid and less leaky, whereas low-T_c lipids tend to leak the encapsulated material

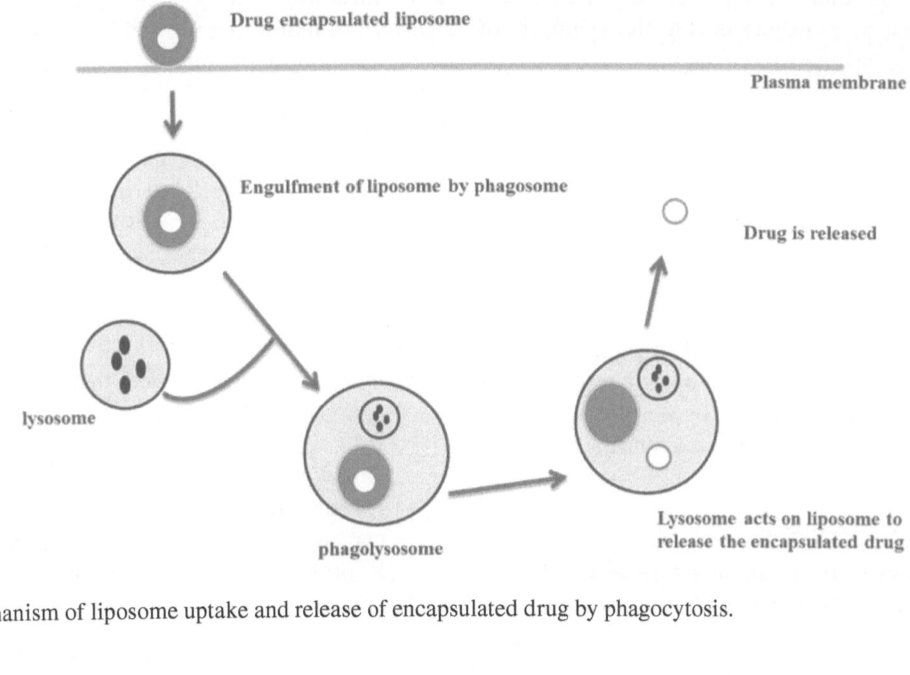

FIGURE 4.5 Mechanism of liposome uptake and release of encapsulated drug by phagocytosis.

at physiological temperature. The high-T_c lipids are less recognized by the RES than low-T_c lipids, which influences the interaction of liposomes with our system. Cholesterol can also be added to increase the membrane fluidity at high concentrations, in contrast to low concentrations; it decreases the membrane fluidity but makes the liposome stable and less leaky.

4.6.2 Surface Charge

The mechanism and liposome–cell interaction may be influenced by the composition and density of charge on the liposome surface. By changing the lipid composition, these factors can be altered. Neutral liposomes will not have a significant impact on cells; in such cases, the drugs enter the targeted site extracellularly from the liposome. A high electrostatic charge on the surface of liposome tends to increase the liposome–cell interaction, and negatively charged liposomes can be taken up through coated pit endocytosis. It has been investigated that this will also influence the plasma clearance after systemic administration.

4.6.3 Surface Hydration

The interaction of liposomes with the blood and its contents and uptake by RES can be minimized by adding a small fraction of hydrophilic compounds such as PEG, thereby making it become sterically stabilized. This kind of hydrophilic coating makes the liposomes more stable in the biological environment and provides steric hindrance to opsonic absorption, which tends liposomes to escape from RES uptake.

4.6.4 Size of the Liposome

The fraction of drug cleared by RES is influenced by the size of liposomes. The RES takes up the liposome based on the size of the vesicles. SUV tend to opsonize quickly compared to large liposomes (LUV and MLV). In the application of drug delivery to tumor cells, the small liposomes can extravagate and reach the tumor site more easily than large liposomes. The liposome size can be managed during the synthesis of liposome by sonication or micro-fluidization.

4.6.5 Method of Preparation

The process of preparing liposomes is discussed later (Section 4.7. synthesis of lipososme; here the suitable and appropriate method to prepare liposomes of certain sizes and characteristics will be discussed. The methods of preparing liposomes of various sizes are presented in Table 4.1.

These methods can also be combined to get the desired size of liposome; for example, SUV can be synthesized from MLV or LUV (from other methods) by sonication or the freeze-thaw extrusion method (Sharma and Sharma, 1997).

4.6.6 Stability of Liposomes

There are several barriers in the human system which will destabilize liposomes and affect the pharmacokinetic and pharmacodynamic profiles of encapsulated drugs. Therefore, it is a need to concentrate on the development and storage factors of liposome encapsulation and manufacture so that improved bio-distribution can be achieved. These stability factors are as follows:

- Physical stability
- Chemical stability

TABLE 4.1
Methods of Liposome Formation with Respect to Sizes

Method	Size
Thin film hydration, freeze-drying, sonication, reverse-phase evaporation, freeze-thaw extrusion	MLV
Thin film hydration method, freeze-thaw extrusion, ethanol injection, reverse-phase evaporation	LUV
Ethanol injection, solvent evaporation, sonication	SUV

Popovska (2013).

4.6.6.1 Physical Stability

This involves size, aggregation, leakage of encapsulated drug, and fusion of liposomes during manufacturing, which deteriorates the physical stability of liposome. The liposome can be aggregated or fused to form new structures which tend to increase in size during storage. Increase in size will influence the mode of administration of the liposome. The bilayer's composition and state determine the leakages of the encapsulated drug. A liquid-state bilayer tends to leak drugs more than a gel-state bilayer. There are several techniques which can estimate these parameters, including electron microscopy, Coulter counter, NMR spectroscopy, and light-scattering ultra-centrifugation (Yadav et al., 2017).

4.6.6.2 Chemical Stability

The unsaturated fatty acids present in the phospholipid bilayer are more prone to oxidation; hydrolysis, pH, and the ionic strength of the liposome may also alter the stability of liposome. The two acyl ester bonds in phosphatidylcholine undergo hydrolysis, which enhances the permeability of liposome. The oxidation of phospholipids liberates free radicals, which tend to form cyclic peroxides and hydroxyperoxidases, and the formation of peroxides can be prevented by adding quenchers (beta-carotene), antioxidants (α-tocopherol), and complexing agents (EDTA) (Kulkarni et al., 2011).

4.6.6.3 Advantages

- It can encapsulate both negatively and positively charged macromolecules
- It has improved pharmacokinetic and pharmacodynamic profiles
- It can accommodate large fragments of DNA, which can combat the problems of gene therapy
- It provides targeted and site-specific therapeutic effect

- It protects the encapsulated drug from the physical and chemical environment of our system to avoid unfavorable reactions

4.6.6.4 Disadvantages

- Instability of liposomes due to oxidation and hydrolysis of phospholipids
- Easily recognized and cleared by RES
- Production cost is high
- Unfavorable interactions with our system, i.e., non-specific uptake may occur

4.7 SYNTHESIS OF LIPOSOMES

Liposomes are prepared such that they may have improved stability and efficient drug entrapment. They are synthesized by the following procedure, as shown in Figure 4.6.

- Liposome formation
- Sizing of liposomes
- Drug loading (can be carried out before or during the preparation or after sizing)
- Removal of non-encapsulated drug

One should consider certain parameters which influences formation of liposome. They are physical parameters (vesicle shape and size, surface morphology, surface charge, pH, phase transition), chemical parameters (phospholipid and cholesterol concentration, phospholipid peroxidation and hydrolysis, cholesterol auto-oxidation and osmolality), and biological parameters (therapeutic application, clinical trials) (Popovska et al., 2013).

4.7.1 Liposome Formation

Three different strategies have been utilized to carry out hydration of lipids:

- Mechanical dispersion methods
- Solvent dispersion methods
- Detergent solubilization

4.7.1.1 Mechanical Dispersion Methods

In this method, the lipids are dried followed by the addition of aqueous phase carried out by mechanical means. The methods are as follows:

4.7.1.1.1 Sonication

This is the most widely used technique for the preparation of SUV. The disadvantages of this method are low encapsulation efficiency and the degradation of lipid components and the drug may occur. This method is carried out in two ways: probe sonication and bath sonication. In probe sonication, the tip of the sonicator is directly coupled into the lipid dispersion. It requires high energy input and it leads to rise in the vessel's temperature; therefore, the vessel should be placed in an ice or water bath. Bath sonication doesn't require high energy input, temperature control is easier, and the dispersing material can be protected under inert atmosphere (Akbarzadeh, 2013).

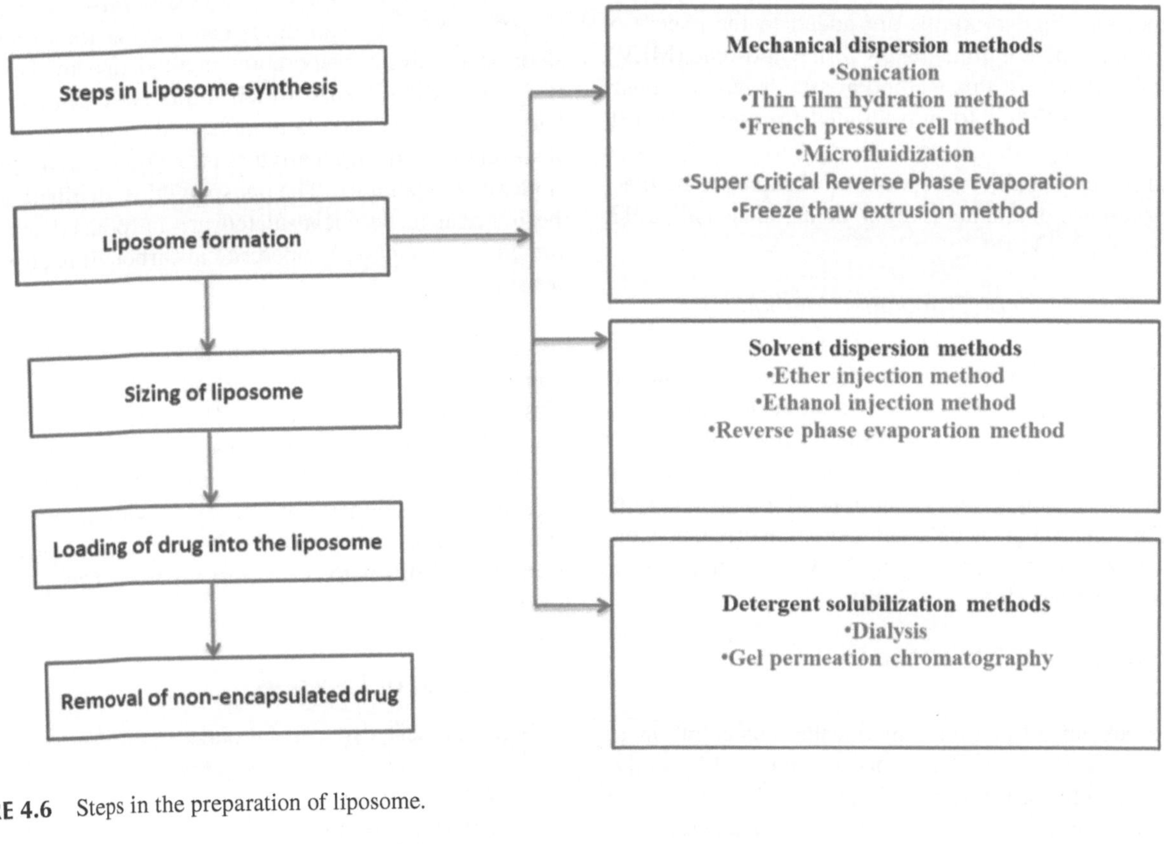

FIGURE 4.6 Steps in the preparation of liposome.

4.7.1.1.2 *Thin-Film Hydration Method*

This method is known as the Bangham method; it is a simple and convenient method for the preparation of MLV. In this method, the phospholipids and cholesterol are dissolved in an organic solvent such as chloroform or ethanol, and on evaporating the solvents under vacuum temperature, lipid films are formed on the flask. For large-scale production, it is desirable to use a rotary evaporator under reduced pressure. The deposited lipid film is then hydrated by adding aqueous buffer solutions at a temperature above lipid transition temperature, and shaking or agitation yields liposome (Laouini et al., 2012).

4.7.1.1.3 *French Pressure Cell Method*

In this method, the MLV are extruded via a small orifice at 20,000 psi at 4°C; it is used to handle unstable materials and it is more effective than sonication methods. It is highly reproducible but it is difficult to achieve desired temperature and allows for only small working volumes (about 50 ml) (Dua et al., 2012).

4.7.1.2 Solvent Dispersion Methods

4.7.1.2.1 *Ether Injection Method*

The phospholipids are dissolved in a diethyl ether/methanol mixture or diethyl ether, and the resulting solution is gradually injected to an aqueous solution of the drug to be encapsulated under reduced pressure at 55–65°C. The liposomes are formed by the removal of ether under a vacuum condition. The main disadvantages of this method are that the material to be encapsulated is exposed to the organic solvent and the population is heterogeneous in size (70 to 200 nm) (Kulkarni et al., 2011).

4.7.1.2.2 *Ethanol Injection Method*

The ethanol and lipid mixtures are added to the excess of aqueous buffer, where it immediately forms liposome (MLV). The main drawbacks of this method are the heterogeneously sized population (30 to 110 nm), diluted liposome, the difficulty in removing ethanol as it forms into zoetrope with water, and the probabaility of bioactive substance to be inactivated in presence of ethanol is high(Akbarzadeh et al., 2013; Popovska et al., 2013).

4.7.1.2.3 *Reverse-Phase Evaporation Method*

This method is used to create inverted micelles by using sonication of mixtures of phospholipids, organic solvents, and the material to be encapsulated. On evaporation of organic solvent under reduced pressure, the resultant inverted micelles are converted to the formation of viscous gel. On removing the organic solvents by a rotary evaporator, the liposomes are shaped. This method gives 65% encapsulation efficiency, but the disadvantage is that there may be DNA strand breakage or protein denaturation (Dua et al., 2012).

4.7.1.3 Detergent Solubilization

4.7.1.3.1 *Dialysis*

Detergents are added at a critical micelle concentration to solubilize lipids; then, the detergents are removed by dialysis and the micelles are formed (liposome LUV). A version of the dialysis system called LIPOPREP (Diachema AG, Switzerland) is available for large-scale removal of detergents. This method is reproducible and the population of liposomes formed is homogenous in size. The drawback is that there may be retention of detergents within the liposome.

4.7.1.3.2 *Gel Permeation Chromatography*

Columns such as sephadex G-50, sephadex G-100, and sepharose 2B-6B can be used to deplete detergent. The liposomes can be percolated through bead spaces rather than penetrating into the pores of the beads. The beads may absorb amphilic lipids; therefore, it is necessary to carry out pre-treatment of column by pre-saturating it with empty liposome suspensions (Akbarzadeh et al., 2013).

4.7.1.3.3 *Industrial Production of Liposomes*

The above-mentioned methods can also be used, but some methods which have greater potential for the large-scale production of liposomes are as follows:

4.7.1.3.3.1 Microfluidization In this method, otherwise known as micro-emulsion, the lipid phase and aqueous phase are injected into the micro-channel. Due to the small channel dimensions and relatively low flow rates in the micro-channel, well-defined mixing can be obtained by interfacial diffusion. Thereby, the size of the liposomes can be controlled by changing the flow rate. MICROFLUIDIZER from Microfluidics Corporation, Massachusetts, can be used to produce 20 gallons/minute of liposomes in the 50- to 200-nm size range. It yields an encapsulation efficiency of about 75% (Laouini et al., 2012; Dua et al., 2012).

4.7.1.3.3.2 Supercritical Reverse-Phase Evaporation (SCRPE) Supercritical fluids can combat the disadvantages of organic solvent evaporation method due to the exposure of drugs to the solvents, which limits large-scale production. They are nontoxic, nonflammable, and recyclable; elimination of solvent from the lipid mixture is easy and can be done under moderate temperature. The phospholipids, distilled water, and the material to be encapsulated are introduced into a given amount of ethanol with supercritical carbon dioxide at reduced pressure.

4.7.1.3.3.3 Freeze-Thaw Extrusion Method Liposomes from the film method are vortexed with the material to be encapsulated, and the resultant MLV are frozen. After several cycles of freeze-thaw and extrusion, the liposomes rupture, which yields SUV, and they fuse to form LUV. This method is used for the synthesis of liposomes with an even size distribution (Samad et al., 2007). Some examples of liposome-formulated drugs with respect to method of preparation are given in Table 4.2.

4.7.2 Sizing of Liposomes

Two approaches are carried out:

- Without sizing step
- With sizing step

TABLE 4.2
Some Liposomal Drugs with Respect to Method of Preparation

Drug	Liposome Formulation	Method of Preparation
Ibuprofen	Egg PC, distearoyl phosphatidylcholine (DMPC), DSPC, stearylamine (SA), dicetylphosphate (DCP), cholesterol (Chol)	Hydration lipidic film
Acetazolamide	Egg PC, type X-E, and cholesterol stearylamine or dicetyl phosphate	Reverse-phase evaporation
Salidroside	Egg PC, cholesterol	Ethanol injection
Plasmid DNA from *Escherichia coli*	Dipalmitoylphosphophatidylcholine (DPPC), DCP, cholesterol (70:20:10 molar ratio)	Heating method
Metronidazole and verapamil hydrochloride drug–cyclodextrin complex	Lecithin	Spray drying

Laouini, 2012; Popovska, 2013.

4.7.2.1 Without Sizing Step

In this case, the liposomes are sized by selecting certain parameters relating to liposome formation and those parameters to be controlled for the formation of acceptable width particle size distributions.

4.7.2.2 With a Sizing Step

The formed liposomes with varying sizes are distributed uniformly, and then fractionated by centrifugation. Gel permeation chromatography can also be used for even particle size distribution (Kulkarni et al., 2011).

4.7.3 Drug Loading

Loading of drugs can be carried out either passively (during liposome formation) or actively (after liposome formation). Hydrophobic and water-soluble drugs can be formulated during vesicle formation based on trapping effectiveness and drug–lipid interactions, while water-soluble drugs with amine function can be actively loaded by employing pH gradients.

4.7.4 Removal of Non-Encapsulated Material

The presence of non-encapsulated drugs may lead to physical instability or other side effects. The non-encapsulated fraction can be eliminated by dialysis, ultra-centrifugation, or gel permeation chromatography (Akbarzadeh et al., 2013).

4.8 LIGAND COUPLING STRATEGIES

Liposomes with ligands or receptors can be synthesized by grafting the ligands to the liposome surface through a spacer molecule and hydrophobic anchor via cross-linking molecules. The choice of hydrophobic anchor depends on the reactive groups presented on the anchor and the ligand, the existence of hetero-functional cross-linking molecules, and the type of chemical bond desired (i.e., stable/unstable, covalent/non-covalent). Because of the reactive amine in its head group, PE, and the presence of various acyl chain lengths with different degrees of unsaturation, PEG lipids are normally incorporated into liposomes; ligands can be coupled before or after the formation of liposome formation or by non-covalent interaction.

4.8.1 Binding of Ligands to the Surface of PEG-Containing Liposomes

The linkage of anchor molecules is done by the application of sulfhydryl reactions. Two thiol-reactive PE derivatives which can be incorporated into liposomes are N-pyridyldithiopropionyl-PE (PDP-PE) and maleimidophenyl butyrate-PE (MPB-PE). PDP-PE can be reduced easily to form free thiol groups at the surface of the liposome. The free thiol groups can then be coupled through thioether bonds to maleimide groups on proteins with the use of a hetero-bifunctional cross-linking reagent like SMPB ((H-succinimidyl-4-)-p-maleimidophenyl) butyrate). The presence of PEG in the lipid bilayer in higher amounts may lead to steric interference with the accessibility of antibodies to the liposomal surface.

4.8.2 Addition of PEG after Coupling of Ligand

To overcome the interference of PEG with the accessibility of antibodies to the surface of the liposome, the PEG can be incorporated to the liposome after the coupling of ligand. This is done by the use of high temperatures; this shows efficient transfer but may destroy ligands. In another method, liposomes containing MB-PE were conjugated first with thiolated antibodies, and then PEG was grafted to the surface of the immune liposomes with the use of PEG-succinyl cysteine (PEG-SC) of various polymer chain lengths. By this method, the ligand and grafted PEG occur on the outer leaflet of the liposomal surface, thereby providing maximal interior space for drug loading.

4.8.3 Incorporation of Ligand–Anchor Conjugates Prior to Liposome Formation

The formation of ligand–anchor conjugates is done by detergent solubilization of conjugates followed by dialysis or by hydration of dry lipid film. By this method, there will be a reduced volume for drug loading as the incorporated ligand may orient towards the interior aqueous space of the liposome.

4.8.4 Non-Covalent Coupling Method

The high-affinity binding of avidin or streptavidin with biotin has been adapted for coupling ligands to the liposomal surface. In one strategy, the liposomes are bound with the avidin or streptavidin, then the ligands are biotinylated, and both the avidin liposomes and biotinylated ligands are incubated. In another strategy, biotinylated ligands are first bound to the target epitope, and this is followed by a chase step using either streptavidin or avidin liposomes (Daraee, 2014).

4.9 LYOPHILIZATION

This is a freeze-drying method to provide long-term stability and storage of liposomes. It involves the removal of water from the liposomal products in the frozen state at extremely low pressures. This is usually applied to dry products that are thermo-labile and would be destroyed by heat-drying. Leakage of entrapped substances may occur during the process of freeze-drying and on reconstitution. This problem can be overcome by the use of trehalose, which is proven to be an excellent cryoprotectant for liposomes. Freeze-driers are available in many sizes, from small laboratory models to large industrial units (Daraee, 2014; Akbarzadeh, 2013).

4.10 CHARACTERIZATION OF LIPOSOMES

This is carried out to check whether the synthesized liposomes are of the desired quality and quantity across batches. The parameters used to characterize the liposomes are as follows:

- Physical parameters
- Chemical parameters
- Biological parameters

4.10.1 Physical Parameters

4.10.1.1 Vesicle Shape and Lamellarity Analysis

The drug release kinetics and encapsulation efficiency depend on the number of lipid bilayers present in the liposome. There are several methods for the determination of the number of that liposomes are made of. One method used to determine the external reactive groups on liposome is carried out by the addition of periodate to the lipid bilayer, which contains phosphotidylglycerol; due to diol oxidation it releases formaldehyde. The released formaldehyde is made to react with chromo-tropic acid, which yields a product that can be measured at 570 nm.

The inside–outside distribution of phospholipids within the membrane can be determined by nuclear magnetic resonance (NMR). In this method, the samples are added with 31P-NMR along with Mn^{2+} (paramagnetic ions which can interact with negatively charged phospholipids), which broadens the 31P-NMR resonance and thereby decreases the quantitative signal. At high concentrations, Mn^{2+} has the ability to penetrate inside the lipid bilayer; accordingly, the inside distribution of phospholipids can be estimated. At low pH or in the presence of any complexing agents (HEPES or TRIS buffer), the penetration of Mn^{2+} will be inhibited; therefore, the presence of outer phospholipids can be measured. The quantitative signal from the NMR can be calculated by the ratio of 31P-NMR signal before and after Mn^{2+} addition, which gives the inner to outer phospholipid ratio.

The presence of amino lipids can be found by the addition of 2,4,6-trinitrobenzensulfonic acid (TNBS) to the lipid bilayer, which contains amines; this results in the formation of a complex that can be detected at 420 nm. The external amino lipids can be estimated as lipid bilayers, which are permeable to TNBS: to determine the presence of inner amino lipids, triton-X is added to the lipid bilayer, which will the lyse the liposome and make the TNBS interact with the inner amino lipids. Other methods, such as the freeze-fracture technique with transmission electron microscopy or the generation of fluorescence signal by the quenching action of N-(7-nitrobenz-2-oxa-1,3-diazol-4-yl) (NBD) by sodium dithionite detected at 450 nm (excitation) and 530 nm (emission), can also be used.

4.10.1.2 Size Analysis

Analysis of liposome size is performed to evaluate the therapeutic application, and it is done by methods such as the following.

By the size-exclusion chromatography (SEC) method, the liposomes are fractionated based on the hydrodynamic size; SEC excludes the liposomes by size from descending order with the help of columns such as sephadex G-50. This method can be also be used in combination with HPLC (High Performance Liquid Chromatography) to improve the resolution of liposome size and reproducibility. Another method by which the size of liposomes can be analyzed is dynamic light scattering (DLS) or photon correlation spectroscopy (PCS). In this method, when the liposomes are suspended in a solvent, they scatter light when they are illuminated by light which differs in index of refraction from that of suspending solvent; the particles which show Brownian motion are exceptions to the scattering effect. PCS measures the time-dependent fluctuations of light scattered from the liposomes suspended in solvent. The main advantage of this method is to evaluate the size of the liposome in its native conformation, but this technique will not provide information on the shape of the liposome. It can estimate small quantities of high-molecular-weight aggregates and large species in the low nanometer to low micrometer range.

A method in which a field is applied perpendicular to the direction of channel flow is known as field-flow fractionation (FFF). There are several types: among them, flow FFE (Free-flow electrophoresis) is used for liposome size analysis, in which it separates vesicles on the basis of hydrodynamic size. The channel wall has a semipermeable membrane that allows only carrier fluid to pass. Two liquids are subjected to the sample: one is the channel flow and the other is cross-flow along the perpendicular direction of channel flow. The sample is injected into the instrument via an injector. After a certain period of time, the channel flow is stopped and cross-flow is allowed to act on the sample. The channel flow slows down the motion of particles present near the channel walls, and cross-flow propels the particles towards the channel wall. The retention time of the particle is proportional to the hydrodynamic diameter, and detectors such as a UV/VIS spectrophotometer can be used. Other methods, such as ultracentrifugation, which separates particles based on the density, atomic fluorescence microscopy (AFM), and Coulter counter, can also be used to analyze the size of liposomes.

4.10.1.3 Zeta Potential

This parameter is used to monitor the surface charge of the liposome, and it is defined as the overall charge that the surface of the particle acquires in a surrounding medium. It is exhibited by any particle existing in a colloidal state. This method is potent for measuring the stability of liposomes. The main principle behind finding the liposomal surface charge is that if the liposome suspension has a large negative or positive zeta potential, the liposomes will not form an aggregate, as they repel each other. If there is a low zeta potential, they tend to flocculate. The zeta potential of a liposome is measured by passing a laser beam through the sample; when this is done, the light is scattered at an angle of 13°. By the application of an electric field to the sample, the particles move in the known strength in the interference pattern and produce scattered light which is proportional to the speed of the particles. The acquired information is passed to the computer via a signal processor, and finally zeta potential is calculated. Zetasizer is commercially used to measure zeta potential-above (+/–) 30 mV is considered to be stable (Huang et al., 2005; Popovska et al., 2013).

4.10.1.4 Stability of Liposomes

Liposome stability has been discussed earlier; the stability of liposomes can be measured by certain techniques, and those techniques are explained at this point.

Ion transport imaging is a technique to determine the K^+ permeability with the help of K^+ indicator dye PBFI (Photophysics of the fluorescent K^+ indicator). To image the K^+ permeability, the liposomes are encapsulated with 100 mM Na^+, 100 mM sucrose, and PBFI placed in a solution containing 100 mM K^+ and 100 mM glucose. The density difference between the inner and outer environment causes flattening of liposome, which indicates that the liposome is impermeable to K^+ and the dye intensity remains stationary. On addition of nigericin, a K^+–H^+ exchanger makes the liposome permeable to K^+; therefore the fluorescence intensity of PBFI has been doubled, which results in the liposome having a deformed spherical shape (Akashi et al., 1996).

4.10.1.5 Leakage Analysis

The leakage of encapsulated hydrophilic molecules can be detected by the application of a fluorescence-quenching system. A dye known as carboxyfluorescein (CF) has been used in this system along with two types of probes used to detect the leakage at different pH values. The CF dye can self-quench at concentrations of >100 mM (>95%) and the sample pH should be less than 7, which is directly proportional to the fluorescence intensity. The liposomes are encapsulated with concentrated solutions of CF. When there is a leakage, the dye is released with an increase in fluorescence intensity as it becomes diluted. Two probes, namely, the polyanionic fluorophore ANTS and the cationic quencher DPX, have been used since ANTS is said to have a pH-independent fluorescence spectrum. Both of these have been co-encapsulated into liposomes as CF dye. On dilution, due to collisional quenching of ANTS and DPX, the decrease in quenching by DPX results in an increase of ANTS fluorescence (Bergstrand, 2003).

4.10.1.6 Encapsulation Efficiency

The efficiency of encapsulation of the drug depends on the amount of drug entrapped by the liposome. The entrapment of drug by the liposome is usually low; this can be improved by active drug loading, where the uncharged drug can cross the lipid bilayer easily, and by making the interior of the liposome charged, so the drug will not escape. To achieve this, the inner part of the liposome is charged by an entrapping low-pH environment and by placing the liposome in a neutral environment which contains the neutral drug.

To find out the percentage entrapment of drug by liposome, separate the drug encapsulated liposome and the free drug from the mixture and to quantify the entrapped drug released from the drug encapsulated liposome.

Minicolumn centrifugation separates the drug loaded liposomes and the drug-free liposomes on the basis of difference in their size. The liposome mixture is subjected to the sephadex column, the column is centrifuged at 2000 rpm for 3 minutes, and the void volumes are expelled. The drug-free liposome is bound to the column and the drug-loaded vesicles are collected at each stage of centrifugation. Ultra-centrifugation can also be used, but the liposomes are aggregated by adding an equal volume of protamine solution to enable improved sedimentation; this is considered to be an effective method for the separation of drug-free and drug-loaded liposomes.

After the separation of drug-loaded vesicles from the drug-free liposomes methanol or triton X-100 is added, which will disrupt the lipid bilayer to release the entrapped drug. Quantification of the released drug can be done by methods such as spectrophotometry or fluorescence spectroscopy methods or FFF. Both separation and determination of encapsulation efficiency can be done by the method of coupling the concentration detector to FFF. The method is purely based on the material of encapsulate and the nature of the drug.

4.10.1.7 In Vitro Drug Release

In vitro drug release can be studied by the dialysis tube diffusion method and the modified Franz vertical diffusion assembly. Aliquots of drug encapsulated liposome suspension are placed inside the dialysis bag, which is called the donor medium, and the receptor compartment contains a buffer such as acetate or phosphate. The entire setup is kept at 37°C and the diffusion starts. Dialysate samples are taken at regular intervals and assayed by HPLC or spectrophotometry (Huang et al., 2005).

4.10.2 Chemical Parameters

The stability and efficiency of liposome may rely on the chemical parameters of the lipid bilayer such as phospholipid and cholesterol concentration, phospholipid peroxidation and hydrolysis, cholesterol auto-oxidation, and osmolarity.

4.10.2.1 Analysis of Lipid

The phospholipid content can be determined by Stewart's method based on the formation of a colored complex between phospholipid and ammonium ferrothiocyanate reagent. In the Bartlett assay, the organic materials present in the liposome are digested by the addition of sulfuric acid at 160°C, and then the phospholipids are oxidized to inorganic phosphates by hydrogen peroxide; interaction with ammonium molybdate leads to the formation of phosphomolybdate, and this is followed by reduction upon interaction with 1,2,6-aminonaphtolsulfonic acid at 100°C. This assay results in the formation of a blue-colored compound that can be detected at 830 nm.

In the ascorbic acid method, phosphomolybdic acid is formed as a result of acid digestion between ammonium molybdate and orthophosphates, and it is reduced by ascorbic acid, which gives a blue-colored compound that can be analyzed at 820 nm. The presence of cholesterol can be determined by the formation of a colored complex in the reaction between cholesterol and a freshly prepared acetic anhydride–concentric sulfuric acid mixture (20:1) at 680 nm.

Lipids can be separated and detected by chromatographic techniques such as HPLC (High Performance Liquid Chromatography), GC (Gas Chromatography), and TLC (Thin Layer Chromatography). An advanced method for the determination of cholesterol, cardiolipin, and 1,2-dioleoyl-sn-glycero-3-phosphocholine simultaneously by isocratic high-performance liquid chromatographic method (column ASTEC diol; 250 mm×4.6 mm, 5-μm particle size; and mobile phase-chloroform–methanol–aqueous ammonium acetate (71:26:3 v/v)) (Popovska et al., 2013).

4.10.3 Biological Parameters

These parameters include biosafety, bio-distribution, animal toxicity, pharmacokinetics, and pharmacodynamics profiles; these can be analyzed by using animal models and clinical trials which are purely based on the therapeutic applications.

4.11 APPLICATIONS

Over several decades, targeted and site-specific drug delivery was achieved by liposome, a nanovesicle that can protect the entrapped drug from enzymatic degradation and provide sustained release of the drug to the target site. The unique properties of liposome have paved the way for the development of novel drug delivery systems which can overcome the disadvantages of other delivery systems. Advancements in liposome formulation, encapsulation efficiency, and drug loading have facilitated a wide variety of applications in various fields. The drug encapsulated liposome can be administered via systemic or topical route, providing flexibility in the drug's mode of administration.

4.11.1 Clinical Applications

4.11.1.1 Cancer Therapy

Several cytotoxic drugs are made to target cancer or malignant cells, but due to their non-specific distribution to the body, they also affect normal cells, which leads to various toxic side effects. In accordance with the concept of sustained release and microencapsulation, the entrapment of cytotoxic drugs such as doxorubicin (DXR), daunorubicin (DNR), and vincristine (VCR) tends to increase the targeted drug delivery and specificity towards the tumor cells, enhance the uptake of liposomes by phagocytic or endocytic cells, and decrease liposomal uptake by the RES. It has been proven that after entrapping the anti-cancer agents within the liposome, the load of cytotoxic drugs is higher in malignant cells than in normal cells as the liposome enters the tumor cells by the EPR effect.

Liposome-formulated anthracyclins show an improved pharmacokinetic profile and higher half-life compared to free drug, which distributes within minutes and shows drug clearance within 24 hours by our system. Anti-cancer effects against solid tumors were increased by the co-administration of TNF (Tumour Necrosis Factor) solution with liposomal drugs. Long-circulating immunoliposome, which has been discussed earlier, is prepared by attaching antibodies at the distal end of the PEG chain within the PEG immunoliposome. This type of liposome is known as a pendant-type immunoliposome, and it shows much higher specificity than ordinary immunoliposomes on both targeting sites of lung endothelial cells and solid tumor tissue. This pendant-type liposome shows higher targeted delivery than ligand-mediated delivery. This targeted delivery is achieved by the liposome by escaping from the gaps between adjacent endothelial cells and openings at the vessel termini by passive convective transport during tumor angiogenesis. Incorporation of fusogenic molecule within the pendant-type immunoliposome may tend to induce endocytosis or liposome fusion towards the target cells (Goyal et al., 2005).

Photodynamic therapy (PDT) is another anti-cancer therapy which shows less side effects compared to chemotherapy and radiotherapy. Combinations of the ether lipids octadecyl phosphocholine and ET-18-OCH3 with both paclitaxel and teniposide showed enhanced in vitro cytotoxic activity against leukemic cells. Liposomal formulation by ether lipids is considered to beneficial in that they provide reduced hemolytic activity, incorporated within the liposome.

4.11.1.2 Gene Therapy

There are numerous genetic disorders that are caused by a lack of enzymes or defective or missing functional genes. These kinds of disorders should be corrected at the genome level rather than by conventional chemotherapies. Gene delivery systems are designed to deliver the genetic material into the patient's cells to synthesize the therapeutic protein.

Successful in vivo gene transfer involves administration of the therapeutic gene where it encounters certain barriers that the gene should escape from and finally enter into the cell nucleus. The gene transfer can be done by direct administration of genes to patients by virally or non-virally mediated strategies. Virus-mediated gene transfer involves the use of viruses such as retrovirus, adenovirus, adeno-associated virus, and herpes virus. Though they are suitable vectors for gene transfer, they possess some potential risks as they tend to mutate easily and show immune and inflammatory responses, as reported in clinical trials.

The drawbacks associated with the use of viral vectors give rise to the development of non-viral gene transfer methods. One such method is liposome-mediated transfer, as it is advantageous that the lack of proteins in lipid complexes makes them relatively non-immunogenic and able to accommodate large base pairs of genetic material, which can overcomeits two of its main disadvantages in the use of viral vectors. The negatively charged genetic material cannot be encapsulated within the liposomes but can be complexed with cationic lipids by electrostatic interactions. The transfection efficiency of DNA liposomes is highly dependent on the composition of lipid, the lipid/DNA ratio, the particle size of the liposome–DNA complex, and the cell lines used.

The lack of cell or organ specificity can be overcome by the use of receptor-mediated gene delivery systems that are able to introduce exogenous DNA to the target. However, not only the nature of the ligands attached onto the carriers but also the overall physicochemical properties of the complexes also contribute to effective cell-selective targeting of plasmid DNA.

4.11.1.3 As Vaccine Adjuvant

The advancement of encapsulation technology in gene delivery has led to the development of DNA liposomes as powerful adjuvants. They provide improved humoral and cell-mediated immunity compared to conventional vaccines. Liposomes are formulated by phosphatidyl serine and the antigens are covalently bound to the liposomal membrane. The antigen release and antibody responses can be improved by encapsulating DNA liposome a second time with alginate lysine microcapsules.

They can also be used to prepare hybrid delivery systems containing two or more components which are destined to present the antigen to antigen-presenting cells (APC), to stimulate immune responses (by incorporating adjuvants), to cause adhesion to the mucosal surface, and to protect the antigen from degradation. These kinds of vaccines are reported to be safe and effective in animal models (Turánek, 2009).

4.11.1.4 Against Parasitic and Fungal Infections

Parasitic diseases such as leishmaniasis and several fungal infections are some of the best-known examples of "Trojan Horse–like" mechanisms, as they normally reside in cells of the mononuclear phagocytic system. Liposomes tend to accumulate in the same cells and eliminate the parasitic infection by the sustained release of a drug. The best liposomal antifungal therapy for humans was found to be done with the encapsulation of amphotericin B. The dosage of amphotericin B is reduced due to its ionosphere and neurotoxicity, and the drug itself is toxic to our system in the case of the free drug, but liposomal formulated amphotericin B renders reduced toxicity and passive targeting. The targeting of liposomes to macrophages can be done by macrophage activation and injection. Some toxins can activate the macrophages, and this can be improved by the use of haptens (which induce immune response when attached to larger particles) (Daraee et al., 2014).

4.11.1.5 Antimicrobial Therapy

Antibiotics can act only against intracellular microbes that can reside within the phagocytic cells. It is easier to target pathogens residing in liver, spleen, and RES by liposomes. Treatment of mycobacterial infections differs from that of other bacterial diseases because of their unique properties, such as their complex lipid-rich envelope protection, which renders anti-mycobacterial therapy ineffective. To be effective, antibiotics must be able to penetrate macrophages where the colonization of mycobacterium occurs, show reduced toxicity, and demonstrate the capability for prolonged therapy at low doses. The mechanism of liposome encapsulated antibiotic uptake and the release of antibiotics to perform their activity have been shown in Figure 4.7. Liposome encapsulation may alter the distribution of antibiotics in our system. Intramuscular administration of gentamicin to rabbits provides more prolonged plasma concentration of drug than does intravenous administration.

Rifabutin encapsulated liposome resulted in a significant enhancement of anti-mycobacterial activity against *Mycobacterium avium* infection compared to free rifabutin. The anti-tubercular activity of rifampin was increased by encapsulating rifampin in a liposome made of egg phosphatidylcholine. The activity of rifampin was further increased by macrophage activator tetrapeptide tuftsin, which was grafted

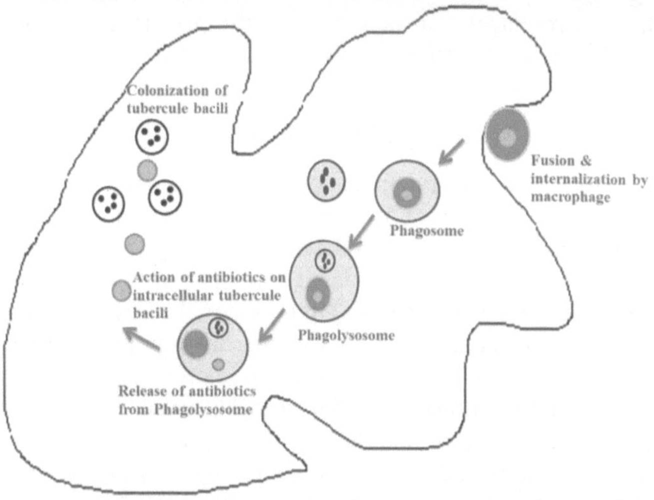

FIGURE 4.7 Engulfment of antibiotic liposome by macrophage and the action of antibiotics on tubercule bacilli.

onto the surface of drug-loaded liposomes and also by the use of combination therapy of other antibiotics.

4.11.2 Therapeutic Applications

4.11.2.1 As Brain Targeting

Liposomes can diffuse across the blood–brain barrier (BBB) and transport the drug encapsulated within them. Several drugs cannot be transported across the so-called tight junctions of the brain. To overcome this, the drugs are encapsulated within cationic liposomes, which can undergo absorptive mediated endocytosis into the brain cells. Reports suggest that liposome coated with mannose or sulfatide (sulfur ester of galactocerebroside) can transport load drug through BBB (Samad et al., 2007).

4.11.2.2 Ocular Applications

Drug administration to the eye fails due to poor drug penetration into the layers of eye as the eye is protected by three highly efficient mechanisms: 1) an epithelial layer that provides a formidable barrier to penetration, 2) tear flow, and 3) blinking reflex. To overcome the limitations of ocular drug delivery, the use of liposome carriers may promote improved drug penetration. These carriers include ganglioside-liposome and wheat germ agglutinin, a lectin which shows high binding affinity towards cornea and ganglioside in terms of corneal adhesion. The use of mucoadhesive polymers such as carbopol 934P and carbopol 1342 can retain the drug-loaded liposome at the cornea.

4.11.2.3 Pulmonary Applications

Due to the solubilization capacity of liposomes, drugs with poor water solubility can be rendered through aerosols, and it is potent against alveolar macrophages. The pulmonary residence times can be prolonged without any allergic or side effects. Aerosolic liposomes can also be administered to *Mycobacterium avium* infections by incorporating anti-mycobacterial drugs. The use of liposomal amikacin against *M. avium-intracellulare* (MAI)-infected alveolar macrophages shows superior killing efficacy in vivo (Kulkarni et al., 2011).

4.11.3 Applications in Food Industry

Liposomes can have applications within the food industry such as the delivery of antimicrobials and food preservative entrapment through encapsulation technology. Food preservatives such as lysozyme can be entrapped within the liposome, and this has been used to preserve cheeses, including Gouda, Emmental, and Edam. As these are highly susceptible to microbial contamination, the addition of chemical preservatives such as nitrate may affect the health of consumers, and the addition of lysozyme during cheese processing tends to lower the antimicrobial activity on binding with casein. liposomeencapsulated lysozyme may improve the antimicrobial activity by the phenomenon of sustained release of lysozyme. Other enzymes such as lipases and proteases have shown increased enzyme activity when they are entrapped within liposome in the production of commercial dairy products.

The addition of antimicrobials may tend to affect the growth of starter cultures, with their delayed release properties; the antimicrobials encapsulated in liposome will not affect the starter cultures and provide antimicrobial activity against the unwanted bacteria. Bacteriocin entrapped in liposome shows improved antilisterial effect of pediocin AcH in beef tallow and muscle slurries compared to unencapsulated bacteriocin.

Nisin, an antibiotic, shows a broad inhibitory spectrum against gram-positive bacteria and food pathogens. Unencapsulated nisin may degrade or lose activity. Encapsulated nisin may eliminate unfavorable and undesired reactions, resulting in long-term and targeted antimicrobial activity. The main advantage of encapsulating antibiotics within the liposome is that this decreases the emergence of antibiotic-resistant strains (Mozafari et al., 2008). The sustained release of nisin by liposome is shown in Figure 4.8.

Biocides such as fungicides, herbicides, and pesticides can also be encapsulated within liposome; this can render their activity safer by causing less damage to other life forms. They can be made present in leaf for longer periods by making the surface of the liposome sticky and not easily washed out.

4.11.4 Cosmetic Applications

The uniqueness of liposomes has led to their use in cosmetics. Lipids are well hydrated and can be used to reduce dryness of the skin, which is a primary cause for aging. Liposome can supply lipids such as linolenic acid which can replenish the skin. The anti-aging cream "Capture," launched by Christian Dior in 1986, was the first liposomal cosmetic product to appear on the market. Liposomes have been also

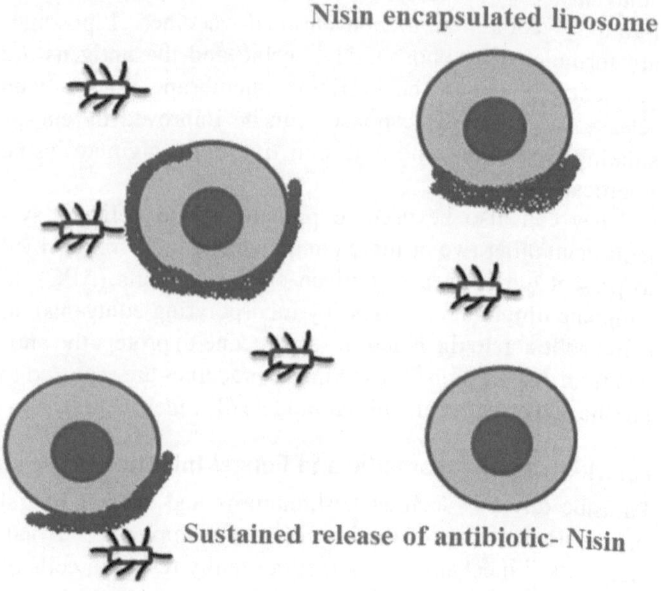

FIGURE 4.8 Sustained release of nisin by liposome encapsulation technology.

TABLE 4.3
Approved Liposomal Drugs on the Market for Human Use (Banerjee, 2015; Allen, 2013)

Drug	Commercial name	Application	Company
Doxorubicin	Doxil™	Kaposi's sarcoma in AIDS	Sequus Pharmaceuticals, Inc., CA
Amphotericin B	Ambisome™	Serious fungal infections	NeXstar Pharmaceutical, Inc., CO
Daunorubicin citrate	Daunoxome™	Kaposi's sarcoma in AIDS	NeXstar Pharmaceutical, Inc., CO
Vincristine sulfate	Marqibo	Acute lymphoblastic leukemia	Talon Therapeutics
Inactivated hepatitis A virus (strain RG SB)	Epaxal	Hepatitis A	Crucell Company, Berna Biotech
Inactivated hemagglutinin of influenza virus strains A and B	Inflexal V	Influenza	Crucell Company, Berna Biotech

TABLE 4.4
Some Liposomal Drugs under Clinical Trial (Allen, 2013)

Drug	Product	Application	Clinical Trial
Annamycin	L-annamycin	Breast cancer	Phase I
Vinorelbine	Alocrest	Advanced breast cancer	Phase I
Cisplatin	Lipoplatin	Colon cancer, gastric tumor	Phase II
Nystatin	Nyotran	Systemic fungal infections	Phase II/III
Cytarabine:daunorubicin	CPX-351	Acute myeloid leukemia	Phase II
Irinotecan HCl:floxuridine	CPX-1	Colorectal cancer	Phase II

used in the treatment of hair loss; minoxidil, a vasodilator, is an active ingredient in products like "Regaine" that can prevent hair loss. Dry skin can be treated by preparation of empty or moisture-loaded liposome which can reduce transdermal water loss. They also enhance the supply of lipids and water to the stratum corneum. Liposome formulated with egg phospholipids exhibited a 1.5-fold increase in skin water content over soya phospholipid formulations. Creams, gels, and ointments can be replaced by the use of liposome pastes containing moisturizers, various extracts, antibiotics, etc. (Laouini et al., 2012).

Some of the liposomal drugs that have been approved or are undergoing clinical trials are represented in Tables 4.3 and 4.4.

4.12 CONCLUSION

The development of liposomes as drug delivery systems has led to scientific and technological advancements. From the day of discovery, it created a milestone in the history of medical sciences. Numerous drugs have been formulated: some have been approved and some are under clinical trials. The uniqueness of liposome has attracted research to develop novel strategies. Controlled release and targeted drug delivery have energized various fields and created a new path which everyone previously thought impossible: moreover, liposomes have emerged as a nano-vehicle to carry genetic elements, thereby eliminating the obstacles in the gene delivery system. Liposomal encapsulation technology should be encouraged to increase innovation and provide new platforms in the future.

REFERENCES

Akashi, K., H. Miyata, H. Itoh, and K. Kinosita. 1996. Preparation of giant liposomes in physiological conditions and their characterization under an optical microscope. *Biophysical Journal* 71, no. 6: 3242–50. doi: 10.1016/S0006-3495(96)79517-6.

Akbarzadeh, A., R. Rezaei-Sadabady, S. Davaran, and S. W. Joo, et al. 2013. Liposome: classification, preparation, and applications. *Nanoscale Research Letters* 8, no. 1: 102. doi: 10.1186/1556-276x-8-102.

Banerjee, K., S. Banerjee, and M. Mandal. 2015. Liposomes as a drug delivery system. In *Biological and pharmaceutical applications of nanomaterials*, edited by Polina Prokopovich, 53–100. CRC Press.

Bergstrand, N. 2003. *Liposomes for drug delivery: From physico-chemical studies to applications*. PhD Thesis.

Daraee, H., M. Kouhi, and A. Akbarzadeh. 2014. Application of liposomes in medicine and drug delivery. *Artificial Cells, Nanomedicine, and Biotechnology* 44, no. 1: 381–91. doi: 10.3109/21691401.2014.953633.

Dua, J. S., A. C. Rana, and A. K. Bhandari. 2012. Liposome: methods of preparation and applications. *Internaional Journal of Pharmaceutical Student Research* 3, no. 2: 14–20. doi: 10.1017/CBO9781107415324.004.

Goyal, P., K. Goyal, and S. Gurusamy. 2005. Liposomal drug delivery systems – clinical applications. *Acta Pharmaceutica (Zagreb, Croatia)* 55, no. 1: 1–25. http://www.ncbi.nlm.nih.gov/pubmed/15907221

Huang, Y. Z., J. Q. Gao, W. Q. Liang, and S. Nakagawa. 2005. Preparation and characterization of liposomes encapsulating chitosan nanoparticles. *Biological & Pharmaceutical Bulletin* 28, no.2: 387–90. doi: 10.1248/bpb.28.387.

Kulkarni, P. R., J. D. Yadav, and K. A. Vaidya. 2011. Liposomes: a novel drug delivery system. *International Journal of Current Pharmaceutical Research* 3, no. 2: 10–18.

Laouini, A., C. Jaafar-Maalej, I. Limayem-Blouza, and S. Sfar, et al. 2012. Preparation, characterization and applications of liposomes: state of the art. *Journal of colloid Science and Biotechnology*, 1 no. 2: 147–168.

Mellman, I, R. Fuchs, and A. Helenius. 1986. Acidification of the endocytic and exocytic pathways. *Annual Review of Biochemistry* 55, no. 1: 663–700.

Mozafari, M. R., C. Johnson, S. Hatziantoniou, and C. Demetzos. 2008. Nanoliposomes and their applications in food nanotechnology. *Journal of Liposome Research* 18, no. 4: 309–27. doi: 10.1080/08982100802465941.

Popovska, O., J. Simonovska, Z. Kavrakovski, and V. Rafajlovska. 2013. International journal of pharmaceutical and an overview : methods for preparation and characterization of liposomes as drug delivery systems. *International Journal of Pharmaceutical and Phytopharmacological Research* 3, no. 3: 182–89.

Samad, A., Y. Sultana, and M. Aqil. 2007. Liposomal drug delivery systems: an update review. *Current Drug Delivery* 4, no. 4: 297–305.

Sharma, A., and U. S. Sharma. 1997. Liposomes in drug delivery: progress and limitations. *International Journal of Pharmaceutics* 154, no. 2: 123–140.

Turanek, J., and Ltd Csnmt Mrss N. M. S Tanger. 2009. Liposomes - Biocompatible Nanoparticles and Their Application for Construction of Drug Delivery Systems and Vaccines. *Nanocon 2009, Conference Proceedings*, 299–303.

Yadav D., K. Sandeep, D. Pandey, and R. K. Dutta. 2017. Liposomes for drug delivery. *Journal of Biotechnology and Biomaterials* 7: 276. doi: 10.4172/2155-952X.1000276.

Section 2

Nano Biotechnology Applications on Food and Agriculture

5 New Insights into the Pharmaceutical and Food-Based Applications of Starch Nanocrystals

Neethu Hari, Supriya Radhakrishnan, Athira S. Dev, and Ananthakrishnan Jayakumaran Nair

CONTENTS

5.1 An Introduction to Nanobiotechnology and Nanoparticles (NPs) 64
5.2 Polymer Nanoparticles 64
5.3 Synthesis of Polymer Nanoparticles 64
 5.3.1 Synthetic Polymer-Based Nanoparticles 65
 5.3.2 Natural Polymer-Based Nanoparticles 65
  5.3.2.1 Protein-Based Nanoparticles 65
  5.3.2.2 Gelatin-Based Nanoparticles 66
  5.3.2.3 Albumin-Based Nanoparticles 66
  5.3.2.4 Casein-Based Nanoparticles 66
  5.3.2.5 Zein-Based Nanoparticles 66
  5.3.2.6 Whey Protein (β-Lactoglobulin and α-Lactalbumin) Nanoparticles 66
 5.3.3 Polysaccharide-Based Nanoparticles 66
  5.3.3.1 Alginate-Based Nanoparticles 66
  5.3.3.2 Chitosan-Based Nanoparticles 67
5.4 Starch Nanocrystals (SNCs) 67
5.5 Synthesis of Starch Nanocrystals 67
5.6 Structural Characterization of Starch Nanocrystals 67
 5.6.1 X-Ray Diffraction (XRD) 67
 5.6.2 Fourier Transform Infrared Spectroscopy (FTIR) 68
 5.6.3 Scanning Electron Microscopy (SEM) 68
 5.6.4 Transmission Electron Microscopy (TEM) 68
5.7 Physical Characterization of Starch Nanocrystals 68
 5.7.1 Swelling Study 68
 5.7.2 Degradation Study 69
 5.7.3 Rheological Properties of Starch Nanocrystals 69
 5.7.4 Solubility of Starch Nanocrystals 69
5.8 Modification of SNCs 69
 5.8.1 Chemical Modifications of SNCs 69
5.9 Grafting 70
 5.9.1 SNCs Modifications Based on the "Graft-To" Strategy 70
 5.9.2 SNCs Modifications Based on the "Graft-From" Strategy 70
5.10 Application of Starch Nanocrystals in Food Industry 70
 5.10.1 Starch Nanocrystals as Packaging Component with Barrier Property 70
 5.10.2 Starch Nanocrystals as Food Additives 71
  5.10.2.1 Fat Replacer 71
  5.10.2.2 Emulsion Stabilizer 71
 5.10.3 Antimicrobial Activity of Starch Nanocrystals 71

5.11 Application of Starch Nanocrystals in Biomedical Field 72
5.11.1 Starch Nanocrystals as Drug Carriers 72
5.11.2 Role of Starch Nanocrystals in Wound Healing 73
5.11.3 Starch Nanocrystals in Tissue Engineering 73
5.11.4 Fluorescently Labeled Starch Nanocrystals as Biosensors 74
5.12 Conclusion 74
References 74

5.1 AN INTRODUCTION TO NANOBIOTECHNOLOGY AND NANOPARTICLES (NPS)

Nanotechnology is an expeditiously progressing field with many potential applications, and their benefits in different science disciplines are expected to increase with time. Nanobiotechnology and nanomaterials impact various aspects of pharmacology and the food system due to their improved physiochemical features, including enhanced surface-to-volume ratio, facile synthetic strategies, tunable size and shape, and better encapsulation and release profiles. Nanomaterials endowed with these additional characteristics provide particular interest in terms of functional foods, nutraceuticals, and packaging products.

The application of nanotechnology in food packaging enables greater food protection through improved physicochemical and biological properties. Nano-engineered biobased food packaging imparts protection against leakage, gas permeation, and invasion of pathogens into foods. Nanomaterials also play an imperative role in food safety applications such as the development of processing aids, quality enhancers, and quality and safety detectors such as nanosensors. The growing importance of and interest in their use has led to the development various fabrication strategies and approaches, which offer a wide variety of futuristic applications. Though the applications of nanomaterials draw a lot of scientific attention, public acceptance of pharmacological and food-based products with nanomaterials incorporated into them will depend on their safety.

5.2 POLYMER NANOPARTICLES

Polymeric nanoparticles are colloidal solid particles or particle dispersions with a nanometer size range of ≈20–1000 nm, and they comprise a polymeric core with an entrapped bioactive component surrounded by a surfactant layer (tween, span, CTAB, etc.) that stabilizes the whole system. Polymeric nanoparticles are highly reliable drug delivery systems used for many target-specific deliveries of potential molecules including drugs, vaccines, proteins, vitamins, nutraceuticals, etc., and act as a promising futuristic medicine capable of revolutionizing the field of medicine and impeding the side effects of conventional chemotherapeutics. Polymeric nanotherapy involves a physical approach to alter and improve the pharmacokinetic and pharmacodynamic properties of therapeutic formulations, thereby enhancing the efficiency of phytomedicines with minimum adverse effects. These advantages of polymer nanoparticles in the medical field are also applicable to the food industry in terms of developing new food products with better quality and shelf life. The polymer nanoparticles can be developed from food-grade synthetic (Cristina and Carlos, 2015) and natural biodegradable polymers (Chang and Chen 2005), which helps in controlled release and targeted delivery of functional compounds such as vitamins and antimicrobial components; this release is caused by specific environmental triggers by reorienting the solution conditions, which results in complete dissolution of the particles or variation in particle porosity (Weiss et al., 2006). Cristina and Carlos (2015) reported numerous additional food-based applications of polymer nanoparticles including improved solubilization and stabilization of hydrophobic natural pigments (e.g., β-carotene) in water; protection of peptides, proteins, and bioactive agents against oxidation, degradation, and photodegradation; sustained release of bioactive compounds and vitamins; and enhanced bioavailability of poorly absorbed minerals.

5.3 SYNTHESIS OF POLYMER NANOPARTICLES

Polymer systems approved by the US Food and Drug Administration (FDA) are commonly used for synthesis of polymer-based nanoparticles and are safe for human use. Several methods for fabricating polymer nanoparticles have been developed and are classified according to whether the particle is formed by a "top-down" or a "bottom-up" approach. The top-down approach mainly involves the direct formation of polymer nanoparticle from a macromolecule or pre-existing polymer, whereas the bottom-up approach tailors nanoparticles by means of polymerization reactions. Moreover, the fabrication techniques depend on the physicochemical and biological properties of the polymer and its active ingredients.

Both natural and synthetic polymer nanoparticles can be synthesized by following top-down methods like the single-emulsification method (solvent diffusion and solvent evaporation), the double-emulsion method (solvent evaporation), nanoprecipitation, salting-out methods (Quintanar-Guerrero et al., 1998; Pinto et al., 2006), evaporation precipitation (Sarkari et al., 2002), and dialysis and electrospray methods. Recently, the emergence of a new generation of polymer nanoparticles used for smart drug delivery was reported, and the wide therapeutic applications of these nanoproducts are expected to overcome various hurdles related to drug delivery and help to minimize unwanted side effects in the body.

These polymer nanomaterials with improved applications are being developed using modern techniques like microelectromechanical systems (MEMS) (Nuxoll, 2013), microfluidic systems (Valencia et al., 2012), electro-dropping system (Choi et al., 2013), microneedle-based system (Wendorf et al., 2011), advanced high-pressure homogenization, interfacial emulsion polymerization, and combined systems. Several other techniques (emulsion, mini-emulsion, micro-emulsion, interfacial polymerization, controlled/living radical polymerization) that come under the bottom-up approach have also been used for the synthesis of polymer nanoparticles. Dried polymeric nanoparticles especially used in inhaler and oral drug formulation for the treatment of pulmonary diseases have been developed using different freeze-drying methods like atmospheric freeze drying, spray freeze drying, vacuum freeze drying, and the lyophilization method.

5.3.1 Synthetic Polymer-Based Nanoparticles

Among the different synthetic polymers, members of the polyester family (e.g., polylactic acid or polylactide [PLA], poly(e-caprolactone) [PCL], poly(glycolic acid) [PGA]) have gained a lot of attention in the various sectors of food and medicine due to their biocompatibility and biodegradability properties (Hans and Lowman, 2002). Poly-beta amino esters (PBAEs) are a class of biodegradable cationic polymers obtained from the co-polymerization of diacrylate and amine molecules (Lynn and Langer, 2000). These polymer-based nanoparticles possess many integral properties for development of site-specific drug (hydrophobic and hydrophilic) and gene delivery systems. It was observed that these nanoparticles act as a stimuli-sensitive (pH, temperature, etc.) dosage system for tumor-targeted delivery of hydrophobic drugs (e.g., paclitaxel, doxorubicin base, etc.); the rapid dissolution of polymer occurs when the pH/temperature of the medium varies, and the system releases its contents when it reaches the acidic tumor microenvironment and endo/lysosome compartments of cells (Shenoy et al., 2005). Similarly, PBAEs are also used for the efficient delivery of therapeutic molecules to the cartilage of joints (Stefano and Polina, 2017). Among the different eco-friendly biopolymers used in medical field, PLA has attracted significant attention. PLA is an inexpensive, stable, biodegradable thermoplastic used in the human body, and this aliphatic polyester is derived from naturally occurring organic acid (lactic acid). Being thermoplastic and biodegradable, PLA has become an attractive material for biological and medical applications especially for regenerative surgery implants and other surgical applications such as subcutaneous sutures. Poly(lactic-co-glycolic acid) is an FDA-approved biodegradable polymer with multiple properties like drug biocompatibility, perfect biodegradation kinetics, mechanical properties, and ease of processing (Jain, 2000). Biodegradable PLGA NPs have been studied for sustained, targeted, and localized delivery of different agents, including drugs, proteins, and peptides (Prabha and Labhasetwar, 2004; Panyam et al., 2003). Since PLGA NPs have the ability to protect DNA from endolysosome degradation, they have recently been used for the transport and delivery of plasmid DNA (Cui et al., 2013; Nafee et al., 2007). Marcos et al., (2014) observed the efficiency of the polycaprolactone polymer nanoparticles in encapsulating progesterone. Poly(caprolactone) (PCL) is a biodegradable, semicrystalline aliphatic polymer with reduced toxicity and high permeability to drugs. Now it has gained wide attention for use as implantable biomaterial in medical applications due to its degradability through the hydrolytic cleavage of ester groups in physiological conditions and its ability to excrete completely from the body once bioresorbed (Natu et al., 2011).

Just as for biomedical applications, synthetic polymers play major roles in food-based applications. Poly(lactic-co-glycolic acid) (PLGA) is a common synthetic polymer used in the food industry approved by the FDA as GRAS (generally recognized as safe) (Jain, 2000). The degradation of PLGA mainly occurs by the process of hydrolysis, resulting in the formation of monomers, lactic acid, and glycolic acid followed by its metabolism through the Krebs cycle (Tice and Cowsar, 1984). PLA polymer matrix loaded with nanocomposites is a very important novel group of materials, with a wide range of possible applications in food packaging.

5.3.2 Natural Polymer-Based Nanoparticles

The usage of natural and biobased components for the fabrication of polymer nanoparticles helps to replace synthetic polymers with naturally available biodegradable components suitable for medicine and food applications. Common natural polymers used for the synthesis of polymeric nanoparticles include polysaccharides, proteins, etc., with various surface charges, hydrophobic-to-hydrophilic ratios, solubilities, site-specific targeting, sustained and tunable release profiles, entrapment efficiencies of bioactive components, and degradation properties. Studies on the new possibilities in pharmaceutical, medicine, and food domains proved that relatively less effort is needed for the introduction of polymer nanoparticles as compared to that of new chemical entities.

5.3.2.1 Protein-Based Nanoparticles

The many unique properties of protein-based nanoparticles—like its highly organized structures and symmetry, biodegradability, potential to specifically act at different interfaces (inside and outside the protein cage and between subunits in macromolecular assembly), and an exemplary size for delivery of vaccines—make them advantageous over other vaccine delivery systems. Protein-based nanoparticles, especially virus-like particles (VLPs) and cage proteins (CPs), exhibit symmetries and physical properties similar to that of viruses and thereby enhance the association of the vaccine components with antigen-presenting cells. It was also revealed that the caged protein nanoparticles have been used as nanocarriers to transport cancer vaccine and increase its interaction with the immune system (Neek et al., 2019). Other commonly used protein nanoparticles include the following:

5.3.2.2 Gelatin-Based Nanoparticles

Gelatin is one of the protein materials that can be used for the development of biodegradable nanoparticles and is obtained from the hydrolysis of collagen. This biocompatible polymer nanoparticle gained much attention in the nanopharmaeceutical area due to its low immunogenicity, and it is considered as a component of various registered blood substitutes. It was reported that the amino acid sidechain of gelatin nanoparticles shows versatility for further modification and is also characterized by additional features like non-toxicity and being easy to cross-link and modify chemically (Kushagri and Abha, 2014). The preparation of biodegradable gelatin nanoparticles was achieved by the two-step method called desolvation (Azimi et al., 2014), and the gelatin nanoparticle so formed can be used for the delivery of Bovine Serum Albumin (BSA). Previous reports showed the ability of gelatin nanoparticles to cross the blood-brain barrier, hence proven as a nanomaterial with high potential to address brain disorders (Azimi et al., 2014).

5.3.2.3 Albumin-Based Nanoparticles

Different properties of albumin like biodegradability, biocompatibility, low immunoreactivity, versatility, and cost make it a proper biomaterial for the synthesis of polymeric nanoparticles. Several techniques are available for the fabrication of albumin nanoparticles, including desolvation, emulsification, and a thermal treatment method (Cristina and Carlos, 2015). Albumin nanoparticles can encapsulate a variety of bioactive components including drugs, phytochemicals, nutraceuticals, and functional food ingredients. Abraxame® is a paclitaxel-loaded albumin nanoparticle currently used for treatment of different cancers like breast, lung, and pancreatic cancer (Elzoghby et al., 2012).

5.3.2.4 Casein-Based Nanoparticles

Food-grade proteins normally offer numerous advantages such as digestibility, price, and good capability to interact with a large number of compounds and nutrients, so they can act as adequate candidates for the preparation of nano- and microparticles. Casein comes under the family of food-grade protein, i.e., milk protein (90%). Rebeca et al. (2015) synthesized casein nanoparticles by the method of coacervation and then stabilized them using lysine followed by spray drying. The synthesized nanoparticle was then used for the oral delivery of folic acid. Owing to the high temperature stability of casein protein, spray drying is suitable to synthesize casein nanoparticles (Elzoghby et al., 2013). The final particle size of casein nanoparticles is affected by the concentration of casein, as shown by Shapira et al. (2010). In a previous study, it was noticed that reassembled casein micelles and casein nanoparticles can be used as a nano-vehicle for ω-3 polyunsaturated fatty acids, which are nutraceuticals extensively used for the enrichment of food (Patricia et al., 2011). Reports based on biomedical applications of casein nanoparticles are also found. For example, a β-casein nanoparticle–based oral drug delivery system was used for treatment of gastric carcinoma (Alina et al., 2012).

5.3.2.5 Zein-Based Nanoparticles

Zein is an important protein present in corn which accounts for 50% of the total protein content. It belongs to the prolamine class with lipophilic amino acid and has been under extensive study due to its hydrophobicity, biocompatibility, and biodegradability properties. Many fabrication strategies like nanoprecipitation, liquid–liquid dispersion, phase separation, and electrospraying are used for the synthesis of zein nanoparticles, for the purpose of loading different active compounds (Pascoli et al., 2018). Wu et al. (2012) synthesized zein nanoparticles with particle sizes ranging from 100 to 700 nm. Here also the concentration of zein is highly correlated to its particle size. The release of active components is possible by means of molecular diffusion from the core and degradation of zein. Zein nanoparticles so formed can be used for the encapsulation of different functional materials like quercetin (Rebeca et al., 2017), peppermint oil (Chen and Zhong, 2015), etc.

5.3.2.6 Whey Protein (β-Lactoglobulin and α-Lactalbumin) Nanoparticles

Whey protein is a combination of globular proteins isolated from whey. This protein is mainly composed of 65% β-lactoglobulin and 25% α-lactalbumin. β-Lactoglobulin is the major whey protein of ruminant species. Nanoparticles based on this protein are widely used in food-based applications since they satisfy different properties like improved sensory properties, enhanced flavor perception, better mouthfeel, etc. Earlier reports on loading efficiency of whey protein nanoparticles explained that the nanoparticles synthesized by cross-linking denatured whey protein by means of pH cycling was used for the loading of aroma (Giroux and Britten, 2011), and in another study whey protein nanoparticles loaded with the aqueous extract of date palm pit were obtained through desolvation by ethanol. The so formed extract-loaded particles are then added to beverages without any change in the sensory properties of the food product (Bagheri et al., 2013).

5.3.3 Polysaccharide-Based Nanoparticles

5.3.3.1 Alginate-Based Nanoparticles

Alginate comes under the category of polysaccharide nanoparticle, it exhibits heterogenous properties like biocompatibility, biodegradability, stability, sustainability, and controlled drug release. Sarei et al. (2013) discussed the possibility of biocompatible alginate nanoparticle to deliver new vaccines. Herein, alginate nanoparticles act as a carrier with adjuvant and their prolonged tunable release properties enable the enhancement of immunogenicity of vaccines. Ionic gelation techniques were used for the development of diphtheria toxoid–loaded alginate nanoparticles followed by their characterization with respect to size and shape, zeta potential, encapsulation efficiency, release profile, and immunogenicity. Saeed et al. (2013) reported the ability of sodium alginate nanoparticles to carry venom-derived peptides, thereby improving its effectiveness. Alginate nanoparticles can also be widely used in the food industry. Alginate nanoparticles loaded with vitamins, minerals, and other functional food ingredients are used for fortification of foods.

5.3.3.2 Chitosan-Based Nanoparticles

Chitosan is a naturally occurring linear biodegradable polysaccharide and is extensively studied as a drug-loading and mucoadhesive polymer formula. Even though chitosan nanoparticles possess the characteristics of chitosan, they attain new properties when they reach to nanometer size; these new properties include high surface-to-volume ratio, enhanced surface energy, high interface, and quantum size effects (Ingle et al., 2008). Recently, a modified chitosan nanoparticle (norcantharidin-associated galactosylated chitosan nanoparticles) was used for hepatocyte-targeted delivery (Wang et al., 2010). Yan-Qiong et al. (2019) studied the enhanced drug-loading capacity of galactosylated-chitosan-triptolide-nanoparticles (GC-TP-NPs) and their targeted delivery to hepatocellular carcinoma. Similarly, a recent study showed that glycine-arginine-glycine-aspartic acid-serine (GRGDS) functionalized chitosan nanoparticles to act as a potential intravenous hemostat for traumatic hemorrhage control in an animal model (Zhang et al., 2018). Chitosan nanoparticles and their derivatives with improved biodegradability and antimicrobial properties are some of the best barriers and film coating materials for protecting the quality of food by enhancing its shelf life. Vitamin C–loaded chitosan nanoparticle was used in food processing due to its continuous and sustained release, and this nanoparticle has the ability to retain its antioxidant properties (Li-Ming et al., 2011).

5.4 STARCH NANOCRYSTALS (SNCs)

Starch is the second most abundant organic polymer in nature, and most plants use it as a source of energy. It is one of the most commonly used biopolymers in industries for various products, because of its total biodegradability, non-toxicity, stability in storage, cost-effectiveness, and feasible fabrication method. Native starch is composed of two types of biomacromolecules: amylose and amylopectin. Amylose is a linear soluble polysaccharide made of α-D-glucose units linked by α-(1→4) linkages with a molecular weight of 105–106 g/mol (Angellier et al., 2005). Amylose is highly resistant to digestion as compared to amylopectin. Amylopectin is a highly branched polymer consisting of about 95% α-(1→4) and 5% α-(1→6) linkages showing molecular weights between 106 and 107 g/mol. The amylopectin content of starch influences the crystalline organization of starch. The conventional model of starch possesses two regions, i.e., crystalline and amorphous lamellae, which results in the formation of crystalline and amorphous growth rings (Jenkins et al., 1994). Its unique semi-crystalline structure favors the production of biobased elements like starch nanocrystals or starch nanoparticles. The platelet-like morphology of the starch crystal, together with its inherent rigidity, powerful interfacial interactions, and percolation network, are responsible for its optimized thermal and mechanical properties, solvent absorption, and excellent barrier properties.

5.5 SYNTHESIS OF STARCH NANOCRYSTALS

Starch nanocrystals are commonly synthesized by the methods of acid hydrolysis. Mild acid treatment of starch granules resulted in the isolation of crystalline regions. This is possible by the dissolution of a region of low lateral order to expose the concentric lamellar structure of the starch granules. This process occurs at temperatures below the gelatinization temperature (Dufresne, 2010). Herein, the amorphous regions undergo more rapid hydrolysis than the crystalline region. The resultant hydrolysis product is starch nanocrystals, possessing highly crystalline and nanoscale morphology. The structure, morphology, size, and final yield of starch nanocrystals are determined by the botanical origin of a starch (crystalline type, relative proportion of amylose and amylopectin, and granule morphology), conditions (type of acid, acid concentration, temperature), and duration of hydrolysis, etc. (Gérard et al., 2002). Shahrodin et al. (2015) observed the high yield of cassava starch nanocrystals (CSN) at different concentrations (2.8 M, 3.0 M, 3.2 M, and 3.4 M) of sulfuric acid (H_2SO_4). Herein, the physical and chemical properties of synthesized CSN was also investigated, and it was found that the hydrolysis process for five consecutive days, resulting in the highest yield of CSN at 3.4 M CSN with a size range of 5–20 nm. In another study, the effect of the duration of hydrolysis on the size of starch nanocrystals was investigated; it was found that almost 24 hours of hydrolysis using sulfuric acid resulted in the formation of starch nanocrystals, and that as the time of hydrolysis increased beyond 24 hours, microparticles were also observed along with nanocrystals and they were separated by means of differential centrifugation (LeCorre et al., 2011a). LeCorre et al. (2012) studied the influence of the botanical origin of starch nanocrystals on final properties (morphological and mechanical properties) of natural rubber nanocomposites. Likewise, a recent study has investigated the effect of the botanical origin and amylose content of initial starch on the final properties like morphology, crystallinity, and rheology of SNCs (LeCorre et al., 2011b).

5.6 STRUCTURAL CHARACTERIZATION OF STARCH NANOCRYSTALS

5.6.1 X-Ray Diffraction (XRD)

One of the most important applications of XRD is the determination of crystalline structures, and this is therefore useful in determining the crystalline structures within the native starch as well as in starch nanocrystals. The crystalline portion of starch granules enables its easy detection with the help of XRD (Namazi and Dadkhah, 2010). Through XRD analysis, it is revealed that starch generally contains three types of crystalline structures: A, B, and C (LeCorre et al., 2012). Earlier, Bogracheva et al. (1998) reported that sometimes B- and A-type structures together constitute C-type structures in such a way that B-type forming the center of the granule is surrounded by A-type structures. Often, modifications could bring about changes in the crystalline forms. In one such attempt, a transformation in crystalline structure from style

A to style B was reported for acetylated starch nanocrystals (ASNs) (Xu et al., 2010). Xu et al. (2014) proposed that the diffraction peaks of corn, barley, and tapioca starches showed a characteristic A-type crystalline structure. In another study. Zheng et al. (2009) used XRD to understand the formation of large domain-like structures in the soy protein reinforced with SNCs. A recent report on XRD analysis of chitosan film modified with β-carotene starch nanocrystals (BSNCs) revealed that the crystallinity of modified film got enhanced with the addition of the BSNCS and showed the presence of strong peaks at 2θ, 17.79° and 28.71° (Hari et al., 2018).

5.6.2 Fourier Transform Infrared Spectroscopy (FTIR)

FTIR is an analytical technique used to understand the various chemical group present in intact starch nanocrystals as well as in modified ones. Recently, FTIR studies helped in the structural analysis of two graft polymers, $PEGME_{550}$ (Thielemans et al., 2006) and $PPGBE_{4000}$ (Labet et al., 2007), using toluene (as a solvent) and 2,4-TDI (as coupling agent). Starch nanocrystals modified by surface grafting using $PTHF_{1400}$, PCL_{10000}, and PCL_{42500} (Labet et al., 2007) were studied using FTIR. Similarly, methyl ethyl ketone solvent in stearate graft polymer used for the modification of starch nanocrystals was analyzed (Thielemans et al., 2006) using FTIR. Also, ASA_{350} graft polymer with toluene solvent used for surface modification was characterized using FTIR (Angellier et al., 2005). FTIR conformation of hydrophobic modifications of starch nanoparticles using nanonoyl octanoyl and decanoyl chloride was done, and vibrations of carbonyl group in 1738–1742 cm^{-1} were observed, which ensured the esterification of long chain fatty acid chlorides to starch, derived from potato and corn (Namazi and Dadkhah, 2010).

5.6.3 Scanning Electron Microscopy (SEM)

Surface morphology details can be revealed with the help of SEM analysis (Figure 5.1). SEM analysis was used to find out the influence of botanic origin on the morphology (Shape) of SNCs, and no correlation between the two was reported (LeCorre et al., 2011b). SEM observation was also used to understand the effect of enzymatic hydrolysis on the morphology of starch nanocrystals. Starch nanocrystals synthesized by the hydrolysis of granular waxy maize starch revealed that the amorphous layer of SNCs is more prone to hydrolysis during the initial stages with the application of alpha-amylase (Foresti et al., 2014). LeCorre et al. (2011b) determined the shape and size of SNCs through SEM analysis before and after acid hydrolysis and reported the largest nanocrystal from potato starch. In addition to this, the morphological study of SNC-based nanocomposite was also done using SEM as well as Atomic Force Microscopy (AFM).

5.6.4 Transmission Electron Microscopy (TEM)

TEM can be used to understand the morphological aspects of starch nanocrystals as well as the effect of different solvents. The use of TEM for structural characterization of SNCs derived from waxy maize granules was first reported in a study conducted by Putaux et al. (2003). Herein, TEM analysis of SNCs revealed that, depending on the origin of the starch, the modified SNCs platelets could have thicknesses between 5 and 7 nm as well as diameters between 30 and 100 nm and concluded that the size variation of starch nanocrystals depends on the botanical origin of the starch. A TEM study of SNCs derived from waxy maize confirmed the size of nanocrystals, which were found to be in range from 70 to 100 nm (Namazi and Dadkhah, 2008). Also, a size range of 40–70 nm was reported for potato starch–based nanocrystals (Chen et al., 2008). However, a TEM study of starch nanocolloids revealed that the particle size rduction (μm to nm) ooccured in colloidal starch due to its transitional properties (Liu et al., 2009). The difference in the diameters of different SNCs was attributed to their crystallinities and shapes. TEM characterization of SNCs derived from waxy maize starch granules reflected lamellar fragmented structure, with stack of elongated elements(Putaux et al., 2003).

FIGURE 5.1 SEM image of starch nanocrystals along with microparticles (16000 ×).

5.7 PHYSICAL CHARACTERIZATION OF STARCH NANOCRYSTALS

5.7.1 Swelling Study

The addition of various solvent molecules for diverse purposes could alter the properties of starch nanocrystals. Swelling study can be used to analyze the effect of various agents on the formation of additional cross-links between different solvent molecules and starch nanocrystals. The swelling of starch nanocrystals is greatly limited by covalent modifications (Ren et al., 2012). The water absorption kinetics of SNCs showed a faster increase in swelling initially, followed by a plateau (Kristo and Biliaderis, 2007). Herein, the reduction in swelling occurred due to leaching of starch nanocrystals at later stages of absorption kinetics (Ghasemolu et al., 2016). In another study it has been reported that in natural rubber-based nanocomposite-incorporated film, the rate of water uptake has a strong relationship with the film, the nanocomposites, and its aqueous environment (LeCorre et al., 2012). According to Kristo and Biliaderis (2007), the difference in the swelling behavior can be attributed to the different filler concentrations. However, a study using unfilled soy protein matrix and nanocomposite filled with pea starch nanocrystals highlighted that there was no change in the rate of water uptake (Zheng et al., 2009) and consequently swelling behavior.

5.7.2 Degradation Study

Kummerer et al. (2011) observed that even though there was no ready biodegradability of starch nanoparticles, the degradability of these nanoparticles is greater than that of their macroscopic counterparts in the aquatic environment. This increased degradation can be attributed to the high surface-to-volume ratio of starch nanoparticles.

5.7.3 Rheological Properties of Starch Nanocrystals

Better understanding of the rheological properties of SNCs is very important to study the effect of the addition of various solvent molecules as part of various processing activities (LeCorre and Angellier, 2014). Morphological comparison of different starch nanocrystals obtained from different starch sources and their crystallinity and rheological properties was first done by LeCorre et al. (2011b). In an attempt to use starch nanocrystals as barrier material for biobased polymers, a high Brookfield viscosity was obtained with the addition of SNCs to coating colors, which were derived from plasticized starch material (LeCorre et al., 2014). However, the relationship between shapes and viscosity was addressed in a study conducted by LeCorre and Angellier (2014). Their work suggested that suspension of starch nanocrystals derived from lower viscosity displayed ill-defined and disc-like morphological structures, whereas solutions of higher viscosity yielded one with square-like structures.

5.7.4 Solubility of Starch Nanocrystals

As previously discussed, starch is a polymer which is composed of amylase and amylopectin, with amylase content being greater (72% and 82%) than amylopectin content (18% to 28%) (Perez and Bertoft., 2010). Amylose with linear glucose units is soluble in hot water, whereas the branched amylopectin is not (Tester et al., 2004). A significant solubility of the amorphous phase of starch was observed earlier by Song et al., (2008). When the bulk starch was converted to starch nanocrystals, its solubility got further enhanced. Acid treatment of starch could even solubilize its highly crystalline residues to stable solutions, although it is otherwise unstable (Angellier et al., 2005). Chemical modification of starch nanocrystals has also been shown to increase solubility as well as polarity (Ning et al., 2011). Acetylated SNCs produced by the interaction between acetic anhydride and hydroxyl group readily enhanced the solubility property in diverse organic solvents such as acetone, toluene, etc. (Xu et al., 2010).

5.8 MODIFICATION OF SNCs

The most predominant form for starch is a concentric semicrystalline multiscale structure. It allows the production of new nano elements: (i) starch nanoparticles produced from gelatinized starch and (ii) starch nanocrystals resulting from the disruption of amorphous domains from semicrystalline granules by acid hydrolysis (Deborah et al., 2010). The limited applications of starch in industry pertain to its reduced solubility and increased viscosity. The starch nanocrystals produced can be modified in many ways, giving them specific functions accordingly. Various chemical modifications can be adopted to enhance the functional properties of starch, which can be furnished with more functional groups. The abundant hydroxyl groups on glucose units in the polymer chain of starch act as sites to be modified with different multifunctional groups. The three most commonly adopted strategies for chemical modification of starch nanocrystals are (i) modification by chemical reaction with small molecules, (ii) grafting onto polymer chains with coupling agents, and (iii) grafting from polymer chains with polymerization of a monomer. The modification strategies must not affect the integrity of the crystalline structure of SNCs. As the chemical modification transforms the surface characteristics of the starch nanocrystals, the performance of nanomaterials will inevitably be affected. Development of strong interfacial adhesion facilitates the transfer of stress and plays a reinforcing role for nanocrystals. The functional modifications of the starch nanocrystals, such as grafting to fluorescent molecules, may expand their application in the field of biomedicine (Ning et al., 2011). The modified starch nanocrystals are promising biomaterials due to their novelty and utilization in foods and food products, cosmetics, and pharmaceuticals, and also as various modified nanocomposites.

5.8.1 Chemical Modifications of SNCs

Commonly adopted chemical modifications of starch are of three types: anionic, cationic, and hydrophobic modifications. The oxidization of hydroxyl groups of starch into carboxyl groups converts starch into anionic form. When cationic groups are introduced onto the backbone, cationic starch is produced. This is mainly done by introducing amino, ammonium, sulfonium, or phosphonium groups by an etherification reaction with the hydroxyl groups of starch (Wang and Xie, 2010). It has been reported that cationic starch nanoparticles can be used as gene delivery vehicles.

The chemical modifications of SNCs involves cross-linking followed by an esterification process, which leads to an increased degree of substitution of SNCs over what is seen for a single esterification modification. This dual modification converts the properties of SNCs to stronger hydrophobicity and lower polarity. The crystalline structure can be maintained as such even after the dual modifications. This type of modification widens the application of starch nanocrystals. These modified SNCs enhance the reinforcement ability for nanocomposites with stronger hydrophobic polymer matrices and rheology modifiers. The modification also enhances the solubility of SNCs in nonpolar solvents such as chloroform, dichloromethane, toluene, etc. The cross-linking did not affect the crystalline structure of SNCs and it was preserved as such (Lili et al., 2016).

Large chain surface modifications of the monocrystalline nanoscale starch are possible using stearic acid chloride and poly (ethylene glycol) methyl ether (PEGME). As per FTIR, X-Ray Photoelectron Spectroscopy (XPS) spectroscopy, contact angle measurements, X-ray diffraction, and DSC analysis, it is confirmed that due to the surface modifications performed no variations will occur to the starch crystalline structure. Due to reduced polar and hydrogen bonding forces, individualization of nanoparticles was shown via TEM. The grafts embedded at the starch surface were also found to crystallize. Thus, it is possible to modify the starch nanocrystal surface with plasticizing chains. These modified SNCs can be used as compatibilized polymer additives, surface-active particles, and co-continuous nanocomposite precursors (Thielemans et al., 2006).

To introduce carboxylic acids, TEMPO (2,2,6,6-tetramethylpiperidine-1-oxyl) oxidation of SNPs, which selectively oxidizes SNPs towards primary hydroxyls (C6 position), has been widely adopted. Kato et al. (2003) suggested that NaOH activates the carbonyls to form intermediates for the second oxidation into carboxylic acids after the hydroxyl groups are oxidized into aldehydes. This oxidation reaction also requires NaBr (sodium bromide) as a co-catalyst and NaClO (sodium hypochlorite) as an oxidation agent.

5.9 GRAFTING

The surface of the starch nanoparticles can be grafted with short grafting agents. Usually this is done using a two-way strategy as follows:

5.9.1 SNCs Modifications Based on the "Graft-To" Strategy

Ning et al. (2011) report on the chemical modifications of starch nanocrystals that caused improved adhesion between the filler and matrix. They also observed that due to the destruction of the three-dimensional networks of the nanomaterials, a sharp decrease in mechanical performance of the composites was observed with modified starch nanocrystals (after chemical modification) when they were utilized as nanofiller in the polymer matrix. The "grafting-to" methods provided grafting of various polymers like aliphatic chloride, poly(caprolactone) (PCL), alkenyl succinic anhydride (ASA), poly(ethylene glycol) methyl ether (PEGME), poly(propylene glycol) mono butyl ether (PPGBE), and poly(tetrahydrofuran) (PTHF) to the surface of starch nanocrystals using different coupling agents like 2,4-touene diisocyanate (2,4-TDI), etc. (Ning et al., 2011).

Yet another method of modification which grafts polymer chains to the starch nanocrystals exists in the reaction between the acid chloride group (–COCl), which includes stearoyl chloride octanoyl, nonanoyl, and decanoyl chloride (Namazi and Dadkhah, 2010), and the hydroxyl group (Thielemans et al., 2006). Under optimized conditions, modified nanocrystals with grafted aliphatic polymers will become more differentiated and moderately larger in size than their unmodified counterparts. It is observed that the polymer chains have taken a crystalline brush-like structure projected outward from the starch surface which enhanced and reinforced nanocrystallinity when stearate polymer was grafted onto the starch nanocrystal surface. (Thielemans et al., 2006). In those studies, the polarity of the starch nanocrystals was decreased and improved dispersion in organic solvents was found when the octanoyl, nonanoyl, and decanoyl polymers were grafted with SNCs. It was also found that the degradation temperature was increased to more than 100°C for SNCs when they are grafted to stearate chains. This is due to the thermal protective crystalline layer formed by the oxygen-poor stearate surface on starch. Similarly, after grafting with octanoyl, nonanoyl, and decanoyl polymers, greater thermal stability has been observed due to the low amount of hydroxyl groups remaining after acylation reaction (Namazi and Dadkhah, 2010). Thus, the "graft-to" strategy, by keeping the crystalline integrity as such, invariably changes the surface properties, thermal properties, physical properties like morphology and size, and chemical properties like polarity and dispersion in aqueous or organic solvents. The reinforcing effect in nanomaterials thus obtained would allow the use of nonpolar polymers as matrices for starch nanocomposite materials, thus stimulating the functional and effective application of starch nanocrystals (Ning et al., 2011).

5.9.2 SNCs Modifications Based on the "Graft-From" Strategy

The major disadvantages of "graft-to" technology are low controllability and grafting efficiency and inability to graft long chain polymers, but these can be overcome by "graft-from" technology. Here propagation of polymeric chain by the polymerization of monomers is induced, and finally the polymer chains are coupled with the starch nanocrystals. Using "graft-from" technique together with ring-opening polymerization (ROP), different polymers have been grafted from the surface of starch nanocrystals. This is mainly done for poly(caprolactone) (PCL) (Chang et al., 2009) and free radical polymerization (FRP) for polystyrene (PS) (Song et al., 2008). Through this, a greater grafting density can be obtained. Here the compatibility between nanocrystals and matrix is enhanced with the grafted chains.

5.10 APPLICATION OF STARCH NANOCRYSTALS IN FOOD INDUSTRY

5.10.1 Starch Nanocrystals as Packaging Component with Barrier Property

The high demand for biologically safe and healthy natural food packaging materials has encouraged the research and development of edible nano-based polymer packaging films. These films are capable of preventing deterioration and loss of food quality. Active and smart packaging forms an innovative concept that could find major applications in the food industry. Incorporation of nanomaterials into packaging materials is a novel packaging strategy which aids in reducing spoilage, ensuring food safety and quality, repairing breaks in packaging, reducing the shortage of food, and improving the health of the people. Polymer nanoparticles used in packaging materials act as carriers of food additives, vitamins, pigments, etc. Hari et al. (2018) reported the development of an active chitosan-based packaging film incorporated with β-carotene-loaded starch nanocrystals and found that these films showed a controlled release of β-carotene for more than one week. Herein, the pigment β-carotene retained its antioxidant activity even after its addition into packaging film. Starch nanoparticles have attracted a lot of interest as a material in barrier films for food packaging. The barrier properties of packaging increase the shelf life of the product inside from deterioration caused by oxidation, photodegradation, spoilage due to microorganisms, etc. In addition to this, barrier packaging mainly focuses on oxygen permeability and water vapor transmission. It has been noticed that a material with good barrier properties to water vapor usually has poor barrier properties to oxygen and vice versa. A decrease in water vapor permeability by the incorporation of maize starch nanoparticles has been reported. Because of the hydrophilic nature of starch nanocrystals, only a few studies have been carried out to learn the barrier properties to oxygen and water vapor. Angelier et al. (2009) observed a significant decrease in permeability upon starch nanocrystal addition, whereas Kristo and Biliaderis (2007) observed that the addition of a lower concentration of starch nanocrystals does not show any significant differences in water vapor permeability as compared to that of film without starch nanocrystals. Meanwhile, a higher concentration of starch nanocrystals leads to a significant decrease in water vapor permeability. Starch nanocrystals were found to reduce oxygen diffusion and permeability through the nanocomposite films upon increased filler content (Angelier et al., 2009).

5.10.2 Starch Nanocrystals as Food Additives

Starch nanocrystals can be used as food additives, as they have a wide range of uses in foods, such as fat replacer, emulsion stabilizer, etc.

5.10.2.1 Fat Replacer

Fat replacers are non-fat substances that act like fat in foods. Some lipid, carbohydrate, or protein-based particles (Lucca and Tepper, 1994) imitate the organoleptic or physical properties of triglycerides. Native and modified starches and starch nanocrystals with smaller particle sizes can act as fat replacers. The particle size of the starches and starch nanocrystals is a prime factor in determining both fat-like taste and mouthfeel (Daniel and Whistler, 1990). Cereal starch nanoparticles developed by the method of ultrasonication were utilized in salad dressing as a fat replacer. Likewise, there are reports on the possibility of using hydrolyzed starch for the development of low-fat mayonnaise (Ma et al., 2006). Earlier reports documented that the smaller size of starch nanocrystals made them an attractive candidate as a fat replacer, and the blending of starch nanocrystals with other components resulted in the formation of a smooth cream-like substance that has properties similar to fat and also helps to replace high-calorie fat, which resulted in the decrease in calories.

5.10.2.2 Emulsion Stabilizer

An emulsion is a colloidal system formed by combining two or more immiscible liquids where one liquid contains a dispersion of the other liquid. They are used profusely in food, cosmetics, biomedicine, and pharmaceutics (Yang et al., 2017). Pickering emulsions (Pickering, 1907) are solid particle-stabilized emulsions used to stabilize droplets against coalescence. Pickering emulsions have gained extensive research interest in many fields due to their unique property of reduced coalescence (Yang et al., 2017), which results in enhanced stabilization, reduced toxicity for in vivo usage, irreversible anchoring, specific inter-particle interactions at the interface, etc. (Chen et al., 2014). These improved properties of emulsions help them to completely block the process of Ostwald ripening, which results in the formation of emulsion with long-term stability (Chen et al., 2014). Studies revealed that environmentally benign particles (cellulose microparticles, cellulose nanoparticles, chitin nanocrystals, starch microparticles, and starch nanoparticles) can replace a number of commonly used particles (hydrophobic silica, clays, carbon nanotubes, latex, and microgels) that have been widely used for stabilizing emulsions. Previous reports established the possibility of using starch nanoparticles as a pickering emulsifier; the increased content of SNPs caused a reduction in the particle sizes of the polymer dispersion and thereby confirmed the key role of the SNPs in the stabilization process (Haaj et al., 2014).

A previous study revealed that oil-in-water emulsions could be stabilized by starch nanocrystals (Li et al., 2012a). The creaming effect of emulsions got inhibited by SNCs due to the formation of a strong inter-droplet network. It was also found that ionization of a sulfate group at the surface of SNCs may contribute the negative charge to SNCs and that the aggregation of SNCs in aqueous dispersion is influenced by the electronic repulsion between the charged SNCs. SNCs used for the development of particle-stabilized emulsifier is synthesized by the acid hydrolysis of waxy maize starch below the gelatinization temperature of starch. The nanoscale crystalline residues developed during the hydrolysis process were separated from starch granules by hydrolyzing amorphous regions (Chen et al., 2014).

5.10.3 Antimicrobial Activity of Starch Nanocrystals

Almost 25% of the world's food supply—approximately 1.3 billion tons of food—is lost (Gustavsson et al., 2011) due to spoilage caused by microbial contamination. Microbial contamination becomes a serious problem when it is related to food products, especially weaning foods (Xiaojia and Huey-Min, 2016). The prime applications of nanotechnology in food safety are the antimicrobial effects of nanoparticles and the identification of spoiling and contaminating food microbes using nanosensors (Nasr, 2015). This will help to extend the shelf life of food. Therefore, nano-based antimicrobials act as an alternative to antibiotics and chemicals used in food and food packaging systems. In our present study on starch nanocrystals loaded with streptomycin (SSNCs), we observed that starch nanocrystals are efficient candidates for loading and sustained release of antibiotics. Thus, the SSNCs are identified as an ideal formulation due to the enhanced drug-carrying capacity and high antibacterial efficacy at a low constant dose against different Gram-positive and Gram-negative pathogenic microorganisms. This is mainly because the starch nanocrystals, which act as a carrier for antibiotics, help to enhance the availability of streptomycin or other food-grade antibiotics like nisin to a larger extent of area and thereby develop a larger zone of inhibition (Figure 5.2). These results are encouraging for the use of antibiotic-loaded starch nanocrystals for the treatment of infections caused by antibiotic-resistant bacteria and also in antimicrobial-based food packaging films.

Moreover, our studies on the synergistic action of antibiotic-loaded starch nanocrystals with other antibiotics showed an enhanced antibacterial activity. The antibiotic selected for the loading purpose was streptomycin. The possible mechanism (Figure 5.3) responsible for this synergistic antimicrobial activity involves two modes of antimicrobial action: the first mode of action (MA I) comprises the combined effect of individual actions performed by SSNCs and antibiotics (e.g., polymyxin B and ciprofloxacin). Herein, the starch nanocrystals act as a carrier for streptomycin and thereby increase the availability of streptomycin/nisin to bacterial cell membrane. After this, the cell membrane will disturb cell functions such as permeability, respiration, and that eventually leads to cell death. The action of antibiotics like polymyxin/ciprofloxacin (described above) is also included. The second possible mode of action (MA II) involves enhancing antimicrobial activity by the interaction of modified SSNCs, i.e., antibiotics (polymyxin/ciprofloxacin), which are adsorbed on the surface of SSNCs (Figure 5.3). In this context, the synergistic antibacterial property of SSNCs with ciprofloxacin and polymyxin B is considered as an alternative and attractive method to combat the increasing spread of drug resistance, and such an approach is likely to provide many potential applications in medical devices and microbe-resistant systems. The combination of antibiotics with drug-loaded starch nanocrystals will help to reduce the concentration of antibiotics, thereby helping to minimize the cost and side effects of antibiotics. For food-based applications, it is possible to use many food-grade antibiotics like bacteriocins (nisin, pediocin), microcins, β-lactams, bacitracins, etc. (Nicholson, 1997) instead of streptomycin.

5.11 APPLICATION OF STARCH NANOCRYSTALS IN BIOMEDICAL FIELD

Eco-friendliness and compatibility with pharmaceutical and biomedical applications for large-scale production of the modified starch nanocrystals lead to their potential applications in the biomedical field. Modified SNCs can be used for the creation of tissue scaffolds or bone-regenerating treatments. Since effective loading of a drug and its and controlled release from the nanoparticles can be accomplished, these can be used for drug delivery applications. Starch particles and nanoparticles/nanocrystals can be used as both matrix and reinforcement and have been widely used for the creation of new biocomposites with various applications in the biomedical field. The major biomedical applications of starch nanoparticles discussed in this section include drug carriers, wound-healing activity, tissue engineering, and fluorescently labeled biosensors.

5.11.1 Starch Nanocrystals as Drug Carriers

It has already been established that cross-linked starch nanoparticles can be used for controlled and sustained drug

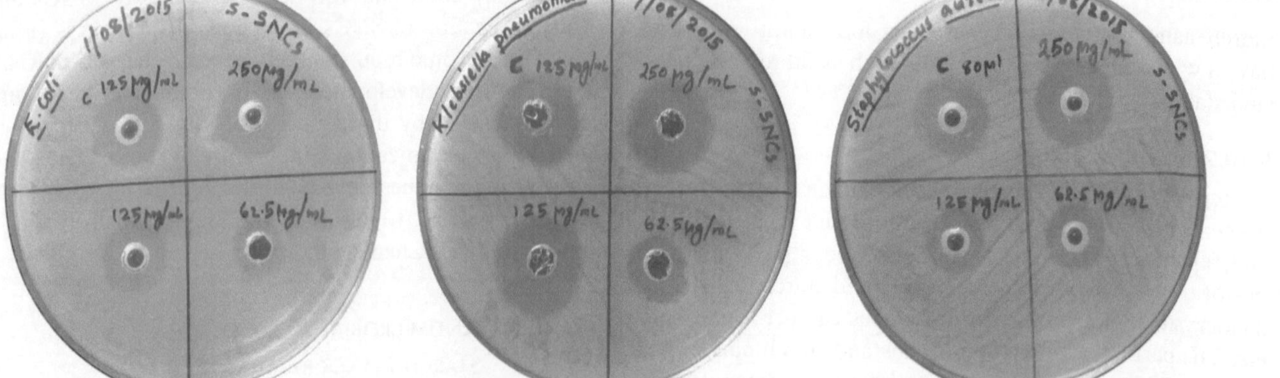

FIGURE 5.2 Antibacterial study of SSNCs with different concentrations (625, 125, and 250 μg/mL) against Gram-positive (*Staphylococcusaureus*) and Gram-negative (*Klebsiella pneumonia* and *Escherichia coli*) organisms.

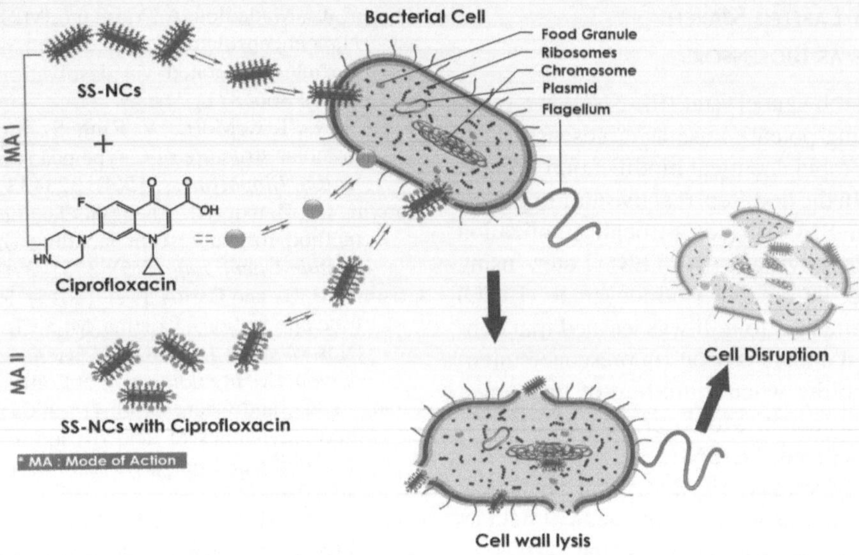

FIGURE 5.3 Possible mechanism of synergistic antibacterial activity by SSNCs and antibiotics (ciprofloxacin).

delivery (Li et al., 2012b), e.g., for *N,N*-bis(acryloyl)crystamine cross-linked nanoparticles exhibited to have controlled drug release with reduction-sensitive properties (Yang et al., 2014). By cross-linking and adding functionalizing moieties with starch-based hydrogels, they can be modified for its use in drug release studies. Li et al. (2014) showed that an in situ hydrogel preparation can be attained through Schiff reaction between starch nanoparticles and polyvinyl amine and later by cross-linking and encapsulating the cancer drug doxorubicin (DOX) to it. Recently it has been accepted that for the controlled delivery of DOX for cancer therapy, acetylated starch nanocrystals (ASNs) using broken rice is a promising candidate (Huaxi et al., 2016). It has been studied that the hollow starch nanoparticles (HSNPs) with gelled starch as the shell form a highly promising platform for cancer therapy (Yang et al., 2017).

5.11.2 Role of Starch Nanocrystals in Wound Healing

Wound healing is the result of innate physiological responses of the body that will soon refurnish the anatomic as well as cellular continuity of the damaged tissue with the normal healthy ones. For the rebuilding of damaged tissue, and also to maintain homoeostasis, a coordinated completion of a series of physio-biochemical events and also security against invading pathogenic microorganisms are necessary (Velnar et al., 2009). Delayed healing often ends up in severe bacterial infections, nutritional deficiencies, and stress (Guo and DiPietro 2010). The demand for novel and rapid wound-healing materials with inherent antimicrobial properties is on the rise. There are selective, biocompatible, and biodegradable polymer materials available that are broadly used for assisted cellular messaging and as scaffolds for wound healing. In this scenario, starch nanocrystals with high biodegradability and bioavailability and low antigenicity and are often a choice for the manufacture of wound management scaffolds (Vijaya et al., 2018). Cross-linked scaffolds made out of starch nanocrystals can be used for sustained and controlled drug delivery in wounds so that it can be rapidly cured. The various antibiotics that are incorporated into the biocompatible starch nanocrystals can be released in a sustained fashion, thereby preventing subsequent infections. SNCs loaded with the drug prove successful in starch biomaterial dressings associated with wound healing.

5.11.3 Starch Nanocrystals in Tissue Engineering

Starch is a commonly used biomaterial and porous starch nanofiber composites can be applied for efficient tissue engineering and regeneration applications extensively due to the formation of macro structures. It is having profound applications in construction of 3D nanofibrous composite scaffolds, which can be used for bone tissue engineering, cellular regeneration, and growth. Modified starch nanocrystals also offer the expected degree of wound healing and tissue regeneration properties. The most common chemically modified forms include cadexomer, oxidized starch (OS), and hydroxyethyl starch (HES) (Raveendran et al., 2017). Cadexomer is a hydrophilic starch cross-linked with epichlorohydrin and is popularly blended with 0.9% (w/w) iodine to form cadexomer-iodine (CI) ointment for wound healing. Here, cadexomer serves as a carrier for iodine. It swells when applied to the wounds, aids in the slow and controlled release of iodine into the wounded tissue, and provides antiseptic properties to it. Chemical modification of starch with oxidizing reagents produces OS, which is an excellent material known for its film-forming capability, due to low viscosity and high stability (Torres et al., 2013). CS-OS nanofiber-based scaffolds were developed for bone tissue regeneration. HES is obtained as a result of hydrolysis and later hydroxyethylation of amylopectin. It is most commonly used as a tissue adhesive which helps in the closure of wounds (Mo et al., 2010).

5.11.4 Fluorescently Labeled Starch Nanocrystals as Biosensors

Fluorescently labeled starch nanocrystal (FL-SNC) is a functional nanomaterial having potential use as a biosensor. It is usually synthesized in a rapid chemical modification process that occurs in two steps. In the first step, reactive amino groups are introduced onto the SNC surface through silanization with 3-aminopropyl triethoxysilane (APTES). Later, using thiourea as intermediate, fluorescein isothiocyanate (FITC) groups were covalently attached to it. It was learned that connected FL-SNC displayed a more obvious fluorescence intensity and higher photostability when a mixture of uncoupled FITC and SNCs was considered. Moreover, considering the cells that could be internalized, FL-SNCs are non-toxic and biocompatible. Thus, polyhydroxyl FL-SNCs can be widely exploited for developing biosensors in the biomedical field as well as biomarkers in food industries (Canxin et al., 2017).

5.12 CONCLUSION

Starch is a natural polymer which can be isolated from available natural sources such as rice, wheat, corn, potato, etc. Due to its outstanding biocompatible characteristics, starch can be used to make new biocomposites and also as both matrix and reinforcement material. This chapter gives an overview of starch in the form of SNCs and its characterization, properties, and applications in the food and biomedical fields. The rheological and morphological characteristics, strong interfacial interactions, swelling property, and strong network organization produced by these nanocrystals are the key factors that contribute to their improved physical, chemical, mechanical, and functional activities. The chemical modifications based on "graft-to" and "graft-from" models enhance the performance of SNCs to a large extent. The modified SNCs loaded with antibiotics showed enhanced antimicrobial activity against different food-borne pathogens. Due to the biocompatibility of chemically modified starches, SNCs can be extensively used in tissue engineering, wound healing, and drug delivery and also as fluorescently labeled biosensors in the biomedical field and as food-based biomarkers in the food and agriculture sector.

REFERENCES

Alina, S., Irit, D., Noa, A., Yehuda, G. A., and Yoav, D. L. (2012). β-Casein nanoparticle-based oral drug delivery system for potential treatment of gastric carcinoma: Stability, target-activated release and cytotoxicity. *Eur. J. Pharm. Biopharm.*, 80(2): 298–305.

Angellier, H., Molina-Boisseau, S., Belgacem, M. N., and Dufresne, A. (2005). Surface chemical modification of waxy maize starch nanocrystals. *Langmuir*, 21: 2425–2433.

Angellier-Coussy, H., Putaux, J.-L., Molina-Boisseau, S., Dufresne, A., Bertoft, E., and Perez, S. (2009). The molecular structure of waxy maize starch nanocrystals. *Carbohydr. Res.*, 344(12): 1558–1566.

Azimi, B., Nourpanah, P., Rabiee, M., and Arbab, S. (2014). Producing gelatin nanoparticles as delivery system for bovine serum albumin. *Iranian Biomed. J.*, 18(1): 34–40.

Bagheri, L., Madadlou, A., Yarmand, M.S., and Mousavi, M. (2013). Nanoencapsulation of date palm pit extract in whey protein particles generated via desolvation method. *Food Res. Int.*, 51(2): 866–871.

Bogracheva, T. Y., Morris, V., Ring, S., and Hedley, C. L. (1998). The granular structure of C-type pea starch and its role in gelatinization. *Biopolymers*, 45(4): 323–332.

Canxin, C., Benxi, W., Zhengyu, J., and Yaoqi, T. (2017). Facile method for fluorescent labelling of starch nanocrystal. *ACS Sustain Chem. Eng.*, 5(5): 3751–3761.

Cristina, M. S., and Carlos, E. A. (2015). Polymeric nanoparticles for food applications, Cristina, M. S, Hongda, C., and Rickey, Y. Y. (Eds.) *Nanotechnology and functional foods: Effective delivery of bioactive ingredients*. First Edition. John Wiley & Sons, Ltd.

Chang, Y. C., and Chen, D. G. H. (2005). Adsorption kinetics and thermodynamics of acid dyeson a carboxymethylated chitosan-conjugated magnetic nano-adsorbent. *Macromol. Biosci.*, 5(3): 254–261.

Chang, P. R., Ai, F., Chen, Y., Dufresne, A., and Huang, J. (2009). Effects of starch nanocrystal-graft-polycaprolactone on mechanical properties of waterborne polyurethane-based nanocomposites. *J. Appl. Polym. Sci.*, 111(2): 619–627.

Chen, H., and Zhong, Q. (2015). A novel method of preparing stable zein nanoparticle dispersions for encapsulation of peppermint oil. *Food Hydrocoll.*, 43: 593–602.

Chen, G., Wei, M., Chen, J., Jin, H., Alian, D., and Peter, R. C. (2008). Simultaneous reinforcing and toughening: New nanocomposites of waterborne polyurethane filled with low loading level of starch nanocrystals. *Polymer.*, 49(7): 1860–1870.

Chen, L., Yunxing, L., Peidong, S., and Yang, C. (2014). Starch nanocrystals as particle stabilisers of oil-in water emulsions. *J. Sci. Food Agri.*, 94(9): 1802–1807.

Choi, D. H., Subbiah, R., Kim, I. H., Han, D. K., and Park, K. (2013). Dual growth factor delivery using biocompatible core-shell microcapsules for angiogenesis. *Small*, 9(20): 3468–3476.

Cui, Y., Xu, Q., Chow, P. K., Wang, D., and Wang, C. H. (2013). Transferrin-conjugated magnetic silica PLGA nanoparticles loaded with doxorubicin and paclitaxel for brain glioma treatment. *Biomaterials*, 34(33): 8511–20.

Daniel, J. R., and Whistler, R. L. (1990). Fatty sensory qualities of polysaccharides (abstract). *Cereal Foods World*, *35*, 825.

Deborah, L. C. -B., Julien, B., and Alain, D. (2010). Starch nanoparticles: A review. *Biomacromolecules*, 11(5): 1139–1153.

Dufresne, A. (2010). Processing of polymer nanocomposites reinforced with polysaccharide nanocrystals. *Molecules*, 15(6): 4111–4128.

Elzoghby, A. O., Samy, W. M., and Elgindy, N. A. (2012). Albumin-based nanoparticles as potential controlled release drug delivery systems. *J. Control. Release*, 157(2): 168–182.

Elzoghby, A. O., Samy, W. M., and Elgindy, N. A. (2013). Novel spray-dried genipin-crosslinked casein nanoparticles for prolonged release of alfuzosin hydrochloride. *Pharm. Res.*, 30(2): 512–522.

Foresti, M. L., Williams, M. D. P., Martínez-García, R., Vazquez, A. (2014). Analysis of preferential action of alpha-amylase from B. licheniformis towards amorphous regions of waxy maize starch. *Carbohydr. Polym.* 102: 80–87.

Gérard, C., Colonna, P., Buléon, A., and Planchot, V. (2002). Order in maize mutant starches revealed by mild acid hydrolysis. *Carbohydr. Polym.*, 48(2): 131–141.

Ghasemolu, M., Gharibzahedi, S. M. T., and Cran, M. (2016). Preparation and characterization of starch nanocrystals, Visakh, P. M., and Long, Y. (Eds.) (60–108), *Starch-based blends, composites and nanocomposites, Green chemistry series*. Cambridge, UK: RSC Publishers.

Giroux, H. J., and Britten, M. (2011). Encapsulation of hydrophobic aroma in whey protein nanoparticles. *Microencapsulation*, 28(5): 337–343.

Guo, S., and DiPietro, L. A. (2010). Critical review in oral biology & medicine: Factors affecting wound healing. *J. Dental Res.*, 89(3): 219–229.

Gustavsson, J., Cederberg, C., Sonesson, U., Robert van, O., and Alexandre, M. (2011). *Global food losses and food waste: Extent causes and prevention*. Rome: Food and Agriculture Organization (FAO) of the United Nations.

Haaj, S. B., Magnin, A., and Boufi, S. (2014). Starch nanoparticles produced *via* ultrasonication as a sustainable stabilizer in Pickering emulsion polymerization. *RSC Adv.*, 4: 42638–42646.

Hans, M. L., and Lowman, A. M. (2002). Biodegradable nanoparticles for drug delivery and targeting. *Curr. Opin. Solid State Mater. Sci.*, 6(4): 319–327.

Hari, N., Stegiya, F., Alakananda, G. R., and Ananthakrishnan, J. N. (2018). Synthesis, characterization and biological evaluation of chitosan film incorporated with β-carotene loaded starch nanocrystals. *Food Pack. Shelf Life*, 16: 69–76.

Huaxi, X., Tao, Y., Qinlu, L., Gao-Qiang, L., Lin, Z., Fengxiang, Y., Yuejiao, C. (2016), Acetylated starch nanocrystals: Preparation and antitumor drug delivery study, *Int. J. Biol. Macromol.* 89: 456–464.

Ingle, A., Gade, A, Pierrat, S., Sonnichsen, C., and Rai, M. (2008). Mycosynthesis of silver nanoparticles using the fungus fusarium acuminatum and its activity against some human pathogenic bacteria. *Curr. Nanosci.*, 4(2): 141–144.

Jain, R. A. (2000). The manufacturing techniques of various drug loaded biodegradable poly(lactide-co-glycolide) (PLGA) devices. *Biomaterials*, 21(23): 2475–2490.

Jenkins, P. J., Comerson, R. E., Donald, A. M., Bras, W., Derbyshire, G. E., Mant, G. R., and Ryan, A. J. (1994). In situ simultaneous small and wide-angle X-ray scattering: A new technique to study starch gelatinization. *J. Polym. Sci. B*, 32(8): 1579–1583.

Kato, Y., Matsuo, R., and Isogai, A. (2003). Oxidation process of water-soluble starch in TEMPO mediated system. *Carbohydr. Polym.*, 51(1): 69–75.

Kristo, E., and Biliaderis, C. G. (2007). Physical properties of starch nanocrystal-reinforced pullulan films. *Carbohydr. Polym.*, 68(1): 146–158.

Kummerer, K., Menz, J., Schubert, T., and Wim, T. (2011). Biodegradability of organic nanoparticles in the aqueous environment. *Chemosphere*, 82(10): 1387–1392.

Kushagri, S., and Abha, M. (2014). Gelatin nanoparticle: Preparation, characterization and application in drug delivery. *IJPSR*, 5(6): 2149–2157.

Labet, M., Thielemans, W., and Dufresne, A. (2007). Polymer grafting onto starch nanocrystals. *Biomacromolecules*, 8(9): 2916–2927.

LeCorre, D., and Angellier-Coussy, H. (2012). Preparation and application of starch nanoparticles for nanocomposite: A review. *React. Funct. Polym.*, 85: 97–120.

LeCorre, D., Bras, J., and Dufresne, A. (2011a). Evidence of micro- and nanoscaled particles during starch nanocrystals preparation and their isolation. *Biomacromolecules*, 12(8): 3039–3046.

LeCorre, D., Bras, J., and Dufresne, A. (2011b). Influence of botanic origin and amylose content on the morphology of starch nanocrystals. *J. Nanoparticle Res.*, 13(12): 7193–7208.

LeCorre, D. Bras, J., and Dufresne. A. (2012). Influence of native starch's properties on starch nanocrystals thermal properties.*Carbohydr. Polym.*, 87(1): 658–666.

LeCorre, D., and Angellier-Coussy, H. (2014). Preparation and application of starch nanoparticles for nanocomposites: A review. *React. Funct. Polym.*, 85: 97–120.

Le Corre, D., Dufresne, A., Rueff, M., Khelifi, B., and Bras, J. (2014). All starch nanocomposite coating for barrier material. *J. Appl. Polym. Sci.*, 131(3): 39826.

Li, C., Sun, P., and Yang, C. (2012a). Emulsion stabilized by starch nanocrystals. *Starch*, 64(6): 497–502.

Li, B.-Z., Wang, L.-J., Li, D., Adhikari, B., and Mao, Z.-H. (2012b). Preparation and characterization of crosslinked starch microspheres using a two-stage water-in-water emulsion method. *Carbohydr. Polym.*, 88(3): 912–916.

Li, Y., Liu, C., Tan, Y., Xu, K., Lu, C., and Wang, P. (2014). In situ hydrogel constructed by starch-based nanoparticles via a schiff base reaction. *Carbohydr. Polym.*, 110: 87–94.

Lili, R., Qian, W., Xiaoxia, Y., Jiang, Z., and Xingguang, S. (2016). Dual modification of starch nanocrystals *via* crosslinking and esterification for enhancing their hydrophobicity. *Food Res. Int.*, 87: 180–188.

Li-Ming, Z., Lu-E, S., Zhi-Liang, Z., Jian-Min, C., Dong-Dong, S., Jie, Y., and Zhen-Xing, T. (2011). Preparation and application of chitosan nanoparticles and nanofibers. *Braz. J. Chem. Eng.* 28(3): 353–362.

Liu, D., Wu, Q., Chen, H., and Chang, P.R. (2009) Transitional properties of starch colloid with particle size reduction from micro to nanometer. *J. Colloid Interface Sci.*, 339(1): 117–124.

Lucca, A., and Tepper, B. J. (1994). Fat replacers and the functionality of fat in foods. *Trends Food Sci. Technol.*, 5(1): 12–19.

Lynn, D. M., and Langer, R. (2000). Degradable poly (beta-amino esters): Synthesis, characterization, and self-assembly with plasmid DNA. *J. Am. Chem. Soc.*, 122(44): 10761–10768.

Ma, Y., Cai, C, Wang, J., and Sun, D.-W. (2006). Enzymatic hydrolysis of corn starch for producing fat mimetics. *J. Food Eng.*, 73(3): 297–303.

Ma, X., Jian, R., Chang, P. R., and Yu, J. (2008). Fabrication and characterization of citric acid-modified starch nanoparticles/plasticized-starch composites. *Biomacromolecules*, 9(11): 3314–3320.

Marcos, B., Sarraga, C., Castellari, M., Kappen, F., Schennink, G., and Arnau, J. (2014). Development of biodegradable films with antioxidant properties based on polyesters containing α-tocopherol and olive leaf extract for food packaging applications. *Food Pack. Shelf Life*, 1(2): 140–150.

Mo, X., Iwata, H., and Ikada, Y. (2010). A tissue adhesive evaluated in vitro and in vivo analysis. *J. Biomed. Mater. Res. A.*, 94(1): 326–332.

Nafee, N., Taetz, S., Schneider, M., Schaefer, U. F., and Lehr, C. M. (2007). Chitosan-coated PLGA nanoparticles for DNA/RNA delivery: Effect of the formulation parameters on complexation and transfection of antisense oligonucleotides. *Nanomedicine*, 3(3): 173–83.

Nasr, N. F. (2015). Applications of nanotechnology in food microbiology. *Int. J. Curr. Microbiol. App. Sci.*, 4(4): 846–853.

Namazi, H., and Dadkhah, A. (2008). Surface modification of starch nanocrystals through ring-opening polymerization of caprolactone and investigation of their microstructures. *Appl. Polym. Sci.*, 110: 2405–2412.

Namazi, H., and Dadkhah. A. (2010). Convenient method for preparation of hydrophobically modified starch nanocrystals with using fatty acids. *Carbohydr. Polym.*,79(3): 731–737.

Natu, M. V., Gaspar, M. N., Ribeiro, C. A., Correia, I. J., Silva, D., de Sousa, H. C., and Gil, M. H. (2011). A poly(μ-caprolactone) device for sustained release of an anti-glaucoma drug. *Biomed. Mater.*, 6(2): 25003.

Neek, M., Kim, T., and Wang, S.-W. (2019). Protein-based nanoparticles in cancer vaccine development. *Nanomedicine*, 15(1): 164–174.

Nicholson, M. D. (1997). The role of natural antimicrobials in food/packaging preservation. *Proceedings of future pack 97*. Wisconsin, Appleton: George O Schroeder associates Inc.

Ning, L., Jin, H., Peter, R. C., Debbie, P. A., and Jiahui, Y. (2011). Preparation, modification, and application of starch nanocrystals in nanomaterials: A review. *J. Nanomater.*, 2011: 13 pg.

Nuxoll, E. (2013). BioMEMS in drug delivery. *Adv. Drug Deliv. Rev.*, 65(11–12): 1611–1625.

Panyam, J., Dali, M. M, Sahoo, S. K, Ma, W., Chakravarthi, S. S., Amidon, G. L., Levy, R. J., and Labhasetwar, V. (2003). Polymer degradation and in vitro release of a model protein from poly(,-lactide-co-glycolide) nano- and microparticles. *J. Control Release*, 92(1–2): 173–187.

Pascoli, M., Renata de, L., and Leonardo, F. F. (2018). Zein nanoparticles and strategies to improve colloidal stability: A mini-review. *Front. Chem.*, 6: 6.

Patricia, Z., Dina, R., and Livney, Y. D. (2011). Re-assembled casein micelles and casein nanoparticles as nano-vehicles for ω-3 polyunsaturated fatty acids. *Food Hydrocoll.*, 25(5): 1270–1276.

Perez, S., and. Bertoft, E. (2010). The molecular structures of starch components and their contribution to the architecture of starch granules: A comprehensive review. *Starch Starke*, 62(8): 389–420.

Pickering, S. U. (1907). CXCVI.–emulsions. *J. Chem. Soc. Trans.*, 91: 2001–2021.

Pinto, R., Neufeld, C., Ribeiro, R. J., and Veiga, F. (2006). Nanoencapsulation I. Methods for preparation of drug-loaded polymeric nanoparticles. *Nanomedicine*, 2(1): 8–21.

Prabha, S., and Labhasetwar, V. (2004). Critical determinants in PLGA/PLA nanoparticle-mediated gene expression. *Pharm. Res.*, 21(2): 354–364.

Putaux, J.-L., Molina-Boisseau, S., Momaur, T., and Dufresne, A. (2003). Platelet nanocrystal resulting from the disruption of waxy maize starch granules by acid hydrolysis. *Biomacromolecules*, 4(5): 1198–1202.

Quintanar-Guerrero, D., Allémann, E., Fessi, H., and Doelker, E. (1998). Preparation techniques and mechanisms of formation of biodegradable nanoparticles from preformed polymers. *Drug Dev. Ind. Pharm.*, 24: 1113–1128.

Raveendran, S., Rochani, A. K., Maekawa, T., and Kumar, D. S. (2017). Smart carriers and nanohealers: A nanomedical insight on natural polymers. *Materials (Basel)*, 10(8): E929.

Rebeca, P., Irene, E., Maite, A., Gonzalez-Navarro, C. J., Gonzalez-Ferrero, C., and Juan, M. I. (2015). Casein nanoparticles as carriers for the oral delivery of folic acid. *Food Hydrocoll.*, 44: 399–406.

Rebeca, P., González-Navarro, C. J, Gamazo, C., Esparza, I., and Juan, M. I. (2017). Zein nanoparticles for oral delivery of quercetin: Pharmacokinetic studies and preventive anti-inflammatory effects in a mouse model of endotoxemia, *Nanomed: Nanotech. Biol. Med.*, 3(11): 103–110,

Ren, L., Jiang, M., Wang, L., Zhou, J., and Tong, J. (2012). A method for improving dispersion of starch nanocrystals in water through crosslinking modification with sodium hexametaphosphate. *Carbohydr. Polym.*, 87(2): 1874–1876.

Saeed, M., Abbas, Z. M., Ali, S., Nasser, M., Saman, S., and Mehrasa, R. B. (2013). Preparation and characterization of sodium Alginate nanoparticles containing ICD-85 (venom derived peptides). *IJIAS.*, 4(3): 534–542.

Sarei, F., Mohammadpour, D. N., Zolfagharian, H., Khaki, P., and Moradi, B. S. (2013). Alginate nanoparticles as a promising adjuvant and vaccine delivery system. *Indian J. Pharm. Sci.*, 75(4): 442–449.

Sarkari, M., Brown, J., Chen, X., Swinnea, S., Williams, R. O., and Johnston, K. P. (2002). Enhanced drug dissolution using evaporative precipitation into aqueous solution. *Int. J. Pharm.*, 243(1–2): 17–31.

Shahrodin, N. S. M., Rahmat, A. R., and Arsad, A. (2015). Synthesis and characterization of cassava starch nanocrystals by hydrolysis method. *Adv. Mater. Res.*, 1113: 446–452.

Shapira, A., Davidson, I., Avni, N., Assaraf, Y. G., and Livney, Y. D. (2010). β-Casein nanoparticle-based oral drug delivery system for potential treatment of gastric carcinoma: Stability, target-activated release and cytotoxicity, *European J. Pharm. Biopharm.*, 80(2): 298–305.

Shenoy, D., Little, S., Langer, R., and Amiji, M. (2005). Poly (ethylene oxide)-Modified poly (β amino Ester) nanoparticles as a pH-sensitive system for tumor-targeted delivery of hydrophobic Drugs: Part I. In vitro evaluations. *Mol. Pharm.*, 2(5): 357–366.

Song, S., Wang, C., Pan, Z., and Wang, X. (2008). Preparation and characterization of amphiphilic starch nanocrystals. *J. Applied Polym. Sci.*, 107(1): 418–422.

Stefano, P., and Polina, P. (2017). Poly-beta-amino-esters nano-vehicles based drug delivery system for cartilage. *Nanomedicine*, 13(2): 539–548.

Tester, R. F., Karkalas, J., and Qi, X. (2004). Starch composition, fine structure and architecture, Starch based blends, composites and nanocomposites. *Cereal Sci.*, 39: 151–165.

Thielemans, W., Belgacem, M. N., and Dufresne, A. (2006). Starch nanocrystals with large chain surface modifications. *Langmuir*, 22(10): 4804–4810.

Tice, T. R., and Cowsar, D. R. (1984). Biodegradable controlled release parenteral systems. *Pharm. Technol. J.* 8: 26.

Torres, F. G. Commeaux, S., and Troncoso, O. P. (2013). Starch-based biomaterials for wound-dressing applications. *Starch*, 65(7–8): 543–551.

Valencia, P. M., Farokhzad, O. C., Karnik, R., and Langer, R. (2012). Microfluidic technologies for accelerating the clinical translation of nanoparticles. *Nature Nanotech.*, 7(10): 623–629.

Velnar, T., Bailey, T., and Smrkolj, V. (2009). The wound healing process: An overview of the cellular and molecular mechanisms. *J. Int. Med. Res.* 37(5): 1528–1542.

Vijaya, S. W., Pallavi, R. W., Satish, D., Aparna, D., Ratnesh, J., and Prajakta, D. (2018). Starch based nanofibrous scaffolds for wound healing applications. *Bioact. Mat.* 3(3): 255–266.

Wang, Y., and Xie, W. (2010). Synthesis of cationic starch with a high degree of substitution in an ionic liquid. *Carbohydr. Polym.*, 80(4): 1172–1177.

Wang, L. F., Pan, S. Y., Hu, H., Miao, W. H., and Xu, X. Y. (2010). Synthesis and properties of carboxymethyl kudzu root starch. *Carbohydr. Polym.*, 80(1): 174–179.

Weiss, J., Takhistov, P., and McClements, D. J. (2006). Functional materials in food nanotechnology. *J. Food Sci.*, 71: R107–R116.

Wendorf, J. R., Ghartey-Tagoe, E. B., Williams, S. C, Enioutina, E., Singh, P., and Cleary, G.W. (2011).Transdermal delivery of macromolecules using solid-state biodegradable micro structures. *Pharmaceutical Res.*, 28(1): 22–30.

Wu, Y. P., Luo, Y. G., and Wang, Q. (2012) Antioxidant and antimicrobial properties of essential oils encapsulated in zein nanoparticles prepared by liquid-liquid dispersion method. *Lwt-Food Sci. Tech.*, 48(2): 283–290.

Xiaojia, H., and Huey-Min, H. (2016). Nanotechnology in food science: Functionality, applicability, and safety assessment. *J. Food Drug Anal.*, 24(4): 671–681.

Xu, Y., Ding, W., Liu, J., Li, Y., Kennedy, J. F., Gu, Q., Shao, S. (2010). Preparation and characterization of organic-soluble acetylated starch nanocrystals. *Carbohydr. Polymers*, 80(4): 1078–1084.

Xu, Y., Sismou, E. N., Grizzard, C., Thomas, M., Pestov, D., Huba, Z., Wang, T., and Bharadwaj, H. L. (2014). Morphological, structural and thermal properties of starch nanocrystals affected by different botanic origins. *Cereal Chem.*, 91(4): 383–388.

Yang, J., Huang, Y., Gao, C., Liu, M., and Zhang, X. (2014). Fabrication and evaluation of the novel reduction-sensitive starch nanoparticles for controlled drug release. *Colloids Surf. B*, 115: 368–376.

Yang, Z., Han, H., Wu, L., Zhang, L., Zhang, L., and Javed, I. M. (2017). Impact of emulsifiers addition on the retrogradation of rice gels during low-temperature storage. *J. Food Qual.* 2017: 7pp.

Yan-Qiong, Z., Yan, S., Ming-Mei, L., Xia, M. M. S., Gu-Jie, M., Chen, Y. M. S., Qiu-Yan, G., Wei-Jie, L. B. S., Xiao-Yue, W. B. S. Na, L., and Thomas, J. W. (2019). Galactosylated chitosan triptolide nanoparticles for overcoming hepatocellular carcinoma: Enhanced therapeutic efficacy, low toxicity, and validated network regulatory mechanisms, *Nanomedicine*, 15(1): 86–97.

Zhang, P., Li, S., Zhang, S., Zhang, X., Wan, L., Yun, Z., Ji, S., Gong, F., Huang, M., Wang, L., Zhu, X., Tan, Y., and Wan, Y. (2018). GRGDS-functionalized chitosan nanoparticles as a potential intravenous hemostat for traumatic hemorrhage control in an animal model. *Nanomedicine*, 14(8): 2531–2540.

Zheng, H., Ai, F., Chang, P. R., Huang, J., and Dufresne, A. (2009). Structure and properties of starch nanocrystals –reinforced soy protein plastics. *Polym. Compos.* 30(4): 474–480.

6 Recent Developments on Nanotechnology in Agriculture

Challenges and Prospects

Yugal Kishore Mohanta, Tapan Kumar Mohanta, and Debasis Nayak

CONTENTS

6.1 Introduction 79
6.2 Nanoparticles and Their Types 80
6.2.1 Inorganic Nanoparticles 80
6.2.2 Organic Nanoparticles 80
6.3 Synthesis of Nanoparticles 81
6.3.1 Physical Route 81
6.3.2 Chemical Route 81
6.3.3 Biological Route 81
6.4 Application of Nanotechnology in the Agriculture and Food Industry 81
6.4.1 Agriculture 82
6.4.2 Nanofertilizers and Nanopesticides 82
6.4.3 Food Processing and Packaging 83
6.5 Conclusion 84
References 84

6.1 INTRODUCTION

Nanoparticles and nanotechnology have emerged as the most prospective and innovative branch of basic science and engineering technology today. Their amalgamation into day-to-day life has provided much-required diversity in every field of science and technology. Starting with textiles, air and water purifiers, detergents, cosmetics, medical equipment, diagnostics, and medicines, nanotechnology has increased in importance in the 21st century (Guo & Sadler, 1999; Hofmann-Amtenbrink, Grainger, & Hofmann, 2015). On the Indian subcontinent, the concept of nanoparticles was demonstrated many centuries ago through Ayurveda. According to the Rasa Shastra, the "bhasma" were synonymous with metallic nanoparticles, which were generally prepared using various concoctions of plant materials (Pal, Sahu, & Haldar, 2014). Various bhasmas of gold, silver, copper, iron, and mercury were used actively to treat various ailments in the ancient period (Rastogi, 2010; Sarkar, Das, & Prajapati, 2010). Partly inspired by the various texts available in Ayurveda, the green synthesis route of nanoparticle synthesis became popular in the 21st century. Although various chemical and physical methods of synthesis are available, their production cost along with the formation of various toxic by-products constitute some major drawbacks (Panigrahi et al., 2004). In comparison to the physical and chemical methods, the green/biological route of synthesis provides quick incentives such as easy availability of a hugely diverse selection of basic raw materials and low costs of production with minimal or no additional reducing or capping agents. Further, since the whole process is ecofriendly, it does not produce any additional toxic by-products (Thakkar, Mhatre, & Parikh, 2010).

The world of nanoparticles is itself highly diverse: based on their composition they are divided into organic and inorganic nanoparticles. As organic nanoparticles are formed from natural resources, they are known to be very biocompatible, facilitating easy delivery with biodegradable qualities. However, inorganic nanoparticles are known for their uniformity and versatility, which is clearly evident from their applicability over a vast range of diagnostic products and instruments (Ealia & Saravanakumar, 2017; Liu, 2006; Mudshinge et al., 2011). Inorganic nanoparticles are highly uniform in their morphology and thus they are easily modulated for various applications, e.g., in cancer cells and in molecular signaling and targeted drug delivery (Bzdek et al., 2011; Jagannathan et al., 2006; Pileni, Tanori, & Filankembo, 1997). Based on the application, the inorganic nanoparticles are synthesized using either "top-down" (breaking of bulk material to nano form) or "bottom-up" (nucleation of nanoparticles at atomic level) techniques (Biswas et al., 2012; Fu et al., 2018; Yamamoto et al., 2013). Inorganic nanoparticles can be synthesized by employing various chemical, physical, and biological entities. For targeted drug delivery, the inorganic/organic nanoparticle can itself act as a drug or various combinations of drugs can be conjugated with the nanoparticles/or nanoemulsion for effective therapeutics. Considerable research has been done at both governmental and private levels to design novel drugs using nanoparticles as a drug delivery approach and to use targeted therapeutics in the fight against cancer. However, other important fields such as the agriculture sector have not received such focus in the implementation of nanotechnology for improving crop conditions and effective pest control.

India being an agricultural county, 70% of the population depends upon the agricultural industry for their primary employment opportunities. However, various environmental and socio-ecological conditions add much complexity to the agrarian economy (Bhatia & Ghanem, 2019; Ferreira et al., 2018). Further, one of the biggest challenges in the agricultural sector is the storage of the products produced. The field of agriculture was lastly focused during the "green revolution," in which various implementations were carried out following the recommendations of M.S. Swaminathan, who is popularly known as the father of the green revolution in India. However, they were done way back in the 1960s, where the main focus was on the use of high-yield crops and fertilizers in the farm sector (Emerick, 2018). Hence, science and technology can provide much-required innovation for overcoming the challenges associated with the improvement of the agro-food industry. In this scenario, research in nanomaterials can have a double bonanza effect where during the production stage the nanoparticles can help in increasing the percentage yield, and after harvesting, nanotechnology can be utilized for storing the harvest for a longer period of time. Hence, the integration of nanomaterial research into the agro-food industry provides viable and vibrant future aspects for the Indian agricultural economy (He, Deng, & Hwang, 2018; Parisi, Vigani, & Rodríguez-Cerezo, 2015).

6.2 NANOPARTICLES AND THEIR TYPES

Nanomaterials or nanoparticles can be simply defined as particles (organic, inorganic) with sizes ranging between 1–100 nm having distinct and diverse properties in relation to their bulk materials (Mohanraj & Chen, 2006). Various research groups have currently inclined towards nanotechnology owing to its unique optical, electrical, and mechanical properties when compared with their bulk materials. Nanobiotechnology is an amalgamation of two diverse branches of material science and biotechnology which has been explored extensively in healthcare, textiles, sanitation, bioremediation, and wound-dressing materials (Couvreur, 2013; Oskam, 2006; Ravishankar Rai & Jamuna Bai, 2011). However, the concept of nanotechnology was provided to the science community way back in 1957 in Caltech University by Prof. Nario Taniguchi, who said, "there is plenty of room at the bottom," in what is regarded as the very first lecture in the history of nanotechnology describing its future endeavors.

Naturally occurring nanoparticles can exist as nanostructures or as composites. The assembly of nanostructures can be from elementary materials or blocks which can have zero dimensions (quantum dots), one dimension (layers), two dimensions (nanotubes, fibers, filaments, springs, columns, needles), or three dimensions (molecules, clusters, fullerenes, rings, metal carbides, powders, grains) (Buzea & Pacheco, 2017; Teo & Sun, 2007). Nanocomposites are generally formed by mixing two types of materials, which can be measurable in micrometers, nanometers, or both. In nanocomposites, the nanometer-sized particles are dispersed in the intragranular or intergranular spaces of the main matrix. Based on the matrix materials used, the nanocomposites can be of ceramic (Al_2O_3/SiO_2; Al_2O_3/CNT), metallic (Fe-Cr/Al_2O_3; Al/CNT; Fe/MgO), or polymeric (Poly vinyl alcohol, polyethylene glycol, poly lactic-co-glycolic acid) form (Glezer, 2011; Njuguna et al., 2014).

Nanoparticles can be further differentiated into organic and inorganic based on the nature of the materials used for their fabrication.

6.2.1 Inorganic Nanoparticles

These are characteristically synthesized by the precipitation of inorganic salts, which are formed by interconnection with various molecules through covalent bonds or metallic bonds, which are naturally very strong forces of attraction. Further, inorganic nanoparticles are known for their uniformity of lattices during their synthesis reaction, where the rate of nucleation has a very prominent role in controlling their shape and size. The primary parameters that directly govern the nucleation process of inorganic nanoparticles are temperature, reaction time, pH, and the presence of surfactants (Ariga et al., 2011; Gasparotto et al., 2012).

6.2.2 Organic Nanoparticles

These are found in nature, where they are assembled in three-dimensional forms. In comparison to inorganic nanoparticles, organic nanoparticles face various challenges where modulation of their size and shape requires precise expertise as they are primarily fabricated by self-assembly of zwitterion particles, with polar and nonpolar regions (Dolez, 2015). Further, due to the weak nature of interactions at the molecular level, the organic nanoparticles are very dynamic in nature, and they can be easily modulated to various forms depending upon the use of surfactants or thermodynamic parameters (Sudha et al., 2018). Using natural and synthetic organic molecules, various emulsions and micelles can be prepared which might present very complex structures in their organic nanoparticles.

Organic nanoparticles are basically synthesized using one of two fundamental approaches: the top-down or the bottom-up method. In the "top-down" approach, as the name suggests, the nanomaterials are prepared by breaking their bulk source materials down to nano size. Hence, the nanoparticles are broken down from the macro to the nano range simply by reducing their natural size to the nanometer size range. There are various top-down techniques such as mechanical milling, microfluidics, and lithography. Mechanical milling is the most popular and cost-effective process of nanoparticle synthesis where the bulk materials are simply crushed using mechanical force to obtain nanomaterials of various shapes, sizes, and forms. However, the mechanical abrasion and contamination with other particles of the abrading product's own nanoparticles remain one of its major drawbacks (Tsuzuki, 2009). Microfluidics and lithography are more advanced forms of the top-down approach where nanoparticles are imprinted using lithography techniques in non-wet substrates using a stone or metal pattern with a smooth surface. The advantage of this approach is that it permits the production of nanoparticles of any shape and size with the encapsulation of macromolecules such as DNA, RNA, and proteins (Vollath, 2008). In the "bottom-up" approach, as the name suggests, the nanoparticles are synthesized at the atomic level, and they aggregate to form the nanomaterials from the precursor materials. The bottom-up techniques basically use a physiochemical methodology using synthetic chemistry and self-organization principles. The advantage of the bottom-up approach is that desired nanoparticles can be synthesized with precise shapes, sizes, and attached molecular groups. Various micelles, liposomes, polymeric conjugates, dendrimers, and polymeric nanoparticles are synthesized using this approach (Sau & Rogach, 2012; Wang & Xia, 2004).

6.3 SYNTHESIS OF NANOPARTICLES

Based on their application nanoparticles are synthesized using various methods. Broadly, the synthesis process is divided into three categories: physical route, chemical route, and biological route.

6.3.1 Physical Route

The physical route is one of the most prominent methods of nanoparticle synthesis where precision and quality of nanomaterials are the primary requisites for the final product. Various methods such as evaporation-condensation, combustion synthesis, arc discharge or plasma discharge, laser/electron beam heating, and laser ablation technique are used in the physical routes (Horikoshi & Serpone, 2013). The advantage of the physical methods is that no toxic chemicals or solvents are used for the synthesis of nanoparticles and uniform nanoparticles are synthesized in this method. However, the use of high pressure, high temperature, and costly high-throughput equipment with very low output is the most prominent drawback of the physical route of nanoparticle synthesis (Altavilla & Ciliberto, 2011).

6.3.2 Chemical Route

The chemical route of synthesis is one of the most prominently used synthesis approaches for the synthesis of nanoparticles. Various organic and inorganic reducing agents such as sodium citrate, sodium borohydride, tollens reagents, polyethylene glycol, etc., are used for the synthesis of nanoparticles based on their chemical requirements. To prevent agglomeration or aggregation of the formed nanoparticles, various capping agents such as poly vinyl alcohol, poly vinylpyrrolidone, polyethylene glycol, poly methacrylic acid, and polymethylmethacrylate are used in combination with various surfactants for generating ideal nanoparticles. Various popularly used techniques such as microemulsion, sol-gel, polymerization, oxidation, UV-initiated photoreduction, microwave-assisted synthesis, irradiation, and sonochemical processing are used for the chemical synthesis of nanoparticles (Hyeon, 2003; Schmidt, 2001). However, chemical synthesis shows some drawbacks such as the use of costly and toxic solvents and reducing and capping agents, the production of hazardous toxic byproducts, and additional purification steps that can increase the overall production cost of the nanoparticles synthesized.

6.3.3 Biological Route

In comparison to the physical and chemical routes of nanoparticle synthesis, the biological route provides a greener and more cost-effective alternative route for nanoparticle synthesis. The biological route provides a diverse range of starting bulk materials such as plants (stem, roots, leaves, flowers, and fruits), bacteria, algae, and fungi which are abundant and easily found in nature (Kharissova et al., 2013). The biological route doesn't require extensive preparation techniques, complex protocols, or extreme conditions of temperature and pressure along with toxic chemicals as reducing agents. The phytochemicals, which are the naturally occurring chemicals present in plants for their natural defense and biological activity, aid themselves as reducing and capping agents during the synthesis period for controlling the shape and size of the nanoparticles formed. Thus, in comparison to the chemical and physical routes, the biological route provides an eco-friendly process with a wide range of raw materials, with no additional use of reducing and capping materials, and with large-scale production capabilities in a cost-effective environment (Nayak et al., 2015).

6.4 APPLICATION OF NANOTECHNOLOGY IN THE AGRICULTURE AND FOOD INDUSTRY

Climate change and various other effects are currently having diverse effects on the agriculture and food industry. Nanoparticles and nanotechnology have come to have applicability in every aspect of human life. From the textile industry to pharmaceuticals and the healthcare industry, nanotechnology is quite popular for its effectiveness. As the food industry has a direct relationship with the surrounding environmental conditions, nanotechnology can help in addressing the large-scale integral limitations faced by the agrarian economy like India (He et al., 2018). The major challenge with the agro-food industry is to tackle the indiscriminate use of pesticides and fertilizers without affecting the environment. Being highly flexible in nature, nanoparticles can be modulated for immediate and efficient use as pesticides, insecticides, and antibacterial agents in packaged food materials (Dasgupta et al., 2015). In the food industry, food spoilage is one of the major factors for loss of economy of the food procured.

Nanotechnology can be utilized in different ways to enhance agricultural production levels and by the package industry to detect food spoilage through sensors or color change detectors that can help in identifying spoiled food materials (García, Forbe, & Gonzalez, 2010; Martins-Lopes, Fernandes, & Martins-Lopes, 2013).

6.4.1 Agriculture

The survival of the human species has a direct connection with agriculture, as it has established its position in the society with its integration for sustainability with relation to the organism's longevity. Agriculture is the foundation of various developing counties, and 70% of the Indian population is engaged in agricultural business. In an agrarian ecosystem, every aspect of agriculture such as preparation of the field, sowing of seeds, supply of clean water, use of fertilizers, insecticides, and pesticides, and after-harvest practices can be detrimentally affected if any one step is disturbed (Bhatia & Ghanem, 2019). Hence, agriculture has the potential to solve various societal issues which can be explored for innovative ecofriendly agricultural practices to have a harmony between every class of the society. Ecological imbalances such as climate change, water pollution, air pollution, soil pollution, pest infestation, and numerous crop diseases continue as a major concern for many generations throughout the world. Therefore, the implementation of various recent innovative farming techniques can be explored for better yield without having any hazardous side effects on human health or the ecosystem.

In the 21st century, the definition of agriculture has changed from ancient times as previously agriculture primarily dealt with production of food grains, whereas presently agriculture means the production of food grains, poultry products, fisheries, medicinal plants, and other cash crops important for textiles, shelter, and horticulture (Sapkota et al., 2019). With increases in population and pollution levels, the agrarian economy is facing direct consequences that result in losses in agricultural yields which lead to huge gaps between demand and supply. Thus, malnutrition and the development of insecticide-resistant species have added to the overall scenario of primarily agrarian economies such as India's (Srivastava et al., 2016). According to a joint report on the state of global food insecurity by World Health Organisation (WHO), the World Food Program (WFP), and the International Fund for Agricultural Development (IFAD) an estimated nearly 870 million people are undernourished, or 12.5% of the global population and 14.9% of those living in developing countries (McGuire, 2013).

In nature, most biological macromolecules (DNA, RNA, proteins) and other dynamic processes occur in the nano range. Hence, scientists and researchers all over the globe are making continuous efforts to improve the scenario in agriculture and its subsidiary outfits for appropriate strategies to defeat the food crisis and malnutrition (Bird et al., 2019; Ngo & Serra-Majem, 2019). Since the chemical organizations of plants and animals are constructed at subatomic levels, nanotechnology can form the connecting linkage between the agro-food industry and nano engineers, agricultural scientists, and molecular biology scientists (Chhipa, 2019; Duhan et al., 2017; Iavicoli et al., 2017). Various research groups have evaluated the effects of metal- and carbon-based nanoparticles for the efficient absorption, accumulation, translocation, and overall effects on growth and development of various plants such as corn, wheat, tomato, alfalfa, soybean, onion, spinach, radish, pumpkin, and cucumber (Nair et al., 2010; Rai et al., 2018). Thus, nanoparticles have shown remarkable results for enhanced germination, photosynthetic activity, and expression of mRNA and protein levels at molecular levels, leading to overall positive changes in gene expression which can have positive impact on the overall growth and yield of the agricultural crops (Hussain et al., 2019; Rico et al., 2011).

6.4.2 Nanofertilizers and Nanopesticides

Soil is one of the most important ecological factors: demand for it has increased 10-fold, leading to loss of natural forest cover as well as agricultural land. The subsequent loss of land resources has had direct consequences on agricultural yield with the extensive use of fertilizers that have deleterious effects on soil health. Soil health is the backbone of the agrarian industry; fertile soil provides essential nutrition to plants, and beneficial nematodes create a food web around the agricultural fields (Feregrino-Perez et al., 2018). The nutrient cycle along with the biological entities present in soil provides the dynamic characteristic for the soil to be called essentially fertile, which means that it allows the absorption and accumulation of essential nutrients for overall development of the agricultural product. Due to the entry of nanotechnology into the agricultural sector, scientists have explored various components such as nanofertilizers, nanopesticides, and nanoagrochemicals for the wholesome development of agricultural products. These nanopesticides and nanofertilizers have proved their efficiency in reducing pest infestations without affecting the biological integrity of the agricultural yield, being non-toxic to human beings and cattle (Elmer, Ma, & White, 2018). Though the use of modern technology aspires toward better agriculture practice and production, different pros and cons should still be seriously taken into account for prolonged adaptation of nanotechnology-based agriculture developments (Raliya et al., 2017) (Figure 6.1).

In case of agricultural nutrient management, instead of single-entity nanoparticles, combinations of various metallic and polymeric nanoformulations are known to impart higher efficacy as they contain various combinations of engineered nanoparticles aided by natural biological entities (Calabi-Floody et al., 2018). The term "nanofertilizer" generally refers to a product that delivers encapsulated substances/nutrients inside nanoporous materials or nanoemulsions using thin polymeric films that can be supplied directly to the standing crop fields. Ideally, nanofertilizers should have the properties to deliver various essential nutrients in a synchronized pattern so that the standing crops can uptake the encapsulated nutrient efficiently without converting prematurely into chemical or gaseous forms (Chen et al., 2018). Nanofertilizers can be

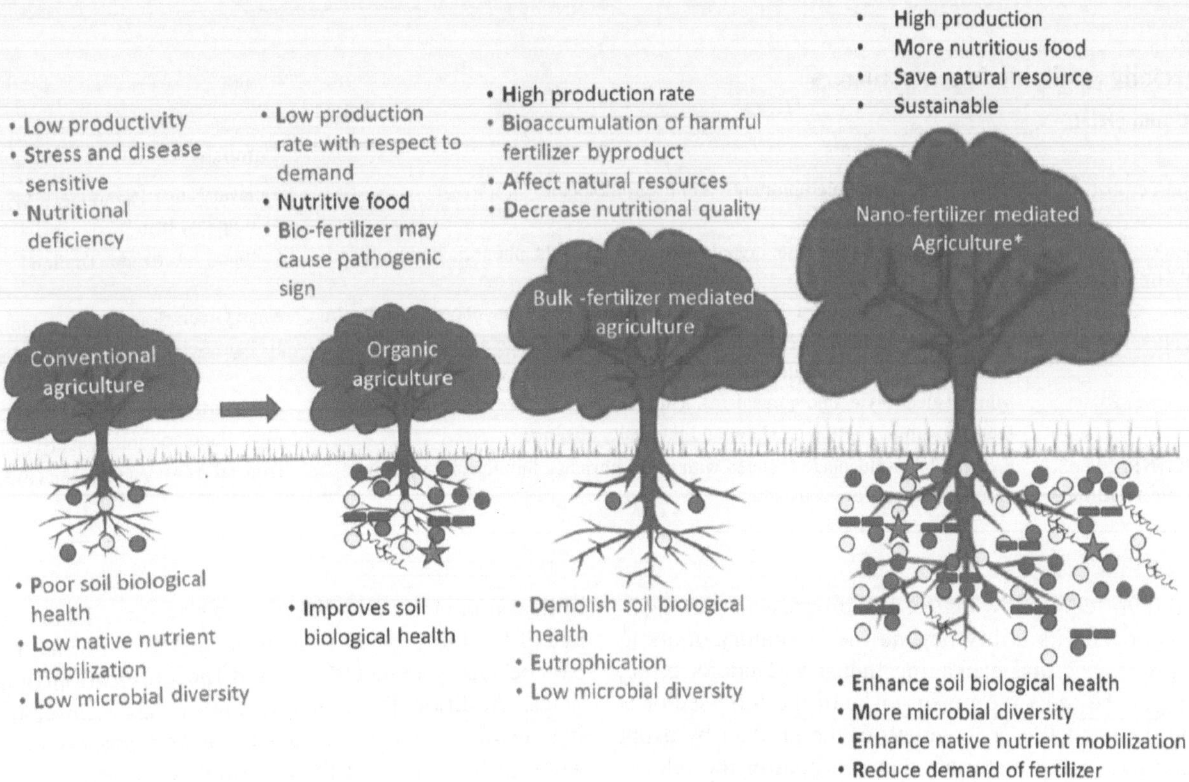

FIGURE 6.1 Comparative analysis of possible pros and cons of the conventional approach with respect to nanotechnology-mediated agriculture production. The impact on the rhizosphere as well as the environment is elucidated. Red text represents potential negative impact of a technology. (*) Note that the influence of the nanofertilizer further depends upon the plant growth and the soil rhizosphere depends upon the nanoparticle type, composition, and exposure concentration. (Reprinted with permission from Raliya et al. (2017). Nanofertilizer for Precision and Sustainable Agriculture: Current State and Future Perspectives. *Journal of Agricultural and Food Chemistry*, *66* (August), 6487–6503. review article. https://doi.org/10.1021/acs.jafc.7b02178). Copyright (2018) American Chemical Society.)

prepared by using the top-down or the bottom-up approach by exploring various organic and inorganic materials that are further loaded with various macro- and micronutrients that can easily be taken up by the plants. One of the important advantages of nanofertilizers is that the nutrients are released slowly and in a controlled manner that can assist in improving the efficacy of the nutrients for more than a month (Juárez-Maldonado et al., 2018).

In developing countries such as India, insects, pests, weeds, and disease cause significant damage to agrarian production levels. According to a study by Dhaliwal et al., 2010, pests cause 25% loss in rice, 10% in wheat, 30–40% in pulses and oilseeds, and more than 50% in cotton (Akoijam, Telem, & Marangmei, 2014; Dhaliwal, Jindal, & Dhawan, 2010). Apart from standing food and cash crops in outside agricultural fields, pests also cause huge losses in storage godowns, bins, and packages, which also significantly damages the quality of the stored food grains. One of the most popular examples of pest management was the use of toxins produced from the bacterium *Bacillus thuringiensis*, which were widely applied against lepidopteran and dipteran species. Currently, various metallic and organic nanoparticles are being used against a wide range of insects, plant pathogens, and pests as pesticides, insecticides, and insect repellants (Barik, Sahu, & Swain, 2008; Goswami et al., 2010; Hayles et al., 2017). According to various research groups, nanoparticles can mediate pest management through direct gene transfer methodology that can be delivered using nanorods, nanoencapsulation, or nanocomposites. Various research groups observed that using nanoemulsions, silica nanoparticles, solid lipid nanoparticles, polyethylene glycol–coated nanoparticles, and metallic nanoparticles such as silver, zinc oxide, and iron oxide nanoparticles were capable of successfully controlling a wide range of agricultural pests and animal ecto-parasites that were potentially harmful for standing food and cash crops (Cătălin Balaure, Gudovan & Gudovan, 2017; Jampílek & Kráľová, 2017; Kah & Hofmann, 2014). Until now, few commercial nanofertilizers have been available in the market, but marketing research is still going on to make the nanofertilizer products more conveyable (Table 6.1).

6.4.3 Food Processing and Packaging

Advances in agricultural practices lead to bumper harvesting of the food grains, which then need proper processing and packaging to aid their longevity. Further, with urbanization and modern lifestyles, the food habits of people have changed drastically, to where people have become accustomed to and prefer packaged and ready-to-use food materials. Along with these changes in food habits, alarming rates of food poisoning

TABLE 6.1
Commercially Available Nanofertilizers

Name of Commercial Product	Content/Ingredients	Company
Biozar Nano-Fertilizer	Organic materials, micronutrients, and macromolecules	Fanavar Nano-Pazhoohesh Markazi Company, Iran
Master Nano Chitosan Organic Fertilizer	Hydrophilic liquid chitosan, organic and salicylic acids, phenolic compounds	Pannaraj Intertrade, Thailand
Nano Green	Natural extracts viz. corn extracts, soybeans, grain, potatoes, coconut, and palm	Nano Green Sciences Inc., India
Nano Max NPK Fertilizer	Multiple organic acids chelated with major nutrients, amino acids, organic carbon, organic micronutrients/trace elements, vitamins, and probiotics	JU Agri Sciences Pvt. Ltd, Janakpuri, India
Nano-Ag Answer R	Mineral electrolyte, microorganisms, and sea kelp	Urth Agriculture, United States
Nano-Gro™	Plant growth promoting regulator and immunity enhancer	Agro Nanotechnology Corp., United States
TAG NANO (NPK, PhoS, Zinc, Cal, etc.) fertilizers	Proteino-lacto-gluconate chelated with micronutrients, humic acid, vitamins, probiotics, and seaweed extracts	Tropical Agrosystem India (P) Ltd, India

have been reported worldwide. Recent progress in nanotechnology has opened its utility beyond the laboratory doors to the food processing and packaging industry (Dudefoi et al., 2018). One of the most prominent tasks of the food sector is increasing the shelf life or longevity of the product by using appropriate packaging materials that can control the release of antimicrobial compounds and decrease exposure to environmental moisture (Bajpai et al., 2018; Hamad et al., 2018). During the processing of food materials, various internal components of the product such as protein, starch, and lipids are subjected to diverse processing conditions such as temperature, pressure, and mechanical properties. Hence, the use of various encapsulated biological entities via nanotechnology can help provide antibacterial and antifungal activities to food packaging materials. Various bioactive components such as docosahexanoic acid (DHA), vitamin B2 (riboflavin), vitamin D, vitamin E, and β-carotene using nanoemulsions and lipid nanoparticles have been shown to improve the shelf life of the stored food items as well as the antimicrobial efficacy of the packaging (Mihindukulasuriya & Lim, 2014).

6.5 CONCLUSION

India is a primarily an agrarian economy, where 70% of the population depends upon agricultural practices for their primary income source. With the increased population rate exacerbated by climate change and high pollution levels in air, water, and soil resources, food and cash crops are seriously affected. With advances in nanotechnology, various organically and inorganically based nanoformulations have shown their applicability in increasing the overall uptake of essential nutritional micro- and macromolecules directly to the plant cells. Further, management of a wide variety of pests, insects, and pathogens in this drug-resistant world has been advanced by research utilizing nanocomposites, nanoemulsions, and nanoformulations in the agro-food industry. The optical properties of nanoparticles are a well-known phenomenon where, due to quantum confinement, the nanoparticles show various optical properties with decreasing size ratio. Hence, the spoilage of food in various Tetra-packaged and -bottled food items can be easily detected through the application of nanoparticles. Additionally, nanotechnology has exhibited promising results in the food packaging and processing industry, where it helps protect the color, texture, aroma, and protein, lipid, and carbohydrate content of the packaged food material by increasing its shelf life. However, proper scanning and research should be conducted at regular intervals to alleviate the side effects of these nanoparticles which have directly or indirectly been incorporated into our daily lifestyles.

REFERENCES

Akoijam, R., Telem, R. S., & Marangmei, L. (2014). Insect pest problems and its changing trends on crop losses. *Environment and Ecology*, *32*(3), 837–841.

Altavilla, C., & Ciliberto, E. (2011). *Inorganic Nanoparticles: Synthesis, Applications, and Perspectives* (pp. 1–17). Altavilla, C., & Ciliberto, E., Editors. New York: CRC Press.

Ariga, K., Ji, Q., McShane, M. J., Lvov, Y. M., Vinu, A., & Hill, J. P. (2011). Inorganic nanoarchitectonics for biological applications. *Chemistry of Materials*, *24*(5), 728–737.

Bajpai, V. K., Kamle, M., Shukla, S., Mahato, D. K., Chandra, P., Hwang, S. K., … Han, Y.-K. (2018). Prospects of using nanotechnology for food preservation, safety, and security. *Journal of Food and Drug Analysis*, *26*(4), 1201–1214. doi: 10.1016/j.jfda.2018.06.011

Barik, T. K., Sahu, B., & Swain, V. (2008). Nanosilica-from medicine to pest control. *Parasitology Research*, *103*(2), 253–258. doi: 10.1007/s00436-008-0975-7

Bhatia, K., & Ghanem, H. (2019). Resource mobilization for agriculture and food security. In Clayton Campanhola and Shivaji Pandey (Eds.), *Sustainable Food and Agriculture* (pp. 523–541). Elsevier. doi: 10.1016/C2016-0-01212-3

Bird, F. A., Pradhan, A., Bhavani, R. V, & Dangour, A. D. (2019). Interventions in agriculture for nutrition outcomes: A systematic review focused on South Asia. *Food Policy*, *82*, 39–49. doi: 10.1016/j.foodpol.2018.10.015

Biswas, A., Bayer, I. S., Biris, A. S., Wang, T., Dervishi, E., & Faupel, F. (2012). Advances in top–down and bottom–up surface nanofabrication: Techniques, applications & future prospects. *Advances in Colloid and Interface Science*, *170*(1–2), 2–27.

Buzea, C., & Pacheco, I. (2017). Nanomaterials and their classification. In Ashutosh Kumar Shukla (Ed.), *EMR/ESR/EPR Spectroscopy for Characterization of Nanomaterials* (pp. 3–45). Springer.

Bzdek, B. R., Zordan, C. A., Luther III, G. W., & Johnston, M. V. (2011). Nanoparticle chemical composition during new particle formation. *Aerosol Science and Technology*, *45*(8), 1041–1048.

Calabi-Floody, M., Medina, J., Rumpel, C., Condron, L. M., Hernandez, M., Dumont, M., & de la Luz Mora, M. (2018). Chapter three – Smart fertilizers as a strategy for sustainable agriculture. In A. Sparks (Ed.), *Advances in Agronomy-* (Vol. 147, pp. 119–157). Academic Press. doi: 10.1016/bs.agron.2017.10.003

Cătălin Balaure, P., Gudovan, D., & Gudovan, I. (2017). 4 – Nanopesticides: A new paradigm in crop protection. In A. M. B. T.-N. P. and S. S. Grumezescu (Ed.) (pp. 129–192). Academic Press. doi: 10.1016/B978-0-12-804299-1.00005-9

Chen, J., Lü, S., Zhang, Z., Zhao, X., Li, X., Ning, P., & Liu, M. (2018). Environmentally friendly fertilizers: A review of materials used and their effects on the environment. *Science of the Total Environment*, *613–614*, 829–839. doi: 10.1016/j.scitotenv.2017.09.186

Chhipa, H. (2019). Applications of nanotechnology in agriculture. In V. Gurtler, A. S. Ball, & S. Soni (Eds.), *Methods in Microbiology* (Vol. 46, pp. 115–142). Academic Press. doi: 10.1016/bs.mim.2019.01.002

Couvreur, P. (2013). Nanoparticles in drug delivery: Past, present and future. *Advanced Drug Delivery Reviews*, *65*(1), 21–23.

Dasgupta, N., Ranjan, S., Mundekkad, D., Ramalingam, C., Shanker, R., & Kumar, A. (2015). Nanotechnology in agro-food: From field to plate. *Food Research International*, *69*, 381–400.

Dhaliwal, G. S., Jindal, V., & Dhawan, A. K. (2010). Insect pest problems and crop losses: Changing trends. *Indian Journal of Ecology*, *37*(1), 1–7.

Dolez, P. I. (2015). Nanomaterials definitions, classifications, and applications. In P. I. Dolez (Ed.), *Nanoengineering* (pp. 3–40). Elsevier. doi: 10.1016/B978-0-444-62747-6.00001-4

Dudefoi, W., Villares, A., Peyron, S., Moreau, C., Ropers, M.-H., Gontard, N., & Cathala, B. (2018). Nanoscience and nanotechnologies for biobased materials, packaging and food applications: New opportunities and concerns. *Innovative Food Science & Emerging Technologies*, *46*, 107–121. doi: 10.1016/j.ifset.2017.09.007

Duhan, J. S., Kumar, R., Kumar, N., Kaur, P., Nehra, K., & Duhan, S. (2017). Nanotechnology: The new perspective in precision agriculture. *Biotechnology Reports*, *15*, 11–23. doi: 10.1016/j.btre.2017.03.002

Ealia, S. A. M., & Saravanakumar, M. P. (2017). A review on the classification, characterisation, synthesis of nanoparticles and their application. In *IOP Conference Series: Materials Science and Engineering* (Vol. 263, p. 32019). IOP Publishing.

Elmer, W., Ma, C., & White, J. (2018). Nanoparticles for plant disease management. *Current Opinion in Environmental Science & Health*, *6*, 66–70. doi: 10.1016/j.coesh.2018.08.002

Emerick, K. (2018). Agricultural productivity and the sectoral reallocation of labor in rural India. *Journal of Development Economics*, *135*, 488–503.

Feregrino-Perez, A. A., Magaña-López, E., Guzmán, C., & Esquivel, K. (2018). A general overview of the benefits and possible negative effects of the nanotechnology in horticulture. *Scientia Horticulturae*, *238*, 126–137. doi: 10.1016/j.scienta.2018.03.060

Ferreira, A. J. D., Guilherme, R., Ferreira, C. S. S., & Oliveira, M. F. L. (2018). Urban Agriculture, a tool towards more resilient urban communities? *Current Opinion in Environmental Science & Health*, *5*, 93–97.

Fu, X., Cai, J., Zhang, X., Li, W.-D., Ge, H., & Hu, Y. (2018). Top-down fabrication of shape-controlled, monodisperse nanoparticles for biomedical applications. *Advanced Drug Delivery Reviews*, *132*, 169–187.

García, M., Forbe, T., & Gonzalez, E. (2010). Potential applications of nanotechnology in the agro-food sector. *Food Science and Technology*, *30*(3), 573–581.

Gasparotto, A., Barreca, D., Maccato, C., & Tondello, E. (2012). Manufacturing of inorganic nanomaterials: Concepts and perspectives. *Nanoscale*, *4*(9), 2813–2825.

Glezer, A. M. (2011). Structural classification of nanomaterials. *Russian Metallurgy (Metally)*, *2011*(4), 263.

Goswami, A., Roy, I., Sengupta, S., & Debnath, N. (2010). Novel applications of solid and liquid formulations of nanoparticles against insect pests and pathogens. *Thin Solid Films*, *519*(3), 1252–1257. doi: 10.1016/j.tsf.2010.08.079

Guo, Z., & Sadler, P. J. (1999). Metals in medicine. *Angewandte Chemie International Edition*, *38*(11), 1512–1531. doi: 10.1002/(SICI)1521-3773(19990601)38:11<1512::AID-ANIE1512>3.0.CO;2-Y

Hamad, A. F., Han, J.-H., Kim, B.-C., & Rather, I. A. (2018). The intertwine of nanotechnology with the food industry. *Saudi Journal of Biological Sciences*, *25*(1), 27–30. doi: 10.1016/j.sjbs.2017.09.004

Hayles, J., Johnson, L., Worthley, C., & Losic, D. (2017). Nanopesticides: A review of current research and perspectives. In A. M. Grumezescu (Ed.), *New Pesticides and Soil Sensors* (pp. 193–225). Academic Press. doi: 10.1016/B978-0-12-804299-1.00006-0

He, X., Deng, H., & Hwang, H. (2018). The current application of nanotechnology in food and agriculture. *Journal of Food and Drug Analysis*, *27*(1), 1–21. doi:10.1016/j.jfda.2018.12.002

Hofmann-Amtenbrink, M., Grainger, D. W., & Hofmann, H. (2015). Nanoparticles in medicine: Current challenges facing inorganic nanoparticle toxicity assessments and standardizations. *Nanomedicine: Nanotechnology, Biology, and Medicine*, *11*(7), 1689–1694. doi: 10.1016/j.nano.2015.05.005

Horikoshi, S., & Serpone, N. (2013). Introduction to nanoparticles. In S. Horikoshi & N. Serpone (Eds.), *Microwaves in Nanoparticle Synthesis: Fundamentals and Applications* (pp. 1–24). Wiley Online Library. doi: 10.1002/9783527648122.ch1

Hussain, I., Singh, A., Singh, N. B., Singh, A., & Singh, P. (2019). Plant-nanoceria interaction: Toxicity, accumulation, translocation and biotransformation. *South African Journal of Botany*, *121*, 239–247. doi: 10.1016/j.sajb.2018.11.013

Hyeon, T. (2003). Chemical synthesis of magnetic nanoparticles. *Chemical Communications*, 8: 927–934.

Iavicoli, I., Leso, V., Beezhold, D. H., & Shvedova, A. A. (2017). Nanotechnology in agriculture: Opportunities, toxicological implications, and occupational risks. *Toxicology and Applied Pharmacology*, *329*, 96–111. doi: 10.1016/j.taap.2017.05.025

Jagannathan, R., Irvin, G., Blanton, T., & Jagannathan, S. (2006). Organic nanoparticles: Preparation, self-assembly, and properties. *Advanced Functional Materials*, *16*(6), 747–753.

Jampílek, J., & Kráľová, K. (2017). Nanopesticides: Preparation, targeting, and controlled release. In A. M. Grumezescu (Ed.), *New Pesticides and Soil Sensors* (pp. 81–127). Academic Press. doi: 10.1016/B978-0-12-804299-1.00004-7

Juárez-Maldonado, A., González-Morales, S., Cabrera-De la Fuente, M., Medrano-Macías, J., & Benavides-Mendoza, A. (2018). Chapter 10 – Nanometals as promoters of nutraceutical quality in crop plants. In A. M. Grumezescu & A. M. B. T.-I. of N. in the F. I. Holban (Eds.), *Handbook of Food Bioengineering* (pp. 277–310). Academic Press. doi: 10.1016/B978-0-12-811441-4.00010-8

Kah, M., & Hofmann, T. (2014). Nanopesticide research: Current trends and future priorities. *Environment International*, *63*, 224–235. doi: 10.1016/j.envint.2013.11.015

Kharissova, O. V, Dias, H. V. R., Kharisov, B. I., Pérez, B. O., & Pérez, V. M. J. (2013). The greener synthesis of nanoparticles. *Trends in Biotechnology*, *31*(4), 240–248.

Liu, W.-T. (2006). Nanoparticles and their biological and environmental applications. *Journal of Bioscience and Bioengineering*, *102*(1), 1–7.

Martins Lopes, C., Fernandes, J. R., & Martins-Lopes, P. (2013). Application of nanotechnology in the agro-food sector. *Food Technology and Biotechnology*, *51*(2), 183–197.

McGuire, S. (2013). WHO, World Food Programme, and International Fund for Agricultural Development. 2012. The State of Food Insecurity in the World 2012. Economic growth is necessary but not sufficient to accelerate reduction of hunger and malnutrition. Rome, FAO. *Advances in Nutrition*, *4*(1), 126–127. doi: 10.3945/an.112.003343

Mihindukulasuriya, S. D. F., & Lim, L.-T. (2014). Nanotechnology development in food packaging: A review. *Trends in Food Science & Technology*, *40*(2), 149–167. doi: 10.1016/j.tifs.2014.09.009

Mohanraj, V. J., & Chen, Y. (2006). Nanoparticles: A review. *Tropical Journal of Pharmaceutical Research*, *5*(1), 561–573.

Mudshinge, S. R., Deore, A. B., Patil, S., & Bhalgat, C. M. (2011). Nanoparticles: Emerging carriers for drug delivery. *Saudi Pharmaceutical Journal*, *19*(3), 129–141.

Nair, R., Varghese, S. H., Nair, B. G., Maekawa, T., Yoshida, Y., & Kumar, D. S. (2010). Nanoparticulate material delivery to plants. *Plant Science*, *179*(3), 154–163. doi: 10.1016/j.plantsci.2010.04.012

Nayak, D., Pradhan, S., Ashe, S., Rauta, P. R., & Nayak, B. (2015). Biologically synthesised silver nanoparticles from three diverse family of plant extracts and their anticancer activity against epidermoid A431 carcinoma. *Journal of Colloid and Interface Science*, *457*, 329–339. doi: 10.1016/j.jcis.2015.07.012

Ngo, J., & Serra-Majem, L. (2019). Hunger and malnutrition. In P. Ferranti, E. M. Berry, & J. R. Anderson (Eds.), *Encyclopedia of Food Security and Sustainability* (pp. 315–335). Oxford: Elsevier. doi: 10.1016/B978-0-08-100596-5.22064-7

Njuguna, J., Ansari, F., Sachse, S., Zhu, H., & Rodriguez, V. M. (2014). Nanomaterials, nanofillers, and nanocomposites: Types and properties. In James Njuguna, K. Pielichowski, & H. Zhu (Eds.), *Health and Environmental Safety of Nanomaterials* (pp. 3–27). Elsevier. doi: 10.1533/9780857096678.1.3

Oskam, G. (2006). Metal oxide nanoparticles: Synthesis, characterization and application. *Journal of Sol-Gel Science and Technology*, *37*(3), 161–164.

Pal, D., Sahu, C. K., & Haldar, A. (2014). Bhasma: The ancient Indian nanomedicine. *Journal of Advanced Pharmaceutical Technology & Research*, *5*(1), 4–12. doi: 10.4103/2231-4040.126980

Panigrahi, S., Kundu, S., Ghosh, S., Nath, S., & Pal, T. (2004). General method of synthesis for metal nanoparticles. *Journal of Nanoparticle Research*, *6*(4), 411–414.

Parisi, C., Vigani, M., & Rodríguez-Cerezo, E. (2015). Agricultural nanotechnologies: What are the current possibilities? *Nano Today*, *10*(2), 124–127.

Pileni, M. P., Tanori, J., & Filankembo, A. (1997). Biomimetic strategies for the control of size, shape and self-organization of nanoparticles. *Colloids and Surfaces A: Physicochemical and Engineering Aspects*, *123*, 561–573.

Rai, P. K., Kumar, V., Lee, S., Raza, N., Kim, K.-H., Ok, Y. S., & Tsang, D. C. W. (2018). Nanoparticle-plant interaction: Implications in energy, environment, and agriculture. *Environment International*, *119*, 1–19. doi: 10.1016/j.envint.2018.06.012

Raliya, R., Saharan, V., Dimkpa, C., & Biswas, P. (2017). Nanofertilizer for precision and sustainable agriculture : Current state and future perspectives. *Journal of Agricultural and Food Chemistry*, *66*(August), 6487–6503. review-article. doi: 10.1021/acs.jafc.7b02178

Rastogi, S. (2010). Building bridges between Ayurveda and Modern Science. *International Journal of Ayurveda Research*, *1*(1), 41–46. doi: 10.4103/0974-7788.59943

Ravishankar Rai, V., & Jamuna Bai, A. (2011). *Nanoparticles and Their Potential Application as Antimicrobials*. A Méndez-Vilas, A., Editor. Mysore: Formatex.

Rico, C. M., Majumdar, S., Duarte-Gardea, M., Peralta-Videa, J. R., & Gardea-Torresdey, J. L. (2011). Interaction of nanoparticles with edible plants and their possible implications in the food chain. *Journal of Agricultural and Food Chemistry*, *59*(8), 3485–3498. doi: 10.1021/jf104517j

Sapkota, T. B., Vetter, S. H., Jat, M. L., Sirohi, S., Shirsath, P. B., Singh, R., … Stirling, C. M. (2019). Cost-effective opportunities for climate change mitigation in Indian agriculture. *Science of the Total Environment*, *655*, 1342–1354.

Sarkar, P. K., Das, S., & Prajapati, P. K. (2010). Ancient concept of metal pharmacology based on Ayurvedic literature. *Ancient Science of Life*, *29*(4), 1.

Sau, T. K., & Rogach, A. L. (2012). *Complex-shaped metal nanoparticles: Bottom-up syntheses and applications*. John Wiley & Sons.

Schmidt, H. (2001). Nanoparticles by chemical synthesis, processing to materials and innovative applications. *Applied Organometallic Chemistry*, *15*(5), 331–343.

Srivastava, P., Singh, R., Tripathi, S., & Raghubanshi, A. S. (2016). An urgent need for sustainable thinking in agriculture–An Indian scenario. *Ecological Indicators*, *67*, 611–622.

Sudha, P. N., Sangeetha, K., Vijayalakshmi, K., & Barhoum, A. (2018). Nanomaterials history, classification, unique properties, production and market. In A. Barhoum & A. S. H. Makhlouf (Eds.), *Emerging Applications of Nanoparticles and Architecture Nanostructures* (pp. 341–384). Elsevier. doi: 10.1016/B978-0-323-51254-1.00012-9

Teo, B. K., & Sun, X. H. (2007). Classification and representations of low-dimensional nanomaterials: Terms and symbols. *Journal of Cluster Science*, *18*(2), 346–357.

Thakkar, K. N., Mhatre, S. S., & Parikh, R. Y. (2010). Biological synthesis of metallic nanoparticles. *Nanomedicine: Nanotechnology, Biology and Medicine*, *6*(2), 257–262.

Tsuzuki, T. (2009). Commercial scale production of inorganic nanoparticles. *International Journal of Nanotechnology*, *6*(5–6), 567–578.

Vollath, D. (2008). An introduction to synthesis, properties and application. *Environmental Engineering and Management Journal*, *7*(6), 865–870.

Wang, Y., & Xia, Y. (2004). Bottom-up and top-down approaches to the synthesis of monodispersed spherical colloids of low melting-point metals. *Nano Letters*, *4*(10), 2047–2050.

Yamamoto, H., Ohnuma, A., Ohtani, B., & Kozawa, T. (2013). Formation of nanoscale reaction field using combination of top-down and bottom-up nanofabricaiton. *Microelectronic Engineering*, *110*, 369–373.

7 An Overview of Mycogenic Nanoparticles

Synthesis and Their Bio-Prospective Applications

Sagarika Mohanta, Sameer Kumar Singdevsachan, Yugal Kishore Mohanta, Sujogya Kumar Panda, and Tapan Kumar Mohanta

CONTENTS

7.1 Introduction 87
7.2 Types of Nanoparticle Synthesized by Using Fungus 89
7.2.1 Gold Nanoparticles 89
7.2.2 Silver Nanoparticles 89
7.2.3 Silicon Nanoparticles/Barium Nanoparticles/Titanium Nanoparticles 91
7.2.4 Cd/Se Nanoparticles 91
7.2.5 Zirconium Nanoparticles 91
7.2.6 Platinum Nanoparticles 91
7.2.7 Bismuth Nanoparticles 91
7.2.8 Magnetite Nanoparticles 91
7.2.9 Strontianite Nanoparticles 92
7.3 Possible Mechanisms for Mycogenic Nanoparticle Synthesis 92
7.4 Factors Affecting Mycogenic Synthesis of Nanoparticles 92
7.5 Synthesis Procedures and Characterizations of Nanoparticles 92
References 94

7.1 INTRODUCTION

Nanotechnology is an interdisciplinary science that exploits the principles of physics, chemistry, biology, materials science, and medicine (Durán and Marcato, 2013; Fakruddin et al., 2012; Krishna et al., 2013; Lyding, 2000; Maine et al., 2014). These fundamental principles are actively connected in the synthesis and fabrication of nanoparticles or nanostructures (Liu et al., 2013; Maab et al., 2012; Yin and Talapin, 2013). In recent years, nanotechnology has been considered a cutting-edge research platform that explores the nanoparticles/materials of prevailing applications in different fields like optics, electronics, catalysis, energy science, agriculture, environmental science, and medicine. The term "nano" signifies any considered parameter that is expressed as a dimension of 10^{-9} times in SI units. The primary perception of nanotechnology was first focused by Richard Feynman in a lecture entitled "There's Plenty of Room at the Bottom" at the American Institute of Technology in 1959. Additionally, the term "nanotechnology" was coined and published by Norino Taniguchi, Tokyo, Japan, in 1974. This new concept of technology gradually expanded to various fields with reliable potential applications. Additionally, bioremediation research initially neglected nano-sized zero-valence metal crystals involved in the process. However, when the mysterious potential of the nanoparticle was discovered, their applications also greatly increased. Nanoparticles are used as metal chips, electronic devices, semiconductor units, and different sophisticated biomedical applications (Gedamu et al., 2012; Goodfellow et al., 2015; Kumar, Abyaneh, et al., 2007; Kumar, Ansary, et al., 2007; Lima and Mano, 2015; Mohanta and Behera, 2014; McDevitt et al., 2016; Ngo et al., 2016; Poudyal et al., 2015). Metal nanoparticles (e.g., silver) are also extensively used as potential antimicrobial agents in many common places (Franci et al., 2015; Padmavathy and Vijayaraghavan, 2008). Additionally, they are used as antimicrobial agents in surgically implanted catheters to reduce infections caused during surgery and have been observed to possess anti-angiogenic, anti-fungal, anti-inflammatory, and anti-permeability activities (Gurunathan et al., 2015, 2009; Samuel and Guggenbichler, 2004; Sheikpranbabu et al., 2009).

Nanotechnology has had a great impact on biology, medicine, food, and agriculture, where it has evolved into a distinct discipline described as "nanobiotechnology" (Chaudhry et al., 2018;

Fakruddin et al., 2012; Jain, 2007; Salata, 2004). With the development and implementation of nanobiotechnology, microbially mediated synthesis of nanoparticles has become a "biological chemistry" approach that is environmentally sustainable and economically achievable (Bar et al., 2009; Iravani, 2011; Raveendran et al., 2003; Sharma et al., 2009; Vigneshwaran et al., 2006a). Overall, microbial nanotechnology is a biologically driven process that bridges nanotechnology and microbial biotechnology (Kalimuthu et al., 2008;Villaverde, 2010). In this context, several microorganisms have been explored for the synthesis of metal nanoparticles including Ag (El-Shanshoury et al., 2011), Au (Thirumurugan et al., 2012), CdS (Abd El-Raheem and El-Shanshoury, 2012), CdSe (S Anil Kumar et al., 2007), PbS (Kowshik et al., 2002), ZnO (Padmavathy and Vijayaraghavan, 2008), and ZnS (Malarkodi and Annadurai, 2013). The biological synthesis of metal nanoparticles by microorganisms has been extensively reviewed by several authors (Gade et al., 2010; Krumov et al., 2009; Vithiya and Sen, 2011). Additionally, intracellular and extracellular nanoparticle biological synthesis has been discussed elsewhere (Deljou and Goudarzi, 2016; Mukherjee et al., 2002; Narayanan and Sakthivel, 2010; Paknikar, 2003; Singh et al., 2016). Some metal sulfide nanoparticles such as ZnS, CdS, and PbS have been synthesized by bacteria, yeasts, and fungi and have been widely applied as semiconductor quantum dots (Durán et al., 2005). The microbial synthesis of metal sulfide nanoparticles was also recently reviewed (Duran and Seabra, 2012).

The biological synthesis of monodispersed metal nanoparticles of different sizes and shapes has been a great challenge in nanotechnology research. Although various physical and chemical methods have been extensively investigated to produce monodispersed nanoparticles, the stability and toxicity of chemicals used are of vital concern. Moreover, the routine use of toxic chemicals and non-polar solvents during the synthesis of nanoparticles limits their applications in the clinical and pharmaceutical fields (Oghabian and Farahbakhsh, 2010). Therefore, updates of hygienic, biocompatible, non-toxic, and eco-friendly methods for synthesis of metal nanoparticles is highly valued. Although biological methods are regarded as safe, worthwhile, sustainable, and environmentally friendly, they also have limitations associated with the culture of microbes, which is time-consuming and difficult. Although it is quite difficult to control size, distribution, shape, crystallinity, and monodispersity, appropriate strain selection along with optimization of conditions such as pH, temperature (growth and reaction incubation), time, metal ion concentration, and active amount of biological material can led to prospective implementation for large-scale production and commercial applications.

In the current thrust of nanoparticle synthesis, fungi have been drawing much attraction from researchers. Fungi are classified as multicellular eukaryotic heterotrophic organisms and have crucial roles in the nutrient cycling of the ecosystem. Interestingly, the reproduction system of fungi is both sexual and asexual and fungi have symbiotic modes of association/relationship with plants and bacteria (Duhan et al., 2017). Moreover, fungi secrete large quantities of enzymes (reductase enzymes), which are precisely involved in the bioreduction and stabilization of nanoparticles; hence large quantities of nanoparticles are produced. Due to the negative charge (electrostatic interactions) and sticky (adhesive substances) nature of the cell surface of the fungal biomass, the metal ions affix on the cell surface of the fungi. As reported, fungi biosynthesize nanoparticles both extracellularly and intracellularly. The extracellular synthesis refers to the immobilization of metal ions in a suitable carrier, whereas intracellular synthesis supports the formation of composite film (Afshar and Sedaghat, 2016). The secondary metabolites as well as enzymes from the fungal cells play crucial roles in converting toxic matter into non-toxic matter, by virtue of which mycogenic nanoparticles are safe and reliable for diverse applications.

Fungi are relatively more valuable than other microorganisms with respect to biological synthesis of nanoparticles in numerous ways:

- The capacity to withstand agitation, flow pressure, and other environmental circumstances in bioreactors of fungal mycelia correlated to bacteria and plants.
- They are fastidious and easy to handle, enabling apparent fabrication of silver nanoparticles.
- Extracellular production of reductive enzymes/proteins.
- Instant application of mycogenic nanoparticles due to extracellular synthesis.

In recent years, there have been several reports available on nanoparticle synthesis using fungi:

Aspergillus terreus
Aspergillus fumigatus
Aspergillus oryzae
Aspergillus niger
Aspergillus flavus
Aspergillus clavatus
Aspergillus conicus
Alternaria alternata
Agaricus bisporus
Acremonium diospyri
Amylomyces rouxii
Colletotrichum sp.
Chrysosporium keratinophilum
Coriolus versicolor
Cladosporium cladosporioides
Chlamydia Mucor Rouxii
Candida utilis
Cryptococcus humicola
Candida glabrata
Candida albicans
Chrysosporium
Epicoccum nigrum
Fusarium semitectum
Fusarium oxysporum
Fusarium solani
Gloeophyllum abietinum

Ganoderma sp.
Neurospora intermedia
Phomosis sp.
Pleurotus
Pleurotus florida
Pleurotus sajor-caju
Pleurotus ostreatus
Penicillium janthinellum
Penicillium verrucosum
Penicillium citrinum
Penicillium fellutanum
Penicillium chrysogenum
Paecilomyces lilacinus
Phoma glomerata
Penicillium nodositatum
Phanerochaete sordid
Phytophthora sp.
Pestalotia sp.
Pycnoporus sanguineus
Phanerochaete chrysosporium
Schizosaccharomyces pombe
Sclerotium rolfsii
Shigella dysenteriae
Trichothecium sp.
Trichoderma versicolor
Trichoderma harzianum
Trichoderma viride
Trichoderma reesei
Verticillium sp.

which have been explored for nanoparticle synthesis and various industrial applications (Ahmad et al., 2005; Balaji et al., 2009; Bhainsa and D'Souza, 2006; Birla et al., 2009; Gada et al., 2008; Kathiresan et al., 2009; Anil Kumar et al., 2007; Philip, 2009; Sanghi and Verma, 2009; Shaligram et al., 2009; S. Shankar et al., 2003; Vigneshwaran et al., 2006a).

7.2 TYPES OF NANOPARTICLE SYNTHESIZED BY USING FUNGUS

7.2.1 Gold Nanoparticles

Fungal species are actively involved in the synthesis of gold nanoparticles. In 2001, (Mukherjee et al. 2001) reported the biological synthesis of gold nanoparticles using eukaryotic microorganism *Verticillium* sp. (AAT-TS-4) of size 20 nm on both the surface and the cytoplasmic membrane of fungal mycelium. These formed nanoparticles are predominantly spherical, with a few triangular or hexagonal ones having well-defined dimensions and good dispersity. Moreover, in intracellular space, *Trichothecium* sp. was found to accumulate Au nanoparticles (Ahmad et al., 2005). *Verticillium luteoalbum* produced spherical particles of <10 nm in diameter as well as spheres and rods when incubated at pH 3.0 and pH 5.0 respectively (Gericke and Pinches, 2006a,b). Fungi are potential candidates of extracellular synthesis of nanoparticles because of their enormous secretory components, which actively involved in the capping and reduction process of nanoparticles. Rapid reduction of Au^{+3} to Au^0 nanoparticles by an endophytic fungus, *Colletotrichum* sp. isolated from the leaves of geranium plant (*Pelargonium graveolens*), produced nanoparticles that were spherical in shape and polydispersed (Shankar et al., 2003a). Glutathiones are potential candidates as capping agents of gold nanoparticles that bind either through free amine group or through cysteine residues (Gole et al., 2001). The fungus *Trichothecium* sp. reduced Au^{+3} to form gold nanoparticles in the range of 8–40 nm (Ahmad et al., 2005). A corresponding finding was also reported in the reduction of $Au^{+3}Cl_4$ to metallic gold by α-NADPH-dependent sulfite reductase (35.6 kDa) and phytochelatin (Table 7.1) (Anil Kumar et al., 2007; Kumar, Abyaneh, et al., 2007).

7.2.2 Silver Nanoparticles

Intracellular and extracellular metabolites of fungi are also highly concerned to produce silver nanoparticles (AgNPs). Biomass of *Verticillium* sp. resulted in the accumulation of silver nanoparticles upon acquaintance to aqueous $AgNO_3$ solution (Mukherjee et al., 2001). Vigneshwaran et al. (2007) also studied the use of *A. flavus* in the aggregation of 8-nm silver nanoparticles on the surface of its cell walls. The white rot fungus *P. chrysosporium* synthesized silver nanoparticles of 50–200 nm on the surface of the mycelium. This mechanism confirmed that the presence of reductase enzymes on the surface of the mycelium is involved in the reduction of ions to silver nanoparticles (Vigneshwaran et al., 2006b). The production of monodispersed silver nanoparticles of 5–25 nm by *A. fumigates* has also been reported (Bhainsa and D'Souza, 2006). Culture filtrate of *F. semitectum* produced silver nanoparticles of 10–60 nm with spherical morphology (Basavaraja et al., 2008). Similarly, *A. niger* sequestered from soil produced spherical Ag nanoparticles of 20 nm by action of nitrate reductase enzyme and quinine during extracellular electron transfer (Gada et al., 2008). In addition, a phytopathogenic fungus isolated from onion, *F. solani* (USM-3790), formed spherical and polydispersed silver nanoparticles of 16.23 nm (Ingle et al., 2008). Spherical silver nanoparticles of 5–40 nm were synthesized by *F. acuminatum* EII. The non-pathogenic *T. asperellum*, a potential biocontrol agent, produced well-defined morphology and stable silver nanoparticles of 13–18 nm (Mukherjee et al., 2008). The fungal filtrate of *P. glomerata* biosynthesized spherical silver nanoparticles (60–80 nm) with moderate anti-bacterial activity against *E. coli*, *Pseudomonas aeruginosa*, and *Staphylococcus aureus* (Birla et al., 2009). The white rot fungus, *C. versicolor*, produced proteins that acted as reducing and capping agents to synthesize silver nanoparticles (Sanghi and Verma, 2009). Rhizospheric fungi from mangrove root-soil *Rhizophora annamalayana* and *P. fellutanum* were found to produce silver nanoparticles of spherical shape and size from 5 to 25 nm (Kathiresan et al., 2009). The synthesis of extracellular silver nanoparticles by the fungus *C. cladosporioides* of polydispersed and spherical particles was reported (Balaji et al., 2009). The biological synthesis and antimicrobial activity of

TABLE 7.1
Different Fungi Exploited for Synthesis of Metal Nanoparticles

Sl. No	Organisms and responsible biomolecules	Type(s) of metal nanoparticle	Size (nm)	Morphology	References
01	*Verticillium* sp.	Au	20±8	Cell wall/spherical, cytoplasmic membrane/quasi-hexagonal	(Mukherjee et al., 2001)
02	*Trichothecium* sp.	Au	ND	ND	(Ahmad et al., 2005)
03	*Verticillium luteoalbum* and isolate 6-3	Au	<10	Spherical, spheres and rods	(Gericke and Pinches, 2006b,a)
04	*Verticillium* sp.	Ag	25±12	Cell wall, cytoplasmic membrane/spherical	(Mukherjee et al., 2001; Senapati et al., 2005)
05	*Aspergillus flavus*	Ag	8. 92	Cell wall	(Vigneshwaran et al., 2007)
06	*Fusarium oxysporum*	Au	20–40	Spherical, triangular	(Mukherjee et al., 2002)
07	*Colletotrichum* sp.	Au	20–40	Spherical	(Shankar et al., 2003)
08	*Trichothecium* sp.	Au	5–200	Triangle, hexagonal	(Ahmad et al., 2005)
09	*Trichoderma asperellum*	Ag	13–18	ND	(Mukherjee et al., 2008)
10	*Trichoderma viride*	Ag	5–40	Spherical, rod-like	(Fayaz et al., 2010)
11	*Fusarium oxysporum*	Ag	5–50	ND	(Mukherjee et al., 2001)
12	*Phanerochaete chrysosporium*	Ag	5–200	Pyramidal	(Vigneshwaran et al., 2006b)
13	*Fusarium solani* USM 3790	Ag	16.23	Spherical	(Ingle et al., 2008)
14	*Fusarium semitectum*	Ag	10–60	Spherical	(Basavaraja et al., 2008)
15	*Fusarium acuminatum* Ell. & Ev.	Ag	5–40	Spherical	(Ingle et al., 2008)
16	*Aspergillus fumigatus*	Ag	5–25	Spherical, triangular	(Bhainsa and D'Souza, 2006)
17	*Coriolus versicolor*	Ag	25–75, 444–491	Spherical	(Sanghi and Verma, 2009)
18	*Aspergillus niger*	Ag	20	Spherical	(Gada et al., 2008)
19	*Phoma glomerata*	Ag	60–80	Spherical	(Birla et al., 2009)
20	*Penicillium brevicompactum*	Ag	58.35±17.88	ND	(Shaligram et al., 2009)
21	*Cladosporium cladosporioides*	Ag	10–100	Spherical	(Balaji et al., 2009)
22	*Penicillium fellutanum*	Ag	5–25	Spherical	(Kathiresan et al., 2009)
23	*Fusarium oxysporum*	Au–Ag	8–14	ND	(Senapati et al., 2005)
24	*Volvariella volvacea*	Au, Ag, Au–Ag	20–150	Spherical, hexagonal	(Philip, 2009)
25	*Fusarium oxysporum*	Si	5–15	Quasi-spherical	(Bansal et al., 2005)
26	*Fusarium oxysporum*	Ti	6–13	Spherical	(Bansal et al., 2005)
27	*Fusarium oxysporum*	Zr	3–11	Quasi-spherical	(Bansal et al., 2004)
28	*Fusarium oxysporum* f.sp. *Lycopersici*	Pt	10–50	Triangles, hexagons, squares, rectangles	(Whiteley et al., 2006)
29	*F. oxysporum*	Magnetite	20–50	Quasi-spherical	(Bharde et al., 2006)
30	*Verticillum* sp.	Magnetite	100–400	Cubo-octahedral	(Bharde et al., 2006)
31	*Fusarium oxysporum*	CdSe	9–15	Spherical	(S Anil Kumar et al., 2007)
32	*Fusarium oxysporum*	$SrCO_3$	ND[1]	Needle morphology	(Rautaray et al., 2004)
33	*Fusarium oxysporum*	$BaTiO_3$	4±1	Quasi-spherical	(Bharde et al., 2006)
34	*Fusarium oxysporum* 66- and 10-kDa protein	Au			(Mukherjee et al., 2002)
35	*Fusarium oxysporum* α-NADPH-dependent sulfite reductase 35.6 kDa and phytochelatin	Au			(Anil Kumar et al., 2007)
36	*Colletotrichum* sp. Glutathiones	Au			(Shankar et al., 2003)
37	*Fusarium oxysporum* α-NADPH-dependent nitrate reductase 44 kDa and phytochelatin	Ag			(Anil Kumar et al., 2007)
38	*Fusarium oxysporum* 24–28 kDa protein	Zr			(Bansal et al., 2004)

(Continued)

TABLE 7.1 (CONTINUED)
Different Fungi Exploited for Synthesis of Metal Nanoparticles

Sl. No	Organisms and responsible biomolecules	Type(s) of metal nanoparticle	Size (nm)	Morphology	References
39	*Fusarium oxysporum* dimeric hydrogenase 44.5 and 39.4 kDa	Pt			(Govender et al., 2010)
40	*Aspergillus niger* nitrate reductase	Ag			(Gada et al., 2008)
41	*Penicillium brevicompactum* compactin	Ag			(Shaligram et al., 2009)
42	*Verticillium* sp. 55- and 13-kDa protein	Magnetite			(Bharde et al., 2006)

[1] not determined

silver nanoparticles by *T. viride* was studied by Fayaz et al. (2010). A biomolecule compactin isolated from the fungal isolate *P. brevicompactum* WA2315 was employed in the biological synthesis of silver nanoparticles of 23 nm (Shaligram et al., 2009). The culture extract of the saprophytic straw mushroom fungus, *Volvariella volvacea*, was involved in the production of gold, silver, and gold–silver nanoparticles (Philip, 2009). It is believed that the formation nanoparticles of different crystal shapes was facilitated by proteins, polysaccharides, and organic acids secreted by the fungus. Recently, in vitro synthesis of silver hydrosol of 10–25 nm was accomplished though the enzymatic route adopting α-NADPH-dependent nitrate reductase (44 kDa) from *F. oxysporum* (Anil Kumar et al., 2007). Durán et al. (2005) explained the involvement of quinine derivates of napthoquinones and anthraquinones as redox centers apart from enzymes in the reduction of silver nanoparticles. Moreover, *F. oxysporum* is also involved in the formation of silver nanoparticles as well as bimetallic gold–silver (Au–Ag) alloy nanoparticles for potential use in the biomedical sector (Table 7.1) (Senapati et al., 2005).

7.2.3 Silicon Nanoparticles/Barium Nanoparticles/Titanium Nanoparticles

F. oxysporum shaped silica and titania nanoparticles through SiF_6^{-2} and TiF_6^{-2} anionic complexes that ultimately facilitated syntheses of crystalline silica by calcinations at 300°C and of titania at room temperature (Bansal et al., 2005). In addition, it produced irregular quasi-spherical morphology of $BaTiO_3$ nanoparticles with an average size of 4 ± 1 nm. The crystalline nature of the tetragonal phase of nanoparticles was confirmed by Selected Area Diffraction (SAED) analysis, which can be useful in microelectronics (Table 7.1) (Mukherjee et al., 2002).

7.2.4 Cd/Se Nanoparticles

Upon incubation with $CdCl_2$ and $SeCl_4$, *F. oxysporum* formed highly luminescent water-soluble CdSe quantum dots with polydispersed spherical morphology and sizes of 9–15 nm (Table 7.1) (Kumar, Ansary, et al., 2007).

7.2.5 Zirconium Nanoparticles

Cationic proteins secreted by *F. oxysporum* produced zirconia nanoparticles when incubated with ZrF_6^{-2} anions. The particles were quasi-spherical in shape (3–11 nm), and the formation of zirconia nanoparticles was found to be facilitated by 24- to 28-kDa protein (Bansal et al., 2004) (Table 7.1).

7.2.6 Platinum Nanoparticles

Platinum nanoparticles with triangular, hexagonal, square, and rectangular morphologies were formed by *F. oxysporum* intercellular mode on the cell wall or membrane and by the extracellular mode in the medium, at sizes of 10–50 nm. It has also been reported that a dimeric hydrogenase enzyme (44.5 and 39.4 kDa) of *F. oxysporum* reduced H_2PtCl_6 to platinum nanoparticles (Table 7.1) (Govender et al., 2010).

7.2.7 Bismuth Nanoparticles

The deposition of bismuth on the periphery of *Yersinia enterocolitica* 8081c cultures with bismuth subsalicylate (BSS) has been reported (Nadeau et al., 1992). *F. oxysporum* was also reported to produce Bi_2O_3 nanocrystals of 5–8 nm (extracellular) with quasi-spherical morphology (Table 7.1) (Uddin et al., 2008).

7.2.8 Magnetite Nanoparticles

The fungus *Verticillum* sp. was actively involved in the formation of iron oxide nanoparticles of 50–400 nm. *F. oxysporum* produced crystalline magnetite nanoparticles with combined salts of $K_3(Fe(CN)_6)$ and $K_4(Fe(CN)_6)$ having quasi-spherical shapes and of 20–50 nm in size (Table 7.1) (Bharde et al., 2006).

7.2.9 Strontianite Nanoparticles

F. oxysporum is the only the source for carbonate ions that reacted with strontium ions to produce strontianite ($SrCO_3$) crystals having needle like quasi-linear superstructures (Table 7.1) (Rautaray et al., 2004).

7.3 POSSIBLE MECHANISMS FOR MYCOGENIC NANOPARTICLE SYNTHESIS

Although extensive research has been undertaken to discover the possible mechanisms of mycogenic nanoparticle synthesis, there are still unanswered questions being incubated in the scientific community (Alghuthaymi et al., 2015; Durán et al., 2005; Meyer, 2008). A number of probable alternative mechanisms exist in fungi, which may superficially explain the reduction process of the metal/metal salts to synthesize nanoparticles. In the extracellular synthesis process, three possible mechanisms are assumed to be followed: (1) nitrate reductase enzymatic action, (2) electron shuttle quinones, or (3) both. Mainly nitrate-dependent reductase and α-NADPH-dependent reductases were reported to be responsible for the synthesis of nanoparticles in fungi. Besides this enzymatic mechanism, the extracellular shuttle quinone also a responsible asset in the nanoparticle synthesis. During intracellular synthesis, the cell wall and sugars present in the cell wall are prone to play a crucial role in the absorption followed with reduction of metal ions by the enzymes within the cell wall, which edge to the aggregation of metal ion and nanoparticle formations (Figure 7.1). It has been studied that metal ($AuCl^{-4}$) ions were trapped by charged amino acids of enzymes present within the fungal cell wall through an electrostatic interaction process (Khandel and Kumar, 2018). Jain et al. (2011) reported that AgNP synthesis using *A. flavus* occurs initially by an unnamed protein (33 kDa) followed by cystein and free amine groups that form capping agents to stabilize the nanoparticles. Several research reports established the involvement of nitrate reductase for extracellular synthesis of NPs (Ingle et al., 2008; Vigneshwaran et al., 2007). The cellular complexity of the fungi encourages more investigation to explore the exact mechanism of nanoparticle synthesis to design a sustainable process for large scale-up (Alghuthaymi et al., 2015).

7.4 FACTORS AFFECTING MYCOGENIC SYNTHESIS OF NANOPARTICLES

As fungi are micro size in nature, they are always in a good symbiotic relationship with hosts and habitats or the open environment in which they are found, and their growth and development are highly influenced by the concerned environmental conditions (Alghuthaymi et al., 2015; Khandel and Kumar, 2018). As the biotic and abiotic factors of the environment affect microbial cellular metabolism, the production of metabolites, viz., enzymes, secondary metabolites, and proteins, is affected by the medium in which they are cultivated. Therefore, the optimization studies of the fungi support better growth as well as product yield enhancement, and some research studies reveal the effects of culture conditions on the mycosynthesis of nanoparticles (Ingle et al., 2008; Klaus et al., 1999). It is noteworthy that mycosynthesis is precisely overwhelmed by incubation conditions, viz., pH, temperature, incubation time, concentration of precursors (metal salts), fungal biomass concentration (Jain et al., 2011; Kumar et al., 2008) (Figure 7.2), and the condition of colloidal interaction. All these common factors control the shape, size, localization, and dispersity of the nanoparticles formed. Moreover, factors such as pH and temperature can be maneuvered for triggering the geometry of nanoparticles (Durán et al., 2005).

7.5 SYNTHESIS PROCEDURES AND CHARACTERIZATIONS OF NANOPARTICLES

The synthesis of nanoparticles using any biological entities is known as biological synthesis or a green (using individual natural

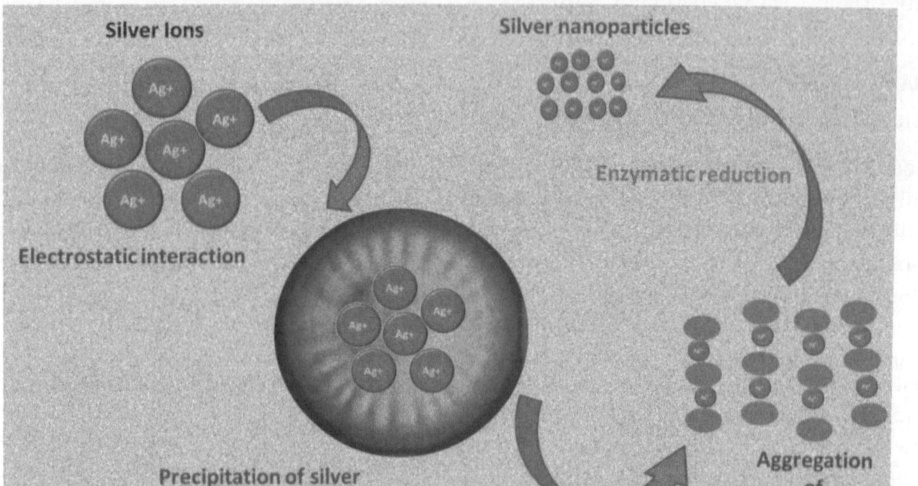

FIGURE 7.1 Mechanism of intracellular synthesis of nanoparticles in fungus (electrostatic interaction between the metal ions and the cell wall enzymes of fungus).

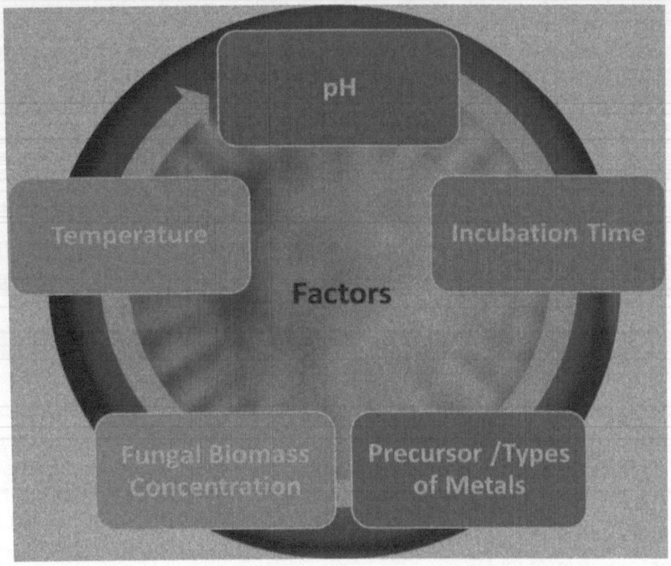

FIGURE 7.2 Factors affecting the mycogenic synthesis of nanoparticles.

TABLE 7.2
Sketch of Characterization Techniques and Properties of Nanoparticles

Characterization techniques	Characterization properties/results
AFM, SEM-HR-SEM, SEM–EDX	Morphology and dispersion of nanoparticles in cells and other matrices/supports, lateral dimensions of nanoparticles, elemental composition
BET (Brunauer–Emmett–Teller)	Accessible surface area
Cryo-TEM	Study complex growth mechanisms, aggregation pathways
DSC (Differential scanning calorimetry)	Size and size distribution
DLS (Dynamica light scattering)	Hydrodynamic size and surface charge, detection of agglomerates
EBSD (Electron backscatter diffraction)	Structure, crystal orientation and phase of materials in SEM, examine microstructures, texture, defects, grain morphology, deformation
EDS/EDX (Energy Dispersive X-ray)	Chemical characterization, investigation of a sample through interactions between light and matter and analyzing X-rays in its particular case
EELS (Electro energy loss spectroscopy)	Energy transformation, change in momentum and ionization potential of an atom, type and quantity of atoms present, chemical state of atoms, collective interactions of atoms with neighbors, bulk plasmon resonance
Electron tomography	Realistic 3D particle visualization, snapshots, video, quantitative information down to the atomic scale
FTIR (Fourier-transform infrared spectroscopy)	Surface composition, ligand binding
HR-TEM/TEM	Electronic core structure, ligand density and arrangement, atomic composition, influence of ligands on nanoparticle shape and size, crystal structure of single particles, differentiation of monocrystalline, polycrystalline, and amorphous nanoparticles
ICP-MS	Elemental composition, size, size distribution, nanoparticle concentration
Liquid TEM	Depict nanoparticle growth in real time, study growth mechanism, single particle motion, super-lattice formation
MALDI (Matrix-assisted laser desorption/ionization)	Surface topography, size of nanoparticle
MFM (Magnetic force microscopy)	Magnetic moments of single nanoparticles, trace magnetic nanoparticles in the interior of cells
NMR	Ligand density and arrangement, electronic core structure, atomic composition, influence of ligands on nanoparticle shape, nanoparticle size
NTA (Nanoparticle tracking analysis)	Nanoparticle size and size distribution
SAXS (Small angle X-ray scattering)	Particle size, size distribution, growth kinetics
TGA (Thermogravimetric analysis)	Mass and composition of stabilizers
TG–DTA	Weight loss and thermal effect during the conversion of precursors to final method oxides in the heat-treatment process

(Continued)

TABLE 7.2 (CONTINUED)
Sketch of Characterization Techniques and Properties of Nanoparticles

Characterization techniques	Characterization properties/results
UV–visible spectrophotometer	Conformation of synthesis, optical properties, size, concentration, agglomeration state, hints on nanoparticle shape
XAS (X-ray absorption spectroscopy)	X-ray absorption coefficient (element-specific)—chemical state of species, interatomic distances, Debye–Waller factors, also for non-crystalline nanoparticles
XPS (X-ray photoelctron spectroscopy)	Electronic structure, elemental composition, oxidation states, ligand binding (surface-sensitive)
XRD	Composition, crystal structure, crystalline grain size

compounds as reducing agent) synthesis method. In the mycosynthesis process, the fungi or fungal mycelia are actively involved in the complete synthesis process. Here, the fungal extracts are used as reducing and stabilizing agents for the synthesis of different nanoparticles using respective precursor molecules.

Characterization of nanoparticles is an indispensable step in material science research to study the different forms or types of synthesized nanoparticles, including their size, shape, and structure. Nanomaterials are of unique structures and sizes which are highly promising for various important biological functions in the food, agriculture, health care, and environmental sectors (Chaudhry et al., 2018; Fortunati and Balestra, 2018; Khandel and Kumar, 2018). Moreover, unique or innovative findings of processes and phenomena of materials at the micro- and nano-scales will bring favorable new circumstances for the assembly of novel nanosystems and nanostructures (Fortunati and Balestra, 2018). The characterization of nanoparticles is furnished by various molecular techniques, viz., UV-visible spectroscopy, Fourier transform infrared spectroscopy (FTIR), dynamic light scattering spectroscopy (DLS), X-ray diffraction technique (XRD), field-emission scanning electron microscopy (FE-SEM), high-resolution transmission electron microscopy (HR-TEM), atomic force microscopy (AFM), energy-dispersive X-ray spectroscopy (EDX), etc. (Table 7.2).

REFERENCES

Abd El-Raheem, R. El-Shanshoury, Elsilk, Sobhy E., Ebeid, Mohamed E., 2012. Rapid biosynthesis of cadmium sulfide (CdS) nanoparticles using culture supernatants of Escherichia coli ATCC 8739, Bacillus subtilis ATCC 6633 and Lactobacillus acidophilus DSMZ 20079T. *African J. Biotechnol.* 11, 7957–7965. doi: 10.5897/AJB11.3708

Afshar, P., Sedaghat, S., 2016. Bio-synthesis of silver nanoparticles using water extract of satureja hortensis L and evaluation of the antibacterial properties. *Curr. Nanosci.* doi: 10.2174/15734137 11666150529202238

Ahmad, A., Senapati, S., Khan, M.I., Kumar, R., Sastry, M., 2005. Extra-/intracellular biosynthesis of gold nanoparticles by an alkalotolerant fungus, trichothecium sp. *J. Biomed. Nanotechnol.* 1, 47–53. doi: 10.1166/jbn.2005.012

Alghuthaymi, M.A., Almoammar, H., Rai, M., Said-, E., Abd-elsalam, K.A. Alghuthaymi, M.A., Almoammar, H., Rai, M., 2015. Myconanoparticles : Synthesis and their role in phytopathogens management. *Biotechnol. Biotechnol. Equip.* 29, 221–236. doi: 10.1080/13102818.2015.1008194

Anil Kumar, S., Abyaneh, M.K., Gosavi, S.W., Kulkarni, S.K., Pasricha, R., Ahmad, A., Khan, M.I., 2007. Nitrate reductase-mediated synthesis of silver nanoparticles from $AgNO_3$. *Biotechnol. Lett.* 29, 439–445. doi: 10.1007/s10529-006-9256-7

Balaji, D., Basavaraja, S., Deshpande, R., Mahesh, D., Prabhakar, B., Venkataraman, A., 2009. Extracellular biosynthesis of functionalized silver nanoparticles by strains of Cladosporium cladosporioides fungus. *Colloids Surf. B Biointerfaces* 68, 88–92. doi: 10.1016/j.colsurfb.2008.09.022

Bansal, V., Rautaray, D., Ahmad, A., Sastry, M., 2004. Biosynthesis of zirconia nanoparticles using the fungus Fusarium oxysporum. *J. Mater. Chem.* 14, 3303–3305. doi: 10.1039/B407904C

Bansal, V., Rautaray, D., Bharde, A., Ahire, K., Sanyal, A., Ahmad, A., Sastry, M., 2005. Fungus-mediated biosynthesis of silica and titania particles. *J. Mater. Chem.* 15, 2583–2589. doi: 10.1039/B503008K

Bar, H., Bhui, D.K., Sahoo, G.P., Sarkar, P., De, S.P., Misra, A., 2009. Green synthesis of silver nanoparticles using latex of Jatropha curcas. *Colloids Surf. A Physicochem. Eng. Asp.* 339, 134–139. doi: 10.1016/j.colsurfa.2009.02.008

Basavaraja, S., Balaji, S.D., Lagashetty, A., Rajasab, A.H., Venkataraman, A., 2008. Extracellular biosynthesis of silver nanoparticles using the fungus Fusarium semitectum. *Mater. Res. Bull.* 43, 1164–1170. doi: 10.1016/j.materresbull.2007.06.020

Bhainsa, K.C., D'Souza, S.F., 2006. Extracellular biosynthesis of silver nanoparticles using the fungus Aspergillus fumigatus. *Colloids Surf. B. Biointerfaces* 47, 160–164. doi: 10.1016/j.colsurfb.2005.11.026

Bharde, A., Rautaray, D., Bansal, V., Ahmad, A., Sarkar, I., Yusuf, S.M., Sanyal, M., Sastry, M., 2006. Extracellular biosynthesis of magnetite using fungi. *Small* 2, 135–141. doi: 10.1002/smll.200500180

Birla, S.S., Tiwari, V. V, Gade, A.K., Ingle, A.P., Yadav, A.P., Rai, M.K., 2009. Fabrication of silver nanoparticles by Phoma glomerata and its combined effect against Escherichia coli, Pseudomonas aeruginosa and Staphylococcus aureus. *Lett. Appl. Microbiol.* 48, 173–179. doi: 10.1111/j.1472-765X.2008.02510.x

Chaudhry, N., Dwivedi, S., Chaudhry, V., Singh, A., Saquib, Q., 2018. Microbial pathogenesis bio-inspired nanomaterials in agriculture and food : Current status, foreseen applications and challenges. *Microb. Pathog.* 123, 196–200. doi: 10.1016/j.micpath.2018.07.013

Deljou, A., Goudarzi, S., 2016. Green extracellular synthesis of the silver nanoparticles using thermophilic bacillus Sp. AZ1 and its antimicrobial activity against several human pathogenetic bacteria. *Indian J. Biotechnol.* 14, 25–32. doi: 10.15171/ijb.1259

Duhan, J.S., Kumar, R., Kumar, N., Kaur, P., Nehra, K., Duhan, S., 2017. Nanotechnology: The new perspective in precision agriculture. *Biotechnol. Rep.* 15, 11–23. doi: 10.1016/j.btre.2017.03.002

Duran, N., Seabra, A., 2012. Microbial syntheses of metallic sulfide nanoparticles: An overview. *Curr. Biotechnol.* 4, 287–296.

Durán, N., Marcato, P.D., 2013. Nanobiotechnology perspectives. Role of nanotechnology in the food industry: A review. *Int. J. Food Sci. Technol.* 48, 1127–1134. doi: 10.1111/ijfs.12027

Durán, N., Marcato, P.D., Alves, O.L., De Souza, G.I.H., Esposito, E., 2005. Mechanistic aspects of biosynthesis of silver nanoparticles by several Fusarium oxysporum strains. *J. Nanobiotechnol.* 3, 8. doi: 10.1186/1477-3155-3-8

El-Shanshoury, A.E.-R.R., ElSilk, S.E., Ebeid, M.E., 2011. Extracellular biosynthesis of silver nanoparticles using Escherichia coli ATCC 8739, Bacillus subtilis ATCC 6633, and Streptococcus thermophilus ESh1 and their antimicrobial activities. *ISRN Nanotechnol.* 2011, 1–7. doi: 10.5402/2011/385480

Fakruddin, M., Hossain, Z., Afroz, H., 2012. Prospects and applications of nanobiotechnology: A medical perspective. *J. Nanobiotechnol.* 10, 1–8. doi: 10.1186/1477-3155-10-31

Fayaz, A.M., Balaji, K., Girilal, M., Yadav, R., Kalaichelvan, P.T., Venketesan, R., 2010. Biogenic synthesis of silver nanoparticles and their synergistic effect with antibiotics: A study against gram-positive and gram-negative bacteria. *Nanomed. Nanotechnol. Biol. Med.* 6, 103–109. doi: 10.1016/j.nano.2009.04.006

Fortunati, E., Balestra, G., 2018. Overview of novel and sustainable antimicrobial nanomaterials for agri-food applications. *Nanomed. Nanotechnol. J.* 2, 115.

Franci, G., Falanga, A., Galdiero, S., Palomba, L., Rai, M., Morelli, G., Galdiero, M., 2015. Silver nanoparticles as potential antibacterial agents. *Molecules* 20, 8856–8874. doi: 10.3390/molecules20058856

Gada, A., Bonde, P., Ingle, A., Marcato, P., Duran, N., Rai, M., 2008. Exploitation of aspergillus niger for synthesis of silver nanoparticles. *J. Biobased Mater. Bioenergy* 2, 243–247.

Gade, A., Ingle, A., Whiteley, C., Rai, M., 2010. Mycogenic metal nanoparticles: Progress and applications. *Biotechnol. Lett.* 32, 593–600. doi: 10.1007/s10529-009-0197-9

Gedamu, D., Paulowicz, I., Jebril, S., Kumar Mishra, Y., Adelung, R., 2012. Procedures and properties for a direct nano-micro integration of metal and semiconductor nanowires on Si chips. *J. Nanotechnol.* 2012. doi: 10.1155/2012/325732

Gericke, M., Pinches, A., 2006a. Microbial production of gold nanoparticles. *Gold Bull* 39, 22–28.

Gericke, M., Pinches, A., 2006b. Biological synthesis of metal nanoparticles. *Hydrometallurgy* 83, 132–140.

Gole, A., Dash, C., Soman, C., Sainkar, S.R., Rao, M., Sastry, M., 2001. On the preparation, characterization, and enzymatic activity of fungal protease–gold colloid bioconjugates. *Bioconjug. Chem.* 12, 684–690. doi: 10.1021/bc0001241

Goodfellow, K.M., Chakraborty, C., Beams, R., Novotny, L., Vamivakas, A.N., 2015. Direct on-chip optical plasmon detection with an atomically thin semiconductor. *Nano Lett.* 15, 5477–5481. doi: 10.1021/acs.nanolett.5b01898

Govender, Y., Riddin, T.L., Gericke, M., Whiteley, C.G., 2010. On the enzymatic formation of platinum nanoparticles. *J. Nanoparticle Res.* 12, 261–271. doi: 10.1007/s11051-009-9604-3

Gurunathan, S., Lee, K.-J., Kalishwaralal, K., Sheikpranbabu, S., Vaidyanathan, R., Eom, S.H., 2009. Antiangiogenic properties of silver nanoparticles. *Biomaterials* 30, 6341–6350. doi: 10.1016/j.biomaterials.2009.08.008

Gurunathan, S., Jeong, J.-K., Han, J.W., Zhang, X.-F., Park, J.H., Kim, J.-H., 2015. Multidimensional effects of biologically synthesized silver nanoparticles in Helicobacter pylori, Helicobacter felis, and human lung (L132) and lung carcinoma A549 cells. *Nanoscale Res. Lett.* 10, 35. doi: 10.1186/s11671-015-0747-0

Hersam, M. C., Guisinger, N. P., Lyding, J. W., 2000. Silicon-based molecular nanotechnology. *Nanotechnology* 11 (2), 70.

Ingle, A., Rai, M., Gade, A., Bawaskar, M., 2008. Fusarium solani: A novel biological agent for the extracellular synthesis of silver nanoparticles. *J. Nanoparticle Res.* 11, 2079–2085. doi: 10.1007/s11051-008-9573-y

Iravani, S., 2011. Green synthesis of metal nanoparticles using plants. *Green Chem.* 13, 2638–2650. doi: 10.1039/C1GC15386B

Jain, K.K., 2007. Applications of nanobiotechnology in clinical diagnostics. *Clin. Chem.* 53, 2002–2009. doi: 10.1373/clinchem.2007.090795

Jain, N., Bhargava, A., Majumdar, S., Tarafdar, J.C., Panwar, J., 2011. Extracellular biosynthesis and characterization of silver nanoparticles using Aspergillus flavusNJP08: A mechanism perspective. *Nanoscale* 3, 635–641. doi: 10.1039/C0NR00656D

Kalimuthu, K., Suresh Babu, R., Venkataraman, D., Bilal, M., Gurunathan, S., 2008. Biosynthesis of silver nanocrystals by Bacillus licheniformis. *Colloids Surf. B. Biointerfaces* 65, 150–153. doi: 10.1016/j.colsurfb.2008.02.018

Kathiresan, K., Manivannan, S., Nabeel, M.A., Dhivya, B., 2009. Studies on silver nanoparticles synthesized by a marine fungus, Penicillium fellutanum isolated from coastal mangrove sediment. *Colloids Surf. B. Biointerfaces* 71, 133–137. doi: 10.1016/j.colsurfb.2009.01.016

Khandel, P., Kumar, S., 2018. Mycogenic nanoparticles and their bio – Prospective applications : Current status and future challenges. *J. Nanostructure Chem.* 8, 369–391. doi: 10.1007/s40097-018-0285-2

Klaus, T., Joerger, R., Olsson, E., Granqvist, C.-G., 1999. Silver-based crystalline nanoparticles, microbially fabricated. *Proc. Natl. Acad. Sci.* 96, 13611–13614. doi: 10.1073/pnas.96.24.13611

Kowshik, M., Vogel, W., Urban, J., Kulkarni, S.K., Paknikar, K.M., 2002. Microbial synthesis of semiconductor PbS nanocrystallites. *Adv. Mater.* 14, 815–818. doi: 10.1002/1521-4095(20020605)14:11<815::AID-ADMA815>3.0.CO;2-K

Krishna, K.V., Ménard-Moyon, C., Verma, S., Bianco, A., 2013. Graphene-based nanomaterials for nanobiotechnology and biomedical applications. *Nanomedicine* 8, 1669–1688. doi: 10.2217/nnm.13.140

Krumov, N., Perner-Nochta, I., Oder, S., Gotcheva, V., Angelov, A., Posten, C., 2009. Production of inorganic nanoparticles by microorganisms. *Chem. Eng. Technol.* 32, 1026–1035. doi: 10.1002/ceat.200900046

Kumar, R., Liu, D., Zhang, L., 2008. Advances in proteinous biomaterials. *J. Biobased Mater. Bioenergy* 2, 1–24. 10.1166/jbmb.2008.204

Kumar, S.A., Abyaneh, M.K., Gosavi, S.W., Kulkarni, S.K., Ahmad, A., Khan, M.I., 2007. Sulfite reductase-mediated synthesis of gold nanoparticles capped with phytochelatin. *Biotechnol. Appl. Biochem.* 47, 191–5. doi: 10.1042/BA20060205

Kumar, S.A., Ansary, A.A., Ahmad, A., Khan, M.I., 2007. Extracellular biosynthesis of CdSe quantum dots by the fungus, Fusarium oxysporum. *Quantum* 3, 190–194. doi: 10.1166/jbn.2007.027

Lima, A.C., Mano, J.F., 2015. Micro/nano-structured superhydrophobic surfaces in the biomedical field: Part II: Applications overview. *Nanomedicine* 10, 271–297. doi: 10.2217/nnm.14.175

Liu, Y., Goebl, J., Yin, Y., 2013. Templated synthesis of nanostructured materials. Chem. Soc. Rev. 42, 2610–2653. doi: 10.1039/C2CS35369E

Kowshik, M., Ashtaputre, S., Kharrazi, S., Vogel, W., Urban, J., Kulkarni, S. K., Paknikar, K. M., 2003. Extracellular synthesis of silver nanoparticles by a silver-tolerant yeast strain MKY3. *Nanotechnology* 14, 95.

Maab, H., Francis, L., Al-saadi, A., Aubry, C., Ghaffour, N., Amy, G., Nunes, S.P., 2012. Synthesis and fabrication of nanostructured hydrophobic polyazole membranes for low-energy water recovery. *J. Memb. Sci.* 423–424, 11–19. doi: 10.1016/j.memsci.2012.07.009

Maine, E., Thomas, V.J., Bliemel, M., Murira, A., Utterback, J., 2014. The emergence of the nanobiotechnology industry. *Nat. Nanotechnol.* 9, 2–5.

Malarkodi, C., Annadurai, G., 2013. A novel biological approach on extracellular synthesis and characterization of semiconductor zinc sulfide nanoparticles. *Appl. Nanosci.* 3, 389–395. doi: 10.1007/s13204-012-0138-0

McDevitt, J.T., De La Garza Ii, R., Newton, T.F., Smith, R., Mahoney, J.J., Hohenstein, J., Gomez, S., Floriano, P.N., Talavera, H., Sloan, D.J., Moody, D.E., Andrenyak, D.M., Kosten, T.R., Christodoulides, N., 2016. Programmable bio-nano-chips for the quantitation of drugs of abuse in oral fluids. *Drug Alcohol Depend.* 140, e140. doi: 10.1016/j.drugalcdep.2014.02.398

McRae, M., Simmons, G., Wong, J., McDevitt, J., 2015. Programmable bio-nano-chip system: A flexible diagnostic platform that learns. *J. Biosens. Bioelectron.* 6, e137. doi: 10.4172/2155-6210.1000e137

Meyer, V., 2008. Genetic engineering of filamentous fungi — Progress, obstacles and future trends. *Biotechnol. Adv.* 26, 177–185. doi: 10.1016/j.biotechadv.2007.12.001

Mohanta, Y.K., Behera, S.K., 2014. Biosynthesis, characterization and antimicrobial activity of silver nanoparticles by Streptomyces sp. SS2. *Bioprocess Biosyst. Eng.* 37, 2263–2269. doi: 10.1007/s00449-014-1205-6

Mukherjee, P., Ahmad, A., Mandal, D., Senapati, S., Sainkar, S.R., Khan, M.I., Parishcha, R., Ajaykumar, P. V., Alam, M., Kumar, R., Sastry, M., 2001. Fungus-mediated synthesis of silver nanoparticles and their immobilization in the mycelial matrix: A novel biological approach to nanoparticle synthesis. *Nano Lett.* 1, 515–519. doi: 10.1021/nl0155274

Mukherjee, P., Senapati, S., Mandal, D., Ahmad, A., Khan, M.I., Kumar, R., Sastry, M., 2002. Extracellular Synthesis of Gold Nanoparticles by the fungus Fusarium oxysporum. *Chembiochem* 3, 461–463. doi: 10.1002/1439-7633(20020503)3:5<461::AID-CBIC461>3.0.CO;2-X

Mukherjee, S., Roy, M., Mandal, B., Dey, G., Mukherjee, P., Ghatak, J., Tyagi, A., Kale, S., 2008. Green synthesis of highly stabilized nanocrystalline silver particles by a non-pathogenic and agriculturally important fungus T. asperellum. *Nanotechnology* 19, 075103.

Nadeau, O.W., Gump, D.W., Hendricks, G.M., Meyer, D.H., 1992. Deposition of bismuth by Yersinia enterocolitica. *Med. Microbiol. Immunol.* 181, 145–152. doi: 10.1007/BF00202054

Narayanan, K.B., Sakthivel, N., 2010. Biological synthesis of metal nanoparticles by microbes. *Adv. Colloid Interface Sci.* 156, 1–13. doi: 10.1016/j.cis.2010.02.001

Ngo, H.T., Wang, H.-N., Fales, A.M., Vo-Dinh, T., 2016. Plasmonic SERS biosensing nanochips for DNA detection. *Anal. Bioanal. Chem.* 408, 1773–1781. doi: 10.1007/s00216-015-9121-4

Oghabian, M.A., Farahbakhsh, N.M., 2010. Potential use of nanoparticle based contrast agents in MRI: A molecular imaging perspective. *J. Biomed. Nanotechnol.* 6, 203–213. doi: 10.1166/jbn.2010.1119

Padmavathy, N., Vijayaraghavan, R., 2008. Enhanced bioactivity of ZnO nanoparticles—An antimicrobial study. *Sci. Technol. Adv. Mater.* 9, 035004. doi: 10.1088/1468-6996/9/3/035004

Philip, D., 2009. Biosynthesis of Au, Ag and Au–Ag nanoparticles using edible mushroom extract. Spectrochim. *Acta Part A Mol. Biomol. Spectrosc.* 73, 374–381. doi: 10.1016/j.saa.2009.02.037

Poudyal, N., Gandha, K., Elkins, K., Ping Liu, J., 2015. Anisotropic SmCo5/FeCo core/shell nanocomposite chips prepared via electroless coating. *AIMS Mater. Sci.* 2, 294–302. doi: 10.3934/matersci.2015.3.294

Rautaray, D., Sanyal, A., Adyanthaya, S.D., Ahmad, A., Sastry, M., 2004. Biological synthesis of strontium carbonate crystals using the fungus Fusarium oxysporum. *Langmuir* 20, 6827–6833. doi: 10.1021/la049244d

Raveendran, P., Fu, J., Wallen, S.L., 2003. Completely "green" synthesis and stabilization of metal nanoparticles. *J. Am. Chem. Soc.* 125, 13940–13941. doi: 10.1021/ja029267j

Salata, O. V, 2004. Applications of nanoparticles in biology and medicine. *J. Nanobiotechnol.* 2, 1–6. doi: 10.1186/1477-3155-2-3

Samuel, U., Guggenbichler, J.P., 2004. Prevention of catheter-related infections: The potential of a new nano-silver impregnated catheter. *Int. J. Antimicrob. Agents* 23, Suppl, 75–78. doi: 10.1016/j.ijantimicag.2003.12.004

Sanghi, R., Verma, P., 2009. Biomimetic synthesis and characterisation of protein capped silver nanoparticles. *Bioresour. Technol.* 100, 501–504. doi: 10.1016/j.biortech.2008.05.048

Senapati, S., Ahmad, A., Khan, M.I., Sastry, M., Kumar, R., 2005. Extracellular biosynthesis of bimetallic Au– Ag alloy nanoparticles. *Small* 1, 517–520. doi: 10.1002/smll.200400053

Shaligram, N.S., Bule, M., Bhambure, R., Singhal, R.S., Kumar, S., Szakacs, G., Pandey, A., 2009. Biosynthesis of silver nanoparticles using aqueous extract from the compactin producing fungal strain. *Process Biochem.* 44, 939–943. doi: 10.1016/j.procbio.2009.04.009

Shankar, S., Ahmad, A., Pasricha, R., Sastry, M., 2003a. Bioreduction of chloroaurate ions by geranium leaves and its endophytic fungus yields gold nanoparticles of different shapes. *J. Mater. Chem.* 13, 1822–1826.

Shankar, S.S., Ahmad, A., Pasricha, R., Sastry, M., 2003b. Bioreduction of chloroaurate ions by geranium leaves and its endophytic fungus yields gold nanoparticles of different shapes. *J. Mater. Chem.* 13, 1822–1826. doi: 10.1039/B303808B

Sharma, V.K., Yngard, R.A., Lin, Y., 2009. Silver nanoparticles: Green synthesis and their antimicrobial activities. *Adv. Colloid Interface Sci.* 145, 83–96

Sheikpranbabu, S., Kalishwaralal, K., Venkataraman, D., Eom, S.H., Park, J., Gurunathan, S., 2009. Silver nanoparticles inhibit VEGF-and IL-1beta-induced vascular permeability via Src dependent pathway in porcine retinal endothelial cells. *J. Nanobiotechnol.* 7, 8. doi: 10.1186/1477-3155-7-8

Singh, P., Kim, Y.-J., Zhang, D., Yang, D.-C., 2016. Biological synthesis of nanoparticles from plants and microorganisms. *Trends Biotechnol.* 34, 588–599.

Thirumurugan, A., Ramachandran, S., Tomy, N.A., Jiflin, G.J., Rajagomathi, G., 2012. Biological synthesis of gold nanoparticles by Bacillus subtilis and evaluation of increased antimicrobial activity against clinical isolates. *Korean J. Chem. Eng.* 29, 1761–1765. doi: 10.1007/s11814-012-0055-7

Uddin, I., Adyanthaya, S., Syed, A., Selvaraj, K., Ahmad, A., Poddar, P., 2008. Structure and microbial synthesis of sub-10 nm Bi_2O_3 nanocrystals. *J. Nanosci. Nanotechnol.* 8, 3909–3913. doi: 10.1166/jnn.2008.179

Vigneshwaran, N., Kathe, A.A., Varadarajan, P. V, Nachane, R.P., Balasubramanya, R.H., 2006a. Biomimetics of silver nanoparticles by white rot fungus, Phaenerochaete chrysosporium. *Colloids Surfaces B Biointerfaces* 53, 55–59. doi: 10.1016/j.colsurfb.2006.07.014

Vigneshwaran, N., Nachane, R.P., Balasubramanya, R.H., Varadarajan, P. V, 2006b. A novel one-pot 'green' synthesis of stable silver nanoparticles using soluble starch. *Carbohydr. Res.* 341, 2012–2018. doi: 10.1016/j.carres.2006.04.042

Vigneshwaran, N., Ashtaputre, N.M., Varadarajan, P. V, Nachane, R.P., Paralikar, K.M., Balasubramanya, R.H., 2007. Biological synthesis of silver nanoparticles using the fungus Aspergillus flavus. *Mater. Lett.* 61, 1413–1418. doi: 10.1016/j.matlet.2006.07.042

Villaverde, A., 2010. Nanotechnology, bionanotechnology and microbial cell factories. *Microb. Cell Fact.* 9, 1–4. doi: 10.1186/1475-2859-9-53

Vithiya, K., Sen, S., 2011. Biosynthesis of nano-particles. *Int. J. Pharm. Sci. Res.* 2(11), 2781.

Whiteley, C.G., Riddin, T.L., Gericke, M., 2006. Analysis of the inter- and extracellular formation of platinum nanoparticles by Fusarium oxysporum f. sp. lycopersici using response surface methodology. *Nanotechnology* 17, 3482.

Yin, Y., Talapin, D., 2013. The chemistry of functional nanomaterials. *Chem. Soc. Rev.* 42, 2484–2487. doi: 10.1039/C3CS90011H

8 Nanonutraceutical Delivery Systems
Application in Food and Feed Industry

Visha Pasuvalingam

CONTENTS

8.1 Introduction 100
8.2 Nutraceuticals 100
8.2.1 Vitamins 100
8.2.2 Bioactive Lipids 101
8.2.3 Probiotics 101
8.2.4 Phytochemical Nutraceuticals 101
8.2.5 Essential Oils 101
8.2.6 Algal Nutraceuticals 101
8.3 Functionality and Bioavailability of Nutraceuticals 101
8.4 Desirable Characteristics of Delivery Systems 101
8.5 Techniques for Nanodelivery of Nutraceuticals 102
8.5.1 Mechanical Processes 102
8.5.2 Chemical Processes 102
8.6 Nanonutraceutical Delivery Materials and Formulations 103
8.6.1 Polysaccharide-Based Nanodelivery Systems 103
8.6.1.1 Chitosan 103
8.6.1.2 Polyethylene Glycol (PEG) 104
8.6.2 Food Protein-Based Nanodelivery Systems 104
8.6.2.1 Milk Proteins and Peptides 104
8.6.3 Peptide Nanotubes 104
8.6.3.1 Gelatin 104
8.6.4 Lipid-Based Nanodelivery Systems 104
8.6.4.1 Liposomes 105
8.6.4.2 Solid Lipid Nanoparticles and Nanostructured Lipid Carriers 105
8.6.4.3 Nanoemulsions 105
8.6.4.4 Micelles 105
8.6.5 Nanodelivery System for Vitamins 106
8.6.6 Phytochemical-Loaded Nanoformulations 106
8.7 Characterization of Nanonutraceutical Delivery Systems 106
8.7.1 Morphology and Particle Size Analysis 106
8.7.1.1 Atomic Force Microscopy 106
8.7.1.2 Tunneling Electron Microscopy 106
8.7.1.3 Scanning Electron Microscopy 106
8.7.1.4 Dynamic Light Scattering 106
8.7.1.5 Zeta Potential 107
8.7.1.6 X-Ray Diffraction 107
8.8 Factors Affecting the Nutraceutical Delivery 107
8.8.1 Nanoparticle Size 107
8.8.2 Surface Characteristics 107
8.8.3 Nature of Surfactants and Building Blocks of Nanoparticles 107
8.8.4 Reticuloendothelial System Clearance 107
8.9 Nanonutrient Uptake and Targeted Delivery 107
8.9.1 Targeting of Nutraceuticals to the Intestine 107
8.9.2 Targeting of Nutraceuticals to the Liver 108
8.9.3 Subcellular Organelle Targeting 108

8.10 Safety Evaluation of Nanodelivery Systems.......... 108
8.11 Current Trends in Marketed Nanoemulsified Nutraceuticals.......... 108
8.12 Challenges and Research Gaps.......... 109
8.13 Ethical and Economical Issues.......... 109
8.14 Future Trends of Nanonutraceuticals.......... 109
8.15 Uniform Regulatory Requirements.......... 110
8.16 Public Perception in Regard to Nanotechnology.......... 110
8.17 Conclusions.......... 110
References.......... 110

8.1 INTRODUCTION

In recent years, nutraceuticals have drawn the attention of researchers as well as consumers due to their ability to prevent and treat a number of diseases like diabetes, cancers, hypertension, cardiovascular disorders, and even neurodegenerative disorders like Alzheimer's disease. Nutraceuticals have inherent health-beneficial potentials like antioxidant properties along with anti-inflammatory, antimicrobial, and immune-potentiating effects, which support their use as therapeutic agents. These bioactive compounds have wide differences in their molecular, physiochemical, and physiological properties. During the incorporation of bioactive compounds into functional foods, solubility, flavor, taste, texture issues, inactivation on exposure to gastrointestinal tract (GIT) environment, and reduced physiological functionality are the most common problems to be encountered. In addition, they are not always easily compatible with food matrix and processing conditions. Thus, the low bioavailability associated with nutraceuticals in the site of action, either due to exposure to harsh conditions during processing and storage or due to their degradation in the GIT, ultimately leading to poor biological functional activity, is one of the major concerns for researchers and nutraceutical scientists today.

In order to address this issue, nanotechnology techniques are being explored and effectively utilized by researchers for the efficient delivery of the nutraceuticals, thus enhancing their biological activity. Various nanoformulations like nanoemulsions (NEs), micelles, nanocapsules, solid lipid nanoparticles (SLNs), nanochelates, liposomes, self-assembled nanoparticles, etc., have been synthesized for the efficient delivery of the nutraceuticals. Designed nanoparticles serve as carriers of bioactive compounds or are bioactive themselves, thus enhancing the stability, delivery, uptake, and bioavailability of the compounds. These nanoformulations ensure targeted delivery and sustained release of the bioactive compounds besides improving its bioavailability and hence the functionality and therapeutic efficacy.

In some cases, scientists have been successful in formulating various nanodelivery modules. Most of these designed nanoparticles have been evaluated in vitro, with meager in vivo studies in animal models. Further extensive in vivo studies still need to be carried out to prove their efficacy and safety. Moreover, industrial scientists and researchers should also ponder about the toxicity potential and safety aspects of these nanoencapsulated nutraceuticals for the human system and the environment. Thus, prior to the release of these bioactive nanoformulations for human use, strict globalized regulations are the need of the hour. Hence, in the coming years, the nanonutraceuticals are sure to revolutionize the food and feed industries.

8.2 NUTRACEUTICALS

Nutraceuticals are components of feed and foods which provide nutritional, functional, and therapeutic benefits to the body. Nutraceuticals include dietary supplements and functional foods. The dietary supplements include vitamins and minerals, herbal elements, plant extracts, and proteins, whereas functional foods consist of omega fatty acids and probiotics (Trottier et al., 2010). Nutraceuticals are proven to enhance human health prevent chronic diseases and even to have therapeutic benefits (Bernal et al., 2011). Numerous therapeutic benefits such as antioxidant activity, immune enhancement, anti-inflammatory, antidiabetic, antiobesity, and cardiovascular protective effects are possessed by nutraceuticals (Slattery et al., 2004). The most common and widely used nutraceuticals are omega-3 fatty acids, vitamins, carotenoids, polyphenols (anthocyanins, curcumin, flavanones, isoflavones, ellagic acid, and proanthocyanidins), and probiotic bacteria (Zaki, 2014). These compounds find significant application in various industries including food/feed supplements, pharmaceuticals, cosmetics, and the dairy industry.

8.2.1 Vitamins

Vitamins are widely consumed functional food supplements required for proper functioning and regulation of cellular metabolism. Vitamins act as cofactors and coenzymes which are involved in basic cellular metabolic reactions. These nutrients are included into a wide range of food products such as cereals, flours, bread, soy milk, salt, and formulated beverages. Food fortification and dietary manipulations to incorporate vitamins are commonly practiced to boost immunity and minimize the risk of chronic diseases. Thirteen different types of vitamins are broadly classified into water- and fat-soluble vitamins based on solubility characteristics. Vitamins A, D, E, and K are the fat-soluble vitamins. Water-soluble vitamins include vitamin C and the B vitamins (thiamine, riboflavin, niacin, pyridoxine, folacin, cyanobalamin, pantothenic acid, and biotin).

8.2.2 Bioactive Lipids

Bioactive lipids play essential roles as signaling and regulatory molecules. In particular, sphingolipids, which form a major component of cell membranes, play a vital role in regulating cell transport across the membrane, cellular growth, senescence, and adhesion (Hannun and Obeid, 2008). Medium-chain triglycerides comprising six to eight carbons help in weight reduction, alleviate a wide range of cardiovascular ailments, and serve as an ideal candidate in preventing Type 2 diabetes. Oleic acid is advocated in the reduction of neurodegenerative diseases such as dementia and Alzheimer's disease and also in the management of colorectal cancer. Polyunsaturated fatty acids such as linolenic and linoleic acids have broad health-promoting beneficial effects such as protection against chronic diseases such as cardiovascular disorders, diabetes, cancer, atherosclerosis, platelet aggregation, and a variety of disorders caused by inflammation (Aluko, 2012).

8.2.3 Probiotics

Probiotics are comprised of myriad live microorganisms that confer health benefits to the host upon consumption in adequate amounts. The normal human alimentary tract harbors around 400 types of probiotic bacteria. Owing to their h ealth benefits, probiotic bacteria have been commonly incorporated into a wide range of dairy products including milk drinks, yogurt, cheese, ice cream, and dairy desserts. Lactobacilli and bifidobacteria are the most commonly used probiotics.

8.2.4 Phytochemical Nutraceuticals

The lipophilic steroid alcohols, phytosterols derived from plants, are chemically similar to cholesterol and exhibit antioxidant, anti-inflammatory, hypocholesterimic, and anticancer properties. These phytochemical nutraceuticals are available as tablets and capsules. The most common among the phytochemicals are isoflavones, sitosterol, genistein, gallic acid, curcumin, β-carotene, and zeaxanthin.

8.2.5 Essential Oils

Essential oils are natural aromatic hydrophobic products extracted from plants and are commonly used in the pharmaceutical, food, and flavoring industries. However, the major obstacles to their application in beverage and agro-food production are related to their very high volatility and high sensitivity to oxidation on exposure to air or UV radiation (Koo et al., 2014).

8.2.6 Algal Nutraceuticals

Currently, various algae have widespread utility in the food/feed supplement industry. Functional polysaccharides present in *Chlorella*, *Spirulina*, and red microalgae contain zeaxanthin, dietary fibers, minerals, polyunsaturated fatty acids (PUFAs), vitamins, and proteins which have health-promoting functions including cholesterol-lowering effects.

8.3 FUNCTIONALITY AND BIOAVAILABILITY OF NUTRACEUTICALS

The adequacy of any nutraceutical product relies mainly upon its active bioavailability, referring to the rate and extent to which it reaches the site of absorption and hence presents an actual functional quantity to the body. Thus, with increasing demand for nutraceuticals, their bioavailability has become a primary concern of the manufacturers, regulators, and users of health-related products (Rapaka and Coates, 2006). Apart from their bioavailability aspects, nutraceutical formulations need to overcome other challenges such as poor water solubility, crystallization, chemical instability, oxidative degradation, and instability in conditions encountered during food processing and storage, before their incorporation into economic food products (Augustin et al., 2012). Further, on oral administration, various characteristics and conditions such as inadequate gastric residence time, loss of activity on exposure to acid/alkaline pH in the GIT, and low permeability and/or solubility within the GI tract limit the activity and health benefits of various nutraceuticals. Big challenges of integrate nutraceuticals are (1) their chemical instability in food products (especially β-carotene, curcumin, vitamins A, D, E, and K) during food processing or storage (e.g., sensitivity to light, heat and oxygen) or within the GI tract (e.g., propensity to be degraded by enzymes, pH) and (2) low water solubility of nutrients leading to low bioavailability. Such crucial factors can affect the expected functionality of nutraceuticals and consequently, the real health benefits may not be appreciated even when nutraceuticals ingested in high amounts. Furthermore, nutraceuticals with distasteful sensory properties (e.g., polyunsaturated fatty acids) can negatively affect the food characteristics and shelf-life when directly incorporated into food products. Similarly, some of the compounds (e.g., vitamin C) have unsavory interactions with other important food components, which adversely affect the appearance, texture, taste, stability, and bioavailability of other important components in the food product.

8.4 DESIRABLE CHARACTERISTICS OF DELIVERY SYSTEMS

Delivery systems should be appropriately fashioned to save value-added nutraceuticals from undesirable cross-interactions with other chemical ingredients and physical entities like pH, light, temperature, or oxygen during nutraceutical processing, storage, and transport. Moreover, potential delivery systems should be promising in terms of protecting the nutraceuticals from normal as well as adverse GI environments (e.g., pro-active digestive enzymes and strong acidity/alkalinity of stomach and intestine) (McClements et al., 2007). Hence, the delivery system should be as below:

- Physiochemically stable for biotic/abiotic stresses while retaining its biofunctionality.
- Augmentative of gastric stability of labile bioactive nutrients.

- Supportive of constant dosage level in systemic circulation.
- With expanded gastric retention time.
- Capable of facilitating lymphatic transport of highly lipophilic compounds (McClements, 2012).

The successful formulation of suitable delivery modules for nutraceuticals needs extensive information about their core properties, use of adequate raw materials, and adept production techniques. Hence, designing appropriate food-grade delivery systems for nutraceuticals having adequate physiochemical stability, technological feasibility, and cost effectiveness involves various challenges. In addition, achieving an acceptable bioavailability for such nutraceuticals is an even more demanding task.

Formulating new delivery systems adapting nanotechnology has been a highly efficient trend for improving the effectiveness of nutraceuticals. The nanotechnology is mainly used to fabricate delivery systems for the least stable and most poorly water-solubilized nutraceuticals. So, one of the major applications of nanotechnology in the food industry is to plan and promote novel functional food components (nutraceuticals) with improved characteristics of high water solubility as well thermal stability and oral bioavailability along with sensory attributes and physiological functionalities. Driven by the ever-increasing consumer awareness and demand for novel food products, both the nutraceutical industry and market have grown in the area of nano-functionalized food fortification. This technology offers great possibilities for commercialization of the bioactive usage by overwhelming the associated restraints and promises a multifold escalation in the utility of this technology in forthcoming years (Augustin and Hemar, 2009). Oral delivery via organic or inorganic polymeric nanoparticles is an attractive route to transfer the therapeutics and bioactive compounds due to its ease of administration and patient compliance for prolonged periods. Moreover, the encapsulation and controlled release of active food ingredients are other significant applications in food industry achievable with nanotechnological approaches. On evaluating the practical advantages of incorporating nanotechnology into the agro-food industry, it has been observed that nanoparticles have novel additional properties that highly deserve to be explored further.

8.5 TECHNIQUES FOR NANODELIVERY OF NUTRACEUTICALS

Generally, two basic approaches to design nanoparticle delivery systems are employed. The first is the **"top-down"** approach, wherein small particles are produced by various size reduction (mechanical) processes, and the next is the **"bottom-up"** approach, wherein the nanoparticle module is synthesized by the self-assembly of smaller biomolecules, viz. lipids and proteins (chemical processes). However, amid the current technological advancement, there is a growing trend to combine **bottom-up** and **top-down** approaches to produce personalized nanoparticle-based delivery systems. The most common mechanical and chemical processes and resulting delivery formulations are described here.

8.5.1 Mechanical Processes

The mechanical process involves techniques that use shear force or particle collisions as the energy source to disintegrate the larger particles into smaller nanoscale aggregates. There are two types of mechanical processes employed in the factory for the nanonization of solid particles, and two types of microfluidic processes for the nanonization of liquids or melts. And as expected, there are contemporary challenges in designing nanoparticle systems using mechanical processes. The challenges include creating high-energy-density "bursts" to inculcate the particles so as to prevent re-aggregation. Very minimal use of chemical additives is an advantage of this process over chemical processes, thereby diminishing concerns regarding the regulations imposed on such formulation components.

8.5.2 Chemical Processes

The chemical process is adopted in those nanoparticle synthesis techniques where chemical reactions or the self-assembly of surfactant and polymers act as primary drivers. In this process, the input of mechanical energy is mostly limited to keeping the suspension fully mixed and to prevent agglomeration and settling down. There are typically five entities involved in the chemical methods: (1) the solute of interest, (2) an internal solvent, (3) an external solvent (mostly water), (4) a surfactant that is dispersed in the external solvent and in some formulations, and (5) a polymer that is soluble in the internal solvent but insoluble in the external one. Three types of internal solvent-based methods have been adopted: (1) lipophilic solvent, (2) amphiphilic solvent, and (3) hydrophilic solvent systems.

The lipophilic solvent method is primarily a process of emulsification/homogenization wherein the presence of a surfactant and/or polymers reduces the energy required for emulsification (by reducing the interfacial tension) and protects the nanodroplets against coalescence. The external solvent is later evaporated through spray drying or lyophilization techniques. SLNs are designed using this lipophilic solvent method. Other lipophilic solvent methods involve the use of self-emulsifying systems that rely on a micro-emulsion phase interaction. The amphiphilic solvent method consists of dissolving the solute in a polar organic (internal) solvent such as acetone or methylene chloride (containing a pre-dissolved lipophilic polymer), and mixing this with an aqueous solution having a surfactant or hydrocolloids. The affinity between the internal and external solvents causes the emulsification to occur spontaneously and, upon addition of more water, the internal solvent diffuses out of the emulsion drop into the aqueous phase, inducing the precipitation of the lipophilic polymer and finally the aggregation of solute in nanoscale. This method has been recently used to produce 20- to 80-nm β-carotene nanoparticles using acetone as the amphiphilic solvent, poly-(lacticco-glycolic) acid as stabilizing polymer, and either Tween® 20 or gelatin as emulsifier. The spontaneous emulsification solvent diffusion method is a small variant of this amphiphilic solvent method.

The hydrophilic (internal) solvent method involves the usage of water-soluble alcohols as the internal solvent. Here, the organic solute and a stabilizing polymer are dissolved in alcohol, and upon mixing with an aqueous solution containing the emulsifier, nanodroplets are spontaneously formed. Soon after emulsification, the alcohol (internal hydrophilic solvent) diffuses into water (without the need of adding external water). As the alcohol diffuses out of the droplet, the solute precipitates in the lipophilic polymer matrix. Another take on the hydrophilic solvent method is the use of polymers or hydrocolloids that could precipitate upon changes of pH, temperature, or electrolyte composition. In recent years, there has been growing interest in using these novel methods to transform globulin proteins and casein micelles into nanocapsules for the delivery of hydrophobic nutraceuticals and hydrophilic minerals.

Among the various chemical processing methods, the most suitable ones, with marked potential related to the distribution of bioactive compounds, are lipid-based delivery systems, which may include SLNs, liposomes, emulsions, and micelles as well as protein- and polysaccharide-based biopolymeric nanoparticles. The specific structural properties of these nano-systems, their low cost, and their non-toxic nature make them suitable carriers in food delivery as compared to others.

Polymer-based delivery systems that trap molecules of interest within networks have been developed extensively for the biomedical and pharmaceutical sectors to protect, transport, and target the released bioactive compounds to the site of action. The major mechanisms involved in the enhanced bioactive molecule absorption by polymeric nanoparticles are as follows:

1. Protection of the bioactive molecule from the harsh environment of the GI tract
2. Prolongation of the residence time in the gut by mucoadhesion
3. Endocytosis of the particles
4. Permeabilizing effect of the polymer

The delivery formulations in nutraceutical food products should minimize the use of synthetic chemical polymers, which can cause side effects; also, being integrated into foods means that there must be no restriction of the quantity of oral intake by consumers. Due to such challenges, only food-grade materials may be used for encapsulation or as core ingredients during nutraceutical delivery. These materials can be obtained from natural sources (bio-based), viz. cellulosic materials, proteins, starch, lipids, and gums or materials with Generally Recognized as Safe (GRAS) status such as cyclodextrins, low-molecular-weight emulsifiers (e.g., Tweens), and mineral salts (Augustin and Hemar, 2009). Moreover, the processing operations must be approved by the regulatory affairs departments of the country in which the nutraceuticals are to be marketed and consumed (McClements, 2012). Different regions of the world have different regulatory agencies, e.g., the European Food Safety Authority (EFSA) in Europe and the Food and Drug Administration (FDA) in the United States.

8.6 NANONUTRACEUTICAL DELIVERY MATERIALS AND FORMULATIONS

The nanoscale systems should be enriched to meet storage conditions as well as to be healthy enough for practical applications, apart from being easy to fabricate and economically viable. Further, their incorporation into final food products should not adversely affect their quality. Some commonly employed delivery systems are as follows:

8.6.1 Polysaccharide-Based Nanodelivery Systems

Polysaccharide-based delivery systems are preferable for nutraceutical applications due to their biocompatible and biodegradable nature and can be easily manipulated to achieve the required properties. Polysaccharides (dietary fibers) have potential health benefits such as cholesterol reduction, cancer prevention, and colonic health improvement. Due to their structural versatility and site-specific digestion properties, polysaccharides are suitable carriers for the targeted and controlled release of nutraceuticals in the GIT. Due to their properties of bio-adhesion onto mucosal surfaces, polysaccharides are widely used for targeting specific organs or cells and prolonging the residence time of bioactive compounds in the intestine. The release of bioactive molecules in the intestine can be controlled by pH changes, transit time, peristaltic pressure, and even enzymatic degradation. Moreover, polysaccharides are considered as a suitable shell under high-temperature processes due to their thermo-stability in comparison to lipid- or protein-based delivery systems, which are very prone to be melted or denatured by a small variation of temperature (Fathi et al., 2012). Polysaccharides including plant-derived carbohydrates (e.g., starch, pectin, and guar gum), animal-derived carbohydrates (e.g., chitosan, chondroitin sulfate), and carbohydrates derived from other sources (e.g., alginate derived from algae) are commonly used during such processes. Sodium alginate and gums are the most common hydrocolloids used for microencapsulation in the food industry. These polymers are usually processed through spray-drying or ionic gelation techniques (Burnside, 2014).

8.6.1.1 Chitosan

Chitosan is the deacetylated form of chitin composed of glucosamine, approved for dietary applications in Japan, Italy, and Finland. It is one of the most widely distributed and used biopolymers due to its cationic, non-toxic, biodegradable, and biocompatible polyelectrolyte nature. It is described as beneficial for improving intestinal absorption capacity of active ingredients, particularly hydrophilic compounds, viz. EGCG, but it is least permeable in the small intestine. Chitosan has been extensively researched for its promising human-based applications, viz. food, cosmetics, biomedical, and pharmaceutical sectors. Due to their subcellular and submicron size, chitosan nanoparticles can invade deep into tissues through fine capillaries and cross the fenestration present in the epithelial lining, which allows efficient delivery of bioactive molecules to the target sites in the body.

8.6.1.2 Polyethylene Glycol (PEG)

One of the most widely accepted materials for the synthesis of polymeric nanoparticles is polyethylene glycol, which may be used as a biodegradable polymer due to the production of the metabolite monomers lactic acid and glycolic acid during its physiological hydrolysis, which is easily processed through the Krebs cycle. Rapid metabolization and low systemic toxicity make this biomaterial useful for the controlled release of bioactive compounds. With enhancement of the circulation time as a steric barrier, polyethylene glycol protects the opsonization of nanoparticles.

8.6.2 Food Protein-Based Nanodelivery Systems

Proteins extracted from animal-derived products and from vegetables are widely used for encapsulation of bioactive substances. These natural polymers exhibit several advantages (e.g., biocompatibility, biodegradability, good amphiphilic, gelation, foaming and water-binding capacity and functional properties) that make them suitable to be used as shell materials in nanofabrication (Nesterenko et al., 2013). However, proteins are difficult to process through electrohydrodynamic technology, mainly because of their complex secondary and tertiary structures, which hinder the entanglement of the polymer and hence the formation of fibers or capsules (Nieuwland et al., 2013). Organic solvents could be used to improve proteins solubility during the electrospinning/electrospraying process; however, these substances limit their use in food and nutraceutical applications. Another strategy to obtain electrosprayed structures from proteins is by blending them with synthetic polymers that are easily spinnable. Thus, diverse proteins including collagen, gelatin, casein, zein, and eggshell membrane protein have been well electrospun with the help of synthetic polymers (Matthews et al., 2002). There is growing interest in developing protein nanocarriers as GRAS drug or nutraceutical delivery devices due to their notable characteristics such as non-antigenicity, high nutritional value, abundance as renewable resources, and unique binding capacity for various drugs or nutraceuticals. In addition, protein nanoparticles can be easily prepared and scaled up during manufacturing. Moreover, protein hydrolysis by digestive enzymes generates bioactive peptides that might exert a number of physiological effects in vivo.

8.6.2.1 Milk Proteins and Peptides

Over the past a few years, milk proteins have attracted intensive research and developmental initiatives to harness and utilize them for delivering health-promoting bioactive compounds in food and drug applications that resulted in broad and novel applications. Among the milk proteins, caseins and whey proteins are the major building materials mostly applied in fabrication of food-grade nano-scaled delivery systems.

8.6.2.1.1 Caseins

The caseins are proline-rich and open-structured proteins with different hydrophobic and hydrophilic domains. Interestingly, most caseins (95%) are naturally self-assembled into casein micelles, which are spherical colloidal particles with diameters in the range of 50–500 nm (average 150 nm). The co-assembly process for casein micelles can be exploited for nanoencapsulation of hydrophobic nutraceuticals, viz. curcumin, vitamin D, and ω-3 polyunsaturated fatty acids.

8.6.2.1.2 Whey Proteins

β-Lactoglobulin and α-lactoglobulin are small globular proteins and constitute the major whey proteins in cow milk. Similarly, bovine serum albumin (BSA), also present in milk and blood serum, is a larger globular protein having alpha-helical structure, 17 disulfide bridges, and a free thiol. Functionally, BSA binds and delivers various small ligands in the blood serum. Although most BSA used in research work is isolated from bovine blood, some is still obtained from milk. Lactoferrin is also a monomeric globular glycoprotein: from the transferrin family, it is involved in nanoencapsulation and has potential antimicrobial activity. Moreover, whey proteins are usually resistant to some proteolytic enzymes and therefore act as candidates for enteric delivery.

8.6.3 Peptide Nanotubes

Novel nanotubes can be synthesized by calcium-induced self-assembly of enzymatically hydrolyzed α-lactalbumin. These peptide nanotubes could be potentially used for nutraceutical delivery due to large surface area, stiffness, and high viscosity. Further, these lactalbumin peptide nanotubes have cavities of 8 nm in diameter that can be used to encapsulate and protect nutraceuticals (vitamins and enzymes) and mask undesirable aromatic compounds. These peptides are considered food-grade materials that facilitate dynamic applications in nanoencapsulating of nutrients, supplements, and pharmaceutical entities.

8.6.3.1 Gelatin

Gelatin is obtained as a denatured protein from collagen by acid or alkaline hydrolysis. It is counted as GRAS material by the FDA and hence has been safely used in pharmaceuticals, cosmetics, and food products. The easy accessibility of its functional groups makes it useful in developing targeted nutraceutical/drug delivery carriers (Huang et al., 2004). Moreover, gelatin is a polyampholyte, having both anionic and cationic groups along with hydrophobic groups. Various methods, viz. desolvation, coacervation-phase separation, emulsification–solvent evaporation, reverse-phase evaporation, and nanoprecipitation, are adopted to prepare gelatin nanoparticles.

8.6.4 Lipid-Based Nanodelivery Systems

Various lipophilic substances such as glycerides, oils, phospholipids, carotenoids, and waxes are also used as carrier materials in microencapsulation (Patel et al., 2012).

They permit barrier creation for the protection of sensitive ingredients against moisture, plus their transport in aqueous media (Nesterenko et al., 2013). The most typical lipid-based carriers are micro/nanoemulsions and liposomes. These delivery modules are especially suitable for water-insoluble bioactive food materials. For instance, liposomes have been effectively used for the protection of lipid-soluble vitamins (A, E, D, and K) (Velikov and Pelan, 2008) and enzymes (Picon et al., 1994). However, these encapsulation systems are in liquid state, which could complicate their handling, processing, and storage. In solid state, SLNs can be obtained. These structures differ from normal emulsions because the core is in the solid state.

8.6.4.1 Liposomes

Liposomes have lipid-bilayered membrane structures composed of phospholipids, which have hydrophilic heads and hydrophobic fatty acid tails assembled into a spheroid structure. Due to their biphasic character, liposomes can serve as carriers for both hydrophilic (in the central aqueous compartment) and hydrophobic (in lipid bilayers) compounds (Langer, 1990). The commonly used methods for nanoliposome synthesis include sonication, extrusion, freeze-thawing, ether injection, and microfluidization. They are typically fabricated using food-grade phospholipids isolated from soy, dairy, or egg sources. Hydrophobic nutraceutical components are incorporated within the non-polar regions formed by the phospholipid tails in the bilayer structures, while hydrophilic components may be incorporated within the interior aqueous core.

Nanoliposomes can be administered parenterally, orally, topically, or nasally (Mozafari et al., 2008) to deliver bioactives such as coenzyme Q_{10}.

8.6.4.2 Solid Lipid Nanoparticles and Nanostructured Lipid Carriers

SLNs are one of the most indispensable lipid-based nanocarriers in nutraceutical and pharmaceutical delivery. SLNs are emulsions created at high temperatures where both dispersed and continuous phases are liquid. As the temperature of the system goes down, the inner core solidifies, generating a solid suspension that can be isolated (Toublan, 2014). Further, spray drying and a fluidized bed can be used to dry the emulsions. SLNs have a similar structure to oil-in-water nanoemulsion but are solid at room temperature. The hydrophobic nutraceuticals can be encapsulated into the solid lipid core for enhanced stability, reduced degradation, and sustained extended release. This minimizes toxicity and improves the target specificity of the compounds to delivery sites. The commonly used lipids in SLNs that have relatively high melting temperatures include fatty acids (stearic acid), triglycerides (e.g., tristearine), waxes (e.g., cetylpalmitate), or a mixture of these (Uner, 2006). The typical surfactants used to stabilize the lipid dispersion include sorbitan esters, phospholipids, bile salts, fatty acid ethoxylates, or a mixture of these components. Curcumin encapsulated in SLN showed a dose-dependent increase in bioavailability (Kakkar et al., 2011). SLNs have some limitations such as low compound loading capacity and leakage during storage. Recently developed nanostructured lipid carriers (NLCs) are smaller and more stable, and have higher loading capacity (Puri et al., 2009; Uner, 2006). More complex lipid mixtures such as a mixture of monoglyceride, diglyceride, or triglyceride with fatty acids of different chain lengths are employed in the hydrophobic core to accommodate more hydrophobic drugs and nutraceuticals such as quercetin and resveratrol to avoid expulsion (Puri et al., 2009; Sanad et al., 2010). SLNs and NLCs are routinely synthesized by high-pressure homogenization, hot/cold homogenization/ultrasonication, phase inversion, and solvent evaporation/emulsification (Das and Chaudhury, 2011). They can be administrated through oral, nasal, parenteral, dermal, ocular, rectal, and pulmonary routes (Iqbal et al., 2012). NLCs have been used for the delivery of the nutraceuticals CoQ_{10} (Junyaprasert et al., 2009), lutein (Lacatusu et al., 2013), and silybin (Jia et al., 2010).

8.6.4.3 Nanoemulsions

The majority of nutraceutical products are either poorly soluble or lipophilic and are significantly influenced by their physico-chemical properties during delivery of their active ingredients. Active components that poorly dissolve in oil or water pose challenges such as route of administration, transport, and reaching their absorption sites, resulting in a poor oral bioavailability. So, to overcome such adverse properties and enhance the bioavailability of bioactive components, the encapsulation of active compounds into nanoemulsion is an alternative method. The microencapsulation and nanoencapsulation of bioactive substances by various techniques, viz. emulsification, freeze-drying, spray drying, coacervation, and nanoprecipitation, have been adopted for encapsulation and drying of bioactive compounds as well as probiotics (Anu Bhushani and Anandharamakrishnan, 2014).

Nanoemulsions are a class of extremely small droplet emulsions that come into sight as transparent or translucent in the range of 50 to 200 nm; they are much smaller than conventional emulsions (1 to 100 μm), which contain a continuous phase, a dispersed phase, and emulsion stabilizer, the emulsifier/surfactant. In comparison to conventional methods – micronizing/milling, spray drying, co-solvent addition and salt formation – the implementation of lipid-based delivery systems, i.e., micro/nanoemulsions and micelles, brings many improvements: (i) high kinetic or thermodynamic stability (significant stability over unstable dispersions, i.e., conventional emulsions and suspensions), (ii) incorporation of both hydrophilic or lipophilic active components into the same nanoemulsion, and (iii) the easiest transportation of the active component through the cell membranes due to the small droplet sizes, resulting in an enhanced nutraceutical concentration in plasma and significantly higher bioavailability.

8.6.4.4 Micelles

Micelles are composed of amphiphilic monomers such as phospholipids and polymers. Micelles are formed when the concentrations of amphiphilic phospholipid reach the critical micelle concentration and temperatures attain the critical micellization temperature (Torchilin, 2007). In polymeric micelles, two monomers having distinct hydrophobicity can be conjugated, and they produce a core–shell micelle structure where the hydrophilic blocks form the micelle shell and the hydrophobic blocks form the micelle core. PEG is a commonly used hydrophilic monomer, and the most widely used core-forming blocks are poly (D, L-lactic acid), poly (caprolactone), poly (propyleneoxide), and poly (laspartic acid) (Gong et al., 2012). Some of the advantages of this method are that micelles enhance the water solubility of hydrophobic compounds, extend their blood circulation time, increase target specificity through enhanced permeability and retention effects, and reduce the toxicity of compounds (Oerlemans et al., 2010).

8.6.5 Nanodelivery System for Vitamins

Hydrophobins (Hyd) used for the nanoencapsulation of nutraceuticals for food enrichment are very interesting, especially for hydrophobic components like vitamin D3. Hyd provide excellent protection to nutraceuticals against degradation and thus are promising nanovehicles for hydrophobic nutraceuticals used in the food and beverage industry.

8.6.6 Phytochemical-Loaded Nanoformulations

Phytochemicals having health booster effects such as plant polyphenols (curcumin, resveratrol, epigallocatechin gallate, etc.) and carotenoids (lycopene, β-carotene, lutein, zeaxanthin, etc.) have gained much attention from researchers, consumers, and food manufacturers because of their extensive applications to lower blood pressure, reduce cancer risk factors, strengthen immune systems, regulate blood sugar concentration, lower cholesterol levels, and serve as antioxidant agents.

Upon oral intake, phytochemicals are identified and treated as xenobiotics and disintegrated by different chemical, enzymatic, and microbial digestion phases in mouth, stomach, and small and large intestine; then, they are absorbed, distributed, and metabolized in cells and tissues and finally excreted via renal, biliary, or pulmonary routes. During the course of metabolism, these compounds or their metabolites may be absorbed or available in the nano/picomolar range in cells and tissues, which may result in inadequate dosages to achieve biological efficacy. Furthermore, unabsorbed phytochemicals reach the large intestine and are metabolized by the local microbiota; this affects their bioefficacy. Hence, there needs to be improvement in nanodelivery modules to overcome these challenges; currently, there is extensive research being conducted on the encapsulation of functional compounds, e.g., curcumin and resveratrol, using nanoemulsions to save these active compounds from digestion. A novel lipid-free nano-CoQ_{10} system formulated and stabilized by various surfactants can be an effective vehicle for improving oral bioavailability of CoQ_{10}. Novel organogel-based nanoemulsions are being developed for oral delivery of poorly soluble nutraceuticals and improve their bioavailability.

8.7 CHARACTERIZATION OF NANONUTRACEUTICAL DELIVERY SYSTEMS

Characterization of nanomaterials/nanoproducts before their incorporation into any food or feed matrix is highly essential. Several characterization techniques are used for this purpose.

8.7.1 Morphology and Particle Size Analysis

Electron microscopes and dynamic light scattering are the most preferred techniques employed for analyzing the size of the nanomaterials. The scanning electron microscope (SEM) and transmission electron microscope (TEM) are mainly used as they have better resolutions in comparison to the optical microscope. Additionally, the surface morphology of the nanomaterials can be analyzed by atomic force microscopy (AFM) (Dasgupta et al., 2015).

8.7.1.1 Atomic Force Microscopy

The surface morphology of a particle is mainly analyzed by atomic force microscopy. Whereas SEM and TEM can image only conducting surfaces and TEM cannot image the magnetic nanomaterials (e.g., Fe-based NMs), AFM has been observed to overcome these limitations. The structural characterization of bioactive molecules, along with the morphology, size, and process dynamics of lipid nanocapsules, can also be analyzed using AFM (Luykx et al., 2008). However, small scratches during sample preparation (smear on slide) and exposure to rough surface when the tip is in direct contact with sample surface can impose errors and difficulties.

8.7.1.2 Tunneling Electron Microscopy

The SEM scans the surface of sample, whereas the TEM penetrates through the sample and generates images of nanomaterials by interacting with the electrons and is best suited to analyze nanosized materials. It provides better resolution than SEM because of the higher energy of the electron beam used (Dasgupta et al., 2015).

8.7.1.3 Scanning Electron Microscopy

SEM is a suitable technique for the evaluation of nanoformulations and has advantages when compared to freeze-fracture transmission electron microscopy, such as lower costs and the opportunity to examine larger areas of dispersed particles in a comparatively short time (Hatziantoniou et al., 2007). SEM is employed for the characterization of nanocapsules, nanoemulsions, and solid lipid nanoparticles.

8.7.1.4 Dynamic Light Scattering

Dynamic light scattering (DLS) is the most common tool for analyzing the size of nanomaterials in suspension. It can measure the particle size down to the 1-nm level, which is not possible using other techniques. Its cost effectiveness and quicker analysis time give additional advantages, but DLS gives a crude indirect measurement and can only measure liquids or gels. Further, nanoformulations in powder form have to be either dissolved or suspended in an inert liquid medium. The use of DLS to evaluate nanoemulsions is well known and has been done in several research studies using peppermint oil NEs (Liang et al., 2012), vitamin E–enriched NEs (McClements, 2012), fisetin NEs (Ragelle et al., 2012), curcumin NEs (Abbas et al., 2014), and β-carotene NEs (Jo and Kwon, 2014).

8.7.1.5 Zeta Potential

Zeta potential is the measure of the overall charge of a particle in a specific medium and gives an indication of the potential stability of a colloidal system. It also explains the electrostatic interaction, mobility, and stability of a colloidal solution (Tiede et al., 2008). Electrostatic repulsion interaction is used to measure and control the stability of the solutions. It explains the reasons for the occurrence of dispersion, aggregation, or flocculation, and can be used to improve the conditions of the colloidal solution (Silva et al., 2012).

8.7.1.6 X-Ray Diffraction

X-ray diffraction (XRD) can be used to determine the basic crystal structure of the material and is a suitable technique to characterize nanopowders (Dasgupta et al., 2015), lipid nanocapsules, albumin protamine-oligonucleotide nanoparticles, albumin–zinc nanoparticles, and lactoglobulin aggregates (Luykx et al., 2008).

8.8 FACTORS AFFECTING THE NUTRACEUTICAL DELIVERY

The development of effective nanoparticle-based target delivery systems requires a clear understanding of various factors influencing the kinetics, absorption, and excretion of nanoformulations in the system (Labhasetwar, 2005).

8.8.1 Nanoparticle Size

As nanoparticle size determines hepatic filtration, tissue extravasation, tissue diffusion, and kidney excretion, size is a key factor in the biodistribution of long-circulating nanoparticles. The recommended size of nanoparticles for biomedical application is 10 to 100 nm. Nanoparticles of <6 nm hydrodynamic size are rapidly eliminated by kidneys (Longmire et al., 2008). Particles smaller than 100 nm are mainly cleared by the liver, while nanoparticles larger than 200 nm are cleared by the spleen (Arruebo et al., 2007).

8.8.2 Surface Characteristics

Nanoparticles are easily recognized by the host immune system. The surface properties of the nanoparticles determine their solubility, stability, and clearance. Nanoparticle hydrophobicity determines the level of binding of opsonins on their surface and subsequent clearance. Hence it becomes necessary to minimize the opsonization by coating nanoparticles with hydrophilic polymers or surfactants such as PEG, polyethylene oxide, poloxamine, polyoxamer, and polysorbate.

8.8.3 Nature of Surfactants and Building Blocks of Nanoparticles

Surfactants, indigestible polysaccharides, proteins, and lipids do not undergo any changes in the mouth. However, digestible polysaccharides undergo hydration, swelling, and digestion by amylase present in the saliva. Digestible protein, lipids, and polysaccharides are digested in the stomach and small intestine by gastric and intestinal enzymes, and the residual components are metabolized by colonic bacteria. Surfactants undergo transformations by surface-active compounds such as bile salts, phospholipids, fatty acids, peptides, or pH-induced changes in the stomach and small intestine, and the residues are likely to be metabolized by bacteria in the colon (Joye et al., 2014).

8.8.4 Reticuloendothelial System Clearance

Nanoparticles have to be functionalized in such a way that they can escape the reticuloendothelial system (RES) and hence have longer circulation time in the bloodstream and a greater chance of reaching the targeted site in required quantities. To prolong the circulation time, coating with polymers such as polyethylene glycol, polyvinyl alcohol (Takeuchi et al., 2000), polyamino acids (Metselaar et al., 2003), and polysaccharides (Haidar et al., 2008) reduces protein adsorption and thereby help to avoid the RES clearance.

8.9 NANONUTRIENT UPTAKE AND TARGETED DELIVERY

In addition to influencing the controlled release of nutraceuticals, nanoparticles also influence the transport of bioactive compounds through enterocytes (by transcellular endocytosis for particles between 20 and 1000 nm) by altering residence time and modulating transportation efficiency and pathways, including the metabolic reactions undergone by such compounds within endothelial cells. For instance, chitosan derivatives have mucoadhesive features and the capability to open tight junctions and enhance paracellular transport. Moreover, the charge of nanoparticles also influences the formation of hydrogen bonds with the mucosal surface, contributing to a momentary retention. Once absorbed from the intestinal lumen, nanoparticles loaded with bioactive nutraceuticals move through the endothelial layer to migrate to tissues into the bloodstream. At this stage, nanoparticles have already been included in endosomes and undergone oxidation, reduction, and hydrolysis reactions as well as conjugations by diverse metabolization routes. However, unlike metabolism of free bioactives, when included in nanoformulations, the bioactive compounds are protected from such reactions, by modulating the exposure to metabolizing enzymes in enterocytes.

8.9.1 Targeting of Nutraceuticals to the Intestine

For intestinal delivery, nutraceuticals (for example, probiotics) could be immobilized into a polymer matrix, which remains intact in the stomach but degrades and dissolves in the intestine. The encapsulation of probiotics has been successfully carried out with alginate, chitosan, carrageenan, gelatin, whey proteins, cellulose acetate phthalate, locust bean gum, and starch by spray drying, freeze-drying, fluidized bed drying, extrusion, and emulsification techniques for targeted intestinal delivery. Biodegradable hydrogels such as amylose, chondroitin sulfate, chitosan, inulin, guar gum, and pectin have also been successfully used to achieve oral colon-targeted delivery. Caseins are easily digestible in the stomach due to their easy accessibility to gastric proteases and provide a target-activated release mechanism (Fox and McSweeney, 2003). Beta casein nanoparticles have been proposed as targeted delivery formulations for gastric cancer treatment (Shapira et al., 2010).

8.9.2 Targeting of Nutraceuticals to the Liver

Particles with high lipid matrix are considered as potential candidates for preventing drug mobilization and precipitation in the gastrointestinal tract. For instance, silybin nanostructured lipid carrier (NLC) formulation and silymarin-loaded NLCs were reported to achieve a prolonged residence time in the blood, selective liver uptake through the chylomicron pathway, and hence targeted delivery to the liver.

8.9.3 Subcellular Organelle Targeting

Subcellular organelle targeting can be achieved by the use of a special targeting moiety chemically conjugated either to the nutraceuticals or to the nanocarriers. Coenzyme Q_{10} and related quinones have been effectively used to decrease mitochondrial oxidative damage. The orally administered ubiquinones are taken up by mitochondria only to a limited extent (Matthews et al., 1998); thus, to overcome this, mitochondrial targeting of ubiquinone can be achieved by conjugation to the membrane-permeable lipophilic triphenylphosphonium cation. Such mitochondrial targeting of antioxidant nutraceuticals using nanocarriers furnishes a powerful tool to enhance the mitochondrial and cellular functioning even during pathological situations (Porteous et al., 2010).

8.10 SAFETY EVALUATION OF NANODELIVERY SYSTEMS

The great potential of nano-based formulations for food applications carries with it serious concerns about their toxicity. Hence the major safety concerns are as follows:

1. Possible cytotoxicity
2. Entry of nanomaterials directly into the tissues on direct exposure to air and their translocation to other surrounding cells
3. Nonbiocompatibility of nanomaterials used in nanoformulations
4. Unclear and unknown effects that may arise due to their nano size on a long-term basis
5. Probable occurrence of alteration in gene expression

These concerns are mainly based on their increased surface-to-volume ratio compared to their micro-sized counterparts and the use of nanofabricated polymers. The potential in vitro and in vivo effects of nano-based formulations cannot be extrapolated from a previous knowledge of bulk analogs. So, it is essential to consider relevant parameters to study toxicity. Nanosize products exhibit unusual properties irrespective of their properties that existed in bulk nature (Roduner, 2006). Nanoparticles can enter any organ in the body and may cause damage to cells due to their increased penetration ability (Cheng et al., 2013). Further, nanoparticles possess different properties in different shapes. Hence, change in a particular shape may result in different properties that may or may not be toxic when used.

These issues about nanofood consumption can be broadly categorized into three domains: low concern, moderate concern, and major concern. An example of the low-concern domain is when processed food with nonbiopersistent nanomaterial is solubilized and digested in the gastrointestinal tract. Similarly, the moderate-concern domain is when a nanofood with nonbiopersistent nanomaterial is not digested but carried across the GI tract. These include improved bioavailability of vitamins/minerals and a higher uptake of food preservatives/colors, leading to exceeding the acceptable daily intake, which might be deleterious to the consumer's health. The major-concern domain is activated when foods include indigestible, insoluble, and potentially biopersistent nanoadditives (e.g., functionalized nanomaterials/metals/metal oxides), which may cause health hazards and have toxicological issues which are not fully known at present (Chaudhry and Castle, 2011).

Hence, safety and toxicity tests are highly mandatory prior to the release of nanoformulations for consumer use. Characterization of the nanostructure itself as an initial step in toxicity assessment is fundamental. For example, determination of materials, structure, size, shape, surface area, charge, solubility, hydrophobicity, stability, and aggregation state is highly essential (Podila and Brown, 2013; Arora et al., 2012).

8.11 CURRENT TRENDS IN MARKETED NANOEMULSIFIED NUTRACEUTICALS

Nanotechnology-based products approved by FDA have entered the market viz., liposomes, encapsulated nanoemulsions, nanoparticles, monoclonal antibody-based products, polymer–protein conjugates, and some polymeric drugs. Unilever has prepared healthier nanoemulsified ice cream without compromising on flavor (Ranjan et al., 2014). Similarly, Aquanova markets its nanotechnology-based carrier micelles to encapsulate bioactives such as vitamins C and E and fatty acid micelles as "NovaSol" and claims that the nanoscale carrier system increases the potency and bioavailability of active ingredients. Aquanova in collaboration with Zyme offers omega-3 fatty acids in 30- to 40-nm nanocapsules, which is 4000 times smaller than the existing product on the market. Aquanova claims improved stability in terms of both temperature and pH of standardized additive concentrations and encapsulated functional compounds (Ranjan et al., 2014). NutraLease, a technology startup company, is initiating research to enhance the bioavailability of functional compounds like lutein; lycopene; β-carotene; vitamins A, D_3, and E; coenzyme Q_{10}; phytosterols; and isoflavones using self-assembled nanoemulsions. These formulations claim to keep the flavor and guard it from temperature, enzymatic reactions, oxidation, and hydrolysis, and they are thermodynamically stable over a broad array of pH values (Silva et al., 2012; Halliday, 2015).

8.12 CHALLENGES AND RESEARCH GAPS

Nanofood technology is one of the fastest-growing fields because of its efficiency and benefits (Dasgupta et al., 2015a). It is the key responsibility of the food processing chain, including researchers and manufacturers, to deliver safe food. Nanofoods are facing controversy and challenges as they attempt to gain acceptance in the marketplace. So far, very few studies have been conducted in this area, and hence it is too early to propose nanofoods as either harmless or hazardous. Many of the recently developed nanodelivery systems can increase the transport of poorly hydrophilic compounds across the GI barrier by enhancing paracellular transport through tight junctions. Certain nanodelivery system components such as surfactants, polymers, and chelating agents increase the gap dimensions, alter barrier integrity, and consequently increase permeability, leading to toxicity. The increased gap disintegrity could stimulate permeability to inflammatory molecules (such as allergens, toxins, and bacteria) damaging colonic epithelium and consequently leading to intestinal inflammatory diseases. Therefore, the action of absorption enhancer used in nanoformulation should be rapid, coinciding with the presence of the nutraceutical at the absorption site, and should cause minimum damage to the gut epithelium.

Currently, nutraceutical industries lack user-friendly, low-cost instruments for the characterization of nanomaterials. This poses a major challenge in developing newer products through nanotechnology platforms. A minor difference in change in pH/temperature/time/ precursors might result in differently sized particles during the synthesis reaction. Hence, size characterization becomes important at every step of nanomaterial synthesis. Similarly, other effects such as their shape, surface chemistry, and interaction with food components have to be thoroughly characterized. Altering any of these factors may result in materials with altered properties and toxicity (Podila and Brown, 2013).

Even though many organizations, including the International Union of Food Science and Technology (IUFoST), the European Federation of Food Science and Technology (EFFoST), and the World Health Organization (WHO), agree on the health benefits of nanonutraceuticals, there is a need for updated data and measurements to ensure the safety of nanofood products. Due to insufficient scientific data on the adverse effects of nanofoods on human and animal health and on the environment (Gutiérrez et al., 2013), there is a serious need for in-depth studies (Ranjan and Dasgupta, 2013) and to implement clear regulatory principles on the use of nanomaterials in the food industry.

8.13 ETHICAL AND ECONOMICAL ISSUES

The major regulatory and ethical issues considered pertaining to nanoformulations are as follows:

- Basic research done by research institutions has to be integrated with industry to commercialize nanotechnology.
- Research institutions should share the technical aspects of the implications of nanotechnology with the public.
- Certification is necessary to provide consumer satisfaction with the safety aspects of nanomaterials.
- Manufacturers should be given separate license for producing nanomaterials, provided that they have all the required instruments and facilities to synthesize nano-based products.
- Proper labeling of nanonutraceutical products must be made mandatory, with revocation of a manufacturer's license for failure to do so.
- Systematic classification and categorization of nanomaterials as framed by the ISO should be established everywhere and followed by manufacturers.
- The FDA and other governmental organizations across the globe should frame strict protocols on synthesizing and characterizing nanomaterials.
- Regulatory acts that govern food sectors, such as the Toxic Substance Control Act (TSCA), Federal Food, Drug, and Cosmetic Act (FDCA), Occupational Safety and Health Act, and Consumer Product Safety Act (CPSA), should regulate and govern the risks of nanotechnology applications in food.

These efforts will promote consumer preference for nanoformulations and boost the nanonutraceutical industry.

8.14 FUTURE TRENDS OF NANONUTRACEUTICALS

The field of food nanotechnology is experiencing significant growth due to the confluence of interests of industry, government, and academia. In the area of nutrient and nutraceutical delivery, there have been important advances made in nanoparticle formulations designed to improve the bioavailability of active compounds. The popularity of nanomaterial-incorporating products is likely to drastically rise over the next 10 to 20 years. However, numerous factors such as uniform regulatory requirements, public perception, and consumer willingness to pay for nanofood products will play an important role in driving the future of nanofoods and their popularity on the global market.

8.15 UNIFORM REGULATORY REQUIREMENTS

Presently, there is a lack of uniform and stringent regulations on food products that contain nanomaterials. Since there are neither any international regulatory agencies nor uniform approaches adopted by countries toward regulation of nanotechnology, many countries have taken divergent approaches for the regulation of nanoproducts, thus making the global safety issues of nanoformulation difficult to quantify. Furthermore, due to the popularity of nanoformulations, consumers might tend to buy products through the Internet that would have not undergone safety tests as per the international regulations prior to marketing.

8.16 PUBLIC PERCEPTION IN REGARD TO NANOTECHNOLOGY

In spite of the many projected benefits of the incorporation of nanoscience into areas such as nutraceutical delivery and food processing, public concern in regard to the safety and environmental effects of nanofoods is proving to be one of the major issues the nanofood industry faces. The lack of understanding and knowledge about the unforeseen risks of nanoscale products interfering with physiological systems in humans and environmental processes of nature is a major challenge to the nanofood industry. Hence, along with research on the effects of nanomaterials on humans, studies and surveys with consumers about their perceptions of nanotechnology have to be promoted. Studies indicate that most Americans perceive that the benefits of nanotechnology outweigh the risks, while European consumers are more skeptical and pessimistic about nanoproducts.

8.17 CONCLUSIONS

The potential benefits of nutraceuticals in nanoscale modules support their application in the food and feed industries. Nanoparticles are promising delivery systems that could contribute to controlled release, fine-tuned kinetics, increased bioavailability, better bioefficacy, and improved stability of bioactive nutrients against environmental and systemic conditions. In the coming years, novel nanomaterials and formulations for nutrient delivery are sure to revolutionize nutraceutical research and manufacturing.

REFERENCES

Abbas, S., M. Bashari, W. Akhtar, W. W. Li, and X. Zhang. 2014. Process optimization of ultrasound-assisted curcumin nanoemulsions stabilized by OSA-modified starch. *Ultrasonics Sonochemistry* 21 no. 4: 1265–1274.

Aluko, R. E. 2012. Bioactive lipids. In *Functional Foods and Nutraceuticals* (First Edition), 23–36, edited by Rotimi E. Aluko. Springer-Verlag, New York. doi:10.1007/978-1-4614-3480-1.

Anu Bhushani, J., and C. Anandharamakrishnan. 2014. Electrospinning and electrospraying techniques: Potential food based applications. *Trends in Food Science & Technology* 38 no. 1: 21–33. 10.1016/j.tifs.2014.03.004

Arora S., J. M. Rajwade, K. M. Paknikar. 2012. Nanotoxicology and *in vitro* studies: The need of the hour. *Toxicology and Applied Pharmacology* 258 no. 2: 151–165.

Arruebo, M., R. Fernández-Pacheco, M. Ricardo Ibarra, and J. Santamaría. 2007. Magnetic nanoparticles for drug delivery. *Nanotoday* 2 no. 3: 22–32.

Augustin, M. A., and Hemar, Y. 2009. Nano- and micro-structured assemblies for encapsulation of food ingredients. *Chemical Society Review* 38 no. 4: 902–912. doi: 10.1039/b801739p

Augustin, M. A., L. Sanguansri, Z. Shan, R. Weerakkody, and D. Y. Yung. 2012. Milk protein-based microencapsulated bioactive for improving the nutritive value of foods. IDF World Dairy Summit, Cape Town, South Africa, 4–8 November 2012.

Bernal, J., J. A. Mendiola, E. Ibanez, and A. Cifuentes. 2011. Advanced analysis of nutraceuticals. *Journal of Pharmaceutical and Biomedical Analysis* 55: 758–774.

Burnside, E. 2014. Hydrocolloids and gums as encapsulating agents. In *Microencapsulation in the Food Industry*, edited by A. G. Gaonkar, N. Vasisht, A. R. Khare, and R. Sobel, 241–250. San Diego, CA: Elsevier.

Das, Surajit, and Anumita Chaudhury. 2011. Recent advances in lipid nanoparticle formulations with solid matrix for oral drug delivery. *AAPS PharmSciTech* 12 no. 1: 62–76. 10.1208/s12249-010-9563-0

Dasgupta, N., S. Ranjan, D. Mundekkad, et al. 2015. Nanotechnology in agro-food: From field to plate. *Food Research International* 69: 381–400. doi: 10.1016/j.foodres.2015.01.005

Fathi, M., M. R. Mozafari, and M. Mohebbi, 2012. Nanoencapsulation of food ingredients using lipid based delivery systems. *Trends in Food Science and Technology* 23 no. 1: 13–27. doi: 10.1016/j.tifs.2011.08.003

Fox, P. F., and P. L. H. McSweeney. 2003. *Advanced Dairy Chemistry, Proteins Part A*, vol. 1. 3rd ed. New York: Kluwer Academic/Plenum.

Gong, Jian, Meiwan Chen, Ying Zheng, Shengpeng Wang, and Yitao Wang. 2012. Polymeric micelles drug delivery system in oncology. *Journal of Controlled Release* 159 no. 3: 312–323. 10.1016/j.jconrel.2011.12.012

Gutiérrez, F. J., S. M. Albillos, E. Casas-Sanz, et al. 2013. Methods for the nanoencapsulation of B-Carotene in the food sector. *Trends in Food Science and Technology* 32 no. 2: 73–83.

Haidar, Ziyad S., Reggie C. Hamdy, and Maryam Tabrizian. 2008. Protein release kinetics for core–shell hybrid nanoparticles based on the layer-by-layer assembly of alginate and chitosan on liposomes. *Biomaterials* 29 no. 9: 1207–1215. 10.1016/j.biomaterials.2007.11.012

Halliday, J. 2015. EFSA opens the floor on nanotechnology. http://www.foodnavigator.com/Financial-Industry/EFSA-opens-the-floor-on-nanotechnology

Hannun, Y. A., and L. M. Obeid. 2008. Principles of bioactive lipid signalling: Lessons from sphingolipids. *Nature Reviews Molecular Cell Biology* 9 no. 2: 139–150.

Hatziantoniou, S., G. Deli, Y. Nikas, et al. 2007. Scanning electron microscopy study on nanoemulsions and solid lipid nanoparticles containing high amounts of ceramides. *Micron* 38 no. 8: 819–823. doi: 10.1016/j.micron.2007.06.010

Huang, Z. M., Y. Z. Zhang, S. Ramakrishna, and C. T. Lim. 2004. Electrospinning and mechanical characterization of gelatin nanofibers. *Polymer* 45 no. 15: 5361. doi: 10.1002/ps.3371

Iqbal, Md Asif, Shadab Md, Jasjeet Kaur Sahni, Sanjula Baboota, Shweta Dang, and Javed Ali. 2012. Nanostructured lipid carriers system: Recent advances in drug delivery. *Journal of Drug Targeting* 20 no. 10: 813–830. 10.3109/1061186X.2012.71684

Jia, Lejiao, Dianrui Zhang, Zhenyu Li, Cunxian Duan, Yancai Wang, Feifei Feng, Feihu Wang, Yue Liu, and Qiang Zhang. 2010. Nanostructured lipid carriers for parenteral delivery of silybin: Biodistribution and pharmacokinetic studies. *Colloids and Surfaces B, Biointerfaces* 80 no. 2: 213–218. 10.1016/j.colsurfb.2010.06.008

Jo, Y. J., and Y. J. Kwon. 2014. Characterization of B-carotene nanoemulsions prepared by microfluidization technique. *Food Science and Biotechnology* 23 no. 1: 107–13.

Joye, I. J., G. Davidov-Pardo, and D. J. McClements. 2014. Nanotechnology for increased micronutrient bioavailability. *Trends in Food Science & Technology* 40 no. 2: 168–182.

Junyaprasert, Varaporn B., Veerawat Teeranachaideekul, Eliana B. Souto, Prapaporn Boonme, and Rainer H. Muller. 2009. Q10-loaded NLC versus nanoemulsions: Stability, rheology and in vitro skin permeation. *International Journal of Pharmaceutics* 377 no. 1–2: 207–214. 10.1016/j.ijpharm.2009.05.020

Kakkar, V., S. Singh, D. Singla, and I. P. Kaur. 2011. Exploring solid lipid nanoparticles to enhance the oral bioavailability of curcumin. *Molecular Nutrition and Food Research* 55: 495–503.

Koo, S. Y., K. H. Cha, D. G. Song, et al. 2014. Microencapsulation of peppermint oil in an alginate–pectin matrix using a coaxial electrospray system. *International Journal of Food Science and Technology* 49 no. 3: 733–739.

Labhasetwar, V. 2005. Nanotechnology for drug and gene therapy: The importance of understanding molecular mechanisms of delivery. *Current Opinion in Biotechnology* 16 no. 6: 674–680.

Lacatusu, Ioana, Elena Mitrea, Nicoleta Badea, Raluca Stan, Ovidiu Oprea, and Aurelia Meghea. 2013. Lipid nanoparticles based on omega-3 fatty acids as effective carriers for lutein delivery. Preparation and in vitro characterization studies. *Journal of Functional Foods* 5 no. 3: 1260–1269. 10.1016/j.jff.2013.04.010

Liang, R., S. Xu, C. F. Shoemaker, et al. 2012. Physical and antimicrobial properties of peppermint oil nanoemulsions. *Journal of Agricultural and Food Chemistry 60* no. 30: 7548–7555. doi: 10.1021/jf301129k

Longmire, M., P. L. Choyke, and H. Kobayashi. 2008. Clearance properties of nano-sized particles and molecules as imaging agents: Considerations and caveats. *Nanomedicine (London, England)* 3 no. 5: 703–717.

Luykx, D. M. A. M., R. J. B. Peters, S. M. Van Ruth, and H. Bouwmeester. 2008. A review of analytical methods for the dentification and characterization of nano delivery systems in food. *Journal of Agricultural and Food Chemistry* 56: 8231–8247. doi: 10.1021/jf8013926

McClements, D. J. 2012. Advances in fabrication of emulsions with enhanced functionality using structural design principles. *Current Opinion in Colloid and Interface Science* 17 no. 5: 235–245.

McClements, D. J., E. A. Decker, and J. Weiss. 2007. Emulsion-based delivery systems for lipophilic bioactive components. *Journal of Food Science* 72: 109–124.

Metselaar, Josbert M., Peter Bruin, Leo W. T. de Boer, Tom de Vringer, Cor Snel, Christien Oussoren, Marca H. M. Wauben, Daan J. A. Crommelin, Gert Storm, and Wim E. Hennink. 2003. A novel family of L-amino acid-based biodegradable polymer-lipid conjugates for the development of long-circulating liposomes with effective drug-targeting capacity. *Bioconjugate Chemistry* 14 no. 6: 1156–1164. 10.1021/bc0340363

Nesterenko, A., I. Alric, F. Silvestre, et al. (2013). Vegetable proteins in microencapsulation: A review of recent nterventions and their effectiveness. *Crops and Products* 42 no. 1: 469–479.

Nieuwland, M., P. Geerdink, P. Brier, et al. 2013. Food-grade electrospinning of proteins. *Innovative Food Science and Emerging Technologies* 20: 269–275.

Oerlemans, Chris, Wouter Bult, Mariska Bos, Gert Storm, J. Frank W. Nijsen, and Wim E. Hennink. 2010. Polymeric micelles in anticancer therapy: Targeting, imaging and triggered release. *Pharmaceutical Research* 27 no. 12: 2569–2589. 10.1007/s11095-010-0233-4

Patel, A. R., P. C. M. Heussen, J. Hazekamp, et al. 2012. Quercetin loaded biopolymeric colloidal particles prepared by simultaneous precipitation of quercetin with hydrophobic protein in aqueous medium. *Food Chemistry* 133: 423–429.

Picon, A., P. Gaya, M. Medina, et al. 1994. The effect of liposome encapsulation of chymosin derived by fermentation on manchego cheese ripening. *Journal of Dairy Science* 77: 16–23.

Podila, R., and J. M. Brown. 2013. Toxicity of engineered nanomaterials: A physicochemical perspective. *Journal of Biochemical and Molecular Toxicology* 27 no. 1: 50–55. doi: 10.1002/jbt.21442

Porteous, C. M., A. Logan, C. Evans, et al. 2010. Rapid uptake of lipophilic triphenylphosphonium cations by mitochondria in vivo following intravenous injection: Implications for mitochondria-specific therapies and probes *Biochimica et Biophysica Acta* 1800: 1009–1017.

Puri, A., K. Loomis, B. Smith, J. H. Lee, A. Yavlovich, E. Heldman, and R. Blumenthal. 2009. Lipid-based nanoparticles as pharmaceutical drug carriers: From concepts to clinic. *Critical Reviews in Therapeutic Drug Carrier Systems* 26: 523–580.

Ragelle, H., S. C. Manciet, J. Seguin, et al. 2012. Nanoemulsion formulation of fisetin improves bioavailability and antitumour activity in mice. *International Journal of Pharmaceutics* 427: 452–459. doi: 10.1016/j.ijpharm.2012.02.025

Ranjan, S., and N. Dasgupta. 2013. Proposal grant 'NanoToF: Toxicological evaluation for nanoparticles used in food'. PI: Dr. C. Ramalingam, Co-PI: Dr. Ashutosh K. and Dr. Ramanathan K. dbtepromis.nic.in

Ranjan, S., N. Dasgupta, A. Roy, et al. 2014. Nanoscience and nanotechnologies in food industries: Opportunities and research trends. *Journal of Nanoparticle Research* 16 no. 6: 1–23. doi: 10.1007/s11051-014-2464-5

Rapaka Rao, S., and P. M. Coates. 2006. Dietary supplements and related products: A brief summary. *Life Sciences* 78 no. 18: 2026–2032.

Roduner, E. 2006. Size matters: Why nanomaterials are different. *Chemical Society Reviews* 35: 583–592. doi: 10.1039/b502142c

Sanad, Rania A., Nevine Shawky Abdelmalak, Tahany S. Elbayoomy, and Alia A. Badawi. 2010. Formulation of a novel oxybenzone-loaded nanostructured lipid carriers (NLCs). *AAPS PharmSciTech* 11 no. 4: 1684–1694. 10.1208/s12249-010-9553-2

Shapira, Alina, Yehuda G. Assaraf, and Yoav D. Livney. 2010. Beta-casein nanovehicles for oral delivery of chemotherapeutic drugs. *Nanomedicine* 6 no. 1: 119–126.

Silva, H. D., M. Â. Cerqueira, and A. A. Vicente. 2012. Nanoemulsions for food applications: Development and characterization. *Food and Bioprocess Technology* 5: 854–867. doi: 10.1007/s11947-011-0683-7

Slattery, M. L., K. P. Curtin, S. L. Edwards, and D. M. Schaffer. 2004. Plant foods, fiber, and rectal cancer. *American Journal of Clinical Nutrition* 79: 274–281. Springer, Technology. doi: 10.1016/j.tifs.2013.05.007

Tiede, K., A. B. A. Boxall, and S. P. Tear. 2008. Detection and characterization of engineered nanoparticles in food and the environment. *Food Additives and Contaminants—Part A Chemistry, Analysis, Control, Exposure and Risk Assessment* 25: 795–821. doi: 10.1080/02652030802007553

Torchilin, V. P. 2007. Micellar nanocarriers: Pharmaceutical perspectives. *Pharmaceutical Research* 24 no. 1: 1–16. 10.1007/s11095-006-9132-0

Trottier, G., P. J. Bostrom, N. Lawrentschuk, et al. 2010. Nutraceuticals and prostate cancer prevention: A current review. *Nature Review in Urology* 7: 21–30.

Uner, M. 2006. Preparation, characterization and physico-chemical properties of solid lipid nanoparticles (SLN) and nanostructured lipid carriers (NLC): Their benefits as colloidal drug carrier systems. *Die Pharmazie* 61 no. 5: 375–386.

Velikov, K. P., and E. Pelan. 2008. Colloidal delivery systems for micronutrients and nutraceuticals. *Soft Matter* 4 no. 10: 1964–1980.

Zaki, N. M. 2014. Progress and problems in nutraceuticals delivery. *Journal of Bioequivalence & Bioavailability* 6 no. 3: 075–077.

Section 3

Nano Biotechnology in Health Care/Therapeutic Applications

9 Nanomedicine for Wound Healing
Challenges and Future Perspectives

Sourav Das, Papia Basu Thakur, Aparna Harri, Manali Bagade, and Chitta Ranjan Patra

CONTENTS

9.1 Introduction 115
9.1.1 FDA-Approved Therapies 115
9.1.2 Types and Causes of Wounds 116
9.1.2.1 Open Wounds 116
9.1.2.2 Closed Wounds 116
9.1.3 Conventional and Modern Approaches 116
9.1.3.1 Conventional Approach 116
9.1.3.2 Modern Approach 117
9.2 Mechanism of Wound Healing 117
9.2.1 Homeostasis Phase 117
9.2.2 Inflammation Phase 118
9.2.3 Proliferative Phase 118
9.2.4 Remodeling Phase 118
9.3 Diabetic and Mechanical Wound Healing 118
9.4 Various Strategies in Wound Healing 119
9.4.1 Nanomedicinal Approach 119
9.4.1.1 Metal-Based Nanoparticles 119
9.4.1.2 Non-Metal and Metalloid-Based Nanoparticles 123
9.4.1.3 Nanoparticles as Carriers 124
9.4.2 Tissue Engineering Approach 125
9.4.2.1 Nanoscaffolds 126
9.4.2.2 Small Molecule–Based Nanostructures 126
9.4.3 Stem Cell–Based Therapy 126
9.5 Unmet Medical Challenges and Future Perspective 126
Acknowledgment 128
Abbreviations 128
References 128

9.1 INTRODUCTION

Wound healing is an intricate continuum process that requires immense coordination among cells, as well as between cells and the extracellular matrix (Hamdan et al. 2017). Over 6.5 billion people were affected to wound-related diseases in 2017, and the number is increasing day by day (Han and Ceilley 2017). It has become a global burden to humanity. In upcoming years, the number is expected to increase due to the increasing prevalence of other types of diseases such as diabetes, obesity, cancer, AIDS, etc., which passively affect the wound healing process. Generally, the emergence of acute and chronic wounds has led to the development of new treatment strategies. The recent days, elevation of chronic wound (diabetic foot ulcers, venous ulcers, pressure ulcers, arterial ulcers, etc.) in an epidemic magnitude creates an urgent need for alternative treatment strategies which will fulfill the unmet medical challenges (Eming, Martin, and Tomic-Canic 2014). The advent of nanomedicine in the wound healing area offers a tremendous opportunity in order to target the complexity and cell specificity of wounds. In this context, we will emphasize the most recent advances in nanomedicine which have led to the paradigm shift in the wound healing domain.

9.1.1 FDA-Approved Therapies

There are four main Food and Drug Administration (FDA)-approved therapies available in the market for cutaneous wounds including recombinant human platelet-derived growth factor, two dermal substitutes, and finally a human skin equivalent (bioengineered) (Marston et al. 2003, Driver et al. 2015, Smiell et al. 1999, Wei, Kirsner, and Eaglstein 2016).

9.1.2 Types and Causes of Wounds

Common causes of wounds are heat, burn, cuts, injury due to accidents, and diabetic foot. Depending on the time of healing, the wound can be acute or chronic (Eming, Martin, and Tomic-Canic 2014). Unlike acute wounds, which are easily cured, chronic wounds are much more complicated and require more time to heal. There are also different types of wounds depending on the level of contamination, which include clean wound, contaminated wound, infected wound, and colonized wound. Clean wounds involve fewer complications, whereas contaminated wounds result from accidental injury, pathogenic organisms, etc. Similarly, infected wounds start from pathogenic infections causing soreness and redness to the skin. Colonized wounds are a kind of chronic state, where the wound, once created, is difficult to heal (for example, bedsores). Likewise, wounds can also be classified as open (cutaneous) or closed (non-cutaneous).

9.1.2.1 Open Wounds

Open wounds are generally exposed to an external environment. There are different types of open wounds such as incision, abrasion, laceration, etc.

9.1.2.1.1 Incision

An incision is generally a surgical cut made by any sharp object (knife, razor, etc.) on the epidermal tissue. Depending on the incising object, the cut might be deep or shallow (Camps 1952).

9.1.2.1.2 Abrasions

These superficial layered injuries are caused due to rubbing or scraping of the skin owing to frictional forces. They can be subdivided into scratches (caused due to contact with sharp object), grazes (caused due to the mechanical friction), or pressure abrasions (caused mechanically) (Robertson and Hodge 1972).

9.1.2.1.3 Laceration

A laceration is produced due to the tearing of the soft tissue of the skin (Cardany et al. 1976). There are also other reasons like splitting and overstretching. Splitting is due to the crushing of the skin in between two hard-blunt objects. Overstretching of fixed tissue includes loading high weight on the hard tissue, especially the bones. Tearing may happen due to accidental or intentional reasons and can be caused by knives or pieces of broken glass (Camps 1952).

9.1.2.1.4 Other Types of Wounds

Stab wounds occur due to the penetration or piercing of a pointed object such as a nail, particularly by metallic instruments such as knives (Camps 1952). The depths of such wounds are typically more than those in incision (e.g., knife stabbings) (Nolan, Hainsworth, and Rutty 2018). Similarly, gunshot wounds and puncture wounds are caused by bullets (or other projectiles) hitting or pointed objects (like needles or nails) puncturing, respectively.

9.1.2.2 Closed Wounds

Closed wounds (non-penetrating wounds) occur inside the body. They can also be classified as contusions and concussions, hematomas, bruises, crush injuries, etc.

9.1.2.2.1 Contusions and Concussions

While contusions are caused due to a pile-up of dead cells or blood under the skin of swollen bruises, concussions occur with damage to the head tissue without external injury (Kurland et al. 2012, Kimbler, Murphy, and Dhandapani 2011).

9.1.2.2.2 Hematomas

Hematomas occur due to the gathering of blood outside the blood vessels, forming a solid mass inside the tissue. They are continuous and ultimately cause bruises (Pieper and Araoz 2016).

9.1.2.2.3 Bruises

Sometimes a mechanical force acting on the body causes internal injury and hence closed wounds. It may or may not be observed (Camps 1952). Due to rupturing of blood vessels, the blood oozes into the extravascular space.

9.1.2.2.4 Crush Injury

A crush injury occurs generally when a large amount of force is applied over a longer period of time.

9.1.3 Conventional and Modern Approaches

Wound healing is now a big concern in the field of nanomedicine. There are different types of approaches which scientists have been adopting in wound healing, including conventional and modern approaches.

9.1.3.1 Conventional Approach

Although advanced and modern methods have been developed to speed up the wound healing process, conventional therapies using natural sources such as honey, plant extracts, maggots, larvae, etc., are still in use. They are highly therapeutic and have cell-stimulatory, antimicrobial, and anti-inflammatory properties. The advantages of these therapies are mainly simplicity, good efficacy, and, for the most part, affordability (Pereira and Bártolo 2016). Usage of traditional methods is common in various regions all over the world, but mainly Asia, Africa, and Latin America. Several plants, e.g., *Aloe vera*, *Vinca rosea*, etc., are used in different parts of the world for their medicinal applications, especially in wound healing (Atiba et al. 2011). In addition to the herbal products, animal-derived materials like honey and propolis are being used as traditional medicine. Honey, a viscous natural acidic sugar, when applied as natural dressing, plays a crucial role in re-epithelialization (Tan et al. 2009). Propolis (bee glue), a resinous substance, is used to shorten healing time due to its wide range of biological (antimicrobial, antioxidative, immunomodulatory, antiviral, antiseptic, anti-inflammatory, and healing) properties and low toxicity (Toreti et al. 2013, Burdock 1998). Similarly, living organisms like maggots and leeches have been significantly used as a wound healing therapy for their ability to trigger wound debridement. All these compounds have several advantages and disadvantages. For instance, herbal and animal-derived products can vary batch to batch with the change of the seasons, which can sometime create side effects like allergic reactions, infections, etc., (Walgrave, Warshaw, and Glesne 2005). Using advanced technology, these compounds are fabricated into gels, fibers, films, etc., by mixing with polymers. For example, *Aloe vera* has been stacked into alginate hydrogels, nanofibers made of silver nanoparticles, and hydrogel sheets have accommodated honey and propolis in dressing films, which have created additional future possibilities for conventional therapies (Pereira and Bártolo 2016).

9.1.3.2 Modern Approach

Apart from conventional approaches, scientists are using various therapies which are more effective and fruitful in the wound repair process. Some of these approaches are listed below.

9.1.3.2.1 Hyperbaric Oxygen Therapy

This approach is fruitful for complicated or recalcitrant ulcers. This therapy is based on oxygen, which has a pivotal role in angiogenesis and leukocyte and fibroblast function (Goldman 2009). It is generally used in clinical settings, which helps oxygen flow to the wound site and ultimately retard amputation (Santema et al. 2018). In the case of ischemic ulcers, it helps in the process of maintaining oxygen pressure in transcutaneous limbs.

9.1.3.2.2 Negative-Pressure Wound Therapy

This therapy is helpful for both acute and chronic wounds. This process removes the interstitial fluid with an increase in angiogenesis and granular tissue on the wound site and also decreases bacterial infection. Similarly, it escalates lymph clearance, skin perfusion level, tensile strength, and collagen synthesis. The surgical incision management system (SIMS) generally used for post-operative wound complications has a negative-pressure therapy unit which has found applications in orthopedics, cardiothoracic surgery, vascular surgery, etc. (Schurtz et al. 2018, Armstrong et al. 2002).

9.1.3.2.3 Bioengineered Cell Construct

Venous leg ulcer, a widespread chronic non-healing disease, has high morbidity rate, affecting human life significantly. BLCC (bilayered living cellular construct), which is FDA approved, gets remarkable attention nowadays for the treatment of life-threatening diseases like VLU. This substitute encompasses human foreskin–derived neonatal fibroblast in combination with human foreskin–derived neonatal epidermal keratinocytes in the type I collagen matrix, which facilitates wound healing (Stone et al. 2017).

9.1.3.2.4 Vascular Surgery

Venous leg ulcers are one kind of non-healing chronic disease which needs effective treatment (Thakral et al. 2015). No major difference in surgery was observed between the control group and the treated group, which was 4% higher than the control group (van Gent et al. 2006). In many cases, the recurrence of venous leg ulcers after surgery was not observed; this also supports the surgery process (Thakral et al. 2015).

9.1.3.2.5 Dressing Materials

Various types of dressing materials are available now in the market for wound healing (Szycher and Lee 1992). Modern custom-designed dressing materials generally absorb excess liquid at the wound site, which is ultimately helpful for wound healing (Fan et al. 2011). For example, dressing materials made of polymers imitate the biological and physical properties of the native tissue (Pereira and Bártolo 2016).

9.2 MECHANISM OF WOUND HEALING

Wound healing is a highly dynamic processes in animals. The mechanism of wound healing consists of an orderly progression of phases from wound to rapid homeostasis, inflammation, proliferation, and remodeling (Hamdan et al. 2017, Gould et al. 2015). Homeostasis refers to the process that stops the actual bleeding through vasoconstriction. Inflammation, which is a part of the body's immune responses, alerts our body about injury. During inflammatory phase, the body releases different types of cells (fibroblasts, endothelial cells, macrophages, neutrophils, keratinocytes, mast cells, etc.) which are responsible for inflammation, migration, and proliferation (Eming, Krieg, and Davidson 2007). In the proliferative phase, the proper aim is to fill and wrap the wound area. Finally, re-epithelialization takes place by forming layers of new skin replacing the damaged skin, which gains flexibility and strength (Hamdan et al. 2017). The different phases of wound healing are discussed below and shown in Figure 9.1.

9.2.1 Homeostasis Phase

Immediately after damage, homeostasis takes place in order to protect the vascular system and vital organs from injury. It builds a safer place for the invasive cells at the injury site and creates a matrix for them (Robson, Steed, and Franz 2001). In combination with the homeostasis process, coagulation begins with an aim to limit blood loss (Pool 1977). Once the platelet comes into action, different growth factors (platelet-derived growth factor [PDGF], transforming growth factor-β [TGF-β], etc.) are released, which activate and trigger the various cells such as neutrophils, macrophages, fibroblasts, and endothelial cells (Lawrence 1998). Since platelets contain vasoactive agents, vasoconstriction starts, which ultimately leads to edema and results in the inflammatory phase (Richardson 2004).

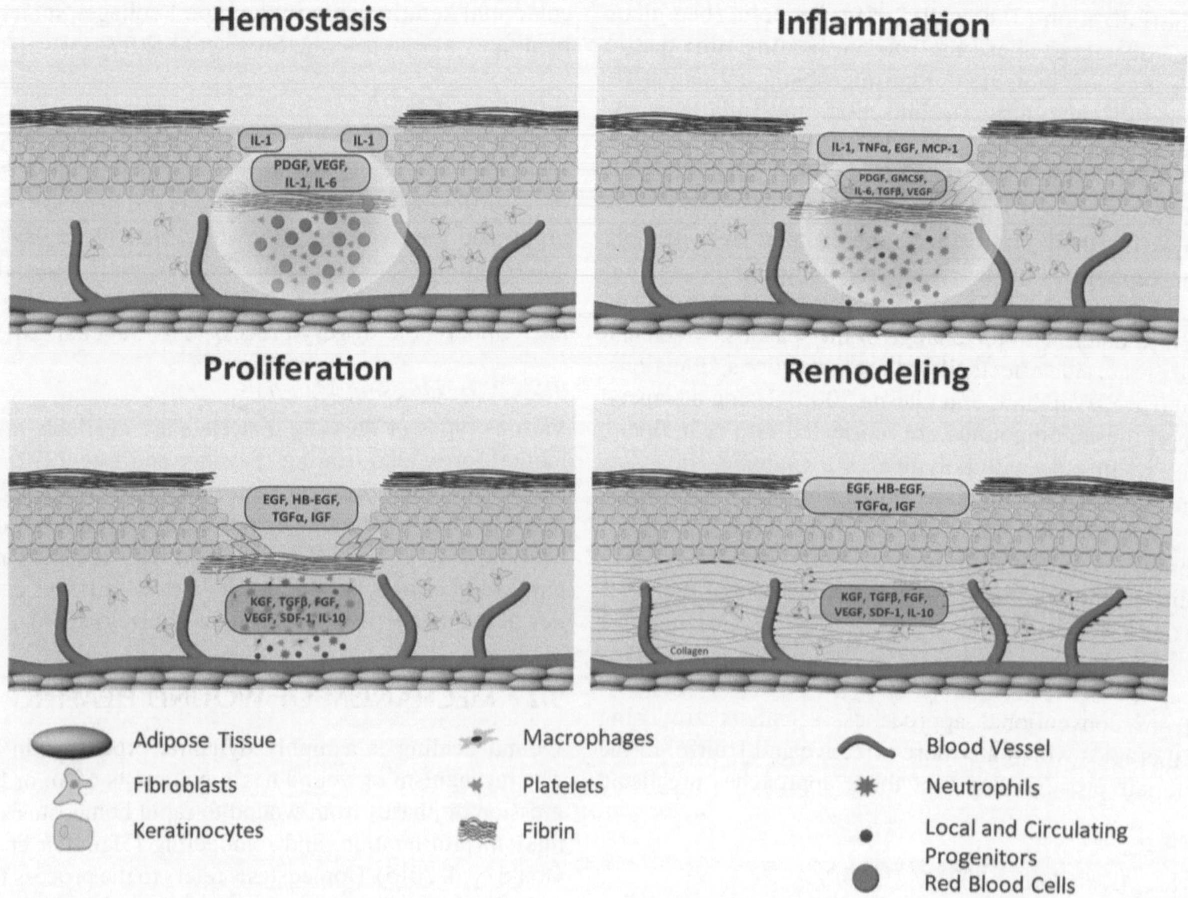

FIGURE 9.1 Phases of cutaneous wound healing depicting the cells and molecules responsible for the regaining of a healthy barrier. (Adapted with permission of Hamdan et al. 2017. Nanotechnology-driven therapeutic interventions in wound healing: Potential uses and applications. *ACS Central Science* 3, no. 3: 163–75; copyright © 2017 American Chemical Society.)

9.2.2 Inflammation Phase

In the inflammatory phase, the area of the injury site immobilizes the wound and prepares it for healing. In this phase, a barrier forms against invading microorganisms. Once the neutrophils come into action, it starts destroying the microorganisms by the mechanism called phagocytosis (Robson, Steed, and Franz 2001). Owing to the change of the surface properties of the neutrophils, they become sticky and attach on the endothelial cells using adhesion and a rolling mechanism (Carmeliet and Jain 2011). Endothelial cells also release chemokines, which also adhere on the site. In a later stage, narrowing of the wound occurs through the connective tissues. These mechanisms are controlled by neutrophils first, followed by macrophages and mast cells (Martin and Leibovich 2005). Lymphocytes come last in the inflammatory stage, which makes a path for the next stages to happen (Hunt, Hopf, and Hussain 2000).

9.2.3 Proliferative Phase

The proliferative phase starts with fibroblast migration followed by collagen deposition, angiogenesis, and granular tissue formation (Witte and Barbul 1997). The fibroblast cells come to the action site and proliferate profusely, which ultimately builds the matrix environment. Migratory fibroblasts are all over the wound area and start producing the collagen. Next, collagens start depositing after an abundant amount of extracellular matrix is accumulated (Clark 1993). During all phases, angiogenesis occurs concurrently as a reparative process. It helps in the proliferation and migration of the endothelial cells. After that, the epithelialization comes into the picture, which leads to the remodeling phase.

9.2.4 Remodeling Phase

In the remodeling phase of wound healing, the new epithelium forms and collagen deposition helps the randomly arranged scar tissue to form a specific alignment (Samuels and Tan 1999). Moreover, the non-elastic nature of the scar tissue potentiates the mimicking characteristics of the tissue undergoing the restoration process. Finally, the scar develops in the injury site through the remodeling forces (O'Kane 2002).

9.3 DIABETIC AND MECHANICAL WOUND HEALING

Delayed and incomplete wound healing is one of the risks for diabetes mellitus patients (Fahey et al. 1991). It is considered to be a life-threatening disease, which ultimately leads to prolonged hospitalization of patients and sometimes amputation, in severe cases (Ramsey et al. 1999). In diabetic patients accompanied with chronic wounds the healing process stops at the inflammatory phase as high blood sugar levels paralyze the blood and tissue defense mechanisms (Tellechea et al. 2016). Also, one of the major problems in the diabetic patients appears to be loss of cutaneous sensation (peripheral insensate neuropathy) and less blood flow in the limbs (Margolis et al. 2000). Generally, there are several reasons including hyperglycemia, hypoxia, chronic inflammation, dysfunction in signaling, circulatory pathways, debilitated neuropathy, etc., which contribute to the retarded wound healing (Baltzis, Eleftheriadou, and Veves 2014). However, it has been reported that the occurrence of the activated macrophages as well as the formation of the lymphatic vessel limits the rate of wound healing in diabetic ulcer patients (Maruyama et al. 2007). The reduction in the number of macrophages and also the lack of endothelial progenitor cell (EPC) mobilization and increasing level of matrix metalloproteinases (MMPs) are reasons for poor wound healing (Liu and Velazquez 2008, Liu et al. 2009). The death rate is high and chances of recurrence of healed ulcers are also higher in diabetic patients. For that reason, it is a compelling need to take adequate actions that can transfer the technologies from bench to bedside and to give meticulous attention in order to avoid precarious disease like diabetic foot ulcers.

Mechanical forces influence the formation of the blood vessel (vascularization) and recreation of tissue in the wound area. It has been reported that applying mechanical forces at the very beginning of healing prevents the normal process of wound healing. In later stages, however, it helps in the wound healing process. So, the process of healing is exquisitely sensitive towards stress (Boerckel et al. 2011). Boerckel et al. (2011) suggested that the application of mechanical stimuli on grown blood vessels induced remodeling with a more robust vascular network. However, the tension which also occurs due to mechanical stress negatively affects the healing mechanism. This has become a matter of concern for plastic surgery patients (Ireton, Unger, and Rohrich 2013). Insufficiency of the tissue owing to the wound dehiscence is one of the major difficulties in plastic surgery which appears due to mechanical stress (Sommerlad and Creasey 1978). Scientists have revealed that enhancement of the tensile strength because of increased level of collagen alignment is the most plausible reason. Nowadays, scientists are using mechanical forces to introduce the scarless condition with minimal risk. They are embracing therapies where mechanical offloading and mechanomodulation are being used, which will eventually give better outcomes in the area of wound healing (Barnes et al. 2018).

9.4 VARIOUS STRATEGIES IN WOUND HEALING

Various treatment strategies which are involved in the wound healing process target different stages of the wound repair. The strategies are categorized into three approaches: the nanomedicinal approach, the tissue engineering approach, and the stem cell (SC)-based approach.

9.4.1 Nanomedicinal Approach

The nanomedicinal approaches include therapies with various metals, non-metals, and metalloids and their applications in wound healing.

9.4.1.1 Metal-Based Nanoparticles

Different metals (gold, silver, copper, terbium, titanium, and zinc) which are involved in the healing process are briefly described below:

9.4.1.1.1 Gold

Colloidal gold nanoparticles (AuNPs) are used in different stages of wound healing, due to their surface, electronic, and optical properties (Eustis and El-Sayed 2006). AuNPs have been reported along with antioxidant agents like epigallocatechin gallate and alpha-lipoic acid in healing diabetic wounds. Similarly, Kim et al. (2015) used hydrocolloid membranes coated with phytochemically stabilized gold nanoparticles (pAuNP-HCM) for therapeutic effects on cutaneous wounds in Sprague Dawley rats. The authors observed that the size of the wound reduced in the treated rats after 5–15 days as compared to other groups. The authors also noticed that on day 5 in the treated groups, there was a decrease in matrix metalloproteinase-1 (MMP-1) and transforming growth factor-1 (TGF-β1) with an increase in collagen, angiopoietin 1, angiopoietin 2, and VEGF expression, which ultimately aided in wound healing. Finally, the authors mentioned that the combined effects of angiogenesis along with the activation of antioxidants and generation of connective tissue helped in the wound healing process, as shown in Figure 9.2. Likewise, Yang et al. (2017) prepared wound dressing materials using electrospun techniques which were resistant against multidrug-resistant (MDR) bacteria. The authors mentioned that the formulation was carried out by binding the AuNPs with the small molecule 6-aminopenicillanic acid, and then the scaffold was prepared by poly(ε-caprolactone) (PCL)/gelatin to evade wound infection by MDR bacteria. The author further added that the as-prepared Au-APA could withstand the MDR bacteria and exhibited tremendous biocompatibility. Additionally, the Au-APA displayed excellent wound healing efficacy in Wistar rats (female) in a 14-day period, as evinced by the lymphocytes and neutrophils (inflammatory cells) present at the wound site. Altogether, the authors shed light on the fruitful clinical application of the homogenous scaffold materials for wound dressing in the near future, as shown in Figure 9.3.

9.4.1.1.2 Silver

Various groups all over the world have been using silver in their formulations for effective therapeutic agents. Likewise, our group also developed a silver metal complex which simultaneously exhibited both wound healing ability and an antibacterial property (Rao et al. 2018). We designed the $Ag_2[Fe(CN)_5NO]$ nanoparticles (silver complex, AgNNPs)

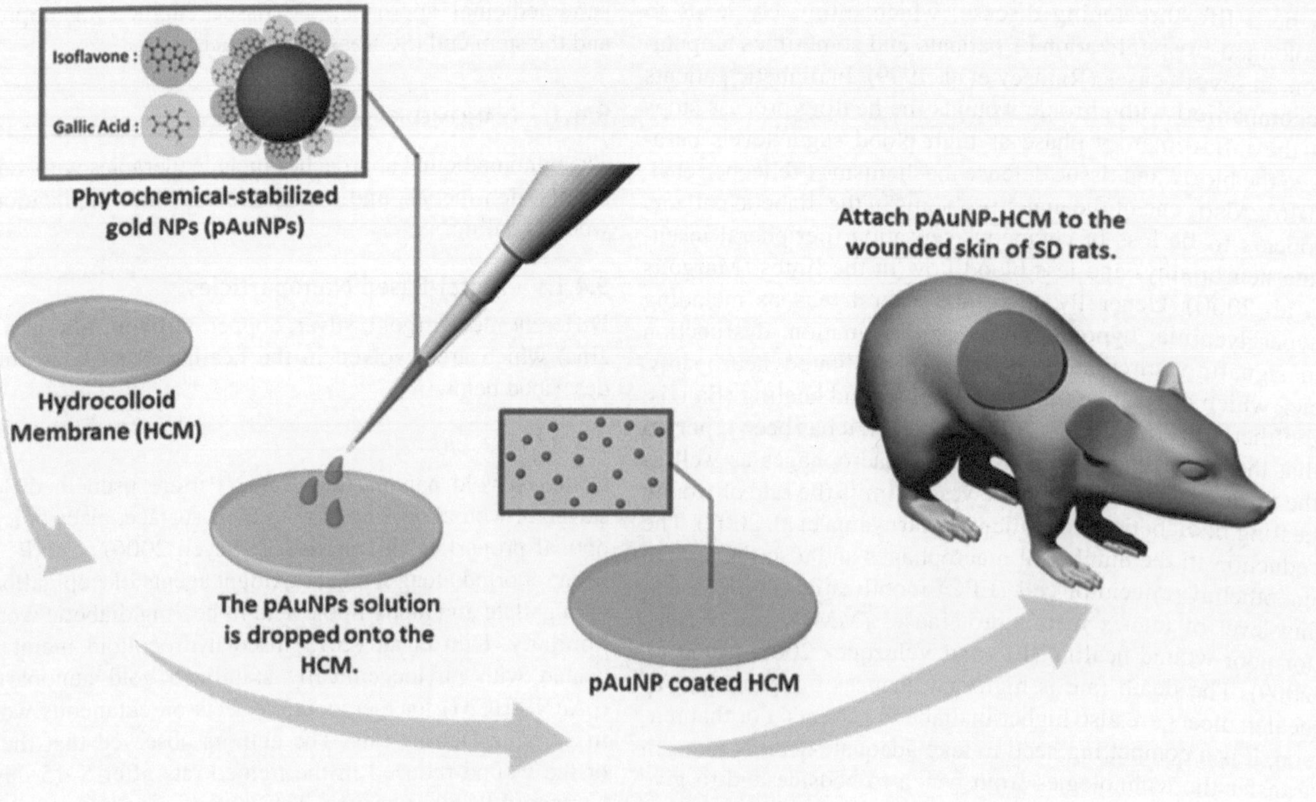

FIGURE 9.2 Illustration of pAuNP-HCM preparation and the skin damage treatment process (non-scalable). (Adapted with permission of Kim et al. 2015. Accelerated healing of cutaneous wounds using phytochemically stabilized gold nanoparticle deposited hydrocolloid membranes. *Biomaterials Science* 3, no. 3: 509–19; copyright © 2014 Royal Society of Chemistry.)

from the precursors silver nitroprusside and silver nitrate. The as-synthesized nanoparticles were thoroughly characterized using several techniques such as DLS, XRD, FTIR, EDAX, Raman, ICP-OES, FESEM, etc. This nanoformulation escalated the wound healing efficacy in the C57/BL6 model by changing the macrophages from M1 to M2. Our group explained that there were two kinds of macrophage activation: classical activation (M1) and alternative activation (M2). The immune-histological studies revealed that the expressions of the arginase-1 (marker of M2) in treated groups were higher than in the untreated group. In fact, the expression of NOS (marker of M1) was observed to be high in untreated as well as vehicle control groups, which clearly stipulated the idea of conversion of the M1 phenotype to M2 after treatment with the nanoformulation. Finally, our group suggested that the conversion of the M1 to M2 decreased the inflammatory response in the mouse model, which eventually enhanced the wound healing process, as shown in Figure 9.4.

9.4.1.1.3 Copper

It has been observed that copper helps in wound healing through formation of skin by synthesis and stabilization of extracellular matrix proteins as well as by angiogenesis. Regarding this, Borkow et al. (2010) introduced copper oxide into wound dressings so the copper ions would release directly and slowly into the wounds. This method helped in wound healing, as copper upregulates integrin, which is a transmembrane receptor of extracellular matrix proteins that transduces cell processes of wound healing. The authors have shown that copper induces PLGF, which activates VEGF through the VEGF receptor involved in the angiogenesis process that assists in wound healing. Presently, with the boon of nanotechnology in healthcare, even copper is transformed from its native form to nanoform with the help of microorganisms, by a process called biosynthesis. This gives better catalytic activity and bioavailability, resulting in increments of antimicrobial activity and reduction in toxicity (Uauy, Olivares, and Gonzalez 1998). Also, copper plays a crucial role in stimulating VEGF for angiogenesis in wound healing and promotes immunity by stimulating interleukin-2. Hence, Tiwari et al. (2014) fabricated copper nanoparticles by *Pseudomonas aeruginosa* which showed application in wound healing. The authors monitored the wound healing activity of biosynthesized copper nanoparticles in rat through excision model.

9.4.1.1.4 Titanium

Along with different metallic nanoparticles like silver, gold, zinc, and copper nanoparticles, titanium also shows therapeutic potential for wound healing (Oyarzun-Ampuero et al. 2015). In 2013, Archana et al. (2013) developed a new wound dressing material composed of biopolymer chitosan, synthetic polymer poly(N-vinylpyrrolidone) (PVP), and

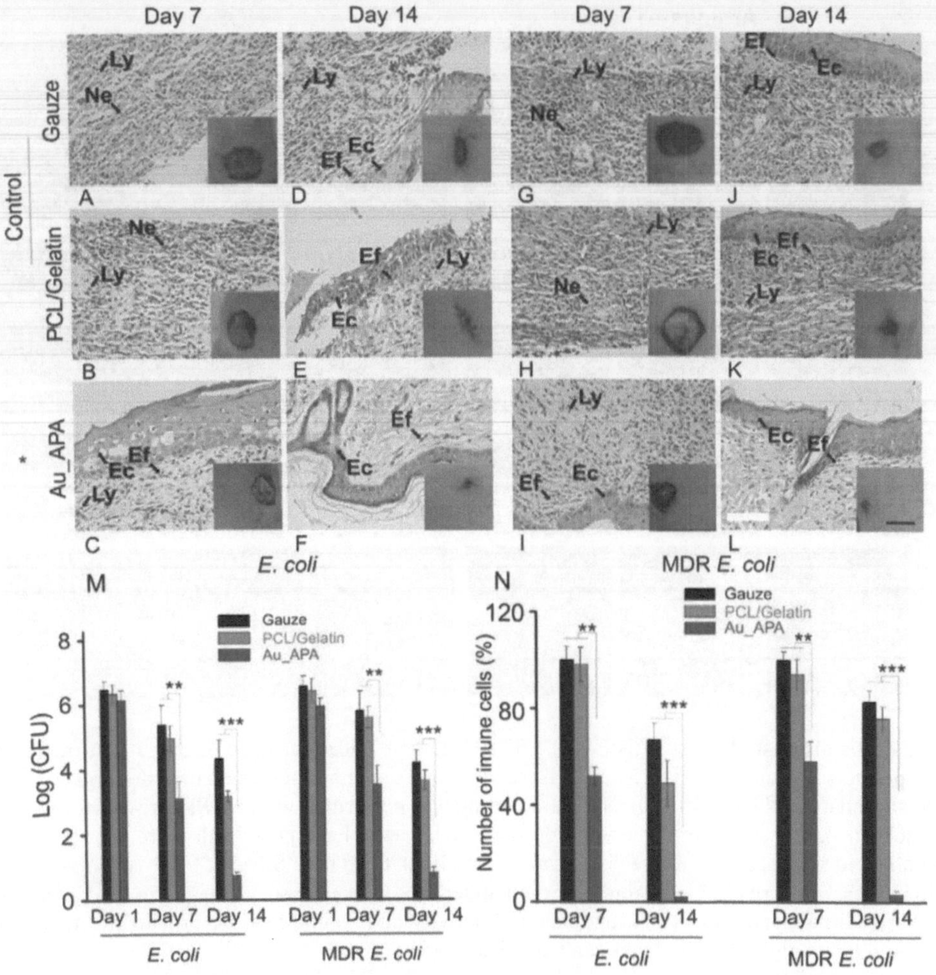

FIGURE 9.3 Animal models of wounds infected by *E. coli* or MDR *E. coli*. Histological images of skin tissues stained by H&E dissected on postoperative day 7 (A–C, G–I) and day 14 (D–F, J–L). (M) Local bacteria count in the wound area and (N) relative to the number of immune cells compared with control group (gauze). Inserted photos show the macroscopic morphologies of wounds. The control groups are treated with wound dressings of gauze (A, D, G, J) and PCL/gelatin nanofibers (B, E, H, K) and the experimental group used Au-APA electrospun nanofibers as wound dressing (C, F, I, L). The presence of Ly and Ne indicates inflammatory response. Ec and Ef are the signal of reepithelization, and they are beneficial for the formation of matured fibrous granulation tissue; ** and *** represent two levels of statistically significant differences (**$P < 0.05$, ***$P < 0.01$). White and black scale bars are 100 µm and 1 cm, respectively (Ly, lymphocyte; Ne, neutrophil; Ec, epithelial cells; Ef, elongated fibroblasts). (Adapted with permission of Yang et al. 2017. Pharmaceutical intermediate-modified gold nanoparticles: Against multidrug-resistant bacteria and wound-healing application via an electrospun scaffold. *ACS nano* 11, no. 6: 5737–45. copyright © 2017 American Chemical Society.)

titanium dioxide (TiO_2) and studied its effects in wound healing. The authors mentioned that the as-prepared nanocomposite demonstrated biocompatibility towards L929 and NIH3T3 fibroblast cells, along with excellent antibacterial activity. The good swelling efficacy with elevated WVTR properties made it effective for wound healing in albino rat model.

9.4.1.1.5 Terbium

Among the lanthanide metals, terbium especially has found a profound application in wound healing. Our group has synthesized pro-angiogenic terbium hydroxide nanorods (THNRs) using a microwave irradiation technique (Nethi et al. 2017). The as-synthesized nanoparticles were thoroughly characterized by several physicochemical techniques like DLS, XRD, TEM, FT-IR, etc. In order to prove the wound healing ability of the nanorods in the in vitro system, we performed the wound healing scratch assay in EA.hy926 cells. The result demonstrated that higher migration of the cells was observed at 8 h upon treatment with the nanorods (dose: 5 µg/mL), as shown in Figure 9.5. Further, the wound healing efficacy was also observed in in vivo mice model (C57/BL6). The wound was created on the dorsal surface of the mice using the punch method. Doses were given in alternative days from day 1 to day 7 with the THNRs (1%). The result demonstrated that upon treatment with the nanorods, there was enhanced wound healing in the THNR-treated group compared to positive control (Silverex-treated) and vehicle control groups, which was confirmed by histopathological analysis of the skin tissue. Also, we found an escalated amount of angiogenesis as well

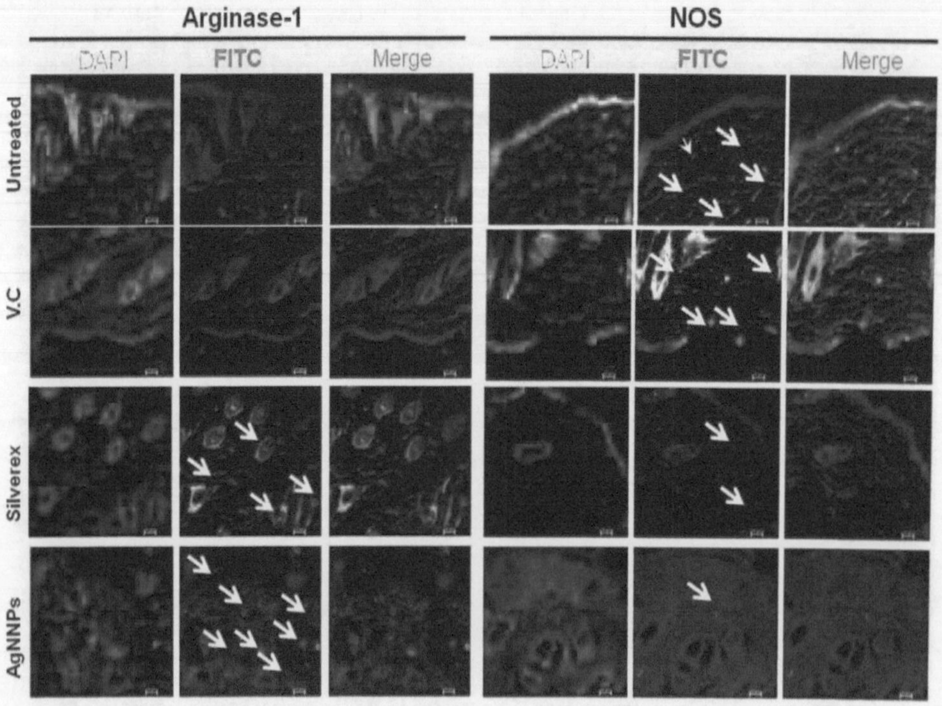

FIGURE 9.4 Immunohistochemistry study of skin tissues of C57/BL6 female mice sacrificed on day 9 post-treatment with AgNNPs. The presence of macrophage polarized forms such as M2 and M1 are identified using antibodies against an Arg-1 and NOS (1:1000), respectively, and FITC-labeled secondary antibody (1:1000). Nuclear staining was performed with DAPI. Arg-1-positive cells depicted as arrows in M2 macrophages present in skin tissues. Similar observations for M1 macrophage polarization are shown with arrows in the stained skin tissues. First row: untreated skin; second row: vehicle control (Vaseline); third row: Silverex 0.2% (w/w) treated; fourth row: AgNNPs 0.1% (w/w)–treated skin tissues, respectively. Cells were observed under 60× objective; scale bar: 5 µm. (Adapted with permission of Rao et al. 2018. Ag_2 [Fe (CN) $_5$NO] nanoparticles exhibit antibacterial activity and wound healing properties. *ACS Biomaterials Science & Engineering* 4, no. 9: 3434–49; copyright © 2018 American Chemical Society.)

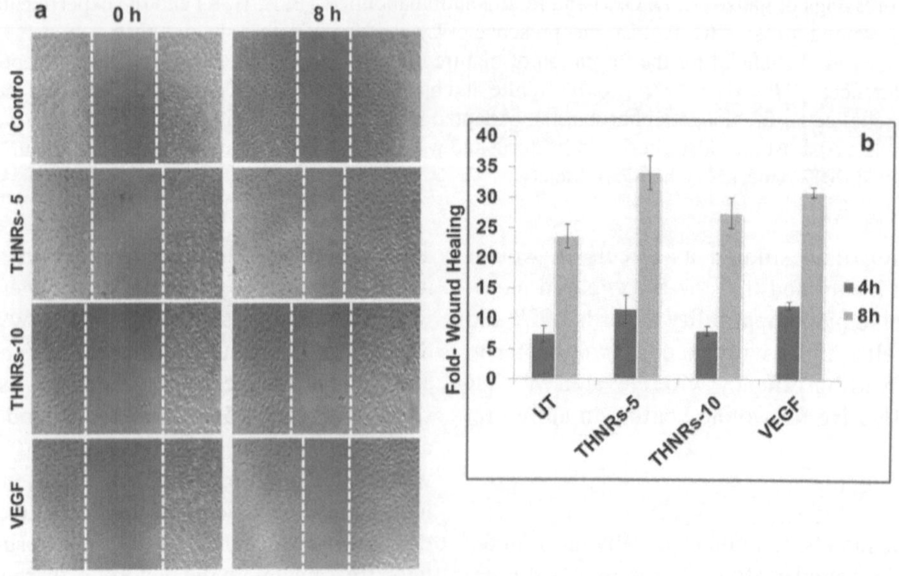

FIGURE 9.5 (a) Determination of endothelial cell (EA.hy 926) migration induced by THNRs (5–10 µg/mL) treatment using wound healing (scratch) assay. (b) Histogram representation of the closure of the scratch area quantified using ImageJ software. (Adapted with permission of Nethi et al. 2017. Pro-angiogenic properties of terbium hydroxide nanorods: Molecular mechanisms and therapeutic applications in wound healing. *ACS Biomaterials Science & Engineering* 3, no. 12: 3635–45; copyright © 2017 American Chemical Society.)

as fibrosis in the tissue section in the treated group, indicating higher wound healing capability of the THNRs, as shown in Figure 9.6.

9.4.1.1.6 *Zinc*

Zinc (Zn) is an essential trace element and is present at less than 50 mg/kg body wt. Besides aiding in growth and development, it also partakes in immune and central nervous system functions. Zn also acts as a co-factor for many enzymes which are involved in DNA repair, transcriptional regulation, metabolic processing, apoptosis, extracellular matrix (ECM) regulation, and antioxidant defense (Lin et al. 2017). ZnO in nanoparticle form has been seen to be most effective in wound healing. It generates ROS, which enhances wound healing by stimulating growth factor–mediated pathways to improve cell migration and adhesion. ZnO has antibacterial and anti-inflammatory activity as well (Sharma, Garg, and Gupta 2016). ZnO can be embedded with any biomaterial and used for wound healing. Similarly, Díez-Pascual et al. (2015) used the versatile casting and solvent mixing method to design nanocomposites. These nanocomposites were prepared using castor oil as the matrix material, and chitosan-modified ZnO was incorporated into it. The authors declared that the nanocomposite is biocompatible, having higher water absorptivity, biodegradability, viscoelasticity, and antibacterial properties as well as superior thermal stability compared to film only. The group mentioned that upon loading, the chitosan modified with ZnONPs eventually increased the oxygen permeability, water vapor transmission rate, etc., of the composites. Further, the CS-ZnO (loading ≤5.0 wt %) exhibited no toxicity against human dermal fibroblast cells, which corroborated the biocompatible nature of the formulation. The authors also claimed that treatment with CO/CS-ZnO manifested enhanced wound healing ability through collagen deposition and re-epithelialization in an in vivo model. Finally, the authors shed light on the future application of the biocompatible CO/CS-ZnO as a wound healing material in the biomedical field.

9.4.1.2 Non-Metal and Metalloid-Based Nanoparticles

Carbon (non-metal) is the most abundant element in the Earth's crust (Cotton and Wilkinsion 1972). Graphene, which is made of carbon, has occupied a significant position in various biological usages. Different groups all over the world are using graphene due to its interesting properties. Additionally, among various metalloids silicon is considered to be one of the most important elements in biomedical application. Herein, we discuss the role of silicon and graphene in wound healing applications.

9.4.1.2.1 *Silicon*

Scientists have been using various materials like collagen, alginate, gelatin, and polyurethane for making different forms of wound dressing materials like foams, hydrogels, films, etc. Ori et al. (2017) showed that a novel formulation involving SiO_2 nanoparticles and a biocompatible material, PVP, accelerated wound healing in murine models. The authors mentioned that cytotoxicity studies of the dressing materials demonstrated no cytotoxicity to murine fibroblasts L929. The group also claimed that the nanocomposite promoted wound closure by a number of mechanisms like allowing efficient diffusion of SiO_2 materials into the wound bed, as PVP is

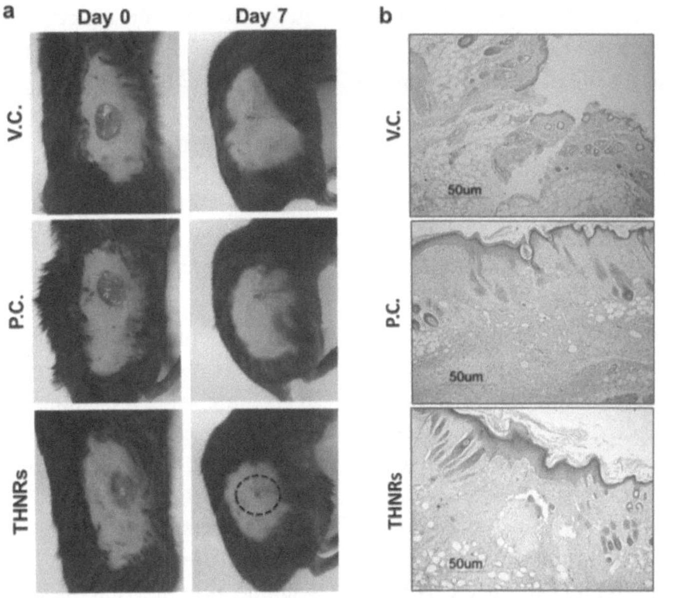

FIGURE 9.6 (a) In vivo wound healing study of THNRs (1%) in C57/BL6 punch biopsy mouse model. (b) Histological analysis of isolated skin sections around the wound area stained with hematoxylin and eosin (H&E). V.C., vehicle control; P.C., positive control (Silverex). (Adapted with permission of Nethi et al. 2017. Pro-angiogenic properties of terbium hydroxide nanorods: Molecular mechanisms and therapeutic applications in wound healing. *ACS Biomaterials Science & Engineering* 3, no. 12: 3635–45; copyright © 2017 American Chemical Society.)

an efficient drug delivery system. Further, the migration of keratinocyte as well as the proliferation and differentiation of epithelial cells ameliorated vascularization at the wound site.

9.4.1.2.2 Graphene

Graphene, due to its unique physicochemical properties (like high surface area and good electrical-thermal conductivity) and biological properties (high drug loading efficacy, biocompatibility, biodegradability, etc.), has gained much interest in the field of biomedicine in recent years. Graphene also has therapeutic applications like the promotion of angiogenesis, cancer therapy, antibiotic therapy, etc. Tissue engineering has also used graphene nanomaterial alone or in combination with other inorganic nanoparticles to obtain nanocomposites for multiple biological applications (Yang et al. 2013). For its role in pro-angiogenesis, graphene has been studied extensively by researchers to evaluate its role in wound healing and tissue repair. In this context, Lu et al. (2012) modeled a wound dressing material consisting of graphene chitosan (CS) and *poly(vinyl alcohol)* (PVA) composites. They adopted the scotch tape method to stack a few layers of graphene, and then the assembly was electrospun and doped with CS-PVA nanofibers. On applying the membrane on artificially created wounds of rabbit and mice models, the group demonstrated that the membrane showed significant wound healing activity. Simultaneously, the nanocomposite exhibited strong antimicrobial activity against *E. coli*, *Agrobacterium* sp., and yeast. This proved that the compound was suitable for direct application for healing of wounds as it prevented infection at the wound site.

9.4.1.3 Nanoparticles as Carriers

The development of nanoparticles acting as carriers in the wound healing area advances modern therapies. Various antibiotics, growth factors, and nucleic acid materials are delivered using nanoparticles to the target site in order to potentiate the healing process.

9.4.1.3.1 Antibiotics Containing Nanoparticles

An antibiotic is an antimicrobial agent that kills bacteria or stops their reproduction. An injured tissue is a target for invasion by microbes, as it provides a favorable environment for microbial growth. The microbes can infect the exposed tissue, and consequently the wound fails to heal. Nanobiotics are nanomaterials with intrinsic antimicrobial activity (Huh and Kwon 2011) that sometimes are conjugated with antimicrobial agents to increase the effectiveness of antibiotics (Gu et al. 2003). The plausible pathways by which antimicrobial nanoparticles cause damage to microbes includes ROS production, membrane disruption, DNA damage, inhibition of enzymatic activities inside the cell, suppression of RNA and protein synthesis, hindrance of biofilm formation, etc. (Li et al. 2008, Kang et al. 2007, Rabea et al. 2003, Huh and Kwon 2011, Pelgrift and Friedman 2013). They have better accumulation and retention efficiency, which is beneficial for sustained and controlled release; localized and uniform distribution in target tissues; enhanced cellular internalization; and reduced side effects. For example, Chakraborty et al. (2010) prepared carboxymethyl chitosan nanoparticles tagged with folic acid by covalent linkage through 2,2-(ethylenedioxy) bis-(ethylamine) (EDBE) for use against vancomycin-resistant *Staphylococcus aureus* (VRSA) and vancomycin-sensitive *Staphylococcus aureus* (VSSA). The authors revealed that the biocompatible, biodegradable nanoparticles acted like a Trojan horse in order to deliver the drug into the target site.

9.4.1.3.2 Growth Factors

A growth factor is one kind of naturally occurring protein or steroid hormone which can promote proliferation, healing and helping in cellular differentiation or growth. They harmonize all cellular processes in wound healing by producing cytokines that attract cells into the wound site, resulting in their proliferation and ECM deposition (Hantash et al. 2008, Barrientos et al. 2008). The growth factors which are normally used to accelerate wound healing are TGF-β (infiltrates cells into the wound area), EGF, PDGF (assisting cell migration, activating the generation of fibroblasts and epithelial cells), FGF, and VEGF (helps in formation of new blood vessels) (Singer and Clark 1999). Delivery systems have been configured in different ways and designed in divergent manners using degradable and non-degradable natural polymers. Generally, poly(lactic acid) (PLA) and poly(lactic-co-glycolic acid) (PLGA) are biocompatible and biodegradable polymers.

9.4.1.3.3 Nanotherapies Using Nucleic Acids

In order to knock down or escalate the target gene which actively participates in a regeneration process, gene-activated matrix therapy plays a vital role (Wang, Ma, and Gao 2014). The aforesaid therapy can come in two different ways: gene therapy and tissue engineering. Gene therapy generally increases the therapeutic efficacy in wound healing. Targeted genes are delivered through divergent techniques like use of viral and non-viral vectors, a direct injection method (plasmid DNA), electroporation of naked or colloidal DNA, gene gun, etc. Scientists are nowadays using matrices made of nanofiber meshes, which are helpful in wound healing. Shi et al. (2018) prepared biodegradable nanospheres containing VEGF-A and PDGF pDNAs and PLGA matrix. The authors proclaimed that the formulation accelerated angiogenesis, collagen deposition, re-epithelialization, and wound closure.

RNAi or siRNA therapy involves the delivery of RNAi to a target site for silencing the expression of any protein-producing gene that can cause disease or amplify a diseased condition. It is used to silence the expression of genes for molecules like MMP-9, which degrades ECM and impairs tissue reorganization during the healing process (Castleberry et al. 2016). Nanoparticle carriers in use include polymer carriers like miscelles (Kim et al. 2014), nanogels (De Backer et al. 2015), dendrimers (Serramía et al. 2015), quantum dots (Subramaniam et al. 2012), liposome nanoparticles (Yang et al. 2014), etc. These technologies are biocompatible and offer more selective targeting and better retention capacity and effectiveness than current RNA delivery systems. Randeria et al. (2015) synthesized gold nanoparticle–conjugated ganglioside-monosialic acid-3 synthase (GM3S)-thiolated siRNA. The authors mentioned that the siRNA downregulated the expression of the GM3 gene. Also, the authors mentioned that GM3S SNA at 5 nM concentrations reduced both the protein and GM3S mRNA at a significant rate (80%), and dose-dependent knockdown was observed in immortalized mouse KCs. In fact, complete elimination of the GM3 ganglioside expression was found at the 72-h time point upon treatment with the GM3S SNA which was confirmed, by confocal microscopy analysis. The in vitro wound healing scratch assay demonstrated the better knockdown ability of the GM3S in primary mKCs cultured in serum-free CnT07 medium compared to GM3S$^{-/-}$ primary mKCs at the 60-h time point. The group mentioned that almost complete migration was noticed with the GM3S SNA-treated group in the hyperglycemic condition. Additionally, significant proliferation was observed with GM3S SNA-treated mKCs compared to GM3S$^{-/-}$ mKCs at 5 days, as shown in Figure 9.7.

9.4.2 Tissue Engineering Approach

This modern approach to tissue regeneration involves construction of scaffolds using nanomaterials and the formation of small molecule–based nanostructures which repair and regenerate injured tissue.

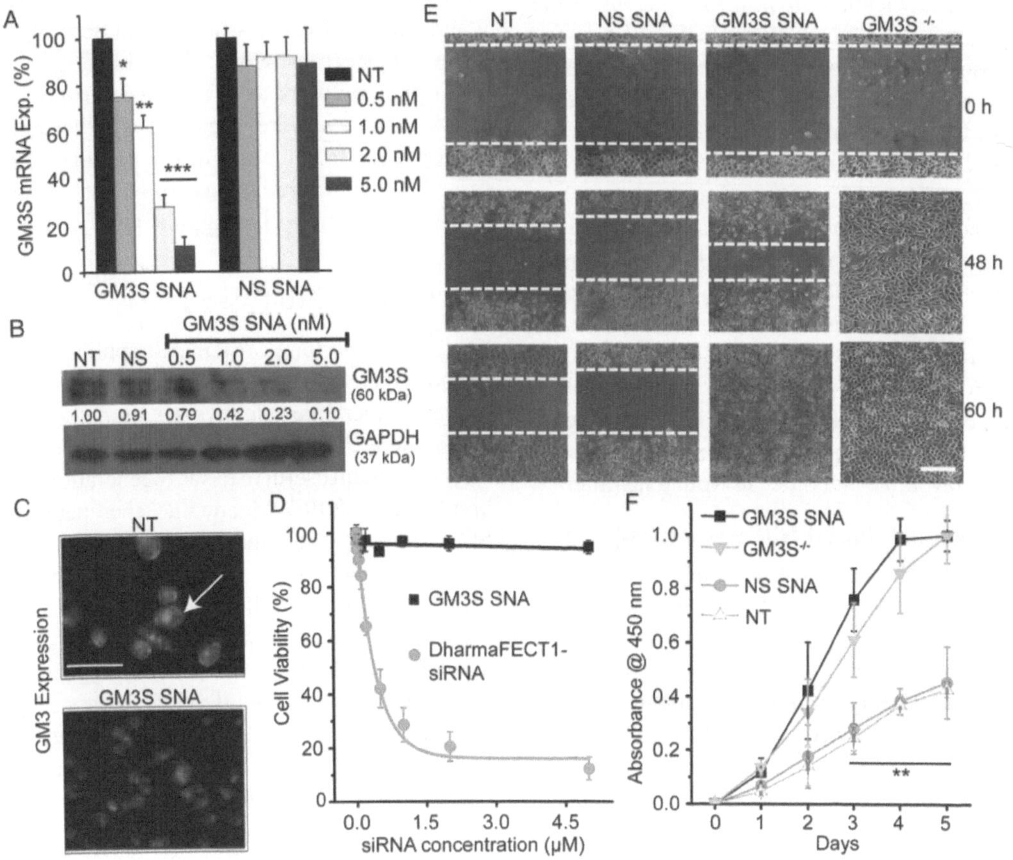

FIGURE 9.7 In vitro characterization of GM3S SNA. (A) GM3S mRNA and (B) protein expression as quantified by RT-PCR and Western blotting after GM3S SNA and NS SNA (0.5, 1.0, 2.0, and 5.0 nM) were incubated with immortalized mKCs in complete medium. GM3S SNA led to a dose-dependent decrease in GM3S mRNA and protein expression as low as 20% of that of control-treated cells. A significant reduction in GM3S mRNA from NS SNA-treated controls was noted with as little as 0.5 nM GM3S SNA. (C) Confocal immunofluorescence shows elimination of GM3 in KCs treated with (lower) GM3S SNA relative to (upper) NT (arrow; green, GM3; red, propidium iodide-stained nuclei). (Scale bar: 50 μm.) (D) Trypan blue exclusion assay showed excellent cell viability after 48 h of treatment with GM3S SNA treatment vs. the identical siRNA sequence coupled with DharmaFECT1 transfection agent. (E) In vitro wound closure (scratch assay) of primary mKCs in serum-free medium with excess glucose (total of 18 mM) were pretreated for 48 h with GM3S SNA, NS SNA, or NT; scratched; and monitored for wound closure. mKCs from GM3S KO mice (GM3S–/–) were studied as a positive control. (Scale bar: 120 μm.) (F) In water-soluble tetrazolium-1 assays, GM3S SNAs increased proliferation of primary mKCs within 3 d with a course similar to that of GM3S–/– mKCs. All data are represented as means ± SDs. All studies were performed in triplicate at least three times. *$P < 0.05$; **$P < 0.01$; ***$P < 0.001$. (Adapted with permission of Randeria et al. 2015. siRNA-based spherical nucleic acids reverse impaired wound healing in diabetic mice by ganglioside GM3 synthase knockdown. *Proceedings of the National Academy of Sciences* 112, no. 18: 5573–8; Copyright © 2018 National Academy of Sciences.)

9.4.2.1 Nanoscaffolds

One of the foremost advances in nanomedicine which is now being used in wound healing is the usage of nanoscaffolds. There are several techniques used to make nanoscaffolds: self-assembly, electrospinning, and phase separation. Among these techniques, electrospinning can provide both degradable (collagen, chitosan) and non-degradable (PLA, PVA) nanofibrous sheets. Shahverdi et al. (2014) fabricated a hybrid nanoscaffold with silk fibroin (SF) and PLGA using the electrospinning technique. This hybrid membrane was porous with high tensile strength that stimulated the attachment and proliferation of fibroblast cells. The authors mentioned that the attachment of SF with PLGA remarkably increased the proliferation of the LC2C cells under in vitro conditions. The group also showed that upon treatment in an animal model, the scaffold effectively reduced the wound area. Scientists are now focusing on various aspects such as nanotopography, which deals with the surface properties of nanoscaffolds. In the wound healing process, it plays a vital role by altering the migration, proliferation rate of cells, generation of ECM, etc. For example, Kim et al. (2012) developed silicon wafer–conjugated polystyrene nanogrooves in order to study the effect of nanotopography on the migration of cells in wound closure. Concomitantly, there are various well-suited characteristics which nanoscaffolds should possess, including biocompatibility, regulated porosity, and permeability (Chawla et al. 2014). The nanoscaffolds should have other important features like the ability to mimic the fibrous nature of ECM and to provide the substratum for various cells in wound healing (Le et al. 2013). Nevertheless, it is important for several factors to be taken into consideration in order to prepare nanoscaffolds to get favorable outcomes in wound healing.

9.4.2.2 Small Molecule–Based Nanostructures

Wound healing requires the influence of a number of biological factors. Using nanotechnology, these factors can be engineered to reach the target sites that influence the process of wound healing in a prominent way. Some small molecules that take part in wound healing are used at the nanoscale level for accelerating the process (Tocco et al. 2012).

9.4.2.2.1 Thrombin

Thrombin (also called factor II or factor IIa) is a serine protease that converts fibrinogen into fibrin in order to form fibrin clots, which helps in activating platelets during blood coagulation. It also regulates homeostatic response and participates in inflammation. Thrombin promotes chemotaxis, agglomerates the lymphocytes, neutrophils, and monocytes, and propagates the endothelial, fibroblast, and epithelial cells. In this way, thrombin is used in wound management and has found topical application over a long period (Carney et al. 1992). Ziv-Polat et al. (2010) conjugated thrombin to maghemite (γ-Fe_2O_3) nanoparticles to check its efficacy in wound repair. The authors cited that the composite showed considerable improvement in skin tensile strength, along with an adhesion property of skin grafts with further reduction in wound dehiscence, which assisted in post-cosmetic surgery recovery in patients. Histopathological analysis revealed that formations of the granulation tissue along with moderate inflammation were observed in all samples. Further, the bound thrombin group displayed lower surgical scarring and less granulation tissue compared to the untreated group, which supported the superior healing activity of the bound thrombin. Additionally, the authors stated that only the free thrombin group showed wide scars as well as narrow scars at 8 days post-operation, as shown in Figure 9.8.

9.4.2.2.2 NO (Nitric Oxide)

Besides promoting vasodilation of blood vessels, NO promotes proliferation and migration of keratinocytes, fibroblasts, endothelial cells, etc., to the wound site in different stages of wound healing mechanisms (Rodenwaldt, Pohl, and de Wit 2007). Freidman et al. (2008) fabricated silane-based hydrogel nanoparticles as a vehicle for delivery of NO to the wound site for accelerating wound healing. The authors used tetramethyl orthosilicate, chitosan, polyethylene glycol, and sugar glass network to form a structural matrix, within which NO_3 molecules were sheathed by thermal reduction. This homogenous formulation of the nanoparticles allows for sustained and controlled release of NO at the target site, evincing anti-infective properties against pathogens.

9.4.3 Stem Cell–Based Therapy

The broad repertoire of stem cell applications in multiple fields includes their usage in the wound healing domain. Their unique characteristics like effective secretome, differentiation ability, and moderate immunogenicity with efficient self-renewal property make them useful in tissue regeneration (Cerqueira, Pirraco, and Marques 2016). Meanwhile, therapeutic approaches using SCs are being used remarkably to induce re-epithelialization in chronic wounds. Some of the cell types that have been used in therapeutics are umbilical cord derived MSCs, mesenchymal stem cells from bone marrow (BM-MSCs), adipose-derived stromal cells (ADSCs), endothelial progenitor cells taken from bone marrow, placenta-derived stem cells, bone marrow–derived mononuclear cells (BMMNCs), etc. (Ojeh et al. 2015, Blanpain et al. 2004). Ma et al. (2011) attached BM-MSCs with a nanofiber scaffold made of collagen/PLGA, which increased the wound healing potency. The group proclaimed that the involvement of BM-MSCs ensured high collagen synthesis with an increase in epidermal edge in rat model. Further, the group demonstrated the expression of keratin 10 as an indication of the involvement of BM-MSCs in various stages of the wound repair procedure, evincing the potentiality of the composite. In spite ofseveral progressions in stem cell therapy, no treatments have received FDA approval in the nanomedicinal field for wound healing (Ojeh et al. 2015).

9.5 UNMET MEDICAL CHALLENGES AND FUTURE PERSPECTIVE

Wounds of any sort – chronic, acute, or diabetic – relentlessly affect large numbers of people worldwide. Current researchers have found various limitations and minimal success in

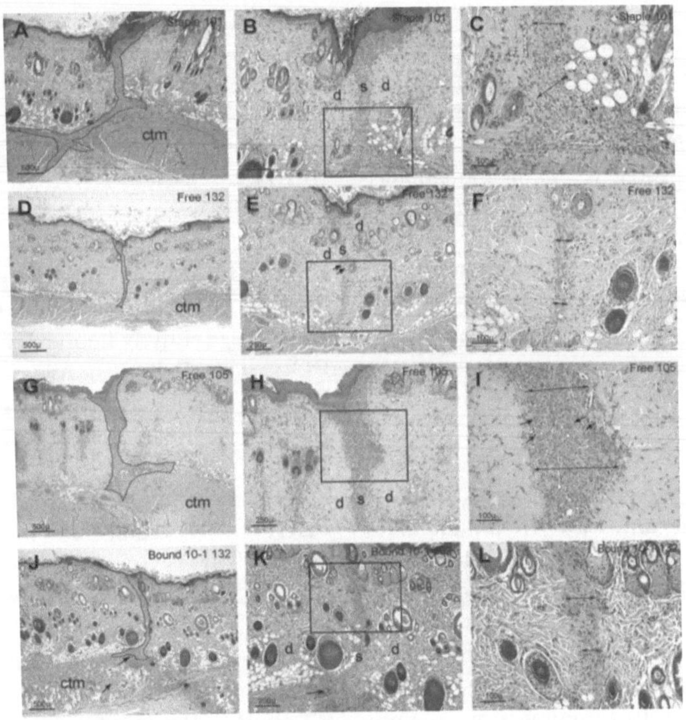

FIGURE 9.8 Hematoxylin and eosin staining of incisional skin wounds at 8 days post-operation. Note: In the hematoxylin-eosin staining procedure, the hematoxylin stains blue the nucleus, the acidic regions of the cytoplasm, and the cartilage matrix. Eosin stains pink the basic regions of the cytoplasm and the collagen fibers. A–C: Control wound closed with staples only; D–F: Wound treated with free thrombin and then closed with staples; G–I: Replicate wound treated with free thrombin and then closed with staples; J–L: Wound treated with bound thrombin and then closed with staples. A: Low-power view of a control sample. The surgical gap is closed by granulation tissue (traced with a black line) (ctm–cutaneous trunci muscle). B: Medium-power view of the center of the sample. The granulation tissue in the scar (s) has a markedly higher cellular density than the flanking dermis (d). The boxed area is shown in C. C: High-power view of the scar. Double-headed arrows indicate the width of the scar. This amount of scarring is considered moderate. D: Low-power view of a sample treated with free thrombin. The surgical gap is closed by granulation tissue (traced with a black line). E: Medium-power view of the center of the sample. The granulation tissue in the scar is moderately more cellular than the flanking dermis. The arrows indicate entrapped hair. The boxed area is shown in F. F: High-power view of the scar. Double-headed arrows indicate the width of the scar. This amount of scarring is considered mild. G: Low-power view of another sample treated with free thrombin. The surgical gap is closed by granulation tissue (traced with a black line). H: Medium-power view of the center of the sample. The granulation tissue in the scar is markedly more cellular than the flanking dermis. The boxed area is shown in I. I: High-power view of the scar. Double-headed arrows indicate the width of the scar. This amount of scarring was considered (high) moderate. Arrows indicate inflammatory cells admixed within the granulation tissue of the scar. J: Low-power view of sample treated with bound thrombin. The surgical gap is closed by granulation tissue (traced with a black line). There is moderate fibrosis in the cutaneous trunci muscle (ctm) region (asterisks) and a few small trichogranulomas (foreign body reaction centered on a free hair shaft, arrows). K: Medium-power view of the center of the sample. The granulation tissue in the scar is moderately more cellular than the flanking dermis. The boxed area is shown in L. An arrow points to a small trichogranuloma. L: High-power view of the scar. Double-headed arrows indicate the width of the scar. This amount of scarring was considered mild. (Adapted with permission of Ziv-Polat et al. 2010. Enhancement of incisional wound healing by thrombin conjugated iron oxide nanoparticles. *Biomaterials* 31, no. 4: 741–7; Copyright © 2009 Elsevier Ltd.)

the case of wound healing. Only a few therapeutic agents have been deemed efficacious by the FDA and are now available in the market. Researchers are now legitimately trying to make new kinds of therapeutic approaches that will deal with advanced stages of disease. In general, there are various factors that have to be taken into consideration for developing such therapies for patients (Hamdan et al. 2017). The colossal challenges that hinder the path into new therapeutic approaches include the following: complexity in patient pathophysiology, insufficient knowledge about the target molecular mechanisms and pathways, optimizing the timing for delivery of therapeutic agents at the target sites, and development of combinational therapy for targeting multiple flawed cellular processes (Senturk et al. 2016). Despite several promising efforts, the challenges still persist due to incomplete knowledge about the behavior of the nanoparticles within the body physiology. For example, various noncovalent interactions of nanomaterials with the blood may cause a number of problems, like preventing the nanoparticles from reaching the target and impeding the flow of blood through blood vessels, which can cause undesired side effects and inflammation (Senturk et al. 2016). So, with comprehensive knowledge about these interactions, scientists are focusing on developing more biocompatible, bioactive nanomaterials to cure diseases.

Additionally, the development of such nanotherapies that will meet the international standards of toxicology and biocompatibility is indispensable. Apart from that, the complications of maintaining the purity of the nanoparticles hinders the overall process. Thus, better synthetic tools and formulations are in incessent demand for the production of the nanomaterials that can be translated to the clinical arena (Gunasekaran et al. 2014). In view of this rising concern, further in-depth studies and solid ambitions are still required in the first place for designing nanomedicines which will get FDA approval in the near future.

ACKNOWLEDGMENT

CRP is thankful to DST, New Delhi, for financial support for this research from Nano Mission-DST (SR/NM/NS-1252/2013; GAP 0570). SD and PBT are thankful to UGC and CSIR, New Delhi, for Senior Research Fellowships and Junior Research Fellowships; respectively. The authors are thankful to the director, CSIR-IICT for his support and encouragement and for his keen interest in this work. The IICT communication number IICT/Pubs./2018/351 dated November 13, 2018, for this manuscript is duly acknowledged.

ABBREVIATIONS

ADSCs	Adipose-derived stem cells
APA	6-Aminopenicillanic acid
AuNPs	Aurum (gold) nanoparticles
bFGF	Basic fibroblast growth factor
BCs	Bacterial cellulose
BLCC	Bilayered living cellular construct
BMMNCs	Bone marrow–derived mononuclear cells
BM-MSCs	Bone marrow–derived mesenchymal stem cells
c-myc	Myelocytomatosis
CS-ZnO	Chitosan-conjugated zinc oxide nanoparticle
DAPI	4′,6-diamidino-2-phenylindole
DLS	Dynamic light scattering
DNA	Deoxyribonucleic acid
EA.hy926	Human endothelial cell line
ECM	Extracellular matrix
EDAX	Energy-dispersive x-ray analysis
EDBE	2,2-(ethylenedioxy) bis-(ethylamine)
EPC	Endothelial progenitor cells
FASEB	Federation of American Societies for Experimental Biology
FESEM	Field emission scanning electron microscopy
FGF	Fibroblast growth factor
FTIR	Fourier-transform infrared spectroscopy
GMS3	Ganglioside-monosialic acid 3 synthase
HCM	Hydrocolloid membranes
HIF-1	Hypoxia-inducible factor1
HIV	Human immunodeficiency virus
HSE	Human skin equivalent
IL2	Interleukin2
ICP-OES	Inductively coupled plasma optical emission spectroscopy
L929	Mouse fibroblast cell line
mRNA	Messenger RNA
MDR	Multidrug-resistant
miRNA	Micro RNA
MMP	Matrix metalloproteinase
MSCs	Mesenchymal stem cells
NIH3T3	Mouse embryo fibroblast cell line
NOS	Nitric oxide synthase
PAuNP	Phytochemically stabilized gold nanoparticles
PCL	Poly(ε-caprolactone)
PDGF	Platelet-derived growth factor
PEG	Polyethylene glycol
PLA	Poly(lactic acid)
PLGA	Poly(lactic-co-glycolic acid)
PLGF	Placenta growth factor
PVA	Poly(vinyl alcohol)
PVP	Poly(N-vinylpyrrolidone)
RBC	Red blood cell
RNA	Ribonucleic acid
ROS	Reactive oxygen species
RT-PCR	Reverse transcriptase polymerase chain reaction
SF	Silk fibroin
siRNA	Small interfering RNA
SIMS	Surgical incision management system
TEM	Transmission electron microscope
TGF-β	Transforming growth factor-β
THNRs	Terbium hydroxide nanorods
TiO_2	Titanium dioxide
TNF-α	Tumor necrotic factor α
V-Akt	Vascular serine/threonine kinase enzyme
VEGF	Vascular endothelial growth factor
VLU	Venous leg ulcers
VSSA	Vancomycin-sensitive *S. aureus*
VRSA	Vancomycin-resistant *S. aureus*
WVTR	Water vapor transmission rate
XRD	X-ray diffraction

REFERENCES

Archana, D., B. K. Singh, J. Dutta, and P. K. Dutta. 2013. In vivo evaluation of chitosan–PVP–titanium dioxide nanocomposite as wound dressing material. *Carbohydrate Polymers* 95, no. 1: 530–39.

Armstrong, D. G., L. A. Lavery, P. Abu-Rumman, et al. 2002. Outcomes of subatmospheric pressure dressing therapy on wounds of the diabetic foot. *Ostomy/Wound Manage* 48, no. 4: 64–8.

Atiba, A., M. Nishimura, S. Kakinuma, et al. 2011. Aloe vera oral administration accelerates acute radiation-delayed wound healing by stimulating transforming growth factor-β and fibroblast growth factor production. *The American Journal of Surgery* 201, no. 6 (June): 809–18. doi: 10.1016/j.amjsurg.2010.06.017

Baltzis, D., I. Eleftheriadou, and A. Veves. 2014. Pathogenesis and treatment of impaired wound healing in diabetes mellitus: New insights. *Advances in Therapy* 31, no. 8: 817–36.

Barnes, L. A., C. D. Marshall, T. Leavitt, et al. 2018. Mechanical forces in cutaneous wound healing: Emerging therapies to minimize scar formation. *Advances in Wound Care* 7, no. 2 (February): 47–56. doi: 10.1089/wound.2016.0709

Barrientos, S., O. Stojadinovic, M. S. Golinko, H. Brem, and M. Tomic-Canic. 2008. Growth factors and cytokines in wound healing. *Wound Repair and Regeneration* 16, no. 5 (September): 585–601. doi: 10.1111/wrr.12205

Blanpain, C., W. E. Lowry, A. Geoghegan, L. Polak, and E. Fuchs. 2004. Self-renewal, multipotency, and the existence of two cell populations within an epithelial stem cell niche. *Cell* 118, no. 5 (September): 635–48. doi: 10.1016/j.cell.2004.08.012

Boerckel, J. D., B. A. Uhrig, N. J. Willett, N. Huebsch, and R. E. Guldberg. 2011. Mechanical regulation of vascular growth and tissue regeneration in vivo. *Proceedings of the National Academy of Sciences* 108, no. 37 (September): E674–80. doi: 10.1073/pnas.1107019108

Borkow, G., J. Gabbay, R. Dardik, et al. 2010. Molecular mechanisms of enhanced wound healing by copper oxide-impregnated dressings. *Wound Repair and Regeneration* 18, no. 2: 266–75. http://cmr.asm.org/

Burdock, G. A. 1998. Review of the biological properties and toxicity of bee propolis (propolis). *Food and Chemical Toxicology* 36, no. 4 (April): 347–63. doi: 10.1016/S0278-6915(97)00145-2

Camps, F. E. 1952. Interpretation of wounds. *British Medical Journal* 2, no. 4787: 770. doi: 10.1136/bmj.2.4787.770

Cardany, C. R., G. Rodeheaver, J. Thacker, M. T. Edgerton, and R. F. Edlich. 1976. The crush injury: A high risk wound. *Journal of the American College of Emergency Physicians* 5, no. 12: 965–70.

Carmeliet, P., and R. K. Jain. 2011. Molecular mechanisms and clinical applications of angiogenesis. *Nature* 473, no. 7347: 298–307.

Carney, D. H., R. Mann, W. R. Redin, et al. 1992. Enhancement of incisional wound healing and neovascularization in normal rats by thrombin and synthetic thrombin receptor-activating peptides. *The Journal of Clinical Investigation* 89, no. 5: 1469–77.

Castleberry, S. A., B. D. Almquist, W. Li, et al. 2016. Self-assembled wound dressings silence MMP-9 and improve diabetic wound healing in vivo. *Advanced Materials* 28, no. 9: 1809–17.

Cerqueira, M. T., R. P. Pirraco, and A. P. Marques. 2016. Stem cells in skin wound healing: Are we there yet? *Advances in Wound Care* 5, no. 4 (April): 164–75. doi: 10.1089/wound.2014.0607

Chakraborty, S. P., S. K. Sahu, S. K. Mahapatra, et al. 2010. Nanoconjugated vancomycin: New opportunities for the development of anti-VRSA agents. *Nanotechnology* 21, no. 10: 105103.

Chawla, R., A. Tan, M. Ahmed, et al. 2014. A polyhedral oligomeric silsesquioxane–based bilayered dermal scaffold seeded with adipose tissue–derived stem cells: In vitro assessment of biomechanical properties. *Journal of Surgical Research* 188, no. 2: 361–72.

Clark, R. A. F. 1993. Regulation of fibroplasia in cutaneous wound repair. *The American Journal of the Medical Sciences* 306, no. 1: 42–8.

Cotton, F. A., and G. Wilkinson. 1972. *Advanced Inorganic Chemistry* (3rd Edn). John Wiley and Sons.

De Backer, L., T. Naessens, S. De Koker, et al. 2015. Hybrid pulmonary surfactant-coated nanogels mediate efficient in vivo delivery of siRNA to murine alveolar macrophages. *Journal of Controlled Release* 217: 53–63.

Díez-Pascual, A. M., and A. L. Díez-Vicente. 2015. Wound healing bionanocomposites based on castor oil polymeric films reinforced with chitosan-modified ZnO nanoparticles. *Biomacromolecules* 16, no. 9: 2631–44.

Driver, V. R., L. A. Lavery, A. M. Reyzelman, et al. 2015. A clinical trial of Integra template for diabetic foot ulcer treatment. *Wound Repair and Regeneration* 23, no. 6: 891–900.

Eming, S. A., T. Krieg, and J. M. Davidson. 2007. Inflammation in wound repair: Molecular and cellular mechanisms. *Journal of Investigative Dermatology* 127, no. 3, (March): 514–25. doi: 10.1038/sj.jid.5700701

Eming, S. A., P. Martin, and M. Tomic-Canic. 2014. Wound repair and regeneration: Mechanisms, signaling, and translation. *Science Translational Medicine* 6, no. 265: 265sr6.

Eustis, S., and M. A. El-Sayed. 2006. Why gold nanoparticles are more precious than pretty gold: Noble metal surface plasmon resonance and its enhancement of the radiative and nonradiative properties of nanocrystals of different shapes. *Chemical Society Reviews* 35, no. 3: 209–17.

Fahey, T. J., A. Sadaty, W. G. Jones, et al. 1991. Diabetes impairs the late inflammatory response to wound healing. *Journal of Surgical Research* 50, no. 4: 308–13.

Fan, K., J. Tang, J. Escandon, and R. S. Kirsner. 2011. State of the art in topical wound-healing products. *Plastic and Reconstructive Surgery* 127 Suppl 1: 44S–59S.

Friedman, A. J., G. Han, M. S. Navati, et al. 2008. Sustained release nitric oxide releasing nanoparticles: Characterization of a novel delivery platform based on nitrite containing hydrogel/glass composites. *Nitric Oxide* 19, no. 1: 12–20.

Goldman, R. J. 2009. Hyperbaric oxygen therapy for wound healing and limb salvage: A systematic review. *PM&R* 1, no. 5: 471–89. doi: 10.1016/j.pmrj.2009.03.012

Gould, L., P. Abadir, H. Brem, et al. 2015. Chronic wound repair and healing in older adults: Current status and future research. *Journal of the American Geriatrics Society* 63, no. 3: 427–38.

Gu, H., P. L. Ho, E. Tong, L. Wang, and B. Xu. 2003. Presenting vancomycin on nanoparticles to enhance antimicrobial activities. *Nano Letters* 3, no. 9: 1261–63.

Gunasekaran, T., T. Haile, T. Nigusse, and M. D. Dhanaraju. 2014. Nanotechnology: An effective tool for enhancing bioavailability and bioactivity of phytomedicine. *Asian Pacific Journal of Tropical Biomedicine* 4, Suppl 1: S1–S7.

Hamdan, S., I. Pastar, S. Drakulich, et al. 2017. Nanotechnology-driven therapeutic interventions in wound healing: Potential uses and applications. *ACS Central Science* 3, no. 3: 163–75.

Han, G., and R. Ceilley. 2017. Chronic wound healing: A review of current management and treatments. *Advances in Therapy* 34, no. 3: 599–610.

Hantash, B. M., L. Zhao, J. A. Knowles, and H. P. Lorenz. 2008. Adult and fetal wound healing. *Frontiers in Bioscience : A Journal and Virtual Library* 13: 51–61.

Huh, A. J., and Y. J. Kwon. 2011. Nanoantibiotics: A new paradigm for treating infectious diseases using nanomaterials in the antibiotics resistant era. *Journal of Controlled Release* 156, no. 2: 128–45.

Hunt, T. K., H. Hopf, and Z. Hussain. 2000. Physiology of wound healing. *Advances in Skin & Wound Care* 13: 6–11.

Ireton, J. E., J. G. Unger, and R. J. Rohrich. 2013. The role of wound healing and its everyday application in plastic surgery: A practical perspective and systematic review. *Plastic and Reconstructive Surgery Global Open* 1, no. 1: e10–9.

Kang, S., M. Pinault, L. D. Pfefferle, and M. Elimelech. 2007. Single-walled carbon nanotubes exhibit strong antimicrobial activity. *Langmuir* 23, no. 17: 8670–73.

Kim, H. N., Y. Hong, M. S. Kim, S. M. Kim, and K. Y. Suh. 2012. Effect of orientation and density of nanotopography in dermal wound healing. *Biomaterials* 33, no. 34: 8782–92.

Kim, H. J., K. Miyata, T. Nomoto, et al. 2014. siRNA delivery from triblock copolymer micelles with spatially-ordered compartments of PEG shell, siRNA-loaded intermediate layer, and hydrophobic core. *Biomaterials* 35, no. 15: 4548–56.

Kim, J. E., J. Lee, M. Jang, et al. 2015. Accelerated healing of cutaneous wounds using phytochemically stabilized gold nanoparticle deposited hydrocolloid membranes. *Biomaterials Science* 3, no. 3: 509–19.

Kimbler, D. E., M. Murphy, and K. M. Dhandapani. 2011. Concussion and the adolescent athlete. *The Journal of Neuroscience Nursing: Journal of the American Association of Neuroscience Nurses* 43, no. 6: 286–90.

Kurland, D., C. Hong, B. Aarabi, V. Gerzanich, and J. M. Simard. 2012. Hemorrhagic progression of a contusion after traumatic brain injury: A review. *Journal of Neurotrauma* 29, no. 1: 19–31. doi: 10.1089/neu.2011.2122

Lawrence, W. T. 1998. Physiology of the acute wound. *Clinics in Plastic Surgery* 25, no. 3: 321–40.

Le, X., G. E. J. Poinern, N. Ali, C. M. Berry, and D. Fawcett. 2013. Engineering a biocompatible scaffold with either micrometre or nanometre scale surface topography for promoting protein adsorption and cellular response. *International Journal of Biomaterials* 2013: 782549. doi: 10.1155/2013/782549

Li, Q., S. Mahendra, D. Y. Lyon, et al. 2008. Antimicrobial nanomaterials for water disinfection and microbial control: Potential applications and implications. *Water Research* 42, no. 18: 4591–602.

Lin, P. H., M. Sermersheim, H. Li, et al. 2017. Zinc in wound healing modulation. *Nutrients* 10, no. 1 (December): 16. doi: 10.3390/nu10010016

Liu, Z. J., and O. C. Velazquez. 2008. Hyperoxia, endothelial progenitor cell mobilization, and diabetic wound healing. *Antioxidants & Redox Signaling* 10, no. 11: 1869–82.

Liu, Y., D. Min, T. Bolton, et al. 2009. Increased matrix metalloproteinase-9 predicts poor wound healing in diabetic foot ulcers. *Diabetes Care* 32, no. 1: 117–9.

Lu, B., T. Li, H. Zhao, et al. 2012. Graphene-based composite materials beneficial to wound healing. *Nanoscale* 4, no. 9: 2978–82. doi: 10.2147/IJN.S38635

Ma, K., S. Liao, L. He, et al. 2011. Effects of nanofiber/stem cell composite on wound healing in acute full-thickness skin wounds. *Tissue Engineering Part A* 17, no. 9–10: 1413–24.

Margolis, D. J., J. Kantor, J. Santanna, B. L. Strom, and J. A. Berlin. 2000. Risk factors for delayed healing of neuropathic diabetic foot ulcers: A pooled analysis. *Archives of Dermatology* 136, no. 12: 1531–35.

Marston, W. A., J. Hanft, P. Norwood, and R. Pollak. 2003. The efficacy and safety of Dermagraft in improving the healing of chronic diabetic foot ulcers. *Diabetes Care* 26, no. 6: 1701.

Martin, P., and S. J. Leibovich. 2005. Inflammatory cells during wound repair: The good, the bad and the ugly. *Trends in Cell Biology* 15, no. 11: 599–607. doi: 10.1016/j.tcb.2005.09.002

Maruyama, K., J. Asai, M. Ii, et al. 2007. Decreased macrophage number and activation lead to reduced lymphatic vessel formation and contribute to impaired diabetic wound healing. *The American Journal of Pathology* 170, no. 4: 1178–91.

Nethi, S. K., A. K. Barui, V. S. Bollu, B. R. Rao, and C. R. Patra. 2017. Pro-angiogenic properties of terbium hydroxide nanorods: Molecular mechanisms and therapeutic applications in wound healing. *ACS Biomaterials Science & Engineering* 3, no. 12: 3635–45.

Nolan, G., S. V. Hainsworth, and G. N. Rutty. 2018. Forces generated in stabbing attacks: An evaluation of the utility of the mild, moderate and severe scale. *International Journal of Legal Medicine* 132, no. 1 (January): 229–36. doi: 10.1007/s00414-017-1702-7

O'Kane, S. 2002. Wound remodelling and scarring. *Journal of Wound Care* 11, no. 8: 296–9.

Ojeh, N., I. Pastar, M. Tomic-Canic, and O. Stojadinovic. 2015. Stem cells in skin regeneration, wound healing, and their clinical applications. *International Journal of Molecular Sciences* 16, no. 10 (October): 25476–501. doi: 10.3390/ijms161025476

Öri, F., R. Dietrich, C. Ganz, et al. 2017. Silicon-dioxide– polyvinylpyrrolidone as a wound dressing for skin defects in a murine model. *Journal of Cranio-Maxillo-Facial Surgery* 45, no. 1: 99–107.

Oyarzun-Ampuero, F., A. Vidal, M. Concha, et al. 2015. Nanoparticles for the treatment of wounds. *Current Pharmaceutical Design* 21, no. 29: 4329–41.

Pelgrift, R. Y., and A. J. Friedman. 2013. Nanotechnology as a therapeutic tool to combat microbial resistance. *Advanced Drug Delivery Reviews* 65, no. 13–14: 1803–15.

Pereira, R. F., and P. J. Bártolo. 2016. Traditional therapies for skin wound healing. *Advances in Wound Care* 5, no. 5 (May): 208–29. doi: 10.1089/wound.2013.0506

Pieper, M. S., and P. A. Araoz. 2016. Chapter 5 – Cardiac tumors: Imaging. In *Clinical Cardio-Oncology*, edited by J. Herrmann, 77–90. Elsevier.

Pool, J. G. 1977. Normal hemostatic mechanisms: A review. *The American Journal of Medical Technology* 43, no. 8: 776–80.

Rabea, E. I., M. E. T. Badawy, C. V. Stevens, G. Smagghe, and W. Steurbaut. 2003. Chitosan as antimicrobial agent: Applications and mode of action. *Biomacromolecules* 4, no. 6: 1457–65.

Ramsey, S. D., K. Newton, D. Blough, et al. 1999. Incidence, outcomes, and cost of foot ulcers in patients with diabetes. *Diabetes Care* 22, no. 3: 382–87.

Randeria, P. S., M. A. Seeger, X. Q. Wang, et al. 2015. siRNA-based spherical nucleic acids reverse impaired wound healing in diabetic mice by ganglioside GM3 synthase knockdown. *Proceedings of the National Academy of Sciences* 112, no. 18 (September): 5573–78. doi: 10.1073/pnas.1505951112

Rao, B. R., R. Kotcherlakota, S. K. Nethi, et al. 2018. Ag2 [Fe (CN) 5NO] nanoparticles exhibit antibacterial activity and wound healing properties. *ACS Biomaterials Science & Engineering* 4, no. 9: 3434–49.

Richardson, M. 2004. Acute wounds: An overview of the physiological healing process. *Nursing Times* 100, no. 4: 50–3.

Robertson, I., and P. R. Hodge. 1972. Histopathology of healing abrasions. *Forensic Science* 1, no. 1: 17–25.

Robson, M. C., D. L. Steed, and M. G. Franz. 2001. Wound healing: Biologic features and approaches to maximize healing trajectories. *Current Problems in Surgery* 38, no. 2: 72–140. doi: 10.1067/msg.2001.111167

Rodenwaldt, B., U. Pohl, and C. de Wit. 2007. Endogenous and exogenous NO attenuates conduction of vasoconstrictions along arterioles in the microcirculation. *American Journal of Physiology-Heart and Circulatory Physiology* 292, no. 5: H2341–8.

Samuels, P., and A. K. W. Tan. 1999. Fetal scarless wound healing. *The Journal of Otolaryngology-Head & Neck Surgery* 28, no. 5: 296–302.

Santema, K. T. B., R. M. Stoekenbroek, M. J. W. Koelemay, et al. 2018. Hyperbaric oxygen therapy in the treatment of ischemic lower- extremity ulcers in patients With diabetes: Results of the DAMO2CLES multicenter randomized clinical trial. *Diabetes Care* 41, no. 1 (January): 112–19. doi: 10.2337/dc17-0654

Schurtz, E., J. Differding, E. Jacobson, C. Maki, and M. Ahmeti. 2018. Evaluation of negative pressure wound therapy to closed laparotomy incisions in acute care surgery. *The American Journal of Surgery* 215, no. 1: 113–5. doi: 10.1016/j.amjsurg.2017.08.009

Senturk, B., G. Uzunalli, R. Mammadov, M. O. Guler, and A. B. Tekinay. 2016. Wound healing applications of nanomaterials. In *Therapeutic Nanomaterials*, edited by M. O. Guler and A. B. Tekinay, 87–117. Wiley.

Serramía, M. J., S. Álvarez, E. Fuentes-Paniagua, et al. 2015. In vivo delivery of siRNA to the brain by carbosilane dendrimer. *Journal of Controlled Release* 200: 60–70.

Shahverdi, S., M. Hajimiri, M. A. Esfandiari, et al. 2014. Fabrication and structure analysis of poly (lactide-co-glycolic acid)/silk fibroin hybrid scaffold for wound dressing applications. *International Journal of Pharmaceutics* 473, no. 1–2: 345–55.

Sharma, S., A. Garg, and A. Gupta. 2016. Healing of wound by tissue engineering and mamoparticles: A review. *International Journal of Innovative Research in Science and Engineering* 2, no. 1: 57–62.

Shi, R., W. Lian, S. Han, et al. 2018. Nanosphere-mediated co-delivery of VEGF-A and PDGF-B genes for accelerating diabetic foot ulcers healing in rats. *Gene Therapy* 25: 425–38.

Singer, A. J., and R. A. F. Clark. 1999. Cutaneous wound healing. *The New England Journal of Medicine* 341, no. 10: 738–46.

Smiell, J. M., T. J. Wieman, D. L. Steed, et al. 1999. Efficacy and safety of becaplermin (recombinant human platelet-derived growth factor-BB) in patients with nonhealing, lower extremity diabetic ulcers: A combined analysis of four randomized studies. *Wound Repair and Regeneration* 7, no. 5: 335–46.

Sommerlad, B. C., and J. M. Creasey. 1978. The stretched scar: A clinical and histological study. *British Journal of Plastic Surgery* 31, no. 1: 34–45.

Stone, R. C., O. Stojadinovic, A. M. Rosa, et al. 2017. A bioengineered living cell construct activates an acute wound healing response in venous leg ulcers. *Science Translational Medicine* 9, no. 371: eaaf8611. doi: 10.1126/scitranslmed.aaf8611.

Subramaniam, P., S. J. Lee, S. Shah, et al. 2012. Generation of a library of non-toxic quantum dots for cellular imaging and siRNA delivery. *Advanced Materials* 24, no. 29: 4014–19.

Szycher, M., and S. J. Lee. 1992. Modern wound dressings: A systematic approach to wound healing. *Journal of Biomaterials Applications* 7, no. 2: 142–213.

Tan, H. T., R. A. Rahman, S. H. Gan, et al. 2009. The antibacterial properties of Malaysian tualang honey against wound and enteric microorganisms in comparison to manuka honey. *BMC Complementary and Alternative Medicine* 9 (September): 34. doi: 10.1186/1472-6882-9-34

Tellechea, A., E. C. Leal, A. Kafanas, et al. 2016. Mast cells regulate wound healing in diabetes. *Diabetes* 65: 2006–19.

Thakral, G., J. La Fontaine, P. Kim, et al. 2015. Treatment options for venous leg ulcers: Effectiveness of vascular surgery, bioengineered tissue, and electrical stimulation. *Advances in Skin and Wound Care* 28, no. 4: 164–72. doi: 10.1097/01.ASW.0000462328.60670.c3

Tiwari, M., K. Narayanan, M. B. Thakar, H. V. Jagani, and J. V. Rao. 2014. Biosynthesis and wound healing activity of copper nanoparticles. *IET Nanobiotechnology* 8, no. 4: 230–7.

Tocco, I., B. Zavan, F. Bassetto, and V. Vindigni. 2012. Nanotechnology-based therapies for skin wound regeneration. *Journal of Nanomaterials* 2012: 4. doi: 10.1155/2012/714134

Toreti, V. C., H. H. Sato, G. M. Pastore, and Y. K. Park. 2013. Recent progress of propolis for its biological and chemical compositions and its botanical origin. *Evidence-Based Complementary and Alternative Medicine: eCAM* 2013, (September): 697390. doi: 10.1155/2013/697390

Uauy, R., M. Olivares, and M. Gonzalez. 1998. Essentiality of copper in humans. *The American Journal of Clinical Nutrition* 67, no. 5: 952S–9S.

van Gent, W. B., W. C. Hop, M. C. van Praag, et al. 2006. Conservative versus surgical treatment of venous leg ulcers: A prospective, randomized, multicenter trial. *Journal of Vascular Surgery* 44, no. 3 (September): 563–71. doi: 10.1016/j.jvs.2006.04.053

Walgrave, S. E., E. M. Warshaw, and L. A. Glesne. 2005. Allergic contact dermatitis from propolis. *Dermatitis* 16, no. 4: 209–15.

Wang, C., L. Ma, and C. Gao. 2014. Design of gene-activated matrix for the repair of skin and cartilage. *Polymer Journal* 46, no. 8: 476.

Wei, E. X., R. S. Kirsner, and W. H. Eaglstein. 2016. End points in dermatologic clinical trials: A review for clinicians. *Journal of the American Academy of Dermatology* 75, no. 1: 203–9. doi: 10.1016/j.jaad.2016.01.052

Witte, M. B., and A. Barbul. 1997. General principles of wound healing. *The Surgical Clinics of North America* 77, no. 3: 509–28.

Yang, K., Y. Li, X. Tan, R. Peng, and Z. J. Liu. 2013. Behavior and toxicity of graphene and its functionalized derivatives in biological systems. *Small* 9, no. 9–10: 1492–503.

Yang, Z. Z., J. Q. Li, Z. Z. Wang, D. W. Dong, and X. R. Qi. 2014. Tumor-targeting dual peptides-modified cationic liposomes for delivery of siRNA and docetaxel to gliomas. *Biomaterials* 35, no. 19: 5226–39.

Yang, X., J. Yang, L. Wang, et al. 2017. Pharmaceutical intermediate-modified gold nanoparticles: Against multidrug-resistant bacteria and wound-healing application via an electrospun scaffold. *ACS Nano* 11, no. 6: 5737–45.

Ziv-Polat, O., M. Topaz, T. Brosh, and S. Margel. 2010. Enhancement of incisional wound healing by thrombin conjugated iron oxide nanoparticles. *Biomaterials* 31, no. 4: 741–47.

10 Recent Advancements on Nano-Based Drug Delivery System for Targeted Cancer Therapy

Arindam Pramanik

CONTENTS

10.1 Introduction 133
10.2 Metallic Nanoparticles 134
10.2.1 Gold Nanoparticles (GNPs) 134
10.2.2 Silver Nanoparticles (Ag NPs) 134
10.2.2.1 Anticancer Activity of Ag NPs on Leukemia 135
10.2.2.2 Anticancer Activity of Ag NPs on Breast Cancer 135
10.2.2.3 Anticancer Activity of Ag NPs on Liver Cancer 135
10.2.2.4 Anticancer Activity of Ag NPs on Lung Cancer 135
10.2.2.5 Anticancer Activity of Ag NPs on Oral and Skin Carcinoma 135
10.2.3 Copper Nanoparticles 135
10.2.4 Zinc Nanoparticles 136
10.2.5 Cobalt Nanoparticles 136
10.3 Organometallic Complexes 136
10.3.1 Metallocenes 137
10.3.2 Metal Arene Complexes 137
10.3.3 Metal–Carbene Complexes 137
10.3.4 Cyclometalated Complexes 137
10.4 Nano-Assisted Drug Delivery 138
10.4.1 Lipid Nanoparticles 138
10.4.2 Polymeric Nanoparticles 138
10.4.3 Dendrimers 138
10.5 Targeted Mechanism 138
10.5.1 Transferrin Receptors 138
10.5.2 Cell-Penetrating Peptides (CPPs) 139
10.5.3 LDL (Low Density Lipoproteins)—LDL Receptor (LDLRs) 139
10.5.4 Integrins and Integrin Ligands 139
10.5.5 Folate Receptors 139
10.6 Conclusion 139
References 139

10.1 INTRODUCTION

According to recent statistics, cancer is the leading cause of death globally, and the number of cases each year is around 10 million (Jemal et al. 2010). Cancer is a disease caused by various environmental and genetic factors (Balmain, Gray, and Ponder 2003), and substantial advancements have led to a better understanding of cancer at the cellular and molecular levels, providing novel strategies and targets for therapy (Praetorius and Mandal 2007). Nevertheless, a lot of research is needed for these advancements to reach an effective stage for functional therapy. Currently, the efficacy of many anticancer drugs is restricted and limited due to the incapability to reach the disease at sufficient doses for the pharmacological effect to occur without causing toxicity or adverse effects to healthy tissues and cells of the body (Ferrari 2005; Peer et al. 2007).

The advancement of technology for controlled synthesis of materials at the nanoscale level has led to the "big revolution" in healthcare and medicine for various diseases (Gil and Parak 2008; Sanvicens and Marco 2008). Nanotechnology offers a rich collection of tools to treat and detect cancer using novel techniques which are capable to bypassing the biological barriers for the purpose of delivering the therapeutic medicine to cancer cells or the diseased tissues involved, at the same time holding the prospect of low costs and specificity (nontoxic to healthy tissue or cells) (Praetorius and Mandal 2007; Baptista et al. 2008; Heath and Davis 2008). The main challenge of nanotechnology is to develop a strategy for molecular therapy which can circulate in the body without being detected by the immune cells and still be able to deliver specifically to the diseased region. As a result, personalized medicine could become a new strategy to combat cancer in the near future, and significant research on this is ongoing.

Among the various classes of nanomaterials, metal nanoparticles (NPs) and organometallic complexes have drawn special interest towards biomedical applications, particularly for cancer therapy mainly due to their easy synthesis protocols and low production costs. Various metallic and their derivative nanoparticles (e.g., gold, silver, copper, zinc, cobalt, etc.) and their organometallic counterparts have been found to have promising anticancer activity on broad classes of cancer. A lot of research is currently focused on utilizing these agents as therapeutics and for their targeted delivery.

10.2 METALLIC NANOPARTICLES

Metallic NPs and their derivatives are found to have broad application in the field of biomedicine including thermal ablation, highly sensitive diagnostic assays (Selvan et al. 2010; Baptista 2009), improved radiotherapy (Huang et al. 2007; Hainfeld, Slatkin, and Smilowitz 2004; Hainfeld et al. 2008), and gene and drug delivery (Yavuz et al. 2009; Han, Ghosh, and Rotello 2007; Thomas and Klibanov 2003). Many of the developed nanoparticle-based strategies can simultaneously be used for diagnostic and therapy applications, that is, as "theranostics," exploring their unique properties of detection within the body for finding the localization of the disease as well as exerting therapeutic effects, allowing a more efficient therapy with a lesser risk in comparison to the old or existing therapies (Xie, Lee, and Chen 2010). The various metals and their derivative nanoparticles are discussed below with their application to types of cancers so far studied.

10.2.1 Gold Nanoparticles (GNPs)

GNPs have the ability to act as an enhancer for radiation dose (for radiotherapy) and can also be used for therapeutic drug carriers. When used as radio sensitizers, GNPs demonstrate natural tumor selectivity and significant efficacy in tumor growth restriction in mice receiving radiotherapy after intravenous injection of GNPs. GNPs can also be utilized as a therapeutic drug carrier, e.g., for bleomycin (BLM), an anticancer drug which was chosen as a model for study by a group of researchers (Ma et al. 2003; Jain et al. 2011). BLM is a natural anticancer drug having potential activity against tumors and has been used as a chemotherapeutic agent in clinical trials (Umezawa et al. 1980; Hecht 1986). However, this drug is restricted due to its non-specificity and toxicity, particularly for pulmonary toxicity (Georgelin et al. 2010). Bleomycin is reported to generate reactive oxygen species (ROS) radicals that cause DNA strand breaks and unwinding (Siu and Malcolm 2005). GNPs can be easily attached to BLM through its sulfate bond and can be used as an ideal drug for combinational therapy. Likewise, many drugs having thiol groups can be conjugated to GNPs to enhance their efficacy and allow better tumor penetration (Figure 10.1).

10.2.2 Silver Nanoparticles (Ag NPs)

The antibacterial and antifungal properties of Ag NPs have been extensively studied and well documented. Recently, Ag NPs has been found to have anticancer activity on various cancer types and have gained a lot of attention for their therapeutic applications against cancer (Goodman et al. 2014). It has been reported that Ag NPs at lower dosages can generate ROS causing DNA damage and chromosomal aberrations (genotoxicity), without causing any significant toxicity (T. Zhang et al. 2014) (Figure 10.2). Some groups of researchers found that up to 10 mg/mL of capped Ag NPs with a size range of 6--80 nm in diameter shows no toxicity against normal human cells (de Lima, Seabra, and Duran 2012). Interesting findings have come up upon studying the anticancer activity of Ag NPs and incorporating them into cancer treatment.

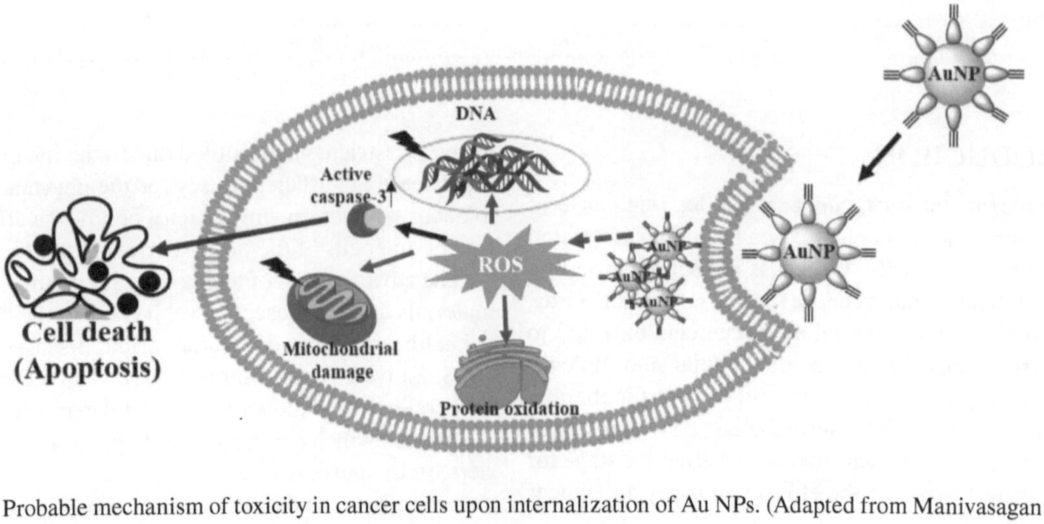

FIGURE 10.1 Probable mechanism of toxicity in cancer cells upon internalization of Au NPs. (Adapted from Manivasagan and Oh 2015.)

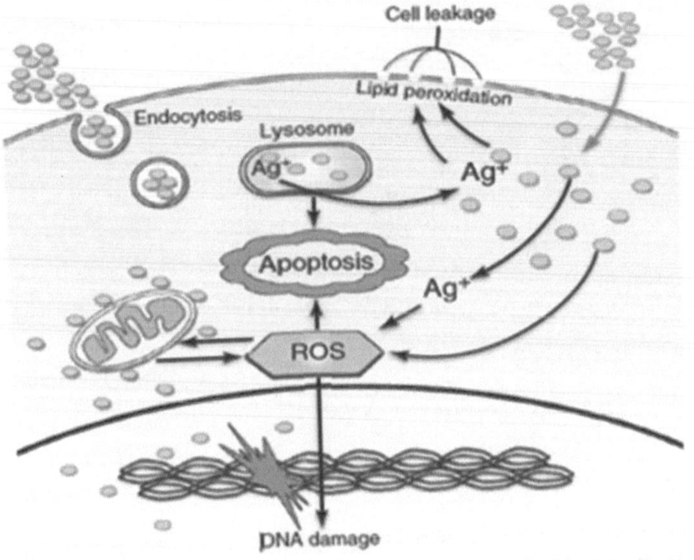

FIGURE 10.2 Possible mechanisms for Ag NP–induced cytotoxicity by ROS generation and apoptosis. (Adapted from Wei et al. 2015.)

Below is a summary of the studies of the effects of Ag NPs on various types of cancers.

10.2.2.1 Anticancer Activity of Ag NPs on Leukemia

Leukemia (Blood cancer) is an abnormal increase of white blood cells due to cancer starting in the bone marrow. Many researchers have found that Ag NPs induced cytotoxic activity against leukemia by inhibiting Jurkat, THP-1, and K562 cell lines. Recently, it was found that poly(N-vinyl-2-pyrrolidone) (PVP)-coated Ag NPs at even low concentrations can successfully kill acute myeloid leukemia (AML) cells (D. Guo et al. 2013). This probably suggests a novel therapeutic strategy for treatment of leukemia in the near future.

10.2.2.2 Anticancer Activity of Ag NPs on Breast Cancer

Dose-dependent anticancer activity has also been found in MCF-7 breast cancer cells with Ag NPs, leading to induction of apoptosis with even nanogram concentration ranges (Franco-Molina et al. 2010). Triple-negative breast cancer cell line MDA-MB-231 (aggressive and poor prognosis) has also been found to be successfully inhibited by this Ag NPs through induction of ROS, DNA fragmentation, and apoptosis-mediated cell death through the caspase 3 pathway. Single-crystalline Ag NPs have also been found to have cytotoxic effects on breast cancer cells through apoptosis-mediated cell death.

10.2.2.3 Anticancer Activity of Ag NPs on Liver Cancer

Cytotoxic studies on a liver cancer cell line (HepG2) have also shown that Ag NP is quite effective in inhibiting the proliferation mainly due to the generation of ROS. It has also been observed by researchers that Ag NPs can cause mitochondrial membrane depolarization, leading to cell death in cancer cells (Sahu et al. 2014). Very interestingly, it is observed that Ag NPs are ~50 times as cytotoxic against liver cancer compared to normal liver cells (murine).

10.2.2.4 Anticancer Activity of Ag NPs on Lung Cancer

Ag NPs have shown substantial amount of cytotoxicity against lung cancers. It is reported that Ag NPs disrupt the mitochondria in a dose-dependent manner in the human alveolar cell line A549, which is primarily caused by induction of ROS, finally leading to apoptosis (Foldbjerg, Dang, and Autrup 2011). Ag NPs are also effective against the H157 (squamous cell lung carcinoma) cell line at low millimolar doses.

10.2.2.5 Anticancer Activity of Ag NPs on Oral and Skin Carcinoma

Some research has also been performed to study the effect of Ag NPs on melanoma cells. It is observed that at low doses, it is cytotoxic towards the HT144 cell line (Nazir et al. 2011). Nuclear targeting peptide-conjugated Ag NPs cause DNA damage by inducing double-strand (ds) breaks, sub-G1 phase cell arrest, and apoptosis in HSC-3 cells (oral squamous cell carcinoma) at low concentrations.

10.2.3 Copper Nanoparticles

Copper NPs are among the nanoparticles synthesized via modern-day technologies that have interesting biological and physiochemical properties (Castro et al. 2014). More specifically, copper NPs can cleave DNA and act as a potential anticancer agent (Fang et al. 2010; Thanh and Green 2010). Copper NPs can also function as effective nano drug delivery systems and control the growth of cancer cells (Pissuwan, Niidome, and Cortie 2011).

Recent attention has been drawn towards the green synthesis of copper NPs from *Sargassum polycystum*, a brown seaweed, using copper sulfate ($CuSO_4$) as the precursor. The resulting NPs were found to have anti-proliferative activity against breast cancer MCF-7 cells at microgram doses. Similarly, green synthesis of Cu NPs was also reported using plants like *Nerium oleander* and *Magnolia kobus* (Gopinath et al. 2014; Lee, Song, and Kim 2013).

Apart from pure copper nanoparticles, derivatives of copper-based nanoparticles have also been studied and found to have active anticancer activity. A few such NPs are CuS, $CuCO_3$, $Cu_3(PO_4)_2$, CuI, CuO, and Cu_2O (Laha et al. 2012, 2014; Pramanik et al. 2015; Nagajyothi et al. 2017; Wang et al. 2012; Guo et al. 2010). It was observed that CuI NPs are more effective in comparison to $Cu(PO_4)_2$ NPs at low doses against cancer cells, although both of them show apoptosis caused due to ROS-mediated DNA damage (Laha et al. 2012). Amorphous and crystalline CuS NPs of similar sizes (≈50–60 nm) are also cytotoxic on cancer cells, although amorphous NPs were more effective than the crystalline ones.

We have used a novel synthetic technique for developing $CuCO_3$ NPs that produce ROS upon internalizing in the cells, causing DNA strand breaks and mitochondria membrane depolarization, resulting in apoptosis-mediated cell death in cancer cells. To further enhance the specificity of the $CuCO_3$ NPs towards cancer cells, folic acid was conjugated, which drives the NPs towards the cancer cell. As a result, folic acid–modified $CuCO_3$ NPs showed more specificity towards tumor cells in vivo, minimizing toxicity to normal cells. During in vivo trials, the survivability of mice increased upon administrating folic acid–conjugated $CuCO_3$ NPs, and tumor volume was reduced (Pramanik et al. 2015).

10.2.4 Zinc Nanoparticles

Zinc functions as a major trace element in the human body and serves as a co-factor for hundreds of mammalian enzymes, playing various vital roles in cellular and sub-cellular activities. Few of such functions of zinc include controlling oxidative stress, replication of DNA, cell repair and apoptosis. ZnO nanoparticles have drawn special attention due to their biocompatibility, selectivity, and enhanced cytotoxicity against cancer cells.

ZnO nanoparticles have a unique UV-ray-absorbing capability. Due to this advantage, ZnO nanoparticles have extensive applications in cosmetic products, e.g., sunscreen cream. ZnO NPs have been thoroughly studied for their anticancer properties.

The electrostatic nature of ZnO nanoparticles is well exploited for its specificity towards cancer cells. At pH above 7 (alkaline), ZnO exists as negatively charged due to the release of protons. At pH less than 7, it exists in its +2 oxidation state. The isoelectric pH of ZnO NPs is 9–10. Thus, at physiological conditions in the blood or tissue (pH 7), it is positively charged, while tumor or cancerous cells have negatively charged anionic phospholipids on their outer membranes (Abercrombie and Ambrose 1962). This results in a selectivity of ZnO NPs towards cancer cells for uptake due to electrostatic attraction between ZnO nanoparticles and cancerous cell outer membranes, leading to specific internalization.

Although it is unclear, the most probable cause for the ZnO cytotoxicity the generation of ROS. It has been observed that cancer cells showed enhanced ROS generation compared to normal cells upon treatment of ZnO NPs (Ostrovsky et al. 2009) (Figure 10.3).

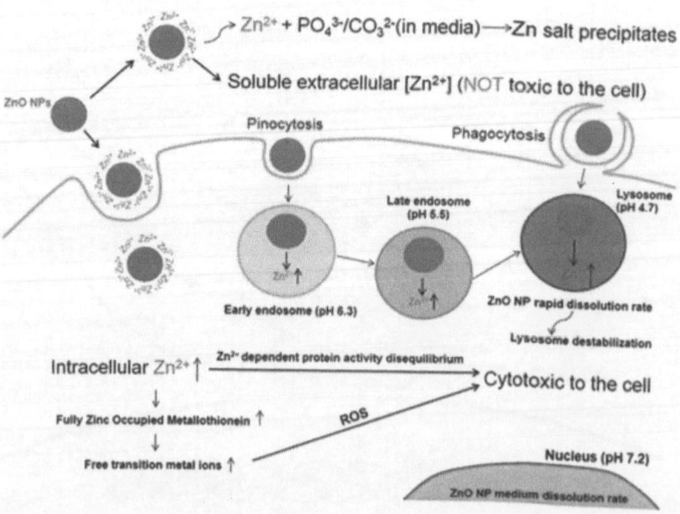

FIGURE 10.3 A schematic representation of the mechanism of cytotoxicity of ZnO nanoparticles. (Adapted from Bisht and Rayamajhi 2016.)

10.2.5 Cobalt Nanoparticles

Phototherapies, a new generation of therapeutic strategies employing photoenergy that includes photodynamic therapy (PDT) and photothermal therapy (PTT), have drawn specific attention towards selectivity and antitumor activity. PDT stands ineffective against hypoxic tumors as it generates reactive oxygen species induced by photosensitizers upon light excitation to kill cancer cells (Fan et al. 2017). PTT, on the other hand, eradicates cancer cells by local hyperthermia, providing advantages such as high efficiency, non-invasiveness, operational simplicity, and remote controllability.

A large number of nanomaterials have been established so far for theranostic applications as imaging-guided PTT. One such agent is phosphide nanoparticles (CoP NPs), which have been established to be a powerful theranostic agent for PTT applications and multimodal imaging. The CoP NPs have potential application as high-contrast triple-modal imaging agent, including photoacoustic (PA), infrared thermal (IRT), and T2-weighted magnetic resonance (MR) imaging. CoP-based NPs show promising application as theranostic agents for cancer therapeutics as well. Mouse breast cancer 4T1 cells, when first incubated with CoP NP dispersions and then irradiated, showed effective anticancer activity of the CoP NP dispersions. Cells receiving both the NPs and laser treatments showed intense increments in IRT imaging signals and significant elevation at temperature to around 57°C, sufficient for inhibiting the cancer cells, whereas without irradiation cytotoxicity was not shown.

Another derivative of cobalt NPs which has been studied for its anticancer activity and cancer specificity is cobalt oxide. The cobalt oxide nanoparticles when coated with phosphonomethyl iminodiacetic acid (PMIDA) show no sign of toxicity on normal oral squamous epithelial cell and human lymphocytes but can potentially kill KB (oral carcinoma) and Jurkat (T-cell lymphoma) cancer cells.

10.3 ORGANOMETALLIC COMPLEXES

The introduction of cisplatin in the 1980s as an anticancer drug led to the incorporation of new generations of platinum drugs (carboplatin and oxaliplatin) as chemotherapeutics. Simultaneously, clinical trials of other platinum drugs and complexes of transition metals as anticancer agents were explored. After this, many bio-organometallic compounds gained interest for potential anticancer therapeutics. There are various classes of organometallic complexes which have been studied for their anticancer activity like metallocenes, metal arene complexes, metal-carbene complexes, etc. Below, detailed light is thrown on this area.

10.3.1 Metallocenes

These are class of compounds where a metal ion is sandwiched between two cyclopentadienyl complexes with an oxidation state of +2 giving the complex aromatic stability (Figure 10.4). Ferrocene is a metallocene of iron $Fe(h_5\text{-}C_5H_5)_2$ that does not show any cytotoxicity; the ferrocenium cation $[Fe(h_5\text{-}C_5H_5)_2]$þ, however, shows a substantial amount of cytotoxicity towards a broad range of cancerous cell lines (Kopf-Maier, Kopf, and Neuse 1984). This may be due to the mechanism of generating ROS inside the cancer cells, which in turn causes DNA damage and hinders the cell membrane integrity. Other researchers have designed a derivative of ferrocenyl iminosugar complex with a fucosidase inhibitor and found this to be significantly effective against melanoma cancer SK-MEL28 and triple-negative breast cancer MDA-MB-231 cell lines (Hottin et al. 2016). Likewise, another complex of ferrocene functionalized with hetero-metallic Pt(II)eguanidine was also found effective for inhibiting cancer cell growth. Very interestingly, it was effective against the cisplatin-resistant colorectal cancer WiDr cell line and breast cancer T-47D cells (Nieto et al. 2015).

Similar to the ferrocene, titanocene complexes have also been found to exhibit activity against cancer cells. $[Ti(h_5\text{-}C_5H_5)_2Cl_2]$ titanocene dichloride contains a cis-dichlorido motif that might be an analog to cisplatin as it forms a similar 1,2 GG intra-strand crosslink in the DNA. However, after a decade of research and clinical trials, the titanocene dichloride was withdrawn mainly due to its solubility issue in water, non-selectivity, and hydrolytic stability, along with a poor understanding of its mechanism of action for exerting cytotoxicity.

10.3.2 Metal Arene Complexes

Organometallic compounds having the base formula $(C_6R_6)_xML_y$ are categorized as metal arene complexes.

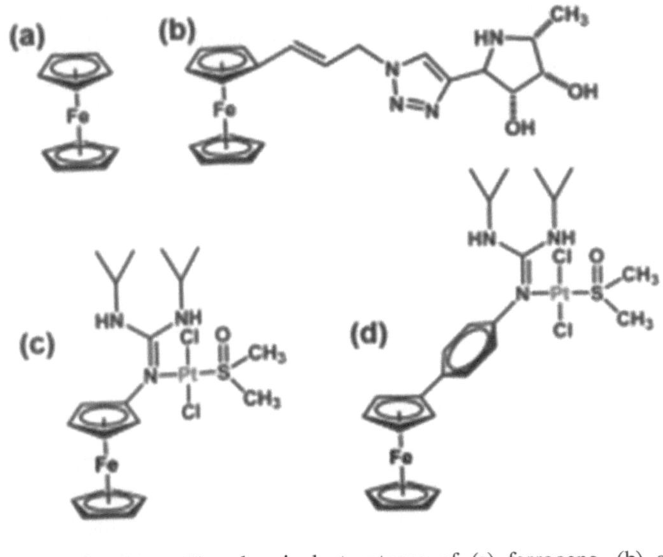

FIGURE 10.4 The chemical structures of (a) ferrocene, (b) a ferrocene-iminosugar conjugate complex, and (c, d) ferrocene-functionalized Pt(II)eguanidine complexes. (Adapted from Zhang and Sadler 2017.)

$(C_6R_6)ML_3$ and $(C_6R_6)_2M$ are among the most common ones. Ruthenium arene complexes have been found to show anti-proliferative activity against many cancer cells mainly due to their binding capability with DNA, which in turn inhibits DNA replication. Likewise, osmium arenes have also been studied for their chemotherapeutic activity and very surprisingly it has been found that their activity is stronger than that of cisplatin in the standard NCI-60 cell line screen, and significantly active during in vivo trials as well. Similarly, iridium and rhodium cyclopentadienyl complexes can cause oxidative stress by increasing the NAD/NADH ratio and impairing cancer cell growth. Interestingly, in the case of Os(II) arene azopyridine complexes, superoxide bursts were detected only in the cancer cells but not in healthy normal cells, thus showing specificity of action. This could be due to a mutation in the DNA of the mitochondria in the cancer cells and their altered function, making them a target of action compared to normal cells.

10.3.3 Metal–Carbene Complexes

Antimicrobial studies have been performed on various N heterocyclic carbine complexes (NHC) of metal, but now interest has been drawn towards studying their anticancer activities. Silver NHC complexes have promising anticancer activity in both in vitro and in vivo models with xenografts of ovarian cancers in mice. Various Au(I) NHC complexes have been found to cause cancer cell death mainly, due to their activity on enzyme thioredoxin reductase. The Au(III) NHC complex can serve a dual purpose: as a fluorescent thiol probe and as a chemotherapeutic agent (Zou et al. 2013). These Au(III) NHC complexes undergo reduction to Au(I)eNHC inside the cells and by the mechanism of the inhibition of enzyme thioredoxin reductase, they show anti-proliferative or cytotoxic activity on a broad range of cancer cells. Likewise, various palladium [Pd(II)] NHC complexes (Figure 10.5) show cytotoxicity against cancer cells, whereas not against normal fibroblast human cell lines. It has also been demonstrated that the Pd(II) complex can inhibit or reduce tumor growth on xenografts of human cancers in mice models.

10.3.4 Cyclometalated Complexes

Cyclometalated Ir(III) complexes can serve for PDT, as they can produce singlet oxygen upon induction with light, which can very effectively kill the cancer cells. A group of researchers

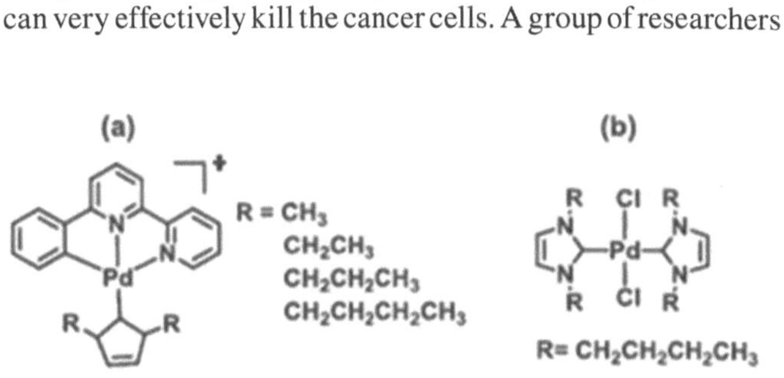

FIGURE 10.5 Chemical structures of palladium(II) NHC complexes. (Adapted from Zhang and Sadler 2017.)

have synthesized cyclometalated Ir(III) complexes with pH-responsive singlet oxygen production and lysosome-specific imaging properties, which make them ideal candidates for PDT, in which they can exert their therapeutic effect and at the same time give real-time assessment for treatment monitoring (He et al. 2015). Cyclometalated Ru(II) complex has been found to internalize into the nuclei of cancer cells within 2 h of incubation. Upon performing a screen on a broad range of cancer cells, it was observed that this Ru(II) complex is more potent against 3D tumor spheroids than the conventional chemotherapeutic drug cisplatin.

10.4 NANO-ASSISTED DRUG DELIVERY

Targeted cancer therapy can distinguish the slight differences between cancer and normal cells. In conventional treatments, the highest tolerable dose of the drug is required for administration, as the body rapidly eliminates the drug (in large amounts). So, conventional methods are very expensive and also exhibit systemic toxicity. Unlike conventional treatments, targeted therapy is very effective without side effects. NPs can deliver the high doses of drug into cancer cells while bypassing normal cells.

Nano-delivery strategies for therapeutic drugs have a lot of advantages over conventional administration. The class of chemotherapeutic agents discussed in the previous section (organometallic complexes) can very effectively carry drugs to be delivered to the cancer cells by the use of nano-delivery vehicles. Currently, there are multiple anticancer therapeutics available for clinical use in the market based on the nano-delivery strategy, such as Vivagel®, Myocet®, Caelyx®, DaunoXome®, DepoCyt®, Doxil®, and VincaXome® (Misra, Acharya, and Sahoo 2010).

10.4.1 Lipid Nanoparticles

Lipid scaffold structure nanomaterials called liposomes were discovered in the 1969. Liposomes are spherically shaped arrangements of phospholipid bilayers of sizes between 30 nm and a few microns (Torchilin 2005). The primary advantage of these liposomes is that they can very effectively encapsulate any hydrophilic or hydrophobic material (drug) and retain them for a long period of time. These nano-carriers provide a lot of additional advantages like biocompatibility, long periods of circulation inside the body, controlled drug release, increased drug stability, high drug-loading capacity, and easy modification for targeting.

10.4.2 Polymeric Nanoparticles

Polymeric nanoparticles of poly (lactic-co-glycolic acid) (PLGA) and chitosan have also been synthesized for drug delivery application. PLGA nanoparticles have also been approved by the Food and Drug Administration (FDA) for drug delivery, and currently Lupron Depot® is available commercially for cancer therapy for clinical use. Likewise, other polymer nanoparticles of chitosan, especially carboxy methyl chitosan, have been successfully used for in vivo trials for delivering anticancer organometallic complexes.

10.4.3 Dendrimers

Dendrimers are repeatedly highly branched 3D polymeric star-like molecules. Poly(amido amine), or PAMAM, is the most well-known dendrimer used for biomedical applications. Dendrimers have been used for synthesis of monodisperse metallic nanoparticles, and for this purpose, PAMAM dendrimers are used for their tertiary amine groups at the branching points within the dendrimer. These nanoparticles range between 1.5 and 10 nm and are known as dendrimer-encapsulated nanoparticles. Dendrimers have been widely used for gene and drug delivery for cancer therapeutics.

10.5 TARGETED MECHANISM

In conventional drug delivery approaches, the therapeutic agent is widely dispersed throughout the body and only a small portion of the therapeutic reaches the diseased part. Cancer-specific targeted drug delivery leads to an increased availability of the drug in the desired area of the body and decreases its non-specific distribution to other healthy organs and tissues, thus minimizing any toxicity or adverse effects.

There are two main approaches for targeted delivery, i.e., active targeting (ligand-based) and passive targeting. In the case of passive targeting, NPs generally accumulate in the tumor region due to the abnormal leaky vasculature of tumor tissue, also known as the enhanced permeation and retention (EPR) effect. It is well studied that the various properties of the nanoparticles such as surface charge, size, and hydrophilic or hydrophobic nature play an important part in determining for passive targeting. Thus, although it is interesting, passive targeting has serious limitations such as inefficient drug diffusion into tumor cells and the lack of EPR effect in some tumors.

In comparison to that, active targeting has many advantages. It is well known that tumor-specific biomarkers are overexpressed on many tumor cells which could be easily targeted using targeting ligands such as peptides, monoclonal antibodies, aptamers, carbohydrates, and other biomolecules (Ulbrich et al. 1997). These ligands, when tagged on the surface of the nanoparticles, can help with internalization into the tumor cells through the mechanism of receptor-mediated endocytosis and then release the cargo (therapeutic) when encountering the acidic pH. A few active targeting approaches using different biological molecules are discussed below.

10.5.1 Transferrin Receptors

Transferrin plays an important role in regulation of iron homeostasis inside the body. The human transferrin receptor 1 (TfR1) is expressed in the nuclei of all cells in the body. Overexpression of TfR1 is found in fast-dividing cancer cells, while it is very low or underexpressed in normal cells. Tumor cells need more iron uptake, and thus 90% of cancer cells overexpress this receptor.

10.5.2 Cell-Penetrating Peptides (CPPs)

CPPs are 5–30 short peptides with a positive charge. This property of this peptide helps its penetration into the cell membrane. This peptide has been found to actively deliver nanoparticles to cancer cells when conjugated to their surface (Levy et al. 1988).

10.5.3 LDL (Low Density Lipoproteins)—LDL Receptor (LDLRs)

Lipoproteins function to drive lipids and cholesterol in the blood. Apolipoprotein B-100, which covers about half of the LDL surface area, targets the LDL to LDL receptors (LDLRs) in some tissues, like ovaries, liver, and adrenal glands. Apolipoprotein B-100 are unrecognized by the phagocytic cells of the immune system. LDLRs are found to be overexpressed in various cancers like colon, lung, adrenal adenoma, acute myelogenous leukemia, pancreatic, brain, and prostate cancers. This fact makes them ideal candidates for targeting cancer cells.

10.5.4 Integrins and Integrin Ligands

Overexpression of integrins on cancer cells makes them a suitable receptor for a targeted delivery strategy in tumor tissues. It has been found that the angiogenesis process of cancer is associated with integrins αvβ3, αvβ5, and α5β; integrins αvβ6 and α6β4 are overexpressed on tumor cells (Sirova et al. 2013). Targeting of αvβ3/αvβ5 integrins, which are overexpressed in tumors, is done using the Arg-Gly-Asp tripeptide (RGD).

10.5.5 Folate Receptors

Folate receptor–based targeting has been explored throughout the world, as it is a low-cost and effective strategy for targeting cancer cells. Folic acid (vitamin B9) is a precursor for DNA synthesis, particularly for synthesis of purines and pyrimidine in all living cells. It is well documented that although the folate receptor is expressed in all cells, cancer cells in particular overexpress them, as they are fast-dividing/replicating cells, making this receptor a suitable and promising route to specifically deliver anticancer drugs through an NP strategy (Pramanik et al. 2015).

10.6 CONCLUSION

Nanotechnology has opened a new door for developing an effective, powerful, and novel strategy to treat and diagnose cancer. In vivo trials of various metal and organometallic nano-complexes as therapeutic agents are developing and moving towards preclinical trials. Nonetheless, the developing strategies and technologies still have to go through a lot of screening and further get approval by the US Food and Drug Administration and its counterparts in other countries. Before reaching that stage, there are still a lot of questions to be answered regarding these novel therapeutic tools, like their toxicity in real scenarios, bio-distribution inside the body, pharmacokinetics, and safety measures regarding their administration. Are these metal and organometallic complexes biocompatible, and can a targeted nano-delivery strategy successfully and effectively minimize any non-specific toxicity or adverse effects inside the body? It is extremely crucial to evaluate the interaction of these NPs with the immune system and other biological molecules, while at the same time killing the diseased cancer cells.

Nanoparticles provide a lot of advantageous over conventional drug delivery like high drug carrying capacity, stability, the capability to deliver both hydrophobic and hydrophilic complexes, controlled drug release, etc. Nanoparticle-assisted targeted delivery also helps in reducing the therapeutic dose required to kill the cancer cells because of its specific drug distribution to diseased cells only. The ligand (biomolecule) chosen for the targeting has to be wisely chosen for each cancer type. Therefore, it is essential to thoroughly design the delivery strategy with plenty of in vitro experiments and in vivo trials before moving on to initial clinical trials with various cancers to come to a rational conclusion. Hopefully, in the next decade, there will be a new generation of low-cost anticancer therapies based on novel metal and organometallic nanoparticles which could revolutionize human healthcare and significantly reduce the mortality rate for cancer.

REFERENCES

Abercrombie, M, and E J Ambrose. 1962. "The Surface Properties of Cancer Cells: A Review." *Cancer Research* 22 (June). United States: 525–48.

Balmain, Allan, Joe Gray, and Bruce Ponder. 2003. "The Genetics and Genomics of Cancer." *Nature Genetics* 33 Suppl (March). United States: 238–44. doi: 10.1038/ng1107.

Baptista, Pedro Viana. 2009. "Cancer Nanotechnology – Prospects for Cancer Diagnostics and Therapy." *Current Cancer Therapy Reviews* 5: 80–8. doi: 10.2174/157339409788166733.

Baptista, Pedro, Eulalia Pereira, Peter Eaton, Goncalo Doria, Adelaide Miranda, Ines Gomes, Pedro Quaresma, and Ricardo Franco. 2008. "Gold Nanoparticles for the Development of Clinical Diagnosis Methods." *Analytical and Bioanalytical Chemistry* 391 (3). Germany: 943–50. doi: 10.1007/s00216-007-1768-z.

Bisht, Gunjan, and Sagar Rayamajhi. 2016. "ZnO Nanoparticles: A Promising Anticancer Agent." *Nanobiomedicine* 3 (January). SAGE Publications Ltd STM: 9. doi: 10.5772/63437.

Castro, Laura, Ma Luisa Blázquez, Jesús ángel Muñoz, Felisa González González, and Antonio Ballester. 2014. "Mechanism and Applications of Metal Nanoparticles Prepared by Bio-Mediated Process." *Reviews in Advanced Sciences and Engineering* 3: 199–216.

Fan, Wenpei, Bryant Yung, Peng Huang, and Xiaoyuan Chen. 2017. "Nanotechnology for Multimodal Synergistic Cancer Therapy." *Chemical Reviews* 117 (22). United States: 13566–638. doi: 10.1021/acs.chemrev.7b00258.

Fang, Jun, Linliang Yu, Pei Gao, Yuguang Cai, and Yinan Wei. 2010. "Detection of Protein-DNA Interaction and Regulation Using Gold Nanoparticles." *Analytical Biochemistry* 399 (2). United States: 262–67. doi: 10.1016/j.ab.2009.11.013.

Ferrari, Mauro. 2005. “Cancer Nanotechnology: Opportunities and Challenges.” *Nature Reviews Cancer* 5 (3). England: 161–71. doi: 10.1038/nrc1566.

Foldbjerg, Rasmus, Duy Anh Dang, and Herman Autrup. 2011. “Cytotoxicity and Genotoxicity of Silver Nanoparticles in the Human Lung Cancer Cell Line, A549.” *Archives of Toxicology* 85 (7). Germany: 743–50. doi: 10.1007/s00204-010-0545-5.

Franco-Molina, Moises A, Edgar Mendoza-Gamboa, Crystel A Sierra-Rivera, Ricardo A Gomez-Flores, Pablo Zapata-Benavides, Paloma Castillo-Tello, Juan Manuel Alcocer-Gonzalez, Diana F Miranda-Hernandez, Reyes S Tamez-Guerra, and Cristina Rodriguez-Padilla. 2010. “Antitumor Activity of Colloidal Silver on MCF-7 Human Breast Cancer Cells.” *Journal of Experimental & Clinical Cancer Research* 29 (November). England: 148. doi: 10.1186/1756-9966-29-148.

Georgelin, Thomas, Sophie Bombard, Jean-Michel Siaugue, and Valerie Cabuil. 2010. “Nanoparticle-Mediated Delivery of Bleomycin.” *Angewandte Chemie* 49 (47). Germany: 8897–8901. doi: 10.1002/anie.201003316.

Gil, Pilar Rivera, and Wolfgang J Parak. 2008. “Composite Nanoparticles Take Aim at Cancer.” *ACS Nano* 2 (11). United States: 2200–2205. doi: 10.1021/nn800716j.

Goodman, Amanda M, Yang Cao, Cordula Urban, Oara Neumann, Ciceron Ayala-Orozco, Mark W Knight, Amit Joshi, Peter Nordlander, and Naomi J Halas. 2014. “The Surprising in Vivo Instability of Near-IR-Absorbing Hollow Au–Ag Nanoshells.” *ACS Nano* 8 (4). American Chemical Society: 3222–31. doi: 10.1021/nn405663h.

Gopinath, M, R Subbaiya, M Masilamani Selvam, and D Suresh. 2014. “Synthesis of Copper Nanoparticles from Nerium Oleander Leaf Aqueous Extract and Its Antibacterial Activity.” *International Journal of Current Microbiology and Applied Sciences* 3: 814–18.

Guo, Yuming, Jie Zhang, Lin Yang, Huajie Wang, Feifei Wang, and Zhi Zheng. 2010. “Syntheses of Amorphous and Crystalline Cupric Sulfide Nanoparticles and Study on the Specific Activities on Different Cells.” *Chemical Communications* 46 (20). The Royal Society of Chemistry: 3493–95. doi: 10.1039/C001714K.

Guo, Dawei, Lingying Zhu, Zhihai Huang, Haixia Zhou, Yue Ge, Wenjuan Ma, Jie Wu, et al. 2013. “Anti-Leukemia Activity of PVP-Coated Silver Nanoparticles via Generation of Reactive Oxygen Species and Release of Silver Ions.” *Biomaterials* 34 (32). Netherlands: 7884–94. doi: 10.1016/j.biomaterials.2013.07.015.

Hainfeld, James F, Daniel N Slatkin, and Henry M Smilowitz. 2004. “The Use of Gold Nanoparticles to Enhance Radiotherapy in Mice.” *Physics in Medicine and Biology* 49 (18). IOP Publishing: N309–15. doi: 10.1088/0031-9155/49/18/n03.

Hainfeld, James F, F Avraham Dilmanian, Daniel N Slatkin, and Henry M Smilowitz. 2008. “Radiotherapy Enhancement with Gold Nanoparticles.” *The Journal of Pharmacy and Pharmacology* 60 (8). England: 977–85. doi: 10.1211/jpp.60.8.0005.

Han, Gang, Partha Ghosh, and Vincent M Rotello. 2007. “Multi-Functional Gold Nanoparticles for Drug Delivery.” *Advances in Experimental Medicine and Biology* 620. United States: 48–56.

He, Liang, Yi Li, Cai-Ping Tan, Rui-Rong Ye, Mu-He Chen, Jian-Jun Cao, Liang-Nian Ji, and Zong-Wan Mao. 2015. “Cyclometalated Iridium(iii) Complexes as Lysosome-Targeted Photodynamic Anticancer and Real-Time Tracking Agents.” *Chemical Science* 6 (10). The Royal Society of Chemistry: 5409–18. doi: 10.1039/C5SC01955A.

Heath, James R, and Mark E Davis. 2008. “Nanotechnology and Cancer.” *Annual Review of Medicine* 59 (1). Annual Reviews: 251–65. doi: 10.1146/annurev.med.59.061506.185523.

Hecht, Sidney M. 1986. “The Chemistry of Activated Bleomycin.” *Accounts of Chemical Research* 19 (12). American Chemical Society: 383–91. doi: 10.1021/ar00132a002.

Hottin, Audrey, Amandine Scandolera, Laurent Duca, Daniel W Wright, Gideon J Davies, and Jean-Bernard Behr. 2016. “A Second-Generation Ferrocene-Iminosugar Hybrid with Improved Fucosidase Binding Properties.” *Bioorganic & Medicinal Chemistry Letters* 26 (6). England: 1546–49. doi: 10.1016/j.bmcl.2016.02.017.

Huang, Xiaohua, Prashant K Jain, Ivan H El-Sayed, and Mostafa A El-Sayed. 2007. “Gold Nanoparticles: Interesting Optical Properties and Recent Applications in Cancer Diagnostics and Therapy.” *Nanomedicine* 2 (5). England: 681–93. doi: 10.2217/17435889.2.5.681.

Jain, Suneil, Jonathan A Coulter, Alan R Hounsell, Karl T Butterworth, Stephen J McMahon, Wendy B Hyland, Mark F Muir, et al. 2011. “Cell-Specific Radiosensitization by Gold Nanoparticles at Megavoltage Radiation Energies.” *International Journal of Radiation Oncology, Bology, Physics* 79 (2). United States: 531–39. doi: 10.1016/j.ijrobp.2010.08.044.

Jemal, Ahmedin, Rebecca Siegel, Jiaquan Xu, and Elizabeth Ward. 2010. “Cancer Statistics, 2010.” *CA: A Cancer Journal for Clinicians* 60 (5). United States: 277–300. doi: 10.3322/caac.20073.

Kopf-Maier, P, H Kopf, and E W Neuse. 1984. “Ferricenium Complexes: A New Type of Water-Soluble Antitumor Agent.” *Journal of Cancer Research and Clinical Oncology* 108 (3). Germany: 336–40.

Laha, Dipranjan, Debalina Bhattacharya, Arindam Pramanik, Chitta Ranjan Santra, Panchanan Pramanik, and Parimal Karmakar. 2012. “Evaluation of Copper Iodide and Copper Phosphate Nanoparticles for Their Potential Cytotoxic Effect.” *Toxicology Research* 1 (2). The Royal Society of Chemistry: 131–36. doi: 10.1039/C2TX00009A.

Laha, Dipranjan, Arindam Pramanik, Jyotirindra Maity, Ananda Mukherjee, Panchanan Pramanik, Aparna Laskar, and Parimal Karmakar. 2014. “Interplay between Autophagy and Apoptosis Mediated by Copper Oxide Nanoparticles in Human Breast Cancer Cells MCF7.” *Biochimica et Biophysica Acta* 1840 (1). Netherlands: 1–9. doi: 10.1016/j.bbagen.2013.08.011.

Lee, Hyo-Jeoung, Jae Yong Song, and Beom Soo Kim. 2013. “Biological Synthesis of Copper Nanoparticles Using Magnolia Kobus Leaf Extract and Their Antibacterial Activity.” *Journal of Chemical Technology & Biotechnology* 88 (11). John Wiley & Sons, Ltd: 1971–77. doi: 10.1002/jctb.4052.

Levy, Y, M S Hershfield, C Fernandez-Mejia, S H Polmar, D Scudiery, M Berger, and R U Sorensen. 1988. “Adenosine Deaminase Deficiency with Late Onset of Recurrent Infections: Response to Treatment with Polyethylene Glycol-Modified Adenosine Deaminase.” *The Journal of Pediatrics* 113 (2). United States: 312–17.

de Lima, Renata, Amedea B Seabra, and Nelson Duran. 2012. “Silver Nanoparticles: A Brief Review of Cytotoxicity and Genotoxicity of Chemically and Biogenically Synthesized Nanoparticles.” *Journal of Applied Toxicology* 32 (11). England: 867–79. doi: 10.1002/jat.2780.

Ma, Brigette B Y, Robert G Bristow, John Kim, and Lillian L Siu. 2003. “Combined-Modality Treatment of Solid Tumors Using Radiotherapy and Molecular Targeted Agents.” *Journal of Clinical Oncology* 21 (14). United States: 2760–76. doi: 10.1200/JCO.2003.10.044.

Manivasagan, Panchanathan, and Junghwan Oh. 2015. "Production of a Novel Fucoidanase for the Green Synthesis of Gold Nanoparticles by Streptomyces Sp. and Its Cytotoxic Effect on HeLa Cells." *Marine Drugs* 13 (11). Switzerland: 6818–37. doi: 10.3390/md13116818.

Misra, Ranjita, Sarbari Acharya, and Sanjeeb K Sahoo. 2010. "Cancer Nanotechnology: Application of Nanotechnology in Cancer Therapy." *Drug Discovery Today* 15 (19–20). England: 842–50. doi: 10.1016/j.drudis.2010.08.006.

Nagajyothi, P C, P Muthuraman, T V M Sreekanth, Doo Hwan Kim, and Jaesool Shim. 2017. "Green Synthesis: In-Vitro Anticancer Activity of Copper Oxide Nanoparticles against Human Cervical Carcinoma Cells." *Arabian Journal of Chemistry* 10 (2): 215–25. doi: 10.1016/j.arabjc.2016.01.011.

Nazir, S, T Hussain, M Iqbal, M Mazhar, A G Muazzam, and M Ismail. 2011. "Novel and Cost-Effective Green Synthesis of Silver Nano Particles and Their Invivo Antitumor Properties against Human Cancer Cell Lines." *Journal of Bioscience and Technology* 2: 425–30.

Nieto, Daniel, Sonia Bruña, Ana Ma González-Vadillo, Josefina Perles, Fernando Carrillo-Hermosilla, Antonio Antiñolo, José M Padrón, Gabriela B Plata, and Isabel Cuadrado. 2015. "Catalytically Generated Ferrocene-Containing Guanidines as Efficient Precursors for New Redox-Active Heterometallic Platinum(II) Complexes with Anticancer Activity." *Organometallics* 34 (22). American Chemical Society: 5407–17. doi: 10.1021/acs.organomet.5b00751.

Ostrovsky, Stella, Gila Kazimirsky, Aharon Gedanken, and Chaya Brodie. 2009. "Selective Cytotoxic Effect of ZnO Nanoparticles on Glioma Cells." *Nano Research* 2 (11): 882–90. doi: 10.1007/s12274-009-9089-5.

Peer, Dan, Jeffrey M Karp, Seungpyo Hong, Omid C Farokhzad, Rimona Margalit, and Robert Langer. 2007. "Nanocarriers as an Emerging Platform for Cancer Therapy." *Nature Nanotechnology* 2 (12). Nature Publishing Group: 751–60.

Pissuwan, Dakrong, Takuro Niidome, and Michael B Cortie. 2011. "The Forthcoming Applications of Gold Nanoparticles in Drug and Gene Delivery Systems." *Journal of Controlled Release* 149 (1): 65–71. doi: 10.1016/j.jconrel.2009.12.006.

Praetorius, Natalie P, and Tarun K Mandal. 2007. "Engineered Nanoparticles in Cancer Therapy." *Recent Patents on Drug Delivery & Formulation* 1 (1). United Arab Emirates: 37–51.

Pramanik, Arindam, Dipranjan Laha, Sourav Chattopadhyay, Sandeep Kumar Dash, Somenath Roy, Panchanan Pramanik, and Parimal Karmakar. 2015. "Targeted Delivery of 'Copper Carbonate' Nanoparticles to Cancer Cells in Vivo." *Toxicology Research* 4 (6). The Royal Society of Chemistry: 1604–12. doi: 10.1039/C5TX00212E.

Sahu, Saura C, Jiwen Zheng, Lesley Graham, Lynn Chen, John Ihrie, Jeffrey J Yourick, and Robert L Sprando. 2014. "Comparative Cytotoxicity of Nanosilver in Human Liver HepG2 and Colon Caco2 Cells in Culture." *Journal of Applied Toxicology* 34 (11). England: 1155–66. doi: 10.1002/jat.2994.

Sanvicens, Nuria, and M Pilar Marco. 2008. "Multifunctional Nanoparticles-Properties and Prospects for Their Use in Human Medicine." *Trends in Biotechnology* 26 (8). England: 425–33. doi: 10.1016/j.tibtech.2008.04.005.

Selvan, Subramanian Tamil, Timothy Thatt Yang Tan, Dong Kee Yi, and Nikhil R Jana. 2010. "Functional and Multifunctional Nanoparticles for Bioimaging and Biosensing." *Langmuir* 26 (14). American Chemical Society: 11631–41. doi: 10.1021/la903512m.

Sirova, M, M Kabesova, L Kovar, T Etrych, J Strohalm, K Ulbrich, and B Rihova. 2013. "HPMA Copolymer-Bound Doxorubicin Induces Immunogenic Tumor Cell Death." *Current Medicinal Chemistry* 20 (38). United Arab Emirates: 4815–26.

Siu, L L M, and J Malcolm. 2005. "Pharmacology of Anticancer Drugs." In Tannock IFH, Richard P, Bristrow RG, Harrington L, Editors. *The Basic Science Oncology*, 322–48. McGraw-Hill.

Thanh, Nguyen T K, and Luke A W Green. 2010. "Functionalisation of Nanoparticles for Biomedical Applications." *Nano Today* 5 (3): 213–30. doi: 10.1016/j.nantod.2010.05.003.

Thomas, Mini, and Alexander M Klibanov. 2003. "Conjugation to Gold Nanoparticles Enhances Polyethylenimine's Transfer of Plasmid DNA into Mammalian Cells." *Proceedings of the National Academy of Sciences* 100 (16). National Academy of Sciences: 9138–43. doi: 10.1073/pnas.1233634100.

Torchilin, Vladimir P. 2005. "Recent Advances with Liposomes as Pharmaceutical Carriers." *Nature Reviews Drug Discovery* 4 (2). England: 145–60. doi: 10.1038/nrd1632.

Ulbrich, Karel, Michal Pechar, Jiři Strohalm, Vladimír Šubr, and Blanka Říhová. 1997. "Polymeric Carriers of Drugs for Site-Specific Therapy." *Macromolecular Symposia* 118 (1). John Wiley & Sons, Ltd: 577–85. doi: 10.1002/masy.19971180176.

Umezawa, Yoji, Hajime Morishima, Seiichi Saito, Tomohisa Takita, Hamao Umezawa, Susumu Kobayashi, Masami Otsuka, Masatoshi Narita, and Masaji Ohno. 1980. "Synthesis of the Pyrimidine Moiety of Bleomycin." *Journal of the American Chemical Society* 102 (21). American Chemical Society: 6630–31. doi: 10.1021/ja00541a073.

Wang, Ye, Xiao-Yuan Zi, Juan Su, Hong-Xia Zhang, Xin-Rong Zhang, Hai-Ying Zhu, Jian-Xiu Li, Meng Yin, Feng Yang, and Yi-Ping Hu. 2012. "Cuprous Oxide Nanoparticles Selectively Induce Apoptosis of Tumor Cells." *International Journal of Nanomedicine* 7. New Zealand: 2641–52. doi: 10.2147/IJN.S31133.

Wei, Liuya, Jingran Lu, Huizhong Xu, Atish Patel, Zhe-Sheng Chen, and Guofang Chen. 2015. "Silver Nanoparticles: Synthesis, Properties, and Therapeutic Applications." *Drug Discovery Today* 20 (5). England: 595–601. doi: 10.1016/j.drudis.2014.11.014.

Xie, Jin, Seulki Lee, and Xiaoyuan Chen. 2010. "Nanoparticle-Based Theranostic Agents." *Advanced Drug Delivery Reviews* 62 (11). Netherlands: 1064–79. doi: 10.1016/j.addr.2010.07.009.

Yavuz, Mustafa S, Yiyun Cheng, Jingyi Chen, Claire M Cobley, Qiang Zhang, Matthew Rycenga, Jingwei Xie, et al. 2009. "Gold Nanocages Covered by Smart Polymers for Controlled Release with Near-Infrared Light." *Nature Materials* 8 (November). Nature Publishing Group: 935–39.

Zhang, Pingyu, and Peter J Sadler. 2017. "Advances in the Design of Organometallic Anticancer Complexes." *Journal of Organometallic Chemistry* 839: 5–14. doi: 10.1016/j.jorganchem.2017.03.038.

Zhang, Tianlu, Liming Wang, Qiang Chen, and Chunying Chen. 2014. "Cytotoxic Potential of Silver Nanoparticles." *Yonsei Medical Journal* 55 (2). Yonsei University College of Medicine: 283–91. doi: 10.3349/ymj.2014.55.2.283.

Zou, Taotao, Ching Tung Lum, Stephen Sin-Yin Chui, and Chi-Ming Che. 2013. "Gold(III) Complexes Containing N-Heterocyclic Carbene Ligands: Thiol 'Switch-on' Fluorescent Probes and Anti-Cancer Agents." *Angewandte Chemie International Edition* 52 (10). John Wiley & Sons, Ltd: 2930–33. doi: 10.1002/anie.201209787.

11 Recent Nanotechnological Advancements in Delivery of Peptide and Protein Macromolecules

Sunita Dahiya and Rajiv Dahiya

CONTENTS

11.1 Introduction 143
11.2 Peptide and Protein Macromolecules 144
11.2.1 Therapeutic Potential, Inherent Problems, and Delivery Challenges 144
11.2.2 Bioavailability Enhancement Strategies 144
11.2.2.1 Penetration Enhancers 144
11.2.2.2 Enzymatic Inhibitors 144
11.2.2.3 Chemical Modification 144
11.2.2.4 Encapsulation 146
11.3 Peptide and Protein Administration 146
11.3.1 The Invasive Injectable Route 146
11.3.2 The Non-Invasive Routes 146
11.3.2.1 Oral Delivery 146
11.3.2.2 Buccal Delivery 147
11.3.2.3 Nasal Delivery 147
11.3.2.4 Pulmonary Delivery 148
11.3.2.5 Transdermal Delivery 149
11.3.2.6 Ocular Delivery 149
11.3.2.7 Rectal Delivery 149
11.4 Nanotechnology-Based Formulation Technologies for Peptide and Protein Delivery 149
11.4.1 Lipid-Based Nanocarriers 149
11.4.1.1 Liposomes 149
11.4.1.2 Microemulsions and Nanoemulsions 150
11.4.1.3 Solid Lipid Nanoparticles 150
11.4.1.4 Nanostructured Lipid Carriers 150
11.4.1.5 Lipid Nanocapsules 150
11.4.1.6 Lipid Nanotubes 150
11.4.2 Polymeric Nanocarriers 150
11.4.2.1 Polymers as Nanocarrier Materials 150
11.4.2.2 Preparation Techniques for NPs 152
11.5 Market Potential of Peptide and Protein Therapeutics 153
11.6 Future Direction for Peptide and Protein Delivery Research 153
11.7 Outlook and Summary 154
References 154

11.1 INTRODUCTION

While there has been tremendous research into the immediate and modified release delivery of small molecules (Pathak et al., 2008; Jain et al., 2010), specific approaches for the successful administration of peptide- and protein-based drugs still need development. Peptides and proteins are essential natural macromolecules that perform different vital functions in the body and are potential candidates to be employed for therapeutic delivery (Joseph et al., 2017). Their major attributes, including bioavailability, biocompatibility, biodegradability, and low toxicity, enable them to be utilized for nano-sized targeted therapeutics, gene targeting therapies, and pharmaceutically active protein molecules for targeted cellular delivery. With traditional and recent advances in synthetic and chemical biology, numerous tailor-made protein or peptide materials have been developed which have garnered attention over the last three decades (Singh et al., 2018; Kumar et al., 2017). This is also evidenced by human antibodies, chimeric proteins, and new protein scaffolds capable of binding to "undruggable" targets, for effective therapies of human diseases including diabetes, cancer, infection, and inflammatory diseases. This can be best exemplified by human insulin, which was approved in 1982 by the US FDA as the first commercially accessible recombinant therapeutic protein; currently, it is extensively used for the treatment of Type I and Type II diabetes mellitus (Leader et al., 2008). More recently, these macromolecules have attracted attention in the search for effective treatments of different inoperable diseases such as AIDS, carcinoma, dwarfism, and autoimmune disorders and also as diagnostic agents. Presently, the invasive injection route is the most commonly used route for administration of therapeutic peptides and proteins. However, it has its share of drawbacks, such as discomfort, expensiveness, and toxicity. For patient compliance and commercial success, delivery routes that are easier and safer than injections are of key concern. The most pressing concern is finding non-invasive routes like oral, buccal, transdermal, nasal, pulmonary, ocular, and rectal for administration of peptides and proteins; unfortunately, there are still many inherent problems and delivery challenges associated with these macromolecules. To overcome these hurdles, research orientations toward developing different strategies such as permeability enhancement, enzyme inhibition, protein structure modification, and protection by encapsulation (Pisal et al., 2010) have been attempted. Also, recent nanotechnological advancements have led to many successful marketed formulations of these biomacromolecules; many more technologies are under clinical consideration.

11.2 PEPTIDE AND PROTEIN MACROMOLECULES

11.2.1 Therapeutic Potential, Inherent Problems, and Delivery Challenges

Although the late 20th century was impacted by numerous therapeutic peptides and proteins in the health market, the delivery of these macromolecules still faces limitations due to their inherent factors such as large molecular size, hydrophilic nature, and enzymatic cleavage. However, as compared to small-molecule drugs, protein therapeutics offers the advantages of higher specificity, greater activity, and less toxicity (Haschek et al., 2013). Among these, therapies that maintain high specificity need to deal with the retention of the structural complexity of proteins, a further challenge to modification and formulation. Furthermore, their vulnerability to enzymatic degradation, instability due to small half-life, and poor membrane absorptivity produce significant blockades for effective distribution of many therapeutic proteins. These hostile integral characteristics of proteins need to be modulated by manipulating proper delivery schemes, as inappropriate design of protein drug formulation can cause denaturation and aggregation, which can lead to various unwanted pharmacological side effects combined with immunologically unfavorable reactions after oral administration. Regardless of their high potency and biofunctionality, protein therapeutics must conquer the obstacles associated with the high molecular weight, short half-lives, instability, and immunogenicity (Bilati et al., 2005a).

11.2.2 Bioavailability Enhancement Strategies

As successful use of peptides and proteins as drugs faces obstacles and challenges due to inherent problems responsible for their poor bioavailability, several strategies have been evaluated and validated clinically to improve the current limitations of these macromolecules. These include the use of penetration enhancers, enzyme inhibitors, encapsulation, and chemical modification (Yu et al., 2016).

11.2.2.1 Penetration Enhancers

Penetration enhancers are employed to enhance the passage of a pharmacologically active quantity of peptide or protein through the mucosal membranes and can be classified into five categories (Table 11.1). Although the complete mechanism of penetration enhancement is still unclear, some of the prominent ways through which the penetration enhancers carry out their function are by altering the structures/properties of mucosal membranes; by altering the thermodynamic activity of proteins and peptides; and by protecting peptides and proteins from proteolytic activity. Although most of the penetration enhancers act by disordering the membrane integrity, not enough information about the correlation between the efficacy and the degree of change/damage to membrane cells with their use is available (Lee and Yamamoto, 1989). Also, penetration enhancers may cause irreversible undetermined biochemical change at the site of effect. Practically, their chronic use may impart toxicity by causing them to absorb themselves. The efficacy of penetration enhancers is determined by the type and characteristics of the protein and penetration enhancer, the administration site, and the delivery system.

11.2.2.2 Enzymatic Inhibitors

High sensitivity toward enzymes is one of the most significant factors in controlling the bioavailability of therapeutic proteins irrespective of their route of administration. Based on this concept, many protease inhibitors have been investigated and tested for the suppression of proteolytic activity. A suitable enzyme inhibiter is chosen according to the treatment of interest and the site of distribution. Furthermore, long-term use of enzyme inhibiters is undesirable as it may disrupt the protein absorption system, leading to the absorption of unwanted proteins; disturb the digestion of nutritive proteins; and trigger the bodily secretion of protease as a result of feedback regulation (Salamat-Miller and Johnston, 2005).

11.2.2.3 Chemical Modification

This commonly used strategy employs the use of direct structural modifications, such as cyclization (e.g., cyclosporine),

TABLE 11.1
Various Penetration Enhancers and Enzyme Inhibitors

Class of penetration enhancer	Agents	Class of enzyme inhibitor	Agents
Chelators	EDTA, citric acid, salicylates, *N*-acyl derivatives of collagen, and enamines (*N*-amino acyl derivatives of diketones)	Non-specific protease inhibitor	Bacitracin
Surfactants	Sodium lauryl sulfate, polyoxyethylene-9-lauryl ether, and olyoxyethylene-20-cetyl ether	Metalloprotease inhibitors	1,10-Phenanthroline, phosphoramidon
Bile salts	Sodium deoxycholate, sodium glycocholate, sodium taurocholate, and their derivatives like sodium taurodihydrofusidate and sodium glycodihydrofusidate	Aspartylprotease inhibitor	Pepstatylglutamic acid
Fatty acids	Oleic acid, caprylic acid, capric acid, and their derivatives such as acylcarnitines, acylcholines, and mono- and diglycerides	Cysteine proteinase inhibitor	p-Hydroxymercuribenzoate
Non-surfactants	Unsaturated cyclic ureas and 1-alkyl and lalkenylazacycloalkanone derivatives	Serine proteinase inhibitors	Diisopropyl fluorophosphate, aprotinin, soybean trypsin inhibitor
		Aminopeptidase inhibitors	Bestatin, puromycin, cx-aminoboronic acid derivatives

amino acid substitution, conjugation to PEG chains as in PEGylation, glycosylation in oligosaccharides, and lipidization in fatty acids (Bruno et al., 2013). PEGylation and glycosylation are demonstrated to improve absorption through biological membranes, showing their structural stability and bioavailability, improved solubility, decreased dosing frequency due to increased systemic stability, increased efficacy, improved safety profile, and reduced immunogenicity (Gupta and Sharma, 2009). Several commercially used PEGylated interferon products are listed in Table 11.2. The use of PEGylation to develop prodrugs provides an unusual strategy to avoid poor solubility, elaborating the pharmacokinetics and lowering the toxicity of protein drugs (Brooks, 2011). The latest PEGylation technologies permit PEG to be tailored in various sizes and shapes with high site specificity and purity, thereby reducing protein immunogenicity and deactivation during conjugation. Further, peptide stapling promoted a peptide's stability, cellular penetration, and binding affinity by locking the conformation of the peptide through multiple synthetic hydrocarbon backbones (Bruno et al., 2013). Rapid elimination of proteins from the blood circulation, which limits their routine use as drugs, is possibly due to their rapid renal clearance, specifically for those of low molecular weights of 40 kDa or less. This logically means that increasing the molecular weight of low-molecular-weight therapeutic molecules to over 40 kDa may reduce the elimination rate. This principle was utilized to conjugate proteins with water-soluble polymers, resulting in reduced renal clearance and thereby prolonging the survival of the drug in the bloodstream (Harris and Chess, 2003). An approach using another principle is about preventing the protein recognition through structural modifications by combining proteins and polymers. This conjugation is assumed to hinder opsonin and antigen processing cells from recognizing these proteins and possibly increase the short half-life of proteins in vivo by decreasing the immune response mechanism. Additionally, the expulsion of proteins from the general circulation by phagocytosis is prevented (Veronese and Mero, 2008). It is interesting that polymers that conjugate with proteins create a steric barrier that impedes the protein degradation by conflicting its binding with the active sites of proteases (Torchilin, 2007; Roberts et al., 2002). Among different polymers used for protein stabilization, the most preferred is PEG, as it is biocompatible and inexpensive, is approved for internal application by drug

TABLE 11.2
Commercial PEGylated Protein Formulations

Brand Name	Drug	Indication
Adagen®	Adenosine deaminase	Severe combined immunodeficiency disease (SCID)
Cimzia®	Anti-TNF Fab	Rheumatoid arthritis & Crohn's Disease
Macugen®	Anti-VEGFc aptamer	Age-related macular degeneration
Mircera®	Erythropoietin (EPO)	Anemia associated with kidney disease
Neulasta®	Granulocyte-colony stimulating factor (GCSF)	Neutropenia
Oncaspar®	Asparaginase	Leukemia
PEG-INTRON®	IFN-a2B	Hepatitis C
PEGASYS®	IFN-a2A	Hepatitis C
Somavert®	Growth hormone antagonist	Acromegaly

regulatory agencies (Roberts et al., 2002), and improves the properties of these biomacromolecules (Bailon and Won, 2009; Torchilin and Lukyanov, 2003).

11.2.2.4 Encapsulation

Encapsulation of peptides and proteins in polymeric reservoirs is a promising delivery method (Xie et al., 2007) offering protection of the encapsulated protein against the effect of enzymes and controlling the speed and site of protein release, which helps to circumvent undesirable side effects (Tan et al., 2010). Different techniques that have been investigated and used for encapsulating therapeutic proteins are discussed in subsequent sections.

11.3 PEPTIDE AND PROTEIN ADMINISTRATION

11.3.1 The Invasive Injectable Route

The most common delivery methods of therapeutic proteins and peptides are by subcutaneous (s.c.), intramuscular (i.m.), and intravenous (i.v.) injections (Almeida and Souto, 2007; Kumar et al., 2006). Use of i.v. injection for protein administration amplifies problems as it is painful, is poorly abided by patients, and requires experts for its administration. Moreover, therapeutic proteins are rapidly cleared from the bloodstream necessitating repeated and high doses of drugs, which may lead to toxic side effects (Ye et al., 2000). Injection by s.c. route may lead to bioavailability up to 100%, though it depends on the molecular weight of the drug used, the site of injection, muscular activity, and pathological conditions associated with the patient (Crommelin et al., 2003). Thus, easier and harmless alternative routes for protein and peptide administration via non-invasive routes appeared as potential therapeutic requirements. This is demonstrated by far-reaching exploration of peptide and protein administration strategies through various routes employing numerous techniques.

11.3.2 The Non-Invasive Routes

Among primary non-invasive alternatives, the oral, nasal, and pulmonary routes have been investigated extensively so far, although the absorbability of peptides and proteins has demonstrated to be very low in most of the verified non-invasive approaches. Another reason for the low feasibility of these complex macromolecules' delivery through non-invasive routes is that they are expensive. Major challenges to the non-parenteral delivery of these macromolecules are summarized in Table 11.3. Since the route of administration of a drug has a significant impact on its therapeutic results (Benet, 1978), peptide and protein delivery approaches have been the focus of intensive investigations in order to find patient-friendly, trouble-free, and effective procedures of administration.

11.3.2.1 Oral Delivery

Oral delivery is the most favored method of administration due to patient suitability and compliance and avoidance of possible contamination and infections caused by the use of injections (Goldberg and Gomez-Orellana, 2003), and it also can be fine-tuned based on the drug's molecular assembly or weight (Elsayed et al., 2009). Enteric coverings and capsules can be utilized as pharmaceutical methodologies to shield the drug from the acidic environment of the stomach, but the evasion of proteolytic enzymes and captivation of these relatively large molecules may present many more challenges. Among recent tools, modification of the physicochemical properties of macromolecules (Cleland et al., 1993), the addition of novel function to macromolecules, and the use of modified and specialized delivery carriers (Varshochian et al., 2013; Stanwick et al., 2012) are reported to improve the oral bioavailability up to 30%–50% from 1%. The oral route is also ideal for delivering certain peptides, as it simulates physiological pathways to reach to their spot of action; one example is oral insulin, whose absorption by the portal vein would directly lead insulin to the target organs, thus reducing adverse systemic effects (Herrero et al., 2012). Among the numerous probable

TABLE 11.3
Major Challenges Associated with Various Delivery Routes

Delivery route	Surface area	Delivery challenges
Oral	200 m^2	Epithelial cell barrier, protease degradation, extreme pH conditions, low bioavailability
Buccal	50.2 m^2	Lack of dosage form retention, presence of drug metabolizing enzymes (oxidases, reductase, cyclooxygenases, peptidases)
Transdermal	1.8	Stratum corneum barrier, protease degradation, macrophages, limitations in surface area and rate of administration
Pulmonary	80–140	Epithelial cell barrier, protease degradation, alveolar macrophages, limits on the amount of protein delivered Per dose
Ocular	–	Tear turnover, protease degradation, protein binding, corneal irritation possibility, low patient acceptance
Nasal	150	Low bioavailability, protease degradation, variable absorption, possibility of nasal mucosal irritation and pathology
Rectal	200–400	Low patient compliance, limited surface area, elimination of drug during bowel movements

formulation methodologies for oral delivery of proteins are enzyme inhibitors, absorption enhancers, mucoadhesive polymeric systems, novel carrier systems, derivatization or chemical modification of proteins and peptides including encapsulation, and prodrug strategies. Also, some novel functionalities to macromolecules, including endogenous cell carrier systems and cell-penetrating peptides (CPPs), have been reviewed (Bohmova et al., 2018). Most significantly, it is essential that these approaches maintain the biological activity of the proteins. Marketed peptide formulations delivered by oral route are given in Table 11.4.

11.3.2.2 Buccal Delivery

The buccal mucosa provides an excellent alternative route for drug administration, which is attributed to its exceptional accessibility, physical robustness, and avoidance of intestinal and hepatic metabolism. Hence, the buccal mucosa provides a remarkably encouraging alternative route for the delivery of peptides and proteins. The buccal mucosa of humans is more permeable due to the presence of 40 to 50 layers of non-keratinized cells, which is of particular importance for the systemic delivery of hydrophilic protein biomacromolecules (Ciach and Moscicka-Studzinska, 2011). The easy access to the buccal mucosa favors both the administration and reversion (in case of adverse event) of formulations at the site. Additionally, the buccal and sublingual regions have low enzymatic activity; enzymes such as trypsin, chymotrypsin, and pepsin, which are present in gastric and intestinal fluids, are almost absent from the oral mucosa, which accordingly are fit for the delivery of enzyme degradation–susceptible proteins (Patel et al., 2011). Moreover, the buccal mucosa is a good vascularized tissue from which blood vessels drain directly into the jugular vein, enabling those candidates that are able to penetrate the buccal mucosal epithelium rapidly into the systemic circulation, thereby avoiding the hepatic first-pass effect (Sudhakar et al., 2006). The molecular masses of peptides and proteins range from 600 to 100,000 Da, which is bigger than most conventional drugs (Veuillez et al., 2001) and adds to the inherent barrier nature of the buccal mucosa, which can be resolved to some extent by various enhancement strategies to fortify a suitable absorption profile. In 1925, the first attempt was made to use the buccal mucosa for insulin delivery. Since then, many attempts were made to improve the limited permeability of insulin across this membrane to attain improved buccal absorption by adding chemical enhancers or by modifying the physicochemical properties of the biomacromolecules. As a result of multiple attempts at improving the buccal delivery of proteins, a buccal system for this protein (Oral-lyn®) has been developed by Generex Biotech for the treatment of Type I and Type II diabetes mellitus (Pozzilli et al., 2005; Heinemann and Jacques, 2009; Annabestani et al., 2010).

11.3.2.3 Nasal Delivery

Recent advancement in biotechnology, inhalation devices, and targeting motifs has considerably raised research interest in peptide and protein delivery via nasal route. It offers advantages including large surface area, highly vascularized mucosa, porous endothelial membrane, lower enzymatic activity relative to gastrointestinal tract (GIT), and avoidance of first-pass metabolism (Saindane et al., 2013). However, the pattern of deposition and size distribution through the delivery device and nasal clearance mechanisms might pose a significant challenge to protein delivery (Smart, 2012). Following intranasal administration, proteins can be absorbed directly into the systemic circulation, into

TABLE 11.4
Commercial Oral Peptide Formulations

Brand name	Peptide	Indication	Strategy/delivery	Company
Neoral®/Sandimmune®	Cyclosporine	Immunosuppression	SNEDDS, systemic delivery	Novartis AG (Switzerland)
DDAVP® Tablets DDAVP® Melt Minrin®	Desmopressin acetate hydrate	Central diabetes insipidus, primary nocturnal enuresis	Chemical modification, systemic delivery	Ferring Pharmaceuticals (Switzerland)/ generic (e.g., Actavis Labs FL Inc., NJ, United States)
Ceredist® Ceredist OD®	Taltirelin hydrate	Spinocerebellar degeneration	Chemical modification to avoid enzymatic hydrolysis, systemic delivery	Mitsubishi Tanabe Pharma Corporation (Japan)
Linzess® (United States) Constella® (Europe)	Linaclotide	Irritable bowel syndrome, chronic idiopathic constipation	Acts locally on luminal surface of intestinal epithelium	Acatavis, Inc. (United States)/Ironwood Pharma, Inc. (United States)
Vancocin®	Vancomycin HCl	infection	Acts locally by inhibition of cell wall biosynthesis	ANI Pharmaceuticals, Inc. (MN, United States)
Koolistin®	Colistin sulfate	Infection	Acts locally to SOD (selective oropharyngeal decontamination) and SDD (selective decontamination of digestive tract)	Biocon Ltd. (India)
Several brands of lozenges	Tyrothricin	Pharyngitis	Acts locally on the throat	Several

central nervous system (CNS), or across GIT (Rapoport and Winner, 2006). Absorption through nasal mucosa is highly dependent on nasal physiology, permeants' physicochemical properties, and the effectiveness of delivery devices (Ugwoke et al., 2001). About 1.5 to 21 mL of mucus generated by serous and seromucous glands forms a double layer of 2–5 μm in thickness (Arora et al., 2002). Mucus acts as a first-line defense by filtering and entrapping foreign particles. Entrapped particles are expelled by cilia in the epithelial surface. This process might significantly impair protein absorption through nasal mucosa. Moreover, the mucociliary apparatus poses a formidable barrier for nasal absorption. Also, cilia along with mucus layer generate a fluctuating movement to clear drug in approximately 12–15 min from the nasal cavity (Merkus et al., 1998). Besides physiological factors, permeants' physicochemical properties also play a considerable role in amending absorption across nasal mucosa. Proteins over 1 kDa have demonstrated overall bioavailability of 0.5% to 5%. Moreover, particles of 1 μm have been reported to reach blood circulation after intranasal administration (Huang and Donovan, 1998). Importantly, lipophilicity and partition coefficient may alter protein and peptide absorption across the nasal mucosal surface. In addition, the particle size and distribution pattern may also determine nasal protein absorption. The nasal cavity offers high permeability of the nasal epithelium as a site for systemic peptide and protein delivery, which allows a higher-molecular-mass cut-off for permeation of nearly 1000 Da and a rapid drug absorption rate, which might be quite identical to that of i.v. injections in some cases. Nasal delivery has been explored as a probable route to target drugs directly to the brain via olfactory neurons, which is one of the most exciting research areas in pharmaceutical sciences owing to the blood–brain barrier (BBB) posing an unbeatable obstacle for a hefty number of antibiotics, antineoplastic agents, and different drugs active in the CNS. Upon administration of a nasal drug formulation into the nasal cavity, the olfactory mucosa might be reached and the olfactory receptor neurons transport drugs into the brain and/or cerebrospinal fluid (CSF). Nasally delivered commercial products are summarized in Table 11.5.

11.3.2.4 Pulmonary Delivery

The numerous advantages of the pulmonary route for the delivery of proteins include enormous absorptive surface area, high vascularization, thin alveolar epithelial membrane (0.1–0.2 μm), and low enzymatic activity. Despite these advantages, several factors may regulate pulmonary protein and peptide absorption. The central airway epithelium is largely constituted of ciliated columnar cells which expresses tight intercellular junctions that may notably limit paracellular transport of peptides and proteins. The diffusion rate in this region is directly proportional to the concentration gradient and lipid solubility. However, diffusion has also been observed to be influenced by peptides' and proteins' molecular size and ionization power (Pontiroli et al., 1989). Mucus binding and subsequent clearance via the mucociliary apparatus may also reduce concentrations available for absorption. Particle wettability, aggregation, crystallinity, polymorphism, and susceptibility towards enzymatic degradation may also determine the absorption rate of peptides and proteins. Importantly, deposition in the respiratory tract significantly depends on the aerodynamic properties of inhaled particles and breathing patterns (Sturm, 2012; Slomkowski and Gosecki, 2011). Particles of 5 to 9 μm in aerodynamic diameter are deposited by impaction in bronchial airways. Fast breathing patterns

TABLE 11.5
Commercial Nasally Delivered Peptides

Brand name	Active drug form	Molecular weight (Daltons)	Therapeutic indication	Company
Synarel®	Luteinizing hormone releasing hormone (LHRH) agonist, nafarelin	1322	Endometreosis	Hoffmann-La Roche, Inc
LHRH (luteinizing hormone releasing hormone)	Kryptocur®	131000	Cryptorchism	Sanofi-Aventis
Miacalcin®	Calcitonin salmon	3432	Postmenopausal osteoporosis,	Novartis Pharmaceuticals
Fortical®	Calcitonin salmon		hypercalcemia or osteoporosis	Unigene's Laboratories/ Upsher-Smith Laboratories, Inc.
Desmopressin	Desmospray®	10700	Cranial diabetes insipidus or nocturia associated with multiple sclerosis	Ferring
Buserelin	Suprecur®	13000	Stimulation of ovulation in infertility treatment or endometriosis	Sanofi-Aventis
Buserelin	Suprefact®	13000	Advanced prostate cancer or endometriosis	Sanofi-Aventis

may result in deposition of 3- to 6-μm particles by a similar mechanism as described above. Gravitational sedimentation results in the deposition of 1- to 5-μm particles in smaller airways. Importantly, particles with aerodynamic diameters of less than 3 μm might be deposited in respiratory bronchioles. Slow breathing patterns can promote absorption of particles in these ranges. Particles with aerodynamic diameters of 500 nm or less can easily be deposited in alveoli through Brownian diffusion, with chances for small particles to be exhaled significantly. At present, no pulmonary insulin products have reached the market; however, numerous second-generation technologies are under development. For example, a remarkably monodispersed (1–3-μm) formulation of insulin microspheres suitable for deep lung delivery has been developed by Epic Therapeutics, a Baxter Healthcare subsidiary. This formulation is distinctive, as all microspheres are essentially insulin and display excellent chemical stability over time.

11.3.2.5 Transdermal Delivery

Transdermal delivery overcomes drawbacks associated with oral and injectable formulations. To date, the passive delivery of proteins through the skin has been found impracticable due to their macromolecular and hydrophilic nature, which limits their transport through the lipophilic stratum corneum, the outermost skin barrier. The large molecular sizes of peptides and proteins usually require chemical or physical interventions to promote skin permeability transiently (Herwadkar and Banga, 2011). For example, the formation of small pores to allow passage of these macromolecules through the stratum corneum. Although there are many technologies to achieve enhanced permeation, the microneedle system is the most promising. Microneedles are architectured to prick holes only in the stratum corneum, avoiding capillaries and nerve endings at the viable epidermis (below stratum corneum). Several ways to administer peptides and proteins include topical administration after microneedle treatment, infusion through hollow microneedles, or the slow release of encapsulated drugs into biodegradable microneedles (Bruno et al., 2013).

11.3.2.6 Ocular Delivery

The eye is considered to be an effective drug delivery route for protein delivery, posing many advantages (Khafagy et al., 2007). The primary advantage of this route is avoidance of gastrointestinal and hepatic first-pass metabolism which otherwise damages proteins (Lee, 2002). However, many drawbacks limit the widespread use of ocular protein delivery route, such as low bioavailability due to poor eye membrane permeability, especially for hydrophilic macromolecules like proteins. Additionally, the ocular tissue contains many enzymes such as protease and aminopeptidase that can degrade peptides and proteins (Zhou and Po, 1991). Even though the ocular route could be considered for protein delivery, it is not the preferred route for systemic protein administration. Despite early reports with promising results of insulin administration via the ocular route (Bartlett et al., 1994), its wider use for peptide and protein delivery needs more research efforts to tackle the associated challenges.

11.3.2.7 Rectal Delivery

Rectal delivery has been used for both local and systematic treatment and is considered important for protein administration, due to the avoidance of first-pass elimination. On the other hand, it possesses drawbacks such as poor peptide and protein absorption through the rectal epithelium, leading to low bioavailability, along with patient discomfort and/or awkwardness. Rectal absorption of proteins and peptides can be improved by using compounds that increase the permeability of the rectal epithelium such as absorption enhancers or decrease the enzymatic degradation of rectally administrated proteins employing enzyme inhibiters (Yamamoto and Muranishi, 1997; de Boer et al., 1990).

11.4 NANOTECHNOLOGY-BASED FORMULATION TECHNOLOGIES FOR PEPTIDE AND PROTEIN DELIVERY

11.4.1 Lipid-Based Nanocarriers

Due to the membrane affinity of lipid materials, lipidic systems possess the ability to increase the internalization of drugs into the cell. In the case of lipid nanocarriers, the submicron size of the particles is another factor favoring cell internalization, and they have been explored as powerful vehicles for site-specific drug delivery to various organs/tissues/systems such as the lymphatic system, brain, lung, and skin (Matougui et al., 2016). The lipid components of lipid-based nanoformulations consist of phospholipids, cholesterol, triglycerides (Rawat et al., 2008), bile salts, and free fatty acids (Liu et al., 2007). The emergence of lipid-based nanocarriers as potential nanocarriers for macromolecular delivery is due to the absorption-enhancing properties of the lipids, the nanocarrier's ability to improve the drug stability, biocompatibility, and the low cost of the production techniques (Karamanidou et al., 2016; Niu et al., 2016). So far, the inclusion of hydrophilic proteins into nanosystems has been limited by their solubility. However, many innovative strategies have been utilized as promising strategies for the formulation of peptides and proteins as lipid-based nanosystems for drug delivery. Specific benefits of lipid-based nanocarriers are control and targeted drug release, improved pharmaceutical stability, high and enhanced drug content, feasibilities of carrying both lipophilic and hydrophilic drugs, biodegradability of most of the lipids, avoidance of organic solvents, easy scale-up and sterilization, relative cost effectiveness and ease to gain/validate regulatory approval.

11.4.1.1 Liposomes

Liposomes are defined as vesicles with an aqueous core in the inner cavity, surrounded by one or more bilayers of amphiphilic phospholipids (Madni et al., 2014). Their sizes range from 20 nm (if unilamellar) to microns (Ibraheem et al., 2014). Amphiphilic lipids able to self-assembly, such as phospholipids, phosphatidylglycerol derivatives and both saturated and unsaturated fatty acids, are most commonly employed for producing liposomes with possible inclusion of polymers and surfactants into their structures (Elizondo et al., 2011; Swaminathan and Ehrhardt, 2012).

11.4.1.2 Microemulsions and Nanoemulsions

Microemulsions are thermodynamically stable transparent isotropic dispersions of low viscosity, consisting of oil and water stabilized by an interfacial film of surfactant molecules, typically in conjunction with a co-surfactant. Usually, the inner phase either oil (o/w emulsions) or water (w/o emulsions) has sizes ranging from 5 to 100 nm. Hydrophilic drugs such as peptides can be successfully incorporated into the dispersed aqueous phase of w/o microemulsion droplets whereby imparting some protection against enzymatic degradation when administered orally. Lecithin-based microemulsions are current research interest as lecithin is a naturally occurring biological surfactant and a major component of membrane lipids safe to use (Cilek et al., 2005, 2006). Lipid nanoemulsions are fine o/w dispersions, having droplets size range of 50–200 nm and introduced for the purpose of parenteral nutrition. Fatty vegetable oils (10%–20%) such as soy oil or middle chain triacylglycerols are used as lipid phase, phospholipids (0.6%–1.5%) as stabilizer and glycerol (2.25%) as osmolarity regulators. These systems also utilized as carriers for lipophilic drugs and several formulations have been commercialized. These systems offer reduced pain during injection and lower local and systemic side effects but possess limited potential for controlled release due to the small size and the liquid state of the carrier.

11.4.1.3 Solid Lipid Nanoparticles

Solid lipid nanoparticles (SLNs) were developed as an alternative to conventional carriers such as emulsions, liposomes, and polymeric nanoparticles (Jenning et al., 2000a, 2000b; Mehnert et al., 2001). SLNs can be produced to incorporate lipophilic or hydrophilic drugs. SLN offers many advantages including usage of physiological lipids, avoidance of organic solvents, a wide application spectrum (dermal, oral, intravenous), established production methods (such as high-pressure homogenization), improved bioavailability, and protection of sensitive drug molecules from the outer environment (water, light, and even controlled release characteristics). Disadvantages of SLNs include particle growth, unpredictable gelation tendency, unexpected dynamics of polymorphic transitions, and inherently low incorporation rate due to the crystalline structure of the solid lipid (Müller et al., 2002a).

11.4.1.4 Nanostructured Lipid Carriers

A new generation of nanostructured lipid carriers (NLCs) consisting of a lipid matrix with a special nanostructure has been developed for improving drug loading (Müller et al., 2002b). These NLCs can be produced by high-pressure homogenization and the process can be modified to yield lipid particle dispersions with solid structure, enabling better drug accommodation from 30 to 80%. Therefore, NLCs' major merit over SLNs is their ability of being able to incorporate large quantities of drugs as a result of the formation of a less ordered lipid matrix with many imperfections, which is the major reason behind their emergence as a great alternative to SLNs. NLCs are prepared by mixing solid and liquid (oil) lipids (Müller et al., 2005; Radtke et al., 2005). Major application areas of NLCs in pharmaceutics are topical, oral, and parenteral drug delivery. NLCs can also be considered as an alternative to liposomes and nanoemulsions due to their advantages such as ease of preparation, high drug loading capacity, sustained parenteral drug release (Wissing et al., 2004), and lower water content in the final particle suspension, which accordingly minimizes possible drug release during storage.

11.4.1.5 Lipid Nanocapsules

Lipid nanocapsules (LNCs) are biomimetic, synthetic nanocarriers of 20–100 nm in size used for drug delivery and imaging purposes (Huynh et al., 2009). The structure of LNCs consists of an oily core surrounded by a surfactant shell. The selection of oily phase and surfactant is adjusted based on the encapsulated drug's properties (Morille et al., 2010). The aqueous phase is a solution of sodium chloride in deionized water. To encapsulate hydrophilic drugs such as peptides, "aqueous-core lipid nanocapsules" have been developed as an additional tool (Anton et al., 2009) whereby incorporation of hydrophilic molecules can be done in either of two ways: (i) distribution of the hydrophilic molecules among the aqueous phase, (ii) injection of a small volume (not more than 2% v/v) of a very concentrated aqueous solution at the phase inversion temperature, before the oil dilution but after the temperature cycling. This helps reduce the degradation of molecules caused by high temperature exposure during the process. LNCs possess benefits in terms of drug protection against biological degradation with an efficient drug loading and/or sustained release to the site of action.

11.4.1.6 Lipid Nanotubes

The unique nanostructure of hollow, cylindrical, long, and open-ended nanotubes with separated inner and outer surfaces offers unique advantages over other drug delivery systems. These include protection of entrapped drug molecules against denaturation and higher loading efficiency with loading of more than one agent, enabling contrast agents and drugs to be used at the same time. Nanotubes can be organic or inorganic, including carbon nanotubes, silica nanotubes, self-assembling lipid nanotubes, and polymer nanotubes (Prato et al., 2008). Lipid nanotubes can be formed via noncovalent self-assembly of lipid molecules in liquid media, and their structure with biocompatible membrane surfaces and easy surface modification makes them potential nanocarriers for drug delivery. Lipid nanotubes can be employed for topical delivery of drug into the skin, whereas fluorescent lipid nanotubes have found application as drug carriers and biomarkers to track and diagnose treatment effectiveness (Zhou et al., 2007).

11.4.2 Polymeric Nanocarriers

11.4.2.1 Polymers as Nanocarrier Materials

Polymer-based nanocarriers can be used for both hydrophobic (e.g., poly (lactide-co-glycolide)) and hydrophilic (e.g., polysaccharides) polymers. Polymeric NPs are solid colloidal carriers composed of synthetic, semi-synthetic, or natural polymers of 10–1000 nm in size (Vrignaud et al., 2011) and are usually categorized as either nanospheres or nanocapsules. In nanospheres, drug is dispersed in a polymeric matrix, whereas nanocapsules are reservoir systems in which drug is confined within a polymeric shell. Proteins can be encapsulated, adsorbed, or chemically linked to the surface of polymeric NPs by various methods such as emulsion solvent evaporation, nanoprecipitation, ionic gelation/crosslinking, and supercritical fluids (SCFs). Polymeric nanoparticulate carriers find applications for the delivery of peptides and proteins via different delivery routes such as oral, pulmonary, nasal, parenteral, transdermal, and ocular (Patel et al., 2014). Requirements for a polymer to be suitable as a polymeric biomaterial are depicted in Figure 11.1. A list of widely used polymers is presented in Table 11.6. Among various polymers employed, PLGA has been widely explored for the PEGylation in protein formulation (Tobío et al., 2000). PLGA is a synthetic co-polymer composed of a mixture of lactic acid and glycolic acid, the monomers which form respectively PLA

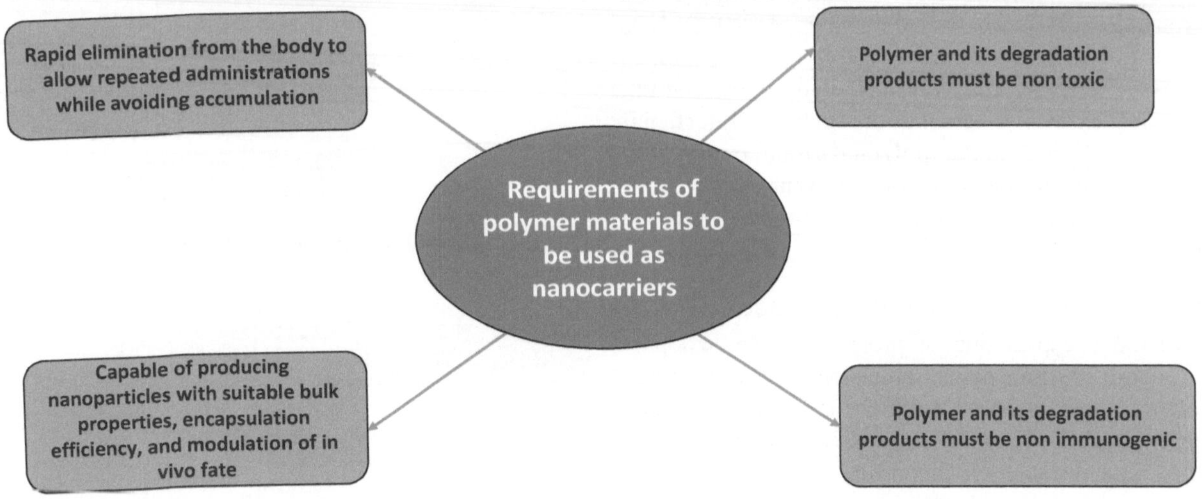

FIGURE 11.1 Requirements of polymer materials as nanocarrier materials.

TABLE 11.6
Polymers Employed as Drug Nanocarriers for Peptides and Protein Delivery

Type	Name	Commercial name/ abbreviation
Synthetic homopolymers	Polylactide	PLA
	Poly(lactide-co-glycolide)	PLGA
	Poly(ε-caprolactone)	PCL
	Poly(*iso*butylcyanoacrylate)	PICBA
	Poly(*iso*hexylcyanoacrylate)	PIHCA
	Poly(*n*-butylcyanoacrylate)	PBCA
	Polyacrylates and polymethacrylates	Eudragit*
Natural polymers	Chitosan	
	Alginate	
	Gelatin	
	Albumin	
	Cellulose derivatives	
Synthetic copolymers	Polylactide-poly(ethylene glycol)	PLA-PEG
	Poly(lactide-co-glycolide)-poly(ethylene glycol)	PLGA-PEG
	Poly(ε-caprolactone)-poly(ethylene glycol)	PCL-PEG
	Poly(hexadecylcyanoacrylate-co-poly(ethylene glycol) cyanoacrylate)	Poly(HDCA-PEGCA)
	Copolymer of methacryloylglycylglycine and *N*-hydroxypropylmethacrylamide	
Colloidal stabilizers	Dextran	
	Pluronic F68	F68
	Poly(vinyl alcohol)	Poly(vinyl alcohol) PVA
	Tween® 20 or Tween® 80	

and PGA. This polymer relies in part of a number of marketed formulations, some of them containing peptide therapeutics (Wang et al., 2016).

11.4.2.2 Preparation Techniques for NPs

The formation of NPs involves the dissolution of the peptide/protein in an aqueous phase and the dissolution of the polymer in an organic solvent. There are several techniques for the NPs' preparation and the main difference among various techniques pertains to the nature of organic solvent in which the polymer is dissolved and the composition of the external aqueous phase (Valencia et al., 2012). NPs with natural polymers are produced using mild techniques including ionic gelation and polyelectrolyte complexation, whereas more extensive techniques such as interfacial polymerization, emulsification–polymerization, emulsification–solvent diffusion, emulsification–solvent evaporation, nanoprecipitation, and supercritical fluids employ synthetic polymers.

11.4.2.2.1 Double Emulsion Solvent Evaporation

Adapted to the encapsulation of proteins within nanoparticles of around 200 nm, double emulsion solvent evaporation (w/o/w) is a widely used method for loading peptides and proteins into NPs. This technique is based on emulsification of an aqueous solution containing the peptide/protein into an organic non-polar solvent, i.e., methylene chloride or ethyl acetate containing the polymer such as PLGA, thereby forming a w/o emulsion which is further emulsified again in a volume of an external aqueous phase containing a surfactant, i.e., polyvinylalcohol (PVA). The remaining solvent is eliminated by evaporation (Blanco and Alonso, 1997). Here, two emulsifications are involved that employ high-energy sources such as homogenization, sonication, or high-speed vortex in the case of small volumes. A schematic view of the procedure is shown in Figure 11.2.

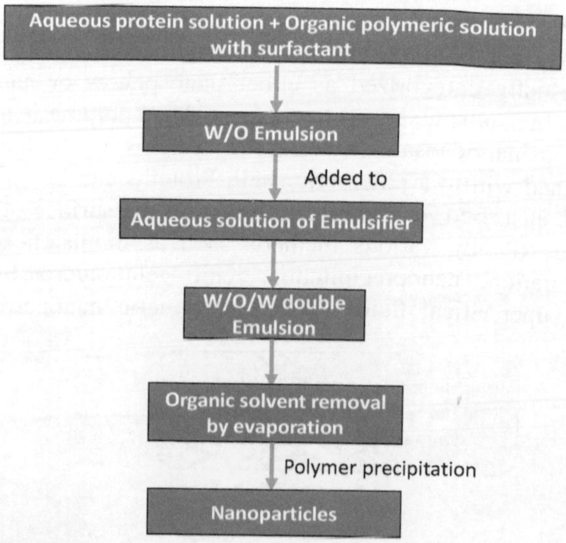

FIGURE 11.2 Flow chart of double emulsion solvent evaporation technique.

11.4.2.2.2 Emulsion Solvent Diffusion

In this method, an aqueous solution of the peptide/protein is emulsified in an organic non-polar phase containing the polymer and surfactant and the resultant emulsion is added to an external polar phase (e.g., mixture of water and ethanol) in which the organic solvent is miscible. The outcome is the polymer precipitation into the polar phase (polymer non-solvents), resulting in the formation of the nanoparticles followed by evaporation to remove the organic solvents (Figure 11.3) (Schubert et al., 2011).

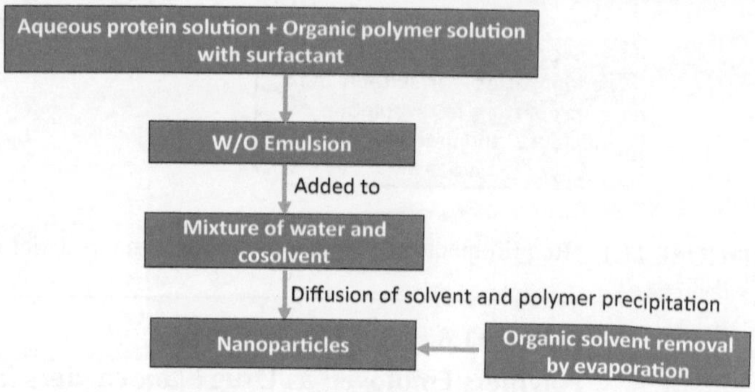

FIGURE 11.3 Flow chart of emulsion solvent diffusion technique.

11.4.2.2.3 Nanoprecipitation

In early phases, this method was limited to encapsulation of hydrophobic molecules (des Rieux et al., 2006) but afterwards modified to accommodate hydrophilic peptides and proteins into PLGA NPs (Bilati et al., 2005b; Bilati et al., 2005c). In this method, the polymer and the proteins are dissolved in a water-miscible organic solvent which is then added dropwise or injected into a dispersing phase in which the polymer is not soluble. The rapid shifting of the solvent into the water causes the nucleation of the polymer, which further aggregates to form nanoparticles, which in turn is followed by an evaporation step to remove solvent traces. The flow diagram of process is depicted in Figure 11.4.

11.4.2.2.4 Ionic Gelation and Crosslinking

The ionic gelation or crosslinking technique is mainly employed for preparing NPs using natural polymers such as sodium alginate, gelatin, and chitosan. This technique is based on the theory about certain charged polysaccharides that can gel in aqueous solution in the presence of small ions and crosslinking agents. The different types of gelling agent may be used depending on the type of polysaccharide, e.g., tripolyphosphate (TPP) is the most commonly crosslinking agent used in the case of chitosan, while calcium salts (calcium chloride, calcium sulfate, or calcium carbonate) are used in the case of alginates. In ionic gelation, chitosan is dissolved in acetic acid in presence/absence of a stabilizer, followed by the addition of polyanion or anionic polymer under mechanical stirring at room temperature (Vandana and Sahoo, 2009). Some of the parameters affecting protein encapsulation by the ionic

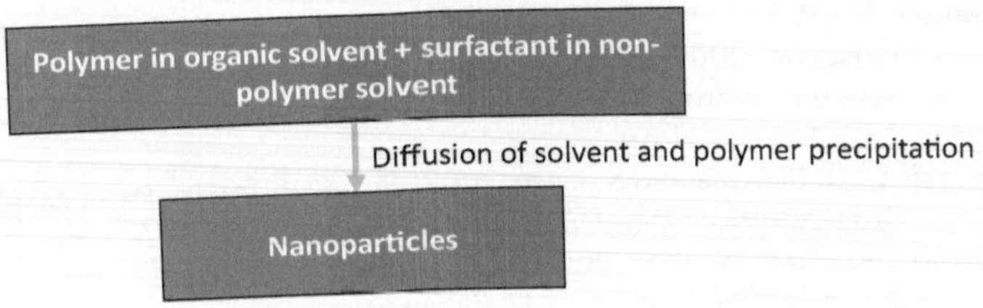

FIGURE 11.4 Flow chart of nanoprecipitation technique.

gelation method are the molecular weight of the polymer, the initial protein concentration, and the polymer concentration used. It is reported that encapsulation efficiency of bovine serum albumin was enhanced around two times by increasing the molecular weight of chitosan from 10 to 210 kDa. Protein and polymer concentrations have demonstrated inverse effects on protein encapsulation efficiency (Xu and Du, 2003). A flow diagram for the ionic gelation technique is depicted in Figure 11.5.

11.4.2.2.5 Polyelectrolyte Complexation

Polyelectrolyte complexes (PECs) result from the mixing of two oppositely charged macromolecules (i.e., polyelectrolytes) to form nanocomplexes (Patwekar et al., 2016). A process flow chart is shown in Figure 11.6. The main parameters influencing the properties of the particles formed are the density of the charges, charge distribution over the polymeric chains, and concentrations of the two polyelectrolytes. The control of the ionic strength and pH of the reaction medium, which influences the degree of ionization, is also fundamental for nanoparticle formation (Hamman, 2010).

11.4.2.2.6 Supercritical Fluids

In this technique, protein and polymer are first dissolved in SCF and the solution is expanded through a nozzle. The SCF is evaporated using a spraying process which leads to precipitation of solute particles (Figure 11.7). One study reported production of insulin-loaded PLA NPs with high product

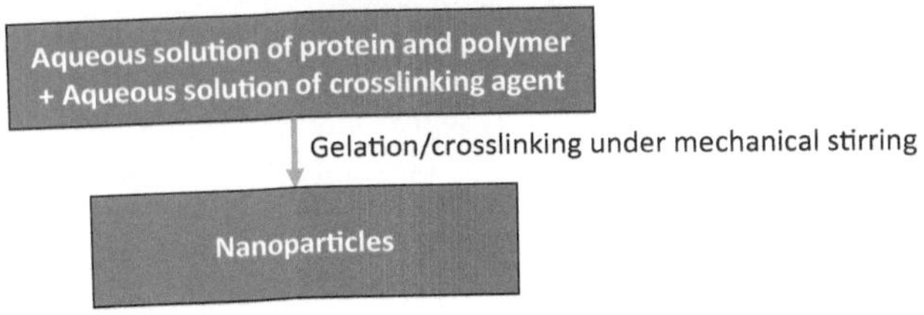

FIGURE 11.5 Flow chart of ionic gelation technique.

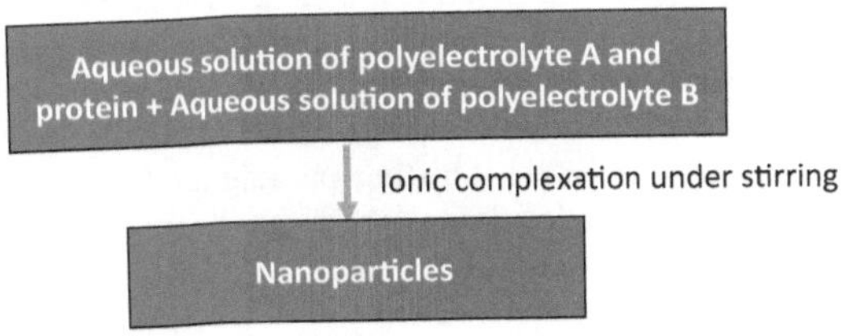

FIGURE 11.6 Flow chart of polyelectrolyte complexation technique.

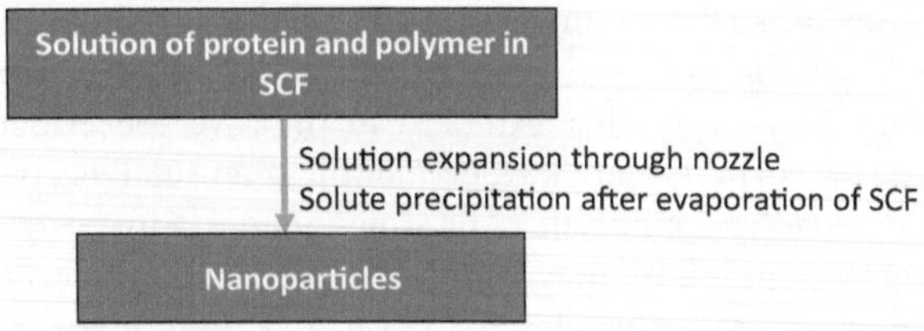

FIGURE 11.7 Flow chart of SCF technique.

yield and maintenance of greater than 80% of the insulin hypoglycemic activity using this method (Elvassore et al., 2001). The major problems observed with this method were the requirement for special equipment, the poor solubility of high-molecular-mass polymers, and strong polar substances in supercritical CO_2 (Marin and Briceno, 2013).

11.5 MARKET POTENTIAL OF PEPTIDE AND PROTEIN THERAPEUTICS

Biotechnological advances have resulted in a greater number of commercially successful peptide and protein therapeutics, and this trend appears to grow in the future. The entire peptide and protein drug market was $47.4 billion in 2006 (Rawat et al., 2008). According to the report of alliedmarketresearch.com, the global oral protein and peptide market accounted for $643 million in 2016, and is anticipated to reach $8,233 million by 2028, registering a compound annual growth rate (CAGR) of 11.7% from 2022 to 2028. The report "Oral Proteins and Peptides Market (3rd Edition, 2018–2030)" at researchandmarkets.com provides a thorough study on the current market prospect for orally administrable protein/peptide-based therapeutics, featuring an extensive analysis on the future potential of this growing market. Despite their clinical and commercial success, oral formulations of protein/peptide-based drugs are associated with inherent drawbacks, such as enzymatic degradation, structural complexities, low bioavailability and high manufacturing costs, thereby limiting their therapeutic potential. This has prompted researchers to develop better technologies in order to improve product stability and therapeutic efficiency. Ongoing innovation is leading toward the exploration of new biological targets, thus strengthening the research channels of various companies attentive to this domain.

11.6 FUTURE DIRECTION FOR PEPTIDE AND PROTEIN DELIVERY RESEARCH

Recently, there have been a myriad of techniques employed for the delivery of peptides and proteins through various non-invasive routes. Among these, insulin delivery via the oral route is of major interest, and many oral delivery strategies for insulin and other protein drugs are under clinical investigation. The latest advancements in biotechnology and newly emerging targeting ligands have significantly boosted interest in developing novel polymeric nanoparticle systems for systemic and local delivery of these biomacromolecules.

Specifically, surface conjugation of polymeric nanoparticles with targeting ligands like antibodies and cell-penetrating peptides may be further explored to improve the effectiveness of the formulation. Surface modification such as coating with mucoadhesive polymers or simultaneous administration of protease inhibitors may significantly enhance the efficacy of polymeric nanoparticles for nasal and pulmonary delivery. Nanotechnological advancements for peptide and protein delivery should enable high loading and entrapment efficiency, protection of structure integrity until it reaches the target site, and release of the encapsulated peptide or protein in a sustained manner to avoid repeated administration. Thus, the development of novel approaches and techniques to enhance the bioavailability of peptides and proteins for non-invasive delivery remains an active field of research. Further studies are still required to develop novel targeted peptide and protein formulations for site-specific and sustained drug delivery employing non-invasive routes of administration.

11.7 OUTLOOK AND SUMMARY

Recent technologies that can be potentially used for peptide and protein delivery are discussed. Each of the technologies related to polymeric materials and various delivery routes have specific advantages and limitations. In spite of voluminous research attempts, the full exploitation of therapeutic peptides and proteins as pharmaceuticals for successful deliveries in therapies to combat dreadful diseases still requires in-depth understanding of their physicochemical properties, including molecular size, immunogenicity, and pharmaco-profiles. The route by which a peptide or protein is delivered depends upon its intended site of action and the limitations imposed by certain physiological processes, such as first-pass hepatic metabolism. From the technological viewpoint, the usage of organic solvent-free and energy-free approaches is desirable, but practically, only some of the lipid-based and polymer-based nanocarriers systems can be produced using such mild preparation techniques. Further, a major limitation of the lipid-based nanosystems is their limited protein-loading capacity, and that of polymeric nanoparticles is low stability and controlled release capacity. To summarize, a delivery system which accommodates peptide or protein therapeutic agents that can entrap efficiently, release appropriately, and preserve sufficiently is yet to be discovered despite many breakthroughs in the field. The use of a rational design approach for ultra-advanced nanocarriers as delivery systems can be seen as a future aspiration for the outstanding non-invasive delivery of potential therapeutic peptides and proteins.

REFERENCES

Almeida, A. J. and E. Souto. 2007. Solid lipid nanoparticles as a drug delivery system for peptides and proteins. *Advanced Drug Delivery Reviews* 59(6):478–490.

Annabestani, Z., S. Sharghi, S. Shahbazi, et al. 2010. Insulin buccal spray (Oral-Lyn) efficacy in type 1 diabetes. *Iranian Journal of Diabetes and Lipid Disorders* 9:1–4.

Anton, N., P. Saulnier, C. Gaillard, E. Porcher, S. Vrignaud and J. P. Benoit. 2009. Aqueous-core lipid nanocapsules for encapsulating hydrophilic and/or lipophilic molecules. *Langmuir* 25(19):11413–11419.

Arora, P., S. Sharma and S. Garg. 2002. Permeability issues in nasal drug delivery. *Drug Discovery Today* 7(18):967–975.

Bailon, P. and C. Y. Won. 2009. PEG-modified biopharmaceuticals. *Expert Opinion on Drug Delivery* 6(1):1–16.

Bartlett, J. D., A. Turner-Henson, J. A. Atchison, T. W. Woolley and D. J. Pillion. 1994. Insulin administration to the eyes of normoglycemic human volunteers. *Journal of Ocular Pharmacology* 10(4):683–690.

Benet, L. Z. 1978. Effect of route of administration and distribution on drug action. *Journal of Pharmacokinetics and Biopharmaceutics* 6(6):559–585.

Bilati, U., E. Allémann and E. Doelker. 2005a. Strategic approaches for overcoming peptide and protein instability within biodegradable nano- and microparticles. *European Journal of Pharmaceutics and Biopharmaceutics* 59(3):375–388.

Bilati, U., E. Allemann and E. Doelker. 2005b. Development of a nanoprecipitation method intended for the entrapment of hydrophilic drugs into nanoparticles. *European Journal of Pharmaceutical Sciences* 24(1):67–75.

Bilati, U., E. Allemann and E. Doelker. 2005c. Nanoprecipitation versus emulsion-based techniques for the encapsulation of proteins into biodegradable nanoparticles and process-related stability issues. *AAPS PharmSciTech* 6(4):E594–E604.

Blanco, M. D. and M. J. Alonso. 1997. Development and characterization of proteinloaded poly(lactide-co-glycolide) nanospheres. *European Journal of Pharmaceutics and Biopharmaceutics* 43(3):287–294.

Bohmova, E., D. Machova, M. Pechar, R. Pola, K. Venclikova, O. Janouskova, and T. Etrych. 2018. Cell-penetrating peptides: a useful tool for the delivery of various cargoes into cells. *Physiological Research* 67(2). Czech Republic: S267–S279.

Brooks, K. 2011. Advances in drug delivery. *Contract Pharma* (06.06.2011).

Bruno, B. J., G. D. Miller and C. S. Lim. 2013. Basics and recent advances in peptide and protein drug delivery. *Therapeutic Delivery* 4(11):1443–1467.

Ciach, T. and A. Moscicka-Studzinska. 2011. Buccal iontophoresis: an opportunity for drug delivery and metabolite monitoring. *Drug Discovery Today* 16:361–366.

Cilek, A., N. Celebi, F. Tirnaksiz and A. Tay. 2005. A lecithin-based microemulsion of rh-insulin with aprotinin for oral administration: investigation of hypoglycemic effects in non-diabetic and STZ-induced diabetic rats. *International Journal of Pharmaceutics* 298(1):176–185.

Cilek, A., N. Celebi and F. Tirnaksiz. 2006. Lecithin-based microemulsion of a peptide for oral administration: preparation, characterization, and physical stability of the formulation. *Drug Delivery* 13(1):19–24.

Cleland, J. L., M. F. Powell and S. J. Shire. 1993. The development of stable protein formulations: a close look at protein aggregation, deamidation, and oxidation. *Critical Reviews™ in Therapeutic Drug Carrier Systems* 10(4):307–377.

Crommelin, D. J., G. Storm, R. Verrijk, L. de Leede, W. Jiskoot and W. E. Hennink. 2003. Shifting paradigms: biopharmaceuticals versus low molecular weight drugs. *International Journal of Pharmaceutics* 266(1–2):3–16.

de Boer, A. G., E. J. van Hoogdalem, C. D. Heijligers-Feijen, J. Verhoef and D. D. Breimer. 1990. Rectal absorption enhancement of peptide drugs. *Journal of Controlled Release* 13(2–3):241–246.

des Rieux, A., V. Fievez, M. Garinot, Y. J. Schneider, V. Préat. 2006. Nanoparticles as potential oral delivery systems of proteins and vaccines: a mechanistic approach. *Journal of Controlled Release* 116(1):1–27.

Elizondo, E., E. Moreno, I. Cabrera, et al. 2011. Liposomes and other vesicular systems: structural characteristics, methods of preparation, and use in nanomedicine. *Progress in Molecular Biology and Translational Science* 104:1–52.

Elsayed, A., M. A. Remawi, N. Qinna, A. Farouk and A. Badwan. 2009. Formulation and characterization of an oily-based system for oral delivery of insulin. *European Journal of Pharmaceutics and Biopharmaceutics* 73(2):269–279.

Elvassore, N., A. Bertucco and P. Caliceti. 2001. Production of insulin loaded poly(ethylene glycol)/poly(l-lactide) (PEG/PLA) nanoparticles by gas antisolvent techniques. *Journal of Pharmaceutical Sciences* 90(10):1628–1636.

Goldberg, M. and I. Gomez-Orellana. 2003. Challenges for the oral delivery of macromolecules. *Nature Reviews Drug Discovery* 2(4):289–295.

Gupta, H. and A. Sharma. 2009. Recent trends in protein and peptide drug delivery systems. *Asian Journal of Pharmaceutics* 3(2):69–75.

Hamman, J. H. 2010. Chitosan based polyelectrolyte complexes as potential carrier materials in drug delivery systems. *Marine Drugs* 8(4):1305–1322.

Harris, J. M. and R. B. Chess. 2003. Effect of pegylation on pharmaceuticals. *Nature Reviews Drug Discovery* 2(3):214–221.

Haschek, W. M., C. G. Rousseaux and M. A. Wallig. 2013. *Haschek and Rousseaux's handbook of toxicologic pathology*, 3rd edition. Cambridge: Academic Press.

Heinemann, L. and Y. Jacques. 2009. Oral insulin and buccal insulin: a critical reappraisal. *Journal of Diabetes Science and Technology* 3(3):568–584.

Herrero, E. P., M. J. Alonso and N. Casaba. 2012. Polymer-based oral peptide nanomedicines. *Therapeutic Delivery* 3(5):657–668.

Herwadkar, A. and A. K. Banga. 2011. Transdermal delivery of peptides and proteins. Chapter 4. In: *Peptide and protein delivery*, ed. C. Van Der Walle, 69–86. Cambridge: Academic Press.

Huang, Y. and M. D. Donovan. 1998. Large molecule and particulate uptake in the nasal cavity: the effect of size on nasal absorption. *Advanced Drug Delivery Reviews* 29(1–2):147–155.

Huynh, N. T., C. Passirani, P. Saulnier, J-P. Benoit. 2009. Lipid nanocapsules: a new platform for nanomedicine. *International Journal of Pharmaceutics* 379:201–209.

Ibraheem, D., A. Elaissari and H. Fessi. 2014. Administration strategies for proteins and peptides. *International Journal of Pharmaceutics* 477(1–2):578–589.

Jain, A., S. Dahiya, K. Pathak. 2010. Development and characterization of piracetam-aerosil R 972 adsorbates for controlled delivery of highly water soluble drug. *Indian Drugs* 47(7):33–39.

Jenning, V., K. Mäder and S. H. Gohla. 2000a. Solid lipid nanoparticles (SLNTM) based on binary mixtures of liquid and solid lipids: a 1H-NMR study. *International Journal of Pharmaceutics* 205(1–2):15–21.

Jenning, V., A. F. Thünemann and S. H. Gohla. 2000b. Characterisation of a novel solid lipid nanoparticle carrier system based on binary mixtures of liquid and solid lipids. *International Journal of Pharmaceutics* 199(2):167–177.

Joseph, M., H. M. Trinh and A. K. Mitra. 2017. Peptide and protein-based therapeutic agents. Chapter 7. In: *Emerging nanotechnologies for diagnostics, drug delivery and medical devices: micro and nano technologies*, ed. A. K. Mitra, K. Cholkar and A. Mandal, 145–167. New York: Elsevier.

Karamanidou, T., V. Bourganis, O. Kammona and C. Kiparissides. 2016. Lipid-based nanocarriers for the oral administration of biopharmaceutics. *Nanomedicine* 11(22):3009–3032.

Khafagy, el-S., M. Morishita, Y. Onuki and K. Takayama 2007. Current challenges in non-invasive insulin delivery systems: a comparative review. *Advanced Drug Delivery Reviews* 59(15):1521–1546.

Kumar, T. R., K. Soppimath and S. K. Nachaegari. 2006. Novel delivery technologies for protein and peptide therapeutics. *Current Pharmaceutical Biotechnology* 7(4):261–276.

Kumar, S., R. Dahiya, S. L. Khokra, R. Mourya, S. V. Chennupati, S. Maharaj. 2017. Total synthesis and pharmacological investigation of cordyheptapeptide A. *Molecules* 22(6):682.

Leader, B., Q. J. Baca and D. E. Golan. 2008. Protein therapeutics: a summary and pharmacological classification. *Nature Reviews Drug Discovery* 7(1):21–39.

Lee, H. J. 2002. Protein drug oral delivery: the recent progress. *Archives of Pharmacal Research* 25(5):572–584.

Lee, V. H. L. and A. Yamamoto. 1989. Penetration and enzymatic barriers to peptide and protein absorption. *Advanced Drug Delivery Reviews* 4(2):171–207.

Liu, J., T. Gong, C. Wang, Z. Zhong and Z. Zhang. 2007. Solid lipid nanoparticles loaded with insulin by sodium cholate-phosphatidylcholine-based mixed micelles: preparation and characterization. *International Journal of Pharmaceutics* 340(1–2):153–162.

Madni, A., M. Sarfraz, M. Rehman, et al. 2014. Liposomal drug delivery: a versatile platform for challenging clinical applications. *Journal of Pharmacy and Pharmaceutical Sciences* 17(3):401–426.

Marin, E., M. I. Briceno and C. Caballero-George. 2013. Critical evaluation of biodegradable polymers used in nanodrugs. *International Journal of Nanomedicine* 8:3071–3090.

Matougui, N., L. Boge, A. C. Groo, et al. 2016. Lipid-based nanoformulations for peptide delivery. *International Journal of Pharmaceutics* 502(1–2):80–97.

Mehnert, M. and K. Mader. 2001. Solid lipid nanoparticles: production, characterization and applications. *Advanced Drug Delivery Reviews* 47(2–3):165–196.

Merkus, F. W., J. C. Verhoef, N. G. Schipper and E. Marttin. 1998. Nasal mucociliary clearance as a factor in nasal drug delivery. *Advanced Drug Delivery Reviews* 29(1–2):13–38.

Morille, M., T. Montier, P. Legras, et al. 2010. Long-circulating DNA lipid nanocapsules as new vector for passive tumor targeting. *Biomaterials* 31(2):321–329.

Müller, R. H., M. Radtke and S. A. Wissing. 2002a. Solid lipid nanoparticles (SLN) and nanostructured lipid carriers (NLC) in cosmetic and dermatological preparations. *Advanced Drug Delivery Reviews* 554(S1):S131–S155.

Müller, R. H., M. Radtke and S. A. Wissing. 2002b. Nanostructured lipid matrices for improved microencapsulation of drugs. *International Journal of Pharmaceutics* 242(1–2):121–128.

Müller, R. H., E. B. Souto and M. Radtke. 2005. Nanostructured lipid carriers: a novel generation of solid lipid drug carriers. *Pharmaceutical Technology Europe* 17(4):45–50.

Niu, Z., I. Conejos-Sanchez, B. T. Griffin, C. M. O'Driscoll, M. J. Alonso. 2016. Lipid based nanocarriers for oral peptide delivery. *Advanced Drug Delivery Reviews* 106(PtB):337–354.

Patel, V. F., F. Liu and M. B. Brown. 2011. Advances in oral transmucosal drug delivery. *Journal of Controlled Release* 153(2):106–116.

Patel, A., M. Patel, X. Yang and A. K. Mitra. 2014. Recent advances in protein and peptide drug delivery: a special emphasis on polymeric nanoparticles. *Protein and Peptide Letters* 21(11):1102–1120.

Pathak, D., S. Dahiya and K. Pathak. 2008. Solid dispersion of meloxicam: factorially designed dosage form for geriatric population. *Acta Pharmaceutica* 58(1):99–110.

Patwekar, S. L., A. P. Potulwar, S. R. Pedewas, M. S. Gaikwad, S. A. Khan and A. B. Suryawanshi. 2016. Review on polyelectrolyte complex as novel approach for drug delivery system. *International Journal of Pharmacy and Pharmaceutical Research* 5(4):97–109.

Pisal, D. S., M. P. Kosloski and S. V. Balu-Iyer. 2010. Delivery of therapeutic peptides and proteins. *Journal of Pharmaceutical Sciences* 99(6):2557–2575.

Pontiroli, A. E., M. Alberetto, A. Calderara, E. Pajetta and G. Pozza. 1989. Nasal administration of glucagon and human calcitonin to healthy subjects: a comparison of powders and spray solutions and of different enhancing agents. *European Journal of Clinical Pharmacology* 37(4):427–430.

Pozzilli, P., S. Manfrini, F. Costanza, et al. 2005. Biokinetics of buccal spray insulin in patients with type 1 diabetes. *Metabolism* 54(7):930–934.

Prato, M., K. Kostarelos and A. Bianco. 2008. Functionalized carbon nanotubes in drug design and discovery. *Accounts of Chemical Research* 41(1):60–68.

Radtke, M., E. B. Souto and R. H. Müller. 2005. Nanostructured lipid carriers: a novel generation of solid lipid drug carriers. *Pharmaceutical Technology Europe* 17(4):45–50.

Rapoport, A. and P. Winner. 2006. Nasal delivery of antimigraine drugs: clinical rationale and evidence base. *Headache* 46(S4):S192–S201.

Rawat, M., D. Singh, S. Saraf and S. Saraf. 2008. Lipid carriers: a versatile delivery vehicle for proteins and peptides. *Yakugaku Zasshi* 128(2):269–280.

Roberts, M. J., M. D. Bentley and J. M. Harris. 2002. Chemistry for peptide and protein PEGylation. *Advanced Drug Delivery Reviews* 54(4):459–476.

Saindane, N. S., K. P. Pagar and P. R. Vavia. 2013. Nanosuspension based in situ gelling nasal spray of carvedilol: development, in vitro and in vivo characterization. *AAPS PharmSciTech* 14(1):189–199.

Salamat-Miller, N. and T. P. Johnston. 2005. Current strategies used to enhance the paracellular transport of therapeutic polypeptides across the intestinal epithelium. *International Journal of Pharmaceutics* 294(1–2):201–216.

Schubert, S., J. T. Delaney Jr and U. S. Schubert. 2011. Nanoprecipitation and nanoformulation of polymers: from history to powerful possibilities beyond poly(lactic acid). *Soft Matter* 7:1581–1588.

Singh, S., K. K. Chandrul and R. Dahiya. 2018. Selective cytotoxic activity of synthetic natural cyclopeptides on HCT11 & B16F10 cells. *International Journal of Pharmaceutical Sciences and Drug Research* 10(6):480–486.

Slomkowski, S. and M. Gosecki. 2011. Progress in nanoparticulate systems for peptide, proteins and nucleic acid drug delivery. *Current Pharmaceutical Biotechnology* 12(11):1823–1839.

Smart, J. D. 2012. Nasal and buccal drug delivery: management forum conference. *Therapeutic Delivery* 3(7):819–822.

Stanwick, J. C., M. D. Baumann and M. S. Shoichet. 2012. Enhanced neurotrophin-3 bioactivity and release from a nanoparticle-loaded composite hydrogel. *Journal of Controlled Release* 160(3):666–675.

Sturm, R. 2012. Theoretical models of carcinogenic particle deposition and clearance in children's lungs. *Journal of Thoracic Disease* 4(4):368–376.

Sudhakar, Y., K. Kuotsu and A. K. Bandyopadhyay. 2006. Buccal bioadhesive drug delivery–a promising option for orally less efficient drugs. *Journal of Controlled Release* 114(1):15–40.

Swaminathan, J. and C. Ehrhardt. 2012. Liposomal delivery of proteins and peptides. *Expert Opinion on Drug Delivery* 9(12):1489–1503.

Tan, Mei Lin, P. F. M. Choong, and C. R. Dass. 2010. Recent developments in liposomes, microparticles and nanoparticles for protein and peptide drug delivery. *Peptides* 31(1):184–193.

Tobio, M., A. Sanchez, A. Vila, I. Soriano, C. Evora, J. L. Vila-Jato, and M. J. Alonso. 2000. The role of PEG on the stability in digestive fluids and in vivo fate of PEG-PLA nanoparticles following oral administration. *Colloids and Surfaces B Biointerfaces* 18(3–4): 315–323.

Torchilin, V. P. and A. N. Lukyanov. 2003. Peptide and protein drug delivery to and into tumors: challenges and solutions. *Drug Discovery Today* 8(6):259–266.

Torchilin, V. P. 2007. Targeted pharmaceutical nanocarriers for cancer therapy and imaging. *AAPS Journal* 9(2):E128–E147.

Ugwoke, M. I., N. Verbeke and R. Kinget. 2001. The biopharmaceutical aspects of nasal mucoadhesive drug delivery. *Journal of Pharmacy and Pharmacology* 53(1):3–21.

Valencia, P. M., O. C. Farokhzad, R. Karnik and R. Langer. 2012. Microfluidic technologies for accelerating the clinical translation of nanoparticles. *Nature Nanotechnology* 7(10):623–629.

Vandana, Mallaredy, and Sanjeeb K. Sahoo. 2009. Optimization of physicochemical parameters influencing the fabrication of protein-loaded chitosan nanoparticles. *Nanomedicine* 4(7). Future Medicine: 773–785.

Varshochian, R., M. Jeddi-Tehrani, A. R. Mahmoudi, et al. 2013. The protective effect of albumin on bevacizumab activity and stability in PLGA nanoparticles intended for retinal and choroidal neovascularization treatments. *European Journal of Pharmaceutical Sciences* 50(3–4):341–352.

Veronese, F. M. and A. Mero. 2008. The impact of PEGylation on biological therapies. *BioDrugs* 22(5):315–329.

Veuillez, F., Y. N. Kalia, Y. Jacques, J. Deshusses and P. Buri. 2001.Factors and strategies for improving buccal absorption of peptides. *European Journal of Pharmaceutics and Biopharmaceutics* 51(2):93–109.

Vrignaud, S., J-P. Benoit and P. Saulnier. 2011. Strategies for the nanoencapsulation of hydrophilic molecules in polymer-based nanoparticles. *Biomaterials* 32(33):8593–8604.

Wang, Y., W. Qu and S. H. Choi. 2016. FDA's regulatory science program for generic PLA/PLGA-based drug products. *American Pharmaceutical Reviews.*

Wissing, S. A., O. Kayser and R. H. Müller. 2004. Solid lipid nanoparticles for parenteral drug delivery. *Advanced Drug Delivery Reviews* 56(9):1257–1272.

Xie, Jingwei, and Chi Hwa Wang. 2007. Encapsulation of proteins in biodegradable polymeric microparticles using electrospray in the Taylor cone-jet mode. *Biotechnology and Bioengineering* 97:1278–1290.

Xu, Y. and Y. Du. 2003. Effect of molecular structure of chitosan on protein delivery properties of chitosan nanoparticles. *International Journal of Pharmaceutics* 250(1):215–226.

Yamamoto, A. and S. Muranishi. 1997. Rectal drug delivery systems Improvement of rectal peptide absorption by absorption enhancers: protease inhibitors and chemical modification. *Advanced Drug Delivery Reviews* 28(2):275–299.

Ye, Q., J. Asherman, M. Stevenson, E. Brownson and N. V. Katre. 2000. DepoFoamTM technology: a vehicle for controlled delivery of protein and peptide drugs. *Journal of Controlled Release* 64(1–3):155–166.

Yu, M., J. Wu, J. Shi and O. C. Farokhzad. 2016. Nanotechnology for protein delivery: overview and perspectives. *Journal of Controlled Release* 240:24–37.

Zhou, X. H. and A. Li Wan Po. 1991. Peptide and protein drugs: I. Therapeutic applications, absorption and parenteral administration. *International Journal of Pharmaceutics* 75(2–3):97–115.

Zhou, Y., M. Kogiso, C. He, Y. Shimizu, N. Koshizaki, T. Shimizu. 2007. Fluorescent nanotubes consisting of CdS-Embedded bilayer membranes of a peptide lipid. *Advanced Materials* 19(8):1055–1058.

12 Nanotechnology
The Hope for New Generation and Its Therapeutic Applications

Dnyaneshwar Rathod, Avinash Kharat, Varsha Wankhade, and Mahendra Rai

CONTENTS

12.1 Introduction 159
12.2 Nanoparticles and Their Applications in Biomedical Science 160
12.2.1 AgNPs 160
12.2.2 Gold Nanoparticles 161
12.2.3 Selenium Nanoparticles 161
12.2.4 Zinc Nanoparticles 162
12.2.5 Chitosan Nanoparticles (ChNPs) 163
12.2.6 Titanium Nanoparticles 163
12.3 Conclusion 164
References 164

12.1 INTRODUCTION

Nanobiotechnology is a combination between nanotechnological and biological materials which is used for the development of novel biological agents at the nano scale (1–100 nm) that can be used in biomedical science. Nanoparticles are structures at the nano level (less than 100 nm) which possess specific biological activity resulting from their chemical and physical properties and maximum surface area to volume ratio. Therefore, nanoparticles can serve as potential candidates, especially in the field of medicine (nanomedicine). The metallic nanoparticles are the most promising candidate because they contain significant antimicrobial potential because of their wide surface area, which increases the scientific attention given to them by in the fight against multidrug-resistant microorganisms (Rai et al., 2009). The fact that the properties of nanoparticles have high constancy plays an important role in biomedical science applications. Therefore, there is a need to synthesize and optimize well-designed nanoparticles with fundamental properties like good stable size, shape, etc. (Wu and Gao, 2018; Bharti et al., 2019).

It is well known that nanotechnology is an interdisciplinary research area which includes engineering, biology, chemistry, and medicine and which shows maximum possibility for the early detection and management of diseases like cancer (Cai et al., 2008). Since the great discovery of nanotechnology, a huge amount of work has been applied in the biomedical field to detect and evaluate novel nanoparticles synthesized from different salt and inorganic compounds, like gold, silver, titanium, selenium, zinc, chitosan, etc., due to their chemical, biological, and physical properties (Burdusel et al., 2018). The authors also mentioned the remarkable devotion oriented to the biomedical applications of silver nanoparticles (AgNPs), which have attracted attention worldwide as unconventional antimicrobial agents. Although there is inadequate evidence about the toxicity and in vivo biological activity of AgNPs, these nanoparticles have been used for a long time as antimicrobial agents for different purposes in the health industry, cosmetics, food storage, pharma, and some environmental applications (Burdusel et al., 2018). Jiang et al. (2018) reviewed that nanomaterials have received more attention because of their prominent biomedical applications and biological characteristics. Moreover, metal oxide nanoparticles showed especially promising and far-ranging prospects for the biomedical field, especially for anti-bacterial action, anticancer drug/gene delivery, cell imaging, biosensing, and so on.

Nanoparticles have wide applications, like in the agricultural, medical, and pharma industries. They show wide applications in many fields, but the medical application of nanotechnology is one of the best application areas, because of its excellent and thrilling properties (Bharti et al., 2019). However, the utilization of heavy metals for the synthesis of nanoparticles may cause toxicity in the human body; therefore; heavy metals cannot be used directly in the drug delivery system. A combination of nanoparticles with a biocompatible biomaterial may help to develop a new and remarkable compound which can be beneficial in the drug delivery system. Nanomedicine is one of the vigorous and quickly developing fields of nanotechnology that deals with the different potential applications (Murali et al., 2018). The authors also explained that nanoscale structures have a size almost like that of many biological molecules, which show different physical and chemical properties in comparison with small molecules or mass materials that show an extensive range of medical applications within the areas of imaging, diagnosis, and therapy. Environmental conditions play an important role in maintaining the biological activities of nanoparticles and therefore also play a significant role in the designing of these materials (Solanki et al., 2008). Zinc is an essential trace element that widely exists in all body tissues, notably the brain, muscle, bone, and skin. Zinc acts as an important component of different enzyme systems, takes part in metabolism, and plays vital roles in protein and nucleic acid synthesis, hematopoiesis, and neurogenesis.

Communicable disease–causing vectors like viruses, bacteria, fungi, parasites, etc., are responsible for roughly 15 million deaths per year worldwide, by infecting severe diseases which may cause respiratory system and immunodeficiency problems and failure (Singh et al., 2017). Just viral infections alone play a crucial role in worldwide health problems; they not only cause harm to health but also create economic burdens on patients. The beneficial biomedical applications of biologically synthesized nanoparticles with non-toxic materials are being considered as an important area of research. These biologically synthesized nanoparticles (silver, gold, selenium, zinc, etc.) are attractive sources which play very important roles in biomedical science. Therefore, many research groups are involved in the study of their uses in medical science. Hence, in the present chapter we discuss biomedical applications of different nanoparticles like AgNPs, gold nanoparticles (AuNPs), ZnO, selenium nanoparticles (SeNPs), etc.

12.2 NANOPARTICLES AND THEIR APPLICATIONS IN BIOMEDICAL SCIENCE

12.2.1 AgNPs

AgNPs have been used as an antimicrobial agent for many purposes, but the exact mode of action is not yet definable. It was proved that silver nanoparticles can act as a promising candidate that can penetrate into bacteria and kill them, by structurally changing the cell membrane and increasing cell permeability, which leads to cell death (Ge et al., 2014). The size and concentration of silver nanoparticles influence the antimicrobial activity. Generally, the maximum concentration of nanoparticles shows maximum antimicrobial activity, whereas small particles can act as better bactericidal killers at minimum concentrations. It is also reported that the size, shape, and concentration of nanoparticles play an important role when AgNPs act as an active antimicrobial agent. Anasane et al. (2015) studied the antifungal activity of biosynthesised silver nanoparticles against fungi responsible for superficial mycoses like *Malassezia furfur, Trichophyton rubrum, Candida albicans*, and *C. tropicalis*. They observed that the maximum antifungal activity of AgNPs was found to be highest for *T. rubrum*, whereas the minimum activity was observed for *M. furfur* and *C. albicans*. From their results, they concluded that synthesized AgNPs can be used as a promising agent against fungi which cause superficial mycoses.

For many decades, silver compounds have been widely used for sanitation and in injury healing purposes because of their promising antibacterial activity (Rigo et al., 2013). Considering the advantages of silver, a number of products have been developed in order to cure chronic wounds. The combination of silver nanoparticles with wound-dressing bands enhanced the healing in superficial burn wounds but not too much variance was observed in deep burn wounds (Chen et al., 2013). Other authors summarized that the silver nanoparticles were found to be a catalyst agent with the wound-dressing, band which increased its efficacy. Moreover, Huang et al. (2007) noticed that silver nanoparticles impregnated in wound-dressing band significantly decreased the time required for wound healing when compared with silver sulfadiazine, and with no side effects. Nanotechnology provides great opportunities to develop wound-healing treatments. The nanoparticles open the new era for the development of novel materials that can be used in advanced medical technology. The silver nanoparticle shows significant biological properties, like anti-inflammatory, antiviral, and antibacterial activities with minimum bacterial resistance (Gunasekaran et al., 2012).

Viruses are considered the most dangerous disease vector and as leading causes of disease and death worldwide (Naidu et al., 2016). They also reviewed that the principal action underlying the viral inhibitory action of AgNPs is not yet completely resolved, but the AgNPs are known as a broad-spectrum agent against different viral strains like HIV-1, hepatitis B virus, respiratory syncytial virus, herpes simplex virus type 1, and monkey pox virus. It was observed that silver nanoparticles showed maximum antiviral activity as compared to silver ions, because AgNPs dissociate into Ag^0 (atomic) and Ag^+ (ionic) clusters, whereas silver salts release only Ag^+ ion (Ge et al., 2014). Lara et al. (2010) studied the antiviral activity of AgNPs of 30–50 nm against HIV-1 isolates by using indicator cells in which infection was quantified by a luciferase-based assay. They also observed inhibition of all strains by AgNPs, which show comparable antiviral efficacies against T-tropic, M-tropic, dual-tropic, and resistant isolates. The final 50% concentration of silver nanoparticles (IC_{50}) was observed in the range from 0.44 to 0.91 mg/mL, which showed maximum inhibition against viral strains used. The authors also concluded that AgNP can be used against HIV activity at an early stage of viral replication, most likely as a viricidal agent or as an inhibitor of viral entry. The antiviral activity of AgNPs and chitosan composites was screened by comparing the $TCID_{50}$ (Median Tissue Culture Infectious Dose) ratio of viral suspensions treated with the composites to untreated suspensions against H1N1 influenza A virus (Mori et al., 2013). While screening, the authors observed that if the concentration of AgNPs increased, antiviral activity against H1N1 influenza A virus increased; however, chitosan alone did not show antiviral activity. Moreover, the influence of the size of silver nanoparticles on antiviral activity was also noticed. They also stated that the smaller AgNPs exhibit stronger antiviral activity. They concluded that AgNP/Ch composites interacting with viruses showed antiviral activity.

12.2.2 Gold Nanoparticles

Gold nanoparticles play an extraordinary role due to their unique optical properties at the nanoscale, biocompatibility, and rich surface chemistry, all of which attract the attention of researchers working in the field of biomedical science. We know that colloidal gold has been used since ancient times in different therapeutic applications, and to learn more about gold nanoparticles, the physical characteristics of nanosized gold have been the subject of a burst of research regarding their synthesis and biomedical application activity. Gold nanoparticles have received the attention of scientists for over a century and are now enormously used in different field such as biomedical science, chemistry, and engineering (Giljohann et al., 2010). The authors also reviewed the multivalent surface of gold nanoconjugates as a possible way to utilize different therapeutic agents while also focusing on and defending molecules on the surface of a single nanoparticle.

The compatibility of gold nanoparticles means that gold possesses beneficial characteristics in the nano-dimension such as unique optical, electronic, and catalytic properties, biocompatibility, non-immunogenicity, antimicrobial properties, and facile surface chemistry (Kumar and Roy, 2016). They also discussed that synthesized gold nanoparticles have the ability to exhibit potential uses in the field of medicine, whether in special applications in diseased cells circulating in the blood, or in immunization against injury caused by pathogens like viruses and in the early diagnosis and treatment of cancer.

Nanotechnology is an interdisciplinary area with extensive ability in the management of cancer, including molecular diagnosis and drug delivery. The consistency in progress for the application of nanotechnology in cancer and its role to management of cancer holds promise for specializations in oncology in which genetic and protein biomarkers are used in treatment of cancer based on the molecular profile of patients (Cai et al., 2008). The AuNPs synthesized by using extract of *Couroupita guianensis* flower were screened for their anticancer activity by using MTT assay, DNA fragmentation, apoptosis, and comet assay for DNA damage (Geetha et al., 2013). The nanoparticles exhibited remarkable cytotoxicity against HL-60 cells. Further they discussed that, notwithstanding the extensive application of AuNPs, much research has reported the cytotoxic effects of gold nanoparticles. As per the authors' statement, this was the first report of gold nanoparticles synthesized from *C. guianensis*-based gold nanoparticles and their anticancer activity. Rattanata et al. (2015) tested the anticancer activity of gold nanoparticles conjugated with gallic acid with different concentrations against two adenocarcinoma cell lines: one well-differentiated and the other moderately differentiated. In their study they observed that the combination of gold nanoparticles and gallic acid inhibited the proliferation of cancer cells as compared to gallic acid alone. They concluded that the activity of gold nanoparticles can be improved by using gallic acid, and that this can be used an alternative medicine in cancer treatments.

The anticancer activity and mechanism of action of gold nanoparticles was tested by using microscopic monitoring, MTT assay, flow cytometric analysis of cell death, and real-time expression analysis (Kajani et al., 2016). Remarkable anticancer activities were observed for nanoparticles synthesized by aqueous and ethanolic extracts. The gold nanoparticles synthesized from ethanolic extract exhibited maximum anticancer activity as compared to AuNPs prepared from aqueous extracts. Biogenically synthesized gold nanoparticles exhibited maximum anticancer activity against HepG2 and A549 cell lines (Rajeshkumar, 2016). Further, the biogenic method for the synthesis of gold nanoparticles was found to be easy and eco-friendly; in summary, this is a new advancement in drug delivery for cancer treatment and innovation in biomedical applications.

12.2.3 Selenium Nanoparticles

Nanotechnology (especially nanobiotechnology) is a thrilling branch of studies affecting different fields of human activity. One of the most interesting applications for nanoparticles is in the field of nanomedicine. Current research in nanotechnology shows the potential for virus technology to change or convert into an entirely new direction. Nowadays, microorganisms are becoming multiresistant against available drugs. Similarly, limitations of tumor therapies are being observed. Therefore, there is an urgent need for alternative safe and effective therapy to combat clinical multidrug-resistant (MDR) bacteria and fungi, along with chronic diseases including cancer.

Selenium is very significant trace element which plays a significant role in the regulation of key cellular functions as selenoproteins (Rayman, 2012). It is an inorganic compound which is found in the form of trigonal and monoclinic crystalline phase. Research have been carried out to examine the application of selenium, and it was observed that Se supplementation can reduce the incidence of different kinds of cancers (Ahmad et al., 2015). Many geographical studies have been noticed that people taking low concentrations of Se in their diets may have maximum cancer mortality rates (Rayman, 2005). However, Se at higher concentrations in the diet is useful as a prooxidant with well-established properties of inhibiting cancer cell growth and also killing them (Aboul-Fadl, 2005). Therefore, the applications of Se compounds in cancer therapy have been strengthened, as proven by much research showing that Se compounds decrease the threat of different kinds of cancers, like exocrine gland, prostate, lung, colon, and liver cancers (Srivastava and Mukhopadhyay, 2013; Beheshti et al., 2013; Gao et al., 2014). The biosynthesized SeNPs from *Streptomyces bikiniensis* (Ess_amA-1) were screened for anticancer activity and it was observed that SeNPs encourage cell death of Hep-G2 and MCF-7 human cancer cells (Ahmad et al., 2015). Moreover, selenium sulfide is an effective antifungal agent, and it has been used in anti-dandruff hair shampoos for treatment of scalp fungal infections (Eswarapriya and Jegatheesan, 2015).

Zonaro et al. (2015) studied the biological preparation of selenium and tellurium nanoparticles by using *Stenotrophomonas maltophilia* and *Ochrobactrum* sp. bacterial strains isolated from polluted sites. The synthesized nanoparticles were screened for biofilm eradication and antimicrobial activity against bacteria such as *Escherichia coli*, *Pseudomonas aeruginosa*, and *Staphylococcus aureus*. During screening, it was observed that selenium nanoparticles had antibacterial activity potential against selected bacterial strains at very low concentrations. They suggested that the probable exploitation of both selenium and tellurium nanoparticles can be used to develop an antimicrobial agent against bacteria and to eradicate biofilms. The biogenically synthesized selenium nanoparticles were screened for their antibacterial activity against nosocomial infections causing bacteria such as *E. coli*, *P. aeruginosa*, *Klebsiella pneumonia*, *Salmonella typhimurium*, *S. aureus*, and *Bacillus subtilis* (Hariharan et al., 2012). The authors observed that selenium nanoparticles showed significant antimicrobial activity against selected bacteria. In their summary, they stated that the application of selenium nanoparticles as an antibacterial agent against virulent pathogens may contribute a higher advantage to decrease contamination in the hospital environment. The SeNPs synthesized by using *Stenotrophomonas maltophilia* and *Bacillus mycoides* showed strong antibacterial activity at very low concentrations against clinical isolates like *Pseudomonas aeruginosa*, whereas they did not inhibit the growth of *Candida albicans* and *C. parapsilosis* (Cremonini et al., 2016). The antibacterial activity of selenium nanoparticles was screened against different foodborne pathogens such as *E. coli*, *S. aureus*, *Salmonella*, and *Listeria monocytogenes* (Nguyen et al., 2017).

Wang et al. (2005) synthesized melatonin-selenium nanoparticle (MT-Se), a novel complex, by combination of selenium nanoparticles and melatonin medium. Further, the synthesized selenium nanoparticles were used to determine the protective potential of MT-Se against immunological liver injury in mice induced by bacillus Calmette-Guérin (BCG)/lipopolysaccharide (LPS). They observed that MT-Se expressively decreased an increase in Tumour Necrosis Factor alpha (TNF-α) and Interleukin 1 (IL-1) levels in the liver and inhibited the production of TNF-α and IL-1 by peritoneal macrophages. They summarized that the melatonin-selenium nanoparticle exhibited a hepatic defensive activity on immunological liver injury in mice. Nguyen et al. (2017) studied the cytotoxicity of SeNPs against the Caco-2 cell line. They performed an MTT assay and observed that adherent Caco-2 cells were slightly influenced by SeNPs at 20 and 40 mg/mL for 24 h, whereas dose-dependent cytotoxicity of SeNPs was not noticed. They also noted no significant variance in cell viability between control and samples treated with different concentrations of SeNPs.

12.2.4 Zinc Nanoparticles

Different types of material can be used to synthesize nanoparticles, but the cytotoxicity of nanoparticles in living cells is considered as a limitation to the exploration of nanoparticles in biomedical science. However, combination therapy with nanoparticle-based drugs is attracting the interest of researchers for use in cancer treatment. Many studies (FDA) have reported that ZnO nanoparticles (ZnONPs) are considered as generally recognized as safe (GRAS) substances (Rasmussen et al., 2010). ZnO nanoparticles are used in the cosmetics industry to prepare sunscreens and facial creams because they absorb UV irradiation and offer optical transparency (Nohynek et al., 2008). They have shown broad-spectrum antibacterial activity, acting as a promising antibacterial compound to manage pathogenic bacteria (Colon et al., 2006). Similarly, ZnO nanoparticles are well known as an antibacterial agent against different pathogenic bacteria (Nagarajan and Kuppusamy, 2013; Bhumi and Savithramma, 2014). Zhou et al. (2006) reported that ZnO nanowires shows biodegradable potential; they dissolve into ions that can be absorbed by the body and act as a nutritional part. Characteristics like favorable band gap, electrostatic charge, surface chemistry, and potentiation of redox-cycling cascades of ZnO nanoparticles show that ZnONPs can be used in biomedical science. Because of these wide applications of ZnO nanoparticles, they have remained in high demand for physicists, chemists, and biologists.

Current studies have suggested that ZnO nanoparticles are potential candidates as cholesterol biosensors and dietary modulators and for hydrolase activity to control diabetes. Moreover, they also showed significant anticancer activity against selected cell lines (Wang et al., 2009; Rasmussen et al., 2010). It was anticipated that ZnONPs having a smaller size range of 4–20 nm were found to be a promising agent against cancer cell lines that can be utilized to minimize the harmful effects on normal body cells. Different studies have proven that biogenically synthesized zinc oxide nanoparticles exhibited effective cytotoxic effect against different cancer cell lines (Sivaraj et al., 2014; Priyadharshini et al., 2014; Namvar et al., 2015). In 2015, Vinardell and Mitjans reviewed that nanoparticles have attracted much attention because of their applications in cancer therapy. Different studies have reported that metal oxide nanoparticles can increase cytotoxicity activity against cancer cell lines, but it was also observed that they did not harm normal cells. In addition, a few studies also noticed the anticancer activity of ZnONPs alone and in combination with different therapies like photocatalytic therapy or some anticancer drugs. Singh et al. (2011) reported the biological synthesis of zinc oxide nanoparticles by using latex *Calotropis procera*.

Given their exceptional characteristics like compatibility, catalytic cytotoxicity, eco-friendliness, and high specialization, zinc oxide nanoparticles are a promising anticancer agent (Bisht and Rayamajhi, 2016). It is well known that zinc is an important element within the human body and a co-factor for more than 300 mammalian enzymes which play crucial roles in balancing vital cellular development such as oxidative stress, DNA replication, repair, cell cycle progression, and apoptosis. Therefore, it is proved that a decrease or disruption in the zinc balance in cancer cells may cause a deleterious effect toward health. Zinc oxide nanoparticles synthesized from *Tabernaemontana* showed significant cytotoxic effect against MCF-7 breast cancer cell lines (Sivaraj et al., 2014). They also observed that the maximum concentration of zinc oxide nanoparticles effectively inhibited the growth of cells by more than 97%. Another study checked how zinc oxide nanoparticles interact with a critical tumor suppressive pathway centered around p53, which is one of the most important known tumor suppressors that defends cells from developing cancer phenotypes through its control over major pathways like apoptosis, senescence, and cell cycle progression (Ng et al., 2011). Authors also observed that the p53 pathway was activated in BJ cells (skin fibroblasts) after ZnO nanoparticle treatment, with a concomitant reduction in cell numbers. The above observations indicated that cellular responses like apoptosis in the presence of ZnONPs need p53 as the molecular master switch towards programmed cell death. Similarly, it is also predicted that in cells without robust p53, protective response can be tipped towards carcinogenesis when cells are stimulated by a DNA damage-inducing agent such as ZnO nanoparticles. Biosynthesized zinc oxide nanoparticles were evaluated for their in vitro cytotoxic activity and cellular effects on murine cancer cell lines (Namvar et al., 2015). Authors found that the ZnONPs exhibited cytotoxicity against murine cancer cell lines. However, no toxicity was observed against a normal mouse fibroblast (3T3) cell line after treatment with ZnONPs. They concluded that biosynthesis and the therapeutic potential of ZnONPs could be new path for further research, mainly in nanomedicine for the treatment of cancer.

12.2.5 Chitosan Nanoparticles (ChNPs)

Chitosan is used in medicine and pharmacology, especially applied in the form of physical or chemical hydrogels. Chitosan is considered as an antibacterial biopolymer and can be combined with other compounds which can be used in different industries like cosmetics, food, and packaging materials. The quaternary ammonium chitosan (QAC)-chitosan combination was evaluated for its antibacterial efficacy against Gram-positive and Gram-negative bacteria, and significant inhibition against them was observed (Omidi and Kakanejadifard, 2019). Divya et al. (2017) screened chitosan nanoparticles against clinical pathogens like *K. pneumoniae*, *E. coli*, *S. aureus*, and *P. aeruginosa*. The minimum inhibitory concentration was also evaluated and compared with chitosan and chitin activity. They concluded that ChNPs showed significant antimicrobial activity against selected bacteria.

Zhang et al. (2016) reported that ChNPs can be used in drug delivery systems to improve their bioavailability and anticancer activity. When they compared chitosan nanoparticles with free mifepristone (MIF), they observed an increase in anticancer activity in many cancer cell lines. They concluded that mifepristone–chitosan nanoparticles can be improved as a powerful delivery system for MIF to develop its anticancer activity and bioavailability. Chitosan coating on alkaloid is used to develop chitosan nanoparticles in order to increase the property for better drug delivery system. The anticancer activity was performed against A549 lung cancer cell lines, and it was observed that ChNPs were capable of targeting in the drug delivery system (Swarnalatha et al., 2015). The authors also noticed that the nano encapsulated with alkaloid inhibited the cell growth of A549 lung adenocarcinoma cell lines when compared with control.

Calcium alginate hydrogel combined with chitosan nanoparticles can be used against inflammation and neovascularization in the treatment of wounds (Wang et al., 2018). This combination has also been observed to exhibit significant antibacterial activity. Through the modulation of generation of reactive oxygen species, it increased synthesis and secretion of IL-6 in vascular endothelial cells, which indicates its efficacy for pro-inflammatory initiation. It was also found to catalyze the vascular endothelial cell invasion, metastasis, and neovascularization to quicken the injury healing. Karri et al. (2016) studied the application of chitosan nanoparticles impregnated into scaffold, which showed that a combination of nanoparticles with scaffold treated wounds remarkably quickly as compared with wound control and placebo scaffold groups. They also analyzed and predicted that complete epithelialization with thick granulation tissue formation was observed in combination, whereas an absence of compact collagen deposition in the placebo scaffold group and the presence of inflammatory cells in the control group was observed. Therefore, it was concluded that the synergistic combination of curcumin, chitosan, and collagen can be a promising agent to treat different pathological manifestations of diabetic wounds.

12.2.6 Titanium Nanoparticles

The synthesis of titanium dioxide (TiO_2) and its applications in different fields has been considered as promising. Among all uses, the biomedical application of TiO_2 is one of strong interest and intensive experimental and theoretical studies, because of its photocatalytic properties, outstanding biocompatibility, high stability, and low toxicity (Yin et al., 2013). They also discussed that the recent advances and medical applications of TiO_2 such as photodynamic therapy, which is used in cancer treatment, drug delivery, cell imaging, biosensors for biological assays, and genetic engineering.

Desai and Kowshik (2009) screened TiO_2 nanoparticle photocatalysts that showed photocatalytic activity against common pathogenic bacteria like *E. coli*, *P. aeruginosa*, *K. pneumoniae*, and *S. aureus* under visible light illumination. The titania nanoparticles synthesized by using hydrothermal process were screened for their antimicrobial activity against *E. coli* (Erdural et al., 2008). They noticed that the sample treated under an alkaline condition showed maximum antimicrobial activity as compared with an untreated sample. The TiO_2 nanoparticles synthesized by the sol-gel method showed significant antibacterial activity against *S. aureus* and *E. coli*. They also found the minimum inhibition concentration of TiO_2 nanoparticles to be 30 μg/ml for *S. aureus* and 40μg/ml for *E. coli*. Shilpa Chakra et al. (2017) studied the antibacterial and anticancer activities of a nanocomposite synthesized mixture of zinc oxide and titanium dioxide nanoparticles. It was observed that the synergistic activity of the nanocomposite against bacteria and cancer was maximum when it compared in alone. Photocatalyzed TiO_2 nanoparticles were evaluated for their antibacterial activity against multidrug-resistant methicillin-resistant *S. aureus* and *Clostridium difficile* (Page et al., 2009).

12.3 CONCLUSION

Nanoparticles are a prominent nano-product extensively used in different medical fields like wound dressing, cancer treatment, and pharmacological treatment. Moreover, to enhance the activity of biological compounds, nanoparticles can act as catalysts in order to increase the potential of compounds. It can be concluded that to provide alternative medicines against infectious diseases, researchers should focus on developing a variety of novel nanoparticles with high efficacy. Nanoparticles hold enormous potential as sources of antimicrobial, anticancer, and antiviral treatment. These nanoparticles may open new vistas for the development of new drugs. Although there is huge demand for and present applications of nanoparticles in biomedical science, information is still needed prior to clinical trials regarding delivery and their non-toxicity against normal tissue.

REFERENCES

Aboul-Fadl, T. 2005. Selenium derivatives as cancer preventive agents. *Current Medicinal Chemistry Anti-cancer Agents* 5:637–652.

Ahmad, M.S. Yasser, M.M. Sholkamy, E.N. Ali, A.M. and Mehanni, M.M. 2015. Anticancer activity of bio-stabilized selenium nanorods synthesized by *Streptomyces bikiniensis* strain EssamA-1. *International Journal of Nanomedicine* 10:3389–3401.

Anasane, N. Golinska, P. Wypij, M. Rathod, D. Dahm, H. and Rai, M. 2015. Acidophilic actinobacteria synthesised silver nanoparticles showed remarkable activity against fungi-causing superficial mycoses in humans. *Mycoses* 59(3):157–166. doi :10.1111/myc.12445.

Beheshti, N. Soflaei, S. and Shakibaie, M. et al. 2013. Efficacy of biogenic selenium nanoparticles against *Leishmania major*: In vitro and in vivo studies. *Journal of Trace Elements in Medicine and Biology* 27(3):203–207.

Bharti, S. Kaur, G. Jain, S. Gupta, S. and Tripathi, S.K. 2019. Characteristics and mechanism associated with drug conjugated inorganic nanoparticles. *Journal of Drug Targeting.* doi: 10.1080/1061186X.2018.1561888.

Bhumi, G. and Savithramma, N. 2014. Biological synthesis of zinc oxide nanoparticles from Catharanthus roseus (l.) G. Don. leaf extract and validation for antibacterial activity. *International Journal of Drug Development and Research* 6(1):208–214.

Bisht, G. and Rayamajhi, S. 2016. ZnO nanoparticles: A promising anticancer agent. *Nanobiomedicine* 3:9.

Burdusel, A.C. Gherasim, O. Grumezescu, A.M. Mogoantă, L. Ficai, A. and Andronescu, E. 2018. Biomedical applications of silver nanoparticles:an up-to-date overview. *Nanomaterials* 8:681–705.

Cai, W. Gao, T. Hong, H. and Sun, J. 2008. Applications of gold nanoparticles in cancer nanotechnology. *Nanotechnology, Science and Applications* 1:17–32.

Chen, J. Ouyang, J. Kong, J. Zhong, W. and Xing, M.M. 2013. Photo-cross-linked and pH-sensitive biodegradable micelles for doxorubicin delivery. *ACS Applied Materials and Interfaces* 5(8):3108–3117.

Colon, G. Ward, B.C. and Webster, T.J. 2006. Increased osteoblast and decreased *Staphylococcus epidermidis* functions on nanophase ZnO and TiO2. *Journal of Biomedical Materials Research Part A* 78:595–604.

Cremonini, E. Zonaro, E. Donini, M. et al. 2016. Biogenic selenium nanoparticles: Characterization, antimicrobial activity and effects on human dendritic cells and fibroblasts. *Microbial Biotechnology* 9(6):758–771.

Desai, V.S. and Kowshik, M. 2009. Antimicrobial activity of titanium dioxide nanoparticles synthesized by sol-gel technique. *Research Journal of Microbiology* 4(3):97–103.

Divya, K. Vijayan, S. George, T.K. and Jisha, M.S. 2017. Antimicrobial properties of chitosan nanoparticles: Mode of action and factors affecting activity. *Fibers and Polymers* 18:221–230.

Erdural, B.K. Yurum, A. Bakir, U. and Karaka, G. 2008. Antimicrobial properties of titanium nanoparticles. A. Vaseashta and I.N. Mihailescu (eds.), *Functionalized nanoscale materials devices and system*, pp. 409–414.

Eswarapriya, B. and Jegatheesan, K.S. 2015. Antifungal activity of biogenic selenium nanoparticles synthesized from electronic waste. *International Journal of Pharmacy and Technology Research* 8(3):383–386.

Gao, F. Yuan, Q. Gao, L. et al. 2014. Cytotoxicity and therapeutic effect of irino`tecan combined with selenium nanoparticles. *Biomaterials* 35(31):8854–8866.

Ge, L. Li, Q. Wang, M. Ouyang, J. Li, X. and Xing, M.M.Q. 2014. Nanosilver particles in medical applications: Synthesis, performance, and toxicity. *International Journal of Nanomedicine* 4:9.

Geetha, R. Ashokkumar, T. Tamilselvan, S. Govindaraju, K. Sadiq, M. and Singaravelu, G. 2013. Green synthesis of gold nanoparticles and their anticancer activity. *Cancer Nanotechnology* 4:91–98.

Giljohann, D.A. Seferos, D.S. Daniel, W.L. Massich, M.D. Patel, P.C. and Mirkin, C.A. 2010. Gold nanoparticles for biology and medicine. *Angewandte Chemie – International Edition England* 49(19):3280–3294.

Gunasekaran, T. Nigusse, T. and Dhanaraju, M.D. 2012. Silver nanoparticles as real topical bullets for wound healing. *The Journal of the American College of Clinical Wound Specialists* 3:82–96.

Hariharan, H. Al-Harbi, N.A. Karuppiah, P. and Rajaram, S.K. 2012. Microbial synthesis of selinium nanocomposite using *Saccharomyces cerevisiae* and its antimicrobial activity against pathogens causing nosocomial infection. *Chalcogenide Letters* 9:509–515.

Huang, Y. Li, X. Liao, Z. et al. 2007. A randomized comparative trial between Acticoat and SD-Ag in the treatment of residual burn wounds, including safety analysis. *Burns* 33(2):161–166.

Jiang, J. Pi, J. and Cai, J. 2018. The advancing of zinc oxide nanoparticles for biomedical applications. *Bioinorganic Chemistry and Applications* 2018:1–18. doi: 10.1155/2018/10625

Kajani, A.A. Bordbar, A.K. Esfahani, S.H.Z. and Razmjou, A. 2016 Gold nanoparticles as potent anticancer agent: Green synthesis, characterization, and *in vitro* study. *RSC Advances* 6:63973–63983.

Karri, V.V. Kuppusamy, G. Talluri, S.V. et al. 2016. Curcumin loaded chitosan nanoparticles impregnated into collagen-alginate scaffolds for diabetic wound healing. *International Journal of Biological Macromolecules* 93:1519–1529.

Kumar, P. and Roy, I. 2016. Applications of gold nanoparticles in clinical medicine. *International Journal of Pharmacy and Pharmaceutical Sciences* 8(7):9–16.

Lara, H.H. Ayala-Nuñez, N.V. Ixtepan-Turrent, L. and Rodriguez-Padilla, C. 2010. Mode of antiviral action of silver nanoparticles against HIV-1. *Journal of Nanobiotechnology*8:1.

Mori, Y. Ono, T. Miyahira, Y. Nguyen, V.Q. Matsui, T. and Ishihara, M. 2013. Antiviral activity of silver nanoparticle/chitosan composites against H1N1 influenza A virus. *Nanoscale Research Letters* 8:93.

Murali, K. Neelakandan, M.S. and Thomas, S. 2018. Biomedical applications of gold nanoparticles. *JSM Nanotechnology and Nanomedicine* 6(1):1064.

Nagarajan, S. and Kuppusamy, K.A. 2013. Extracellular synthesis of zinc oxide nanoparticle using seaweeds of Gulf of Mannar, India. *Journal of Nanobiotechnology* 11:39.

Naidu, K.S.B. Govender, P. and Adam, J.K. 2016. Nano silver particles in biomedical and clinical applications: Review. *Journal of Pure and Applied Microbiology* :1–10.

Namvar, R. Rahman, H.S. Mohamad, R. et al. 2015. Cytotoxic effects of biosynthesized zinc oxide nanoparticles on murine cell lines. *Evidence Based Complementary and Alternative Medicine* doi: 10.1155/ 2015/593014.

Ng, K.W. Khoo, S.P. Heng, B.C. et al. 2011. The role of the tumor suppressor p53 pathway in the cellular DNA damage response to zinc oxide nanoparticles. *Biomaterials* 32:8218–8225.

Nguyen, T.H.D. Vardhanabhuti, B. Lin, M. and Mustapha, A. 2017. Antibacterial properties of selenium nanoparticles and their toxicity to Caco-2 cells. *Food Control* 77:17–24.

Nohynek, G.J. Dufour, E.K. and Roberts, M.S. 2008. Nanotechnology, cosmetics and the skin: Is there a health risk? *Skin Pharmacology and Physiology* 21:136–149.

Omidi, S. and Kakanejadifard, A. 2019. Modification of chitosan and chitosan nanoparticle by long chain pyridinium compounds: Synthesis, characterization, antibacterial, and antioxidant activities 208:477–485.

Page, K. Michael Wilson, M. and Parkin, I.P. 2009. Antimicrobial surfaces and their potential in reducing the role of the inanimate environment in the incidence of hospital-acquired infections. *Journal of Materials Chemistry* 19:3819.

Priyadharshini, R.I. Prasannaraj, G. Geetha, N. and Venkatachalam, P. 2014. Microwave-mediated extracellular synthesis of metallic silver and zinc oxide nanoparticles using macroalgae (*Gracilaria edulis*) extracts and its anticancer activity against human PC3 cell lines. *Applied Biochemistry and Biotechnology* 174:2777–2790.

Rai, M.K. Yadav, A. Gade, A.K. 2009. Silver nanoparticles as a new generation of antimicrobials. *Biotechnology Advances* 27:76–82.

Rajeshkumar, S. 2016. Anticancer activity of eco-friendly gold nanoparticles against lung and liver cancer cells. *Journal of Genetic Engineering and Biotechnology* 14:195–202.

Rasmussen, J.W. Martinez, E. Louka, P. and Wingett, D.G. 2010. Zinc oxide nanoparticles for selective destruction of tumor cells and potential for drug delivery applications. *Expert Opinion on Drug Delivery* 7(9):1063–1077.

Rattanata, N. Daduang, S. Wongwattanakul, M. et al. 2015. Gold nanoparticles enhance the anticancer activity of gallic acid against cholangiocarcinoma cell lines. *Asian Pacific Journal of Cancer Prevention* 16:7143–7147.

Rayman, M.P. 2005. Selenium in cancer prevention: A review of the evidence and mechanism of action. *Proceedings of the Nutrition Society* 64(4):527–542.

Rayman, M.P. 2012. Selenium and human health. *The Lancet* 379:1256–1268.

Rigo, C. Ferroni, L. Tocco, I. et al. 2013. Active silver nanoparticles for wound healing. *International Journal of Molecular Sciences* 14:4817–4840.

Shilpa Chakra, C.H. Rajendar, V. Venkateswara Rao, K. and Kumar, M. 2017. Enhanced antimicrobial and anticancer properties of ZnO and TiO_2 nanocomposites. *3 Biotech* 7:89.

Singh, R.P. Shukla, V.K. Yadav, R.S. Sharma, P.K. Singh, P.K. and Pandey, A.C. 2011. Biological approach of zinc oxide nanoparticles formation and its characterization. *Advanced Materials Letters* 2(4):313–317.

Singh, L. Kruger, H.G. Maguire, G.E.M. Govender, T. and Parboosing, R. 2017. The role of nanotechnology in the treatment of viral infections. *Therapeutic Advances in Infectious Disease* 4(4):105–131.

Sivaraj, R. Rahman, P.K.S.M. Rajiv, P. and Venckatesh, R. 2014. Biogenic zinc oxide nanoparticles synthesis using *Tabernaemontana divaricate* leaf extract and its anticancer activity against MCF-7 breast cancer cell lines. *International Conference on Advances in Agricultural, Biological and Environmental Sciences*. doi: 10.15242/IICBE.C1014032.

Solanki, A. Kim, J.D. Lee, K.B. 2008. Nanotechnology for regenerative medicine: Nanomaterials for stem cell imaging. *Nanomedicine* 3:567–578.

Srivastava, N. and Mukhopadhyay, M. 2013. Biosynthesis and structural characterization of selenium nanoparticles mediated by *Zooglea ramigera*. *Powder Technology* 244:26–29.

Swarnalatha, Y. Gunna, G.K. and Jacob, C.M. 2015. Synthesis of alkaloid loaded chitosan nanoparticles for enhancing the anticancer activity in A549 lung cancer cell line. *Der Pharmacia Lettre* 7(6):378–390.

Vinardell, M.P. and Mitjans, M. 2015. Antitumor activities of metal oxide nanoparticles. *Nanomaterials* 5:1004–1021.

Wang, H. Wei, W. Zhang, S.Y. et al. 2005. Melatonin-selenium nanoparticles protects liver against immunological injury induced by bacillus Calmette-Guérin and lipopolysaccharide. *Acta Pharmacologica Sinica* 26(6):745–752.

Wang, H. Wingett, D. Engelhard, M.H. et al. 2009. Fluorescent dye encapsulated ZnO particles with cell specific toxicity for potential use in biomedical applications. *Journal of Materials Science: Materials in Medicine* 20:11–22.

Wang, T. Zheng, Y. Shen, Y. et al. 2018. Chitosan nanoparticles loaded hydrogels promote skin wound healing through the modulation of reactive oxygen species. *Artificial Cells, Nanomedicine and Biotechnology* 46:138–149.

Wu, X. and Gao, S. 2018. Synthesis of shape-controlled layered double hydroxide and its application in drug delivery. *Nanoscience and Nanotechnology Letters* 10:75–80.

Yin, Z.F. Wu, L. Yang, H.G. and Su, Y.H. 2013. Recent progress in biomedical applications of titanium dioxide. *Physical Chemistry Chemical Physics* 15:4844–4858. doi: 10.1039/c3cp43938k.

Zhang, H. Wu, F. Li, Y. et al. 2016. Chitosan-based nanoparticles for improved anticancer efficacy and bioavailability of mifepristone. *Beilstein Journal of Nanotechnology* 7:1861–1870.

Zhou, J. Xu, N. and Wang, Z.L. 2006. Dissolving behavior and stability of ZnO wires in biofluids: A study on biodegradability and biocompatibility. *Advanced Materials* 18:2432–2435.

Zonaro, E. Lampis, S. Turner, R.J. Qazi, S.J.S. and Vallini, G. 2015. Biogenic selenium and tellurium nanoparticles synthesized by environmental microbial isolates efficaciously inhibit bacterial planktonic cultures and biofilms. *Frontiers in Microbiology.* 6(article 584):1–11. doi: 10.3389/fmicb.2015.00584.

13 Origami-Engineered Structures and Their Biomedical Applications

Dipti Mohanta, Snehaprava Mohapatra, Sonali Khuntia, Suchilagna Sahoo, and Pradipta Ranjan Rauta

CONTENTS

13.1 Introduction 167
13.2 Origami Structures and Mathematical Design 167
13.3 Nano-Origami: Art and Function 168
13.4 Biomedical Applications 168
13.4.1 Drug Delivery 168
13.4.2 Microsurgery 168
13.4.3 Stent Grafts 170
13.4.4 Imaging Tools 170
13.4.5 Foldscope: The Folded Microscope 170
13.4.6 Microfluidic Devices 170
13.4.7 Microgripper 170
13.5 Molecular Origami 170
13.6 Challenges/Issues 172
13.7 Conclusion 172
References 172

13.1 INTRODUCTION

For a long period of time, cultures have arisen with innovative techniques that artistically approach, implement, and familiarize the paper craft. The word "origami" evolved in Japan: *ori* stands for "folding" and *kami* means "paper." According to history, classical and traditional origami was first originated in China around 1400 years ago (Kurin 2004). Traditional origami consisted of folding paper according to a pattern into a sculpture without using any tools. It is believed that the origami concept was brought from China to Japan by monks in the sixth century. Then, it (the art of paper folding) become one of the most iconic cultural contributions of Japan and became the Japanese symbol of pleasure and life: the crane. In 1797, it was reported that the first book on origami was published which mentioned 49 techniques of crane folding. Although origami is most closely tied to Japan, it also has roots in China (e.g., ancient Chinese art), Europe (e.g., mathematically based folding applications of Spain and decorative arts and crafts of Germany), and Korea (e.g., ancient Korean art). Traditional origami is categorized by simple folding patterns that were freely accessible from generation to generation. However, modern origami are complex structures that are created by designers (incorporate advanced mathematical theories, modeling) and considered as copyrightable material or intellectual property.

Perhaps one may be acquainted with traditional origami and might attempt to make a crane, a frog, or an airplane. Less familiar is the fact that the modern concept of origami is also inspiring the design of engineering devices and structures around the world nowadays. This transitional field (origami) has currently perceived remarkable successes with many applications in modern biomedical science. Inspired by the traditional/modern art of origami, this approach is very beneficial not only in manufacturing solar panels/transport antennae, telescopes (folding scopes), and applications in space stations (significantly reducing size of transported objects) but also in making use of shape-memory materials and self-assembling robots, along with a ton of biomedical applications (Felton et al. 2014a). This chapter aims to assess the application of origami-based structures in biomedical applications.

13.2 ORIGAMI STRUCTURES AND MATHEMATICAL DESIGN

Origami-inspired biomedical device design is an emergent technology, because of its ability to be arranged from a minimal foldable pattern to a larger volume. Modern origami is mostly studied on the basis of a mathematical principles/modeling called origamics, which includes the elementary principles of design, structural properties, and kinematics

(Sorguç, Hagiwara, and Selçuk 2009; Dukes and Rusinko 2011). The process of testing an origami design is augmented in a cost-effective manner with advanced computational simulation technologies. This origami-based design can transform an idea into reality in very little time using both 2D and 3D prototyping techniques and parametric graphical software (Johnson et al. 2017).

The techniques of folding pattern can be used in designing many structures with better functionality and compact form for easy storage and transport. At present, these principles have already been successfully used to produce a wide variety of materials. Also, the folding helps in reducing manufacturing costs since they require smaller amounts of materials and simplify constructions. So, the advantages of compact origami principles could be applied to improvement in manufacturing biomedical devices. For example, the origami-inspired design of disposable biomedical devices can make them significantly less expensive to manufacture. Also, a compact form of a biomedical device (applying origami principles) can be inserted into the body via a small incision, travel to an area that is inaccessible to large devices, and then install into its functional form at the target area. The fabrication of nanorobots and other micro electromechanical systems have also benefited from origami-based principles and design (Felton et al. 2014b). The key features of origami-inspired biomedical devices includes a simple synthesis procedure, cost-effective and disposable design (avoids contamination), and highly compact adaptable designs to reach remote areas for biomedical applications. The design range varies from the milli- to the nanoscale, depending upon the application to challenging medical problems.

13.3 NANO-ORIGAMI: ART AND FUNCTION

Nanostructured origami is rapidly evolving to construct delicate structures by integrating novel functionality into crystalline nanostructures. The nanomembranes can be easily folded and unfolded while preserving their natural properties. Otherwise, these structures can also be cut out, which is similar to the alternate form of origami known as kirigami, generating phenomenal artistic 3D nano-objects that work under favorable conditions (at specific stimuli) to address a particular technology task (Cavallo and Lagally 2015).

13.4 BIOMEDICAL APPLICATIONS

Origami-inspired biomedical devices can be manufactured in a precise way using modern prototyping procedures. The structures can be made in a compact form using mathematical folding patterns and then organized into functional forms at particular sites. So, this can be an innovative and less invasive method of treatment and delivery, where the compact devices (micro- or nano-sized) can move throughout the body to previously inaccessible target zones. So, the impact of origami can be observed in a wide variety of biomedical devices (Johnson et al. 2017). The majority of origami-inspired designed devices focus on reaching remote areas within the body to execute a procedure such as imaging, biopsy, surgery, etc. (Johnson et al. 2017; Gregory et al. 2015; Miyashita et al. 2015). A few examples of origami-based designs and their biomedical applications are mentioned in Table 13.1. The major biomedical applications consisting of origami approaches are discussed below.

13.4.1 Drug Delivery

The origami-inspired hydrogenation-aided graphene structure can accomplish drug uptake and storage (in compact form) and help in sustainable delivery (Zhu and Li 2014). These graphene origamic nanostructures could be mechanically programmable by applying an external electric field. This nanoscale origami was achieved by cutting graphene atomically into thin structures and using its chemical properties; the folding and unfolding pattern can be manipulated by applying an electric field in a controlled manner. The graphene can act as a building block and can be shaped into several folding structures. The folded nanostructured graphene could possibly work as a drug delivery system. The size of the graphene origami structure was very important, as the increase in size affects the drug delivery, which proved the advantage of nano-based drug delivery system (Zhu and Li 2014; Johnson et al. 2017).

13.4.2 Microsurgery

Self-folding and controllable robots can be used in the body for surgical procedures. In an attempt, one research group at Massachusetts Institute of Technology (MIT) developed a self-folding micro-robot which can achieve tasks and then degenerate into the specified environment (Miyashita et al. 2015). The planner sheet was composed of three layers along with four square electromagnetic coils, and a thermal responsive polymeric film (made up of polyvinyl chloride) was packed by two rigid structural layers of water-soluble material. There are gaps at the front and back sides of the layer to build sufficient self-folding configurations. The robot was transformed from planner to a 3D structure by heating, as folding starts at 65°C (in only 1 min). With help of the electromagnetic coils controlled by a joystick, the robot could be guided to any place inside the body and used for surgical procedures. However, more improvements are required for this system, such as complete easy removal of the robot after use and orientation to the proper target (remote areas).

In another approach, an ingestible, controllable, and biodegradable origami robot was developed by Shuhei et al. This origami robot is guided to the stomach by the oral route, then it is directed to the exact location using an external magnetic field, where it can be used for surgical procedures (removing foreign body/tumor), to patch wounds, or to deliver drugs to infected sites (Miyashita et al. 2016). This self-folding robot was tested in vivo using a central magnet. The origami robot was encapsulated in an ice capsule for easy ingestion, which then melts at stomach pH. This type of miniature robot can be implemented in several minimally invasive gastrointestinal procedures, such as surgery, imaging, etc.

TABLE 13.1
Examples of Origami-Based Designs and Their Biomedical Applications

Use	Materials	Control mechanism	Advantages	References
Drug delivery	Graphene	Folded by hydrogenation and folding/unfolding controlled by external electric field	Drug/molecular mass uptake, storage, and targeted delivery	Zhu and Li (2014)
Drug delivery	Origami structures combining the anticancer drug DOX via intercalation and tagged with quantum dots	Self-unfolding; delivers the drug at desired location; fluorescence properties of QDs were used to investigate biodistribution and pharmacokinetics	Higher cellular uptake in cancer cells and death due to apoptosis	Jiang et al. (2012), Zhao et al. (2012), Zhang et al. (2014)
Drug delivery	DNA origami-based nano box	It opens and closes via specific external stimuli like photoactivated mechanisms	Senses the condition, discovers the target zone, unfolds the structure at particular stimuli, and releases drugs	Kohman et al. (2016), Li et al. (2016)
Biosensor	Origami-based synthetic nanopore with DNA	DNA translocation rate through the channel can also be affected by external stimuli (electric field)	Highly specific and sensitive	Bell et al. (2012)
Biosensor	3D DNA origami nanorobots immobilized on glass fiber tips	Target DNA sequence identified by bioluminescence	Highly specific, sensitive	Torelli et al. (2018)
Bioimaging	Transient binding of regular fluorophore-tagged oligonucleotides onto DNA origami structures	DNA-based point accumulation for imaging in nanoscale topography (DNA-PAINT) scheme and and by integrating the programmability and photostability of nanocrystal QDs	Overcome photobleaching, superior photoproperties	Bhatia et al. (2016)
Microsurgery robot	Aqueous soluble aluminum-coated polyester- and acetone-degradable material	Can be distantly controlled by external magnetic field	Can do many functions like moving, carrying/delivering blocks	Miyashita et al. (2015)
Microsurgery robot	Use of biocompatible and biodegradable robot of an iron magnet, artificial esophagus, and silicone stomach	Robot encapsulated in ice capsule; robot released inside stomach and then movement is regulated by external electric field	Easy to manufacture, cost effective, and less invasive	Miyashita et al. (2016)
Microsurgery self-folding robot	Aqueous soluble paper, polyvinyl chloride	A thermo-responsive contraction film sandwiched by two rigid structural layers of liquid-soluble material; its movement controlled by electromagnetic field	Can be guided to reach remote areas in vivo as well as other biological treatments	Miyashita et al. (2014)
Stent grafts	Ni-rich titanium/nickel (TiNi) shape memory alloy	Packed into a sheath and then guided to the required site, where it is released from sheath; stent is deployed mechanically or thermally	Easy production and less invasive	Kuribayashi et al. (2006)
X-ray imaging cover for the extendable arm	A synthetic paper product called Tyvek	Cover expands and contracts using origami-inspired engineering (Miura-ori folding)	Cost effective and time saving	Johnson et al. (2017), Varrasi (2014)
Foldscope	Paper (waterproof)	Applied for large-scale manufacturing of microscopes, specifically demonstrating bright-field, dark-field, and fluorescence microscopes	Low cost, user friendly, high performance	Cybulski, Clements, and Prakash (2014)
Microfluidic device	Folded from one piece of paper in a photolithographic step	Simple folding within very little time without any tools; detection point can be placed in any layer due to easy unfolding pattern	Simple and easy design, low cost, and less manufacturing time, without any tools	Liu and Crooks (2011)
Microgrippers	Multilayer photolithography polyvinyl alcohol layer, with trilayer joints made up of a polymer	Triggered only at particular cancerous environments due to specific biomolecular interaction carried out by particular enzymes	Autonomous, biocompatible, less invasive, easy production	Bassik et al. (2010)

13.4.3 Stent Grafts

A self-portable origami stent graft was developed by Kuribayashi et al., of Oxford University; it was composed of a single nickel–titanium shape memory alloy (SMA) with multi-folding structure (Kuribayashi et al. 2006). When the foil is attached adhesively, it forms a cylinder (length 44 mm and diameter 25.4 mm). The origami stent was folded using a negative photochemical imprint procedure. This type of stent has a great advantage over the currently used wire mesh–structured stent, which can lead to restenosis. This newly developed origami SMA stent graft was first inserted into a case and then guided to the selected site, where the stent was delivered from the case mechanically or thermally (at body temperature) and used for minimally invasive surgery, such as those necessitating aorta and esophageal stent grafts.

13.4.4 Imaging Tools

The origami-inspired engineering approach has great applications in improving bioimaging techniques. An origami-inspired cover (made up of synthetic paper product) for the long arm of a medical X-ray machine was designed by a GE health care in collaboration with BYU (Brigham Young University) (Varrasi 2014; Johnson et al. 2017). The cover folds and unfolds using an origami-based approach using a Miura-ori folding pattern. The cover can be made easily movable (in any direction) and adjustable according to the long X-ray arm's geometry. The cover with this special folding pattern also solves the sterilization issues, as it keeps the arm sterile during movement. This cover with the Miura-ori folding approach is very cost effective and time saving compared to the previously used plastic drapes, which require manual replacement with sterilization issues. This invention is a great advancement in X-ray imaging technique (Johnson et al. 2017).

13.4.5 Foldscope: The Folded Microscope

The Foldscope, a portable, ultra-low-cost, and versatile microscope, is made mostly out of paper (waterproof) with an origami-based approach. This origami-inspired approach is applied extensively to the making of microscopes, specifically for dark-field, bright-field, and fluorescence microscopes, etc. (Cybulski, Clements, and Prakash 2014).

The optical design in association with origami principles allows the manufacture of a large volume of microscopes from 2D media. These are made from a type of flat paper sheet, and although it's tough to accept as true, the parts (waterproof paper) necessary to make it cost less than one US dollar. The technical parameters corresponding to Foldscope performance are quite remarkable. It reaches 2,000× magnification, is small in size, and also requires no external power, which makes it user friendly.

This cost-effective technical approach to microscopes with significant high-resolution imaging capabilities makes them suitable for diverse applications, especially in developing countries (Cybulski, Clements, and Prakash 2014; Johnson et al. 2017).

13.4.6 Microfluidic Devices

Microfluidics, otherwise known as lab-on-a-chip (LOC) technology, is applied to a number of biomedical applications including drug delivery and tissue engineering. The recent advancements in microfluidic devices using an origami-based approach are very cost effective as well time saving. The origami concept was incorporated to formulate a very simple technique for fabricating 3D paper microfluidic devices (oPAD) (Liu and Crooks 2011). This device was folded from one piece of paper in a photolithographic step, irrespective of the number of layers. Before that, layer-by-layer fabrication (in a sequential manner) was the only option for designing microfluidic devices. The current development reduces the cost and fabrication time by a significant amount. The folding and unfolding can be done in less than one minute without use of any complex tools.

At any layer of oPAD, the detection points can be positioned due to the easily unfolding design. 3D paper microfluidic devices are very useful in the biomedical field due to their cost-effective nature and easy fabrication.

13.4.7 Microgripper

The microgripper is a microscopic device used to grasp and recover microscale objects and biopsies. The microgripper is distantly guided in vivo to the preferred locations. A microgripper was developed by Leong et al. with an origami approach; this microgripper can be activated by chemical actuation or temperature, under relevant conditions. Once it is guided to the desired location, it is activated by heat or chemicals. The grippers consist of six rotationally symmetric digits (each with three polymer joints), which are stemmed in the most successful design (Stern et al. 2009). Magnetic resonance imaging (MRI) or computer tomography can also be applied for guiding these microgrippers in order to avoid premature lodging. The tetherless microgripper denotes the evolution of marginally invasive, biocompatible, self-governing microtools (Stern et al. 2009).

Again, this concept was expanded by Bassik et al. by aiming at specific enzyme substrate biomolecular interactions to activate the microgrippers (Bassik et al. 2010). The biopolymer that pairs with specific proteolytic enzymes secreted from cancer cells is chosen accordingly, so that the microgripper is activated only in cancerous locations. This microgripper is a characteristic example of the evolution of tools and devices that can reconfigure independently at favorable biological environments.

13.5 MOLECULAR ORIGAMI

Nucleic acids (DNA or RNA) are ideal for creating folding patterns due to unique molecular interactions. Structural DNA origami exploits biological or synthetic DNA as designer molecules for the self-assembly of artificial nanostructures.

In nucleic acids, the specific base pairing (Watson–Crick Model) happens between particular nucleotides (A-T/C-G), making it possible to form specific folds that ultimately result in complex origami patterns (Figure 13.1 and Figure 13.2). With research advancements, DNA has been established for exceptional usefulness in fabricating artificial nanostructures with incorporated complexity and programmability. These structures are of two types: static with well-organized physicochemical features and dynamic with the capability to reorganize under suitable conditions (e.g., external stimuli) (Figure 13.2).

In 2006, DNA origami was first introduced by Paul Rothemund, but after that structural DNA nanotechnology has undergone remarkable progress in fabrication and use in various biomedical applications such bioimaging, biosensing, and drug delivery (Arora and de Silva 2018; Ke, Castro, and Choi 2018). The origami-based synthetic nanopore is an example of a biosensor that responds to a specific electric field that

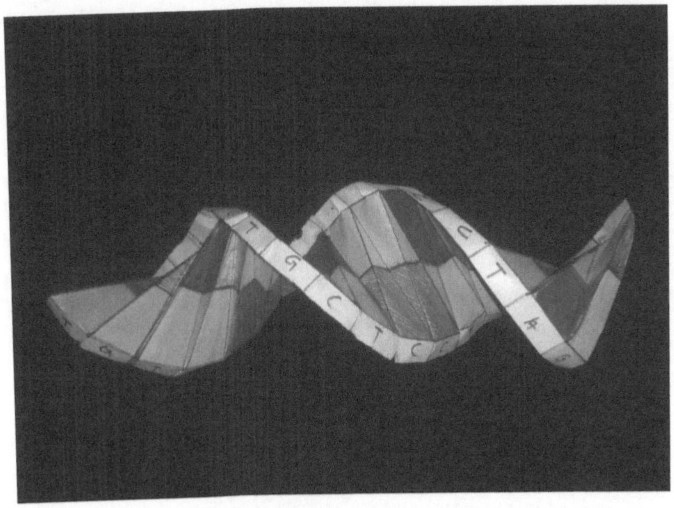

FIGURE 13.1 3D DNA double-helix structure comprised of the specific base pairing (Watson-Crick Model) between particular nucleotides (A-T/C-G), making it possible to form specific folds that ultimately result in complex origami patterns.

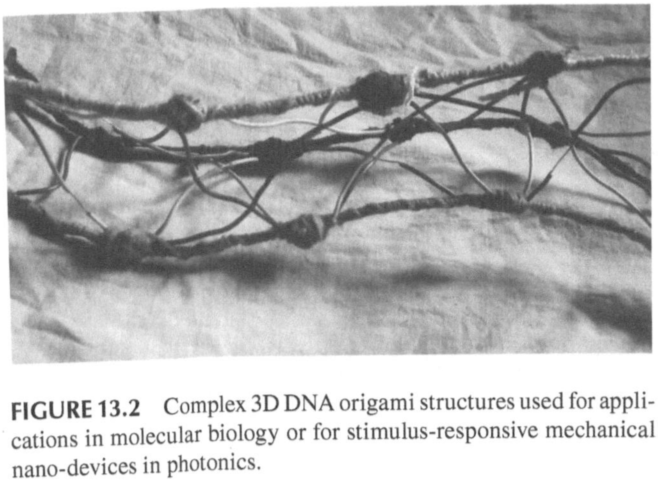

FIGURE 13.2 Complex 3D DNA origami structures used for applications in molecular biology or for stimulus-responsive mechanical nano-devices in photonics.

caused a structural change of the channel. The DNA translocation rate through the channel can also be controlled by voltage, so that this artificial nanopore activity can be regulated via external stimulus (Bell et al. 2012). In another attempt at making biosensors, 3D DNA origami nanorobots were immobilized on glass fiber tips triggered by a target DNA and identified using bioluminescence study (Torelli et al. 2018).

Cellular imaging with DNA probes is routinely implemented in biomedical research. DNA strands are usually labeled or modified with fluorescent (fluorophore-tagged DNA) or radioactive moieties for imaging. Using DNA probes, it is likely to achieve spatial resolution far below the fundamental diffraction limit of so-called super-resolution microscopy. Remarkably, DNA nanotechnology offers unique advantages with unprecedented predictability and programmability that are far better than current microscopy techniques (Ke, Castro, and Choi 2018).

DNA origami also has biomedical applications in drug delivery (Jiang et al. 2012): origami can be used to create structures that carry drug molecules to particular locations inside the body and deliver with high accuracy. The drugs can be loaded to specific shapes or origami cages or bind to the structures via specific biomolecular interactions. Due to specific interaction, the shape/morphology of the structures can be modified in order to provide higher stability. So, these principles can be employed in fabricating nanorobots: a folded origami nanorobot that can be used in target recognition, self- or stimulus-responsive unfolding, and drug delivery. This can be a boon to cancer therapy, whereby the origami nanorobot can recognize and block tumor-linked blood vessels (Li et al. 2018). Similarly, molecular kirigami structures are applied in a wide range of innovative applications, such as kirigami-based bandages (Zhao et al. 2018).

The most common use of static DNA structures in therapeutics is anticancer drug delivery. A DNA origami structure combining the anticancer drug doxorubicin (DOX) via intercalation was fabricated by Ding et al., which showed higher cellular uptake in breast cancer cell lines and resulted in clear observation of apoptotic behavior. These DNA-DOX structures have been proven in vivo, where these structures were tagged with quantum dots (QDs) whose fluorescence properties were used to investigate biodistribution and pharmacokinetics (Jiang et al. 2012; Zhao et al. 2012; Zhang et al. 2014).

DNA structures have been explored for biomedical applications in addition to cancer therapy, e.g., DNA structures for immunostimulant delivery to immune cells. In one study, DNA nanotubes containing CpG sequences were investigated for their effects in spleen cells; the carriers showed no significant toxicity and their immunological behavior functioned as per the design (Schuller et al. 2011). In another study, tetrahedral DNA oligonucleotide nanoparticles (ONPs) were used as vehicles for small-interfering RNA delivery that successfully proved silencing of target genes in vivo (Lee et al. 2012).

Dynamic DNA nanostructures offer another critical advantage: programming the availability of drugs. The DNA origami box can be designed in such a way that it opens and closes via specific external stimuli under favorable condition. This concept was demonstrated in a few studies, where a photoactivated mechanism was used to release drugs from DNA origami (Kohman et al. 2016; Li et al. 2016). Thus, DNA nanorobots can be designed to sense the condition, discover the target zone, unfold the structure, and release payloads.

13.6 CHALLENGES/ISSUES

Issues like biocompatibility, sterilization, and safety are given importance while designing and fabricating any biomedical devices. Many origami-based biomedical devices have met these properties and possess significant advantages over previously used devices. However, in many cases origami-based micro/nanostructures have to be improved in terms of better design and fabrication (input from clinical users), computational design and modeling, and clinical acceptance.

Design and fabrication of origami-based structures focus on many properties such as strain energy (varies with different folding design), rigidity and thickness of materials, and advanced manufacturing methods. Computer-assisted manufacturing methods, including procedures like plasma cutting, abrasive water jet cutting, laser cutting, and incremental sheet forming, are used for producing creases in the material before folding.

There are several procedures available for assembling origami-based structures. Among them, two methods are very promising: self-folding design and smart machines to fold the material into the final product or programmed robots. For self-folding design, there is a requirement for smart material with a self-folding nature to generate the final folded product. Moreover, the existing self-folding technology needs subsystems besides the core material: usually creases or joints, folding mechanisms, a signal regulator, and a power source.

The designing of a crease/joint pattern in an origami structure can be accomplished by advanced software based on mathematical principles. Software like TreeMaker is frequently used by designers to input a tree graph that estimates the ideal shape. The crease pattern can also be modified using TreeMaker's interface. There is other design software called Origamizer that splits a model's surface mesh into flat triangular segments and then assembles into a planar sheet. So, this software forms the surface of the preferred shape (triangular mesh) out of a sheet without overlapping and inserting payloads. Origamizer is very effective, particularly after several repetitions, at accomplishing a successful mesh design. Again, an orthogonal method was created where the structures can be made from a set of existing hinge configurations. The orthogonal structure was achieved in sequential procedures; the grid structures were created from a flat sheet, and then each square block in the grid was assigned one molecule from a set of crease pattern molecules until the orthogonal structure was achieved. This type of designing of crease pattern was very efficient.

The above-mentioned methods are appropriate only for preferred final shapes with specific geometric features. However, these methods are logical, resulting in high precision and efficiency in accomplishing particular tasks.

Clinical adoptability is another important factor to be given priority while designing origami-based biomedical devices. The origami-based devices should be safe, flexible, and more accurate in comparison to existing devices that can contribute to the execution of minimally invasive procedures. These devices must be simple, robust, and trustworthy to use, with consistent performance for rapid acceptance in clinics. More input from clinicians should be given priority, so that the origami-based devices will improve the issues with existing technology.

13.7 CONCLUSION

It is fascinating that the beautiful concept of ancient art with paper folding is now used to produce innovative biomedical devices with remarkable applications. Researchers have discovered the probability of using modern origami designs to develop many biomedical procedures/devices such as cardiac catheterization, stent grafts, disposable sterilization covers, microsurgical devices, microfluidic devices, surgical grippers, and more, especially for drug delivery. Modern origami is still progressing day by day, with improved materials, modeling, designs, fabrication procedures, and folding proficiency. Although the use of origami mathematics in biomedical engineering is intermittent at the current stage, the improvement in evolving algorithms based on the basic concepts of origami is paving the way for future clinical applications. Molecular nano-origami has demonstrated many promising biomedical applications. However, this field requires more improvement in terms of issues like immunogenicity, off-target gene regulation, pharmacokinetics and pharmacodynamics, and cost effectiveness; this will help its increased translation to biomedical applications.

REFERENCES

Arora, Aakriti Alisha, and Chamaree de Silva. 2018. "Beyond the Smiley Face: Applications of Structural DNA Nanotechnology." *Nano Reviews & Experiments* 9 (1–8): 1430976.

Bassik, Noy, Alla Brafman, Aasiyeh M. Zarafshar, Mustapha Jamal, Delgermaa Luvsanjav, Florin M. Selaru, and David H. Gracias. 2010. "Enzymatically Triggered Actuation of Miniaturized Tools." *Journal of the American Chemical Society* 132: 16314–7.

Bell, Nicholas A.W., Christian R. Engst, Marc Ablay, Giorgio Divitini, Caterina Ducati, Tim Liedl, and Ulrich F. Keyser. 2012. "DNA Origami Nanopores." *Nano Letters* 12 (1): 512–17.

Bhatia, Dhiraj, Senthil Arumugam, Michel Nasilowski, Himanshu Joshi, Christian Wunder, Valérie Chambon, Ved Prakash, et al. 2016. "Quantum Dot-Loaded Monofunctionalized DNA Icosahedra for Single-Particle Tracking of Endocytic Pathways." *Nature Nanotechnology* 11 (August). Nature Publishing Group: 1112–9.

Cavallo, Francesca, and Max G. Lagally. 2015. "Nano-Origami: Art and Function." *Nano Today* 10: 538–41.

Cybulski, James S., James Clements, and Manu Prakash. 2014. "Foldscope: Origami-Based Paper Microscope." *PLoS ONE* 9 (6): e98781.

Dukes, Patrick S., and Joseph P. Rusinko. 2011. "A Mathematical Origami Puzzle." *Ohio Journal of School Mathematics* 64:36–42.

Felton, S., M. Tolley, E. Demaine, D. Rus, and R. Wood. 2014a. "A Method for Building Self-Folding Machines." *Science.*

Felton, S, M Tolley, E Demaine, D Rus, and R Wood. 2014b. "Applied Origami. A Method for Building Self-Folding Machines." *Science* 345: 644–46.

Gregory, T. Stan, Kevin James Wu, Jasper Yu, James Brent Box, Rui Cheng, Leidong Mao, Guoyi Tang, and Zion Tsz Ho Tse. 2015. "Magnetohydrodynamic-Driven Design of Microscopic Endocapsules in MRI." *IEEE/ASME Transactions on Mechatronics* 20(6):2691–98.

Jiang, Qiao, Chen Song, Jeanette Nangreave, Xiaowei Liu, Lin Lin, Dengli Qiu, Zhen Gang Wang, et al. 2012. "DNA Origami as a Carrier for Circumvention of Drug Resistance." *Journal of the American Chemical Society* 134 (32): 13396–403.

Johnson, Meredith, Yue Chen, Sierra Hovet, Sheng Xu, Bradford Wood, Hongliang Ren, Junichi Tokuda, and Zion Tsz Ho Tse. 2017. "Fabricating Biomedical Origami: A State-of-the-Art Review." *International Journal of Computer Assisted Radiology and Surgery* 12: 2023–32.

Ke, Yonggang, Carlos Castro, and Jong Hyun Choi. 2018. "Structural DNA Nanotechnology: Artificial Nanostructures for Biomedical Research." *Annual Review of Biomedical Engineering* 20: 377–403.

Kohman, Richie E, Susie S Cha, Heng-Ye Man, and Xue Han. 2016. "Light-Triggered Release of Bioactive Molecules from DNA Nanostructures." *Nano Letters* 16 (4). American Chemical Society: 2781–85.

Kuribayashi, Kaori, Koichi Tsuchiya, Zhong You, Dacian Tomus, Minoru Umemoto, Takahiro Ito, and Masahiro Sasaki. 2006. "Self-Deployable Origami Stent Grafts as a Biomedical Application of Ni-Rich TiNi Shape Memory Alloy Foil." *Materials Science and Engineering A* 419: 131–37.

Kurin, Richard. 2004. "Safeguarding Intangible Cultural Heritage in the 2003 UNESCO Convention: A Critical Appraisal." *Museum International* 56: 66–77. doi: 10.1111/j.1350-0775.2004.00459.x.

Lee, Hyukjin, Abigail K. R. Lytton-Jean, Yi Chen, Kevin T. Love, Angela I. Park, Emmanouil D. Karagiannis, Alfica Sehgal, et al. 2012. "Molecularly Self-Assembled Nucleic Acid Nanoparticles for Targeted In Vivo SiRNA Delivery." *Nature Nanotechnology* 7 (June). Nature Publishing Group: 389–93.

Li, Feiran, Haorong Chen, Jing Pan, Tae-Gon Cha, Igor L Medintz, and Jong Hyun Choi. 2016. "A DNAzyme-Mediated Logic Gate for Programming Molecular Capture and Release on DNA Origami." *Chemical Communications* 52 (54). The Royal Society of Chemistry: 8369–72.

Li, Suping, Qiao Jiang, Shaoli Liu, Yinlong Zhang, Yanhua Tian, Chen Song, Jing Wang, et al. 2018. "A DNA Nanorobot Functions as a Cancer Therapeutic in Response to a Molecular Trigger In Vivo." *Nature Biotechnology* 36 (3): 258–64.

Liu, Hong, and Richard M. Crooks. 2011. "Three-Dimensional Paper Microfluidic Devices Assembled Using the Principles of Origami." *Journal of the American Chemical Society* 133: 17564–66.

Miyashita, Shuhei, Laura Meeker, Michael T. Tolley, Robert J. Wood, and Daniela Rus. 2014. "Self-Folding Miniature Elastic Electric Devices." *Smart Materials and Structures* 23 (9). IOP Publishing: 94005.

Miyashita, Shuhei, Steven Guitron, Marvin Ludersdorfer, Cynthia R. Sung, and Daniela Rus. 2015. "An Untethered Miniature Origami Robot That Self-Folds, Walks, Swims, and Degrades." In *Proceedings – IEEE International Conference on Robotics and Automation*, Seattle.

Miyashita, Shuhei, Steven Guitron, Kazuhiro Yoshida, Shuguang Li, Dana D. Damian, and Daniela Rus. 2016. "Ingestible, Controllable, and Degradable Origami Robot for Patching Stomach Wounds." In *Proceedings – IEEE International Conference on Robotics and Automation*, Stockholm, 909–16.

Schuller, Verena J, Simon Heidegger, Nadja Sandholzer, Philipp C Nickels, Nina A Suhartha, Stefan Endres, Carole Bourquin, and Tim Liedl. 2011. "Cellular Immunostimulation by CpG-Sequence-Coated DNA Origami Structures." *ACS Nano* 5 (12). United States: 9696–702.

Sorguç, A.G., I. Hagiwara, and S.A. Selçuk. 2009. "Origamics in Architecture: A Medium of Inquiry for Design in Architecture." *METU Journal of the Faculty of Architecture* 26 (2): 235–47.

Stern, George M., Christina L. Randall, Timothy G. Leong, Bryan R. Benson, Noy Bassik, and David H. Gracias. 2009. "Tetherless Thermobiochemically Actuated Microgrippers." *Proceedings of the National Academy of Sciences* 106: 703–8.

Torelli, Emanuela, Marisa Manzano, Sachin K. Srivastava, and Robert S. Marks. 2018. "DNA Origami Nanorobot Fiber Optic Genosensor to TMV." *Biosensors and Bioelectronics* 99: 209–15.

Varrasi, J. 2014. "How the Future of Origami Engineering Is Unfolding." https://www.livescience.com/49121-origami-inspired-engineering-is-expanding.html

Zhang, Qian, Qiao Jiang, Na Li, Luru Dai, Qing Liu, Linlin Song, Jinye Wang, et al. 2014. "DNA Origami as an in Vivo Drug Delivery Vehicle for Cancer Therapy." *ACS Nano* 8 (7). American Chemical Society: 6633–43.

Zhao, Yong-Xing, Alan Shaw, Xianghui Zeng, Erik Benson, Andreas M Nyström, and Björn Högberg. 2012. "DNA Origami Delivery System for Cancer Therapy with Tunable Release Properties." *ACS Nano* 6 (10). American Chemical Society: 8684–91.

Zhao, Ruike, Shaoting Lin, Hyunwoo Yuk, and Xuanhe Zhao. 2018. "Kirigami Enhances Film Adhesion." *Soft Matter* 14 (13): 2515–25.

Zhu, Shuze, and Teng Li. 2014. "Hydrogenation-Assisted Graphene Origami and Its Application in Programmable Molecular Mass Uptake, Storage, and Release." *ACS Nano* 8: 2864–72.

Section 4

Biosensors/Nanochip/Nanodevice/Nanocomputing

14 Nanobiosensors and Their Biomedical Applications

Pradipta Ranjan Rauta, Debasis Nayak, and Pankaj Kumar Singh

CONTENTS

14.1 Introduction 177
14.2 Selection and Optimization of Nanobiosensing Technology 178
14.3 Types of Nanobiosensors 178
14.3.1 Label-Free Detection 178
14.3.2 Label-Based Detection 178
14.4 Nanomaterials Used in Biosensing 178
14.4.1 Nanoparticles 178
14.4.2 Electrochemical Biosensors 178
14.4.3 Magnetic Nanobiosensors 179
14.4.4 Acoustic Wave–Based Nanobiosensors 179
14.4.5 Optical Nanobiosensors 179
14.4.6 Nanotubes 179
14.4.7 Nanowires 180
14.4.8 Other Nanomaterials 180
14.5 Nanobiosensors in Biomedical and Diagnostic Applications 180
14.6 Conclusion and Future Perspectives 181
References 182

14.1 INTRODUCTION

At present, monitoring health for early detection of health disorders or disease conditions is a very important component of human health management. Health disorders and diseases are regulated by specific biomolecules, typically called biomarkers. Also, the presence of harmful compounds (e.g., pesticides) in food and water should be detected on a regular basis because of their role in causing disease conditions. So, there is a demand for rapid analysis by efficient biosensing systems with high accuracy in detecting biomolecules. Diagnostic technology that involves biosensors has the potential to perform quick onsite analysis. Biosensors are generally minute transportable devices that are used to identify the presence or even quantitate a range of biological analytes, e.g., biological structures, biomolecules, or microorganisms. Basically, biosensors can be divided into three major parts: a bioreceptor that identifies the analyte by generating a signal, a signal transducer that translates the biological interactions/response into quantifiable analytical signals, and finally a detector that reads the signal (qualitative/quantitative). There has been significant advancement in designing and fabricating biosensors in recent years, and the industry is estimated to cross 20 billion USD in the market in the next two decades. The measuring task of a biosensor is to detect biologically specific molecules including DNA, proteins, antibodies, enzymes, other immunological agents, etc.

The bioreceptor, one of the important elements of biosensors, works as a template for the molecules to be sensed by using biological interactions. For example, antigen can be screened using antibody and vice versa. Similarly, a protein can be detected using its corresponding selective ligand/substrate. The transducer, the second biosensor component, converts the biological interactions/responses (between receptor and analyte) to electrical signals. This electrical system is detected by a detector (third component), which amplifies it into a readable form. Apart from these elements, the immobilization scheme plays an important role in biosensing; it is used to immobilize the bioreceptor so that an accurate reaction is performed in an efficient and feasible manner. This makes the biosensing systems cheaper. Biosensing systems are also affected by physiological conditions such as variations in pH, temperature, the presence of contaminants, etc. (Kissinger 2005).

The use of nanomaterials in biosensors (known as nanobiosensors) adds a significant element into the overall biosensing technology by addressing issues related to specificity and sensitivity in detecting target molecules. The ultra-small size of nanomaterials, which have large surface-to-volume ratios with all vital physicochemical properties, provides a large surface area for biomolecular interaction, resulting in high specificity and sensitivity in detection in comparison to the same material at the bulk scale. The rapid improvement in nanotechnology provides specific and sensitive recognition of analytes by (i) surface modifications of biomaterials on nanocomposites (Mahato et al. 2016), (ii) integration of novel nanofluidics technology into the design of small and user-friendly devices (Noh et al. 2012), and (iii) new and innovative signal amplification strategies using several types of nanomaterials (Mahato et al. 2016). Various nanomaterials that have been discovered on the basis of their novel mechanical, electronic, optical, magnetic, and thermal properties are now being employed in enhanced biological signaling and transduction mechanisms. These widely used nanomaterials are nanoparticles (metal/polymer), nanorods, nanotubes, nanowires, and nanomembranes/films made up of nanocrystalline matter (Jianrong et al. 2004). Research and development in the field of nanobiosensors must remain very diverse and multidisciplinary in order to continue to develop various nanobiosensor prototypes.

14.2 SELECTION AND OPTIMIZATION OF NANOBIOSENSING TECHNOLOGY

The physical and chemical properties of nanomaterials play a significant role in biosensing in terms of sensitive and selective response. The nanofabrication procedure usually targets two crucial jobs: the design and manufacturing of nanoscale adhesive surfaces through integrated circuits and the engineering of surfaces via micromachining (Malik et al. 2013). Techniques used for biosensing purposes mainly follow four basic process: etching/growth, photolithography, surface etching procedures, and chemical interactions. Nanoelectromechanical systems (NEMS) have developed by incorporating highly sensitive electrical and electromechanical properties into several nanomaterials. This technology has thus provided various materials with unique properties due to their nanoscale functionalization. In addition, these devices have been coupled with biological systems and molecules to improve their bio-adhesion features and reactions to a wide range of stimuli. The optimization of nanomaterials is very important while making biosensors for better performance and efficiency as per the desired goal. Moreover, a key strategy in shaping nanomaterials for specific applications involves tuning and engineering their surfaces by micromachining procedures. Biosensing properties using nanomaterials are also affected by several other factors like compatibility, functionalization with other materials, aspect ratios, etc.

14.3 TYPES OF NANOBIOSENSORS

The classification of nanobiosensors is very diverse area, as it depends on several factors such as biomolecule transduction methodology, and the nature and type of nanomaterials. The nature of nanomaterials is one of the most significant criteria for classifying nanobiosensors, as assessing it is critical to the improvement of biosensing. Another way classifying nanobiosensors is based on biomolecule transduction, which can be typically separated into two categories: label-free and label-based.

14.3.1 Label-Free Detection

Label-free detection permits the sensors to directly detect the target molecule without using a secondary molecule. For example, a label-free sensor is capable of sensing directly when the protein binds to the antibody using different unique properties of nanomaterials and electroanalytical transduction technologies (Prasad 2014).

14.3.2 Label-Based Detection

Label-based detection requires a secondary molecule or amplification step for the purpose of biosensing. In most immune-based assays, the detection strategy involves capturing the analyte of interest by utilizing primary antibodies, and then fluorescently labeled secondary antibodies detect the bound analyte indirectly. Enzyme-linked immunosorbent assay (ELISA), which has become a standard diagnostic tool nowadays, detects the occurrence of an antigen or antibody in a sample. Fluorescent detection techniques are mainly built on fluorescent markers that cause fluorescent energy transfer into an optical signal, confirming the binding reaction (Prasad 2014).

The various types of nanobiosensors are discussed in detail below according to their nature and physicochemical properties.

14.4 NANOMATERIALS USED IN BIOSENSING

14.4.1 Nanoparticles

The use of nanoparticles in biosensing has become very popular due to their unique properties such as smaller size, large surface-to-volume ratio, and other favorable physicochemical characteristics (electrochemical, optical, magnetic, acoustic wave-based nature) that influence the binding of target molecule to a large extent.

14.4.2 Electrochemical Biosensors

Electrical biosensors are very popular due to their user-friendly (with reduced size) and cost-effective nature, which enable them to be used as point-of-care diagnostics. The biosensing principle is mainly based on a binding reaction that results in a change in current or voltage. Again, these biosensors can be categorized on the basis of electrical measurement procedures, including voltametric, coulmetric/amperometric, and impedance measurements, etc. Metallic nanoparticles are mostly used in developing electrochemical biosensors, where the electrochemical reaction is carried out very efficiently with reduced nonspecific reactions. In one study, gold nanoparticles were utilized to improve the immobilization of DNA on gold electrodes; this significantly enhanced sensing efficiency by reducing the detection limit (Cai et al. 2001). In other studies, enzyme-conjugated gold nanoparticle–based sensors were established for the detection of glucose, hydrogen peroxide, xanthine, etc. (Crumbliss et al. 1992; Zhao et al. 1996; Xu, Liu, and Ju 2003). Similarly, carbon nanoparticles were used to immobilize the horseradish peroxidase enzyme on gold electrodes, which resulted in faster amperometric response.

In a similar trend, novel photo-electrochemical systems can be designed using nanosized semiconductor crystals tagged with biological molecules such as enzymes or precursors that improve the efficiency of photochemical reactions. With this idea, an enzymatic detection system was designed, where nanocrystalline cadmium sulfide (CdS) was used to immobilize the formaldehyde dehydrogenase onto gold electrodes in order to carry out catalytic oxidation of formaldehyde (Curri et al. 2002). Bioassays based on biotin–streptavidin specificity have also been designed using the above principle (Malik et al. 2013).

14.4.3 Magnetic Nanobiosensors

Magnetic nanoparticles, mostly ferrite-based materials or alloys of transition metals, are used for designing magnetic nanobiosensors. These types of sensors are very beneficial, particularly in biomedical applications (Malik et al. 2013). Integrating the magnetic properties of magnetic nanoparticles in biosensing has made the biosensors more sensitive and effective with wide-ranging applications. Many magnetic bioassay techniques have been developed for specific isolation of magnetically labeled targets with the use of external magnetic fields (Chamorro-Garcia and Merkoçi 2016). Besides this, distinctive equipment such as superconducting quantum interference devices (SQUID) has been utilized for quick detection of biological targets with the help of the super-paramagnetic behavior of magnetic nanoparticles (Y.-T. Chen et al. 2017). Again, these magnetic-based devices are used to screen the specific antigen/antibodies from the biological mixture using the corresponding antibodies/antigens conjugated to magnetic nanoparticles (Wujcik et al. 2014; Kim et al. 2015).

14.4.4 Acoustic Wave–Based Nanobiosensors

There is always the need to amplify the biosensing responses to increase the limits of biodetection. Acoustic wave–based biosensors are mainly used for the amplification of sensing responses using several stimulus-based effects (Durmuş et al. 2015). For example, the conjugated antibody-sol nanoparticles (5–100 nm) that bind themselves to the electrode have been complexed with analytes in way that antibody molecules are attached uniformly over the electrode surface. These highly covered sol-antibody-antigen particles help in detection due to variation in the vibrational frequency of the quartz-based sensing platform (Malik et al. 2013; Chamorro-Garcia and Merkoçi 2016). There are many reports in which other particles, such as platinum, gold, titanium, and cadmium sulfide, are used for designing acoustic wave–based nanobiosensors (Malekzad et al. 2017).

14.4.5 Optical Nanobiosensors

Optical nanobiosensors consist mainly of a target biomolecule coupled to an optical transducing device, such as an optical nanofiber combined to a photometric detection scheme (Damborský, Švitel, and Katrlík 2016; Prasad 2014). This type of biosensing technology provides specific analytical information (qualitative, semiquantitative, or quantitative) using various biomolecular targets (e.g., DNA, protein). This approach has provided exceptional applications that provide insights into intact cell function, thus helping in many studies of molecular functions (such as DNA–protein interactions, protein–protein interactions, apoptosis, etc.). The integration of florescence or surface plasmon resonance into optical fiber has impacted significantly on the development of optically based biosensors with high efficiency (Abdulhalim, Zourob, and Lakhtakia 2008; Damborský, Švitel, and Katrlík 2016). These devices are very helpful in biomedical applications such as medical diagnostics and ultra-high-throughput drug screening. The utilization of nanoparticle-quantifying tags in optical sensors solves many limitations of organic fluorophore-based sensors.

14.4.6 Nanotubes

Nanotubes such as carbon nanotubes (one-dimensional nanomaterials) have been explored in designing biosensors due to their distinctive structure-dependent optoelectronic properties. The other favorable properties of carbon nanotubes are flexible geometric structures (easy functionalization and folding abilities), high aspect ratio (large surface-to-volume ratio), unique electronic conductance, and electron transport properties, all of which can play a significant role in biosensing by influencing the biomolecular interactions, resulting in a quick and sensitive label-free bioelectronic detection (Malik et al. 2013; Sireesha et al. 2018). Both single-walled and multi-walled carbon tubes have been used in designing novel biosensors for better performance. For example, single-walled carbon nanotube (CNT)-based enzyme biosensors have been developed for the detection of glucose with the help of the enzyme glucose oxidase, which is immobilized on nanotubes, offering a quantitative estimation of glucose in several body fluids such as blood, saliva, tears, etc. (Azamian et al. 2002). Higher efficiency has been achieved due to high loading of enzymes (due to large surface-to-volume ratio) and high electrical conductivities. Again, a chemoelectroluminescence effect has been enhanced by pairing CNTs to sensing molecules through controlled and better conductance of charge carriers. CNTs can be easily coupled with biomolecules such as DNA, oligonucleotides, peptides, and proteins due to their remarkable functionalization potentials, which have been employed in designing next-generation biosensors. At present, a series of CNT biosensors have been developed for cancer diagnosis by detecting a wide range of biomarkers with the help of conjugated aptamers, peptides, enzymes, and antibodies (Sireesha et al. 2018). However, issues existing with CNTs, such as controlling the size during fabrication, biocompatibility, maintaining the bioactivity after immobilization, structural stability, and cytotoxicity, are being increasingly focused on for better performance (Sireesha et al. 2018).

14.4.7 Nanowires

Over the past decade, nanowire-based biosensors have already made breakthroughs in biomedical fields due to dramatic increases in sensitivity, reproducibility, and scalability with a relatively simple fabrication process. Nanowires are mainly one-dimensional cylindrical nanostructures with notable electron transport properties that enable them for label-free detection of various species electrically. The versatile nature of nanowires makes them superior than nanotubes in biosensing capability in two major ways: (i) easy and controlled strategy for surface modification (functionalization) and (ii) high biocompatibility of materials during surface modification (Ambhorkar et al. 2018; Malik et al. 2013). These nanowires were mostly made from semiconductor materials, and their surfaces can be easily functionalized by coupling with a number of sensing molecules. The first application of silicon nanowire sensors was to detect proteins in a solution electrically. For example, the detection and isolation of streptavidin from a mixture were achieved by using silicon nanowires coupled with biotin (Malik et al. 2013). Again, the use of nanowires in biosensors has improved the currently used in vivo diagnostic procedures by efficient biodetection of several pathogens or multiple disease biomarker molecules (Ambhorkar et al. 2018). Besides these, nanowire-on-a-chip platforms have been developed for the comprehensive rapid detection, separation, and isolation of biomolecules for practical clinical use. In one such study, nanosized optical fibers coupled with antibodies have been used to detect the presence of toxicants within an individual cell (Malik et al. 2013). Likewise, another study reported the use of Zinc oxide (ZnO) coated on gold electrodes to detect hydrazine using amperometric response (Cullum et al. 2000). However, the use of nanowires in biosensing applications has met several issues such as quality of materials and improvement in electrical conductivities for low-level detection. In one attempt, a complex nanostructure of 200 independent nanowire assemblies was utilized to detect a low-level serum biomarker (for bone cancer) by improved electrical conductivities. Again, other favorable properties of nanowire have been explored to employ in the design of biosensors (Cui and Lieber 2001). For example, the surface resonance ability of nanowires can be incorporated into biosensing probes that can significantly improve the sensitivity at ultra-low label (Huang, Yu, and Zare 2007). Nanowires were also developed through the complementary metal oxide conductor (CMOS) approach, which simplified the process in terms of controlling and regulating the synthesis procedure. This approach has been used for analysis of biofluids such as serum to detect and isolate the various pathogens or proteins of interest (Stern et al. 2007).

14.4.8 Other Nanomaterials

Nanomaterials have revolutionized biosensing technology by improving diagnostic procedures with quick and sensitive detection. Many other nanomaterials are being focused on to be used in biosensing applications. Smart nanomaterials such as quantum dots in conjugation with sensitive dyes have resulted in photochromic, thermochromic, and electrochromic materials that can exhibit ultrasensitive detection (Altintas, Davis, and Scheller 2017). Similarly, many other structures such as nanodots, nanosheets, nanobelts, nanocombs, and nanoribbons have been examined for improving the conventional biosensing procedures (Malik et al. 2013; Mahato et al. 2016). They have significantly impacted the development of a new generation of biosensing systems with higher accuracy.

14.5 NANOBIOSENSORS IN BIOMEDICAL AND DIAGNOSTIC APPLICATIONS

Nanobiosensors have been employed for the qualitative as well as quantitative detection of several biomarkers such as serum antigens, carcinogens, and causative agents of various metabolic disorders. Routine applications in diagnosis are best explained by their use in the detection of diseases or disorders like diabetes, allergic responses, cancer, and several metabolic disorders on the basis of blood analysis at the laboratory desk. There are numerous examples of clinical applications, where these nanobiosensors are used on a regular basis. These applications include improvements in the detection of glucose in diabetic patients (del Barrio et al. 2017; Zhang, Du, and Wang 2015; L. Chen, Hwang, and Zhang 2018; Pickup et al. 2005), detection of HIV-AIDS (Alterman et al. 2001; Lifson et al. 2016; Chamorro-Garcia and Merkoçi 2016), detection of urinary tract bacterial infections (Drummond, Hill, and Barton 2003; Chamorro-Garcia and Merkoçi 2016; Kumar et al. 2016), the diagnosis of cancer and its various stages (Gao et al. 2004; Harisinghani et al. 2003; Gdowski et al. 2014; Sharifi et al. 2019; Salahandish et al. 2018), and detecting other disease biomarkers (Chamorro-Garcia and Merkoçi 2016). There is a numerous research reports on the development of the next-generation synthesis of synthetic and semisynthetic ligands, catalytic electrodes, plastic antibodies, bioinspired nanotechnologies for biosensing, monitoring, therapy and controlled biomolecular interactions, along with in vitro and in vivo devices for biomedical diagnostic and therapeutic applications (Turner 2013; Mahato et al. 2016; Chamorro-Garcia and Merkoçi 2016). Prototypes are being developed for immunosensors, microfluidic sensors, flow-based biosensors, paper inkjet sensors, enzyme sensors, etc., for the detection of several clinically significant specific molecules including pathogen (parasite or bacteria or virus), cancer biomarkers, hormones, heavy metal ions, etc. (Gdowski et al. 2014; Mahato et al. 2016; Chamorro-Garcia and Merkoçi 2016). Aptamer-based disposable nanobiosensors have been explored extensively to be used in diagnostics and therapeutics. These biosensors can integrate numerous biochemical and biophysical processes into low-cost portable devices with quick response, high specificity, and high sensitivity (Wang et al. 2013; Hanif et al. 2018).

Among the existing biosensor types, the electrochemically based readout system is regarded as the most powerful due to fast biomolecular analysis with high sensitivity (Zhu, Chandra, and Shim 2013; Mahato et al. 2016; Chamorro-Garcia and Merkoçi 2016), thereby aiding in rapid therapeutic strategies to reduce patient stress. Electrochemically based biosensing principles have been applied to design new biosensor prototypes. The ultrasensitive detection of viruses, circulating tumor cells, circulating nucleic acids, etc., has also been developed using molecular probes coupled with nanostructures (Sekretaryova, Eriksson, and Turner 2016; Rackus et al. 2015; Green et al. 2016). Recently, smartphone-based nanobiosensors have fascinated the research world, which is designing a new generation of biosensor prototypes (portable sensors) with lots of advantages such as high-resolution cameras, advanced processors, high-end security through fingerprinting, and increased memory. Research on smartphone-based nanobiosensor prototypes performing such tasks as electrochemical sensing, microscopy, lateral flow assays, immunoassays, flow cytometry, and optical detection based on surface plasmon resonance has been reported (Walzik et al. 2015; Roda et al. 2016). Already, there are a few commercialized smartphone-based nanobiosensor prototypes on the market, e.g., iHealth, GENTAG, AliveCor, CellScope, Mobile Assay, etc. (Mahato et al. 2016).

Biochip- and microarray-based approaches have been introduced to construct nanobiosensors, where multiple biological probes are bound to different regions of nanostructures. This approach enables the diagnosis of multiple diseases or biological disorders in a small period of time. However, issues like cross-reactivity (can be addressed using highly specific probe and creative design structures) have to be given priority while designing such types of sensing apparatus (Chamorro-Garcia and Merkoçi 2016; Neethirajan et al. 2018). The introduction of nanomaterials has empowered detecting enzyme systems to be immobilized and reused/recycled. The employment of nano innovations such as microelectromechanical systems (MEMS) and NEMS gives several advantages over existing diagnostic techniques, such as improved biological signaling and transduction mechanisms. Magnetic nanoparticle–based biosensors have been fabricated in a controlled fashion and used for a number of biomedical applications: biosensing, isolation, and separation of several biomolecules related to disease and metabolic disorders (Wen et al. 2014; Parveen, Misra, and Sahoo 2012; Yen, Padmanabhan, and Selvan 2013).

14.6 CONCLUSION AND FUTURE PERSPECTIVES

In the last two decades, biosensing technology has been continuously improved by the development of new and novel techniques capable of performing very specific and sensitive detection (both qualitative and quantitative). The rapid growth of nanotechnology and its introduction into biosensing technology have proved to be significant blessings in biomedical applications such as diagnosis, therapy, etc. Nanobiosensor-based approaches have enabled the creation of quicker, cost-effective, and user-friendly devices. The employment of nanomaterials or specific nanostructures such as nanotubes, nanowires, quantum dots, and hybrid nanostructures has significantly improved the transduction mechanisms and has resulted in ultrasensitive detection. Reproducibility is another positive factor associated with nanobiosensors. The electrochemical transductors are mostly used for commercial purposes due to their favorable properties (simplicity, ease of fabrication, lower cost, robust signal). Right now, nanomaterials are being increasingly considered for the merging of biological and chemical sensors to create multidimensional devices that can provide easy and fast detection with better performance.

Despite the achievement of enhanced transduction and robust detection with nanobiosensors, there are still limitations, which are mainly due to issues with biological probes. Greater understanding of immobilization chemistry and reduction of nonspecific interactions will significantly decrease the limits of detection. The biocompatibility of nanomaterials also needs greater focus during the development of in vivo biosensing applications.

Despite significant advancements in the last decade, the nanobiosensing approach is still being fine-tuned to solve many limitations or issues with existing procedures. Within the next decade, more nanobiosensor-based diagnostic devices that will be able to accomplish thousands of measurements at lower cost and greater efficiency will become available in the market. The integration of biochip technology into nanobiosensing is one of the significant future trends in biomedical diagnostics. Diagnostics based on blood analysis are being given priority at the current stage, as blood molecular patterns will offer sensitive evaluations of disease conditions. Further advancement in nanoscale chemical sensors and molecular electronics will enable the fabrication of microscopic equipment capable of detecting patterns of chemical/target molecules in body fluids such as blood. Given the early success, the future translation into large-volume production and clinical adaptability looks very promising. There is a large requirement for the development of point-of-care devices with simplistic approaches towards detection. The development of non-polymerase chain reaction technologies will also be aided by the use of nanobiosensors. Again, nanotechnology approach can be applied for single-cell analysis to be used in genetic disorder diagnosis. Nanobiosensor-based diagnosis will also play a significant role in linking the diagnosis with treatment and development of personalized medicine. The extensive use of nanobiosensors in the detection of cancer biomarkers will be very helpful for early detection and treatment of cancer. The development of nanotheranostics will offer detection, treatment, and monitoring of the disease condition in vivo. In vivo analysis of biomarkers (in individual living cells) will be made possible by the use of laser-based nanosensors. With their incredible scope, nanotechnologies promise to overcome the limits of existing molecular diagnosis and facilitate point-of care diagnosis, the combination of diagnostics and therapeutics, and the development of personalized medicine. Future work should be focused towards the development of novel, specific, sensitive, rapid, and personalized diagnosis strategies for biomolecular investigation.

REFERENCES

Abdulhalim, Ibrahim, Mohammad Zourob, and Akhlesh Lakhtakia. 2008. "Overview of Optical Biosensing Techniques." In *Handbook of Biosensors and Biochips*, edited by Robert S Marks, David C Cullen, Isao Karube, Christopher R Lowe, and Howard H Weetall, 413–46. John Wiley & Sons, Ltd. doi: 10.1002/9780470061565.hbb040.

Alterman, M, H Sjobom, P Safsten, P O Markgren, U H Danielson, M Hamalainen, S Lofas, et al. 2001. "P1/P1′ Modified HIV Protease Inhibitors as Tools in Two New Sensitive Surface Plasmon Resonance Biosensor Screening Assays." *European Journal of Pharmaceutical Sciences* 13 (2). Netherlands: 203–12.

Altintas, Zeynep, Frank Davis, and Frieder W Scheller. 2017. "Applications of Quantum Dots in Biosensors and Diagnostics." In *Biosensors and Nanotechnology: Applications in Health Care Diagnostics*, edited by Zeynep Altintas, 183–99. John Wiley & Sons, Inc. doi: 10.1002/9781119065036.ch9.

Ambhorkar, Pranav, Zongjie Wang, Hyuongho Ko, Sangmin Lee, Kyo-in Koo, Keekyoung Kim, and Dong-il (Dan) Cho. 2018. "Nanowire-Based Biosensors: From Growth to Applications." *Micromachines (Basel)* 9: 679. doi: 10.3390/mi9120679.

Azamian, Bobak R, Jason J Davis, Karl S Coleman, Claire B Bagshaw, and Malcolm L H Green. 2002. "Bioelectrochemical Single-Walled Carbon Nanotubes." *Journal of the American Chemical Society* 124 (43). American Chemical Society: 12664–5. doi: 10.1021/ja0272989.

del Barrio, Melisa, María Moros, Sara Puertas, Jesús M de la Fuente, Valeria Grazú, Vicente Cebolla, Susana de Marcos, and Javier Galbán. 2017. "Glucose Oxidase Immobilized on Magnetic Nanoparticles: Nanobiosensors for Fluorescent Glucose Monitoring." *Microchimica Acta* 184 (5): 1325–33. doi: 10.1007/s00604-017-2120-8.

Cai, Hong, Chun Xu, Pingang He, and Yuzhi Fang. 2001. "Colloid Au-Enhanced DNA Immobilization for the Electrochemical Detection of Sequence-Specific DNA." *Journal of Electroanalytical Chemistry* 510 (1): 78–85. doi: 10.1016/S0022-0728(01)00548-4.

Chamorro-Garcia, Alejandro, and Arben Merkoçi. 2016. "Nanobiosensors in Diagnostics." *Nanobiomedicine* 3: 1–26. doi: 10.1177/1849543516663574.

Chen, Yi-Ting, Arati G Kolhatkar, Oussama Zenasni, Shoujun Xu, and T Randall Lee. 2017. "Biosensing Using Magnetic Particle Detection Techniques." *Sensors.* 17: 2300. doi: 10.3390/s17102300.

Chen, Longyi, Eugene Hwang, and Jin Zhang. 2018. "Fluorescent Nanobiosensors for Sensing Glucose." *Sensors* 18 (5): 1440. doi: 10.3390/s18051440.

Crumbliss, A L, S C Perine, J Stonehuerner, K R Tubergen, J Zhao, R W Henkens, and J P O'Daly. 1992. "Colloidal Gold as a Biocompatible Immobilization Matrix Suitable for the Fabrication of Enzyme Electrodes by Electrodeposition." *Biotechnology and Bioengineering* 40 (4). United States: 483–90. doi: 10.1002/bit.260400406.

Cui, Yi, and Charles M Lieber. 2001. "Functional Nanoscale Electronic Devices Assembled Using Silicon Nanowire Building Blocks." *Science* 291 (5505): 851–53. doi: 10.1126/science.291.5505.851.

Cullum, B M, G D Griffin, G H Miller, and T Vo-Dinh. 2000. "Intracellular Measurements in Mammary Carcinoma Cells Using Fiber-Optic Nanosensors." *Analytical Biochemistry* 277 (1). United States: 25–32. doi: 10.1006/abio.1999.4341.

Curri, M L, A Agostiano, G Leo, A Mallardi, P Cosma, and M Della Monica. 2002. "Development of a Novel Enzyme/Semiconductor Nanoparticles System for Biosensor Application." *Materials Science and Engineering: C* 22 (2): 449–52. doi: 10.1016/S0928-4931(02)00191-1.

Damborský, Pavel, Juraj Švitel, and Jaroslav Katrlík. 2016. "Optical Biosensors." *Essays in Biochemistry* 60 (1). Portland Press Limited: 91–100. doi: 10.1042/EBC20150010.

Drummond, T Gregory, Michael G Hill, and Jacqueline K Barton. 2003. "Electrochemical DNA Sensors." *Nature Biotechnology* 21 (October). Nature Publishing Group: 1192–99.

Durmuş, N Gözde, Richard L Lin, Mariel Kozberg, Deniz Dermici, Ali Khademhosseini, and Utkan Demirci. 2015. "Acoustic-Based Biosensors." In *Encyclopedia of Microfluidics and Nanofluidics*, edited by Dongqing Li, 28–40. New York: Springer New York. doi: 10.1007/978-1-4614-5491-5_10.

Gao, Xiaohu, Yuanyuan Cui, Richard M Levenson, Leland W K Chung, and Shuming Nie. 2004. "In Vivo Cancer Targeting and Imaging with Semiconductor Quantum Dots." *Nature Biotechnology* 22 (8). United States: 969–76. doi: 10.1038/nbt994.

Gdowski, Andrew, Amalendu P Ranjan, Anindita Mukerjee, and Jamboor K Vishwanatha. 2014. "Nanobiosensors: Role in Cancer Detection and Diagnosis." *Advances in Experimental Medicine and Biology* 807. United States: 33–58. doi: 10.1007/978-81-322-1777-0_4.

Green, Brenda J, Tina Saberi Safaei, Adam Mepham, Mahmoud Labib, Reza M Mohamadi, and Shana O Kelley. 2016. "Beyond the Capture of Circulating Tumor Cells: Next-Generation Devices and Materials." *Angewandte Chemie* 55 (4). Germany: 1252–65. doi: 10.1002/anie.201505100.

Hanif, Aamir, Rabia Farooq, Muneeb U Rehman, Rehan Khan, Sabhiya Majid, and Majid Ahmad Ganaie. 2018. "Aptamer Based Nanobiosensors: Promising Healthcare Devices." *Saudi Pharmaceutical Journal* 27: 312–9. doi: 10.1016/j.jsps.2018.11.013.

Harisinghani, Mukesh G, Jelle Barentsz, Peter F Hahn, Willem M Deserno, Shahin Tabatabaei, Christine Hulsbergen van de Kaa, Jean de la Rosette, and Ralph Weissleder. 2003. "Noninvasive Detection of Clinically Occult Lymph-Node Metastases in Prostate Cancer." *The New England Journal of Medicine* 348 (25). United States: 2491–99. doi: 10.1056/NEJMoa022749.

Huang, Bo, Fang Yu, and Richard N Zare. 2007. "Surface Plasmon Resonance Imaging Using a High Numerical Aperture Microscope Objective." *Analytical Chemistry* 79 (7). United States: 2979–83. doi: 10.1021/ac062284x.

Jianrong, Chen, Miao Yuqing, He Nongyue, Wu Xiaohua, and Li Sijiao. 2004. "Nanotechnology and Biosensors." *Biotechnology Advances* 22 (7): 505–18. doi: 10.1016/j.biotechadv.2004.03.004.

Kim, Nam-Gun, Myeong-Ae Kim, Young-Il Park, Tae-Sung Jung, Seong-Wan Son, ByungJae So, and Hwan-Goo Kang. 2015. "Magnetic Nanoparticle Based Purification and Enzyme-Linked Immunosorbent Assay Using Monoclonal Antibody against Enrofloxacin." *Journal of Veterinary Science* 16 (4). The Korean Society of Veterinary Science: 431–37. doi: 10.4142/jvs.2015.16.4.431.

Kissinger, P T. 2005. "Biosensors—A Perspective." *Biosensors and Bioelectronics* 20: 2512–6.

Kumar, M S, S Ghosh, S Nayak, and A P Das. 2016. "Recent Advances in Biosensor Based Diagnosis of Urinary Tract Infection." *Biosensors & Bioelectronics* 80 (June). England: 497–510. doi: 10.1016/j.bios.2016.02.023.

Lifson, Mark A, Mehmet Ozgun Ozen, Fatih Inci, ShuQi Wang, Hakan Inan, Murat Baday, Timothy J Henrich, and Utkan Demirci. 2016. "Advances in Biosensing Strategies for HIV-1 Detection, Diagnosis, and Therapeutic Monitoring." *Advanced Drug Delivery Reviews* 103 (August): 90–104. doi: 10.1016/j.addr.2016.05.018.

Mahato, Kuldeep, Alisha Prasad, Pawan Kumar Maurya, and Pranjal Chandra. 2016. "Nanobiosensors: Next Generation Point-of-Care Biomedical Devices for Personalized Diagnosis." *Journal of Analytical & Bioanalytical Techniques* 7 (2): 1000e125. doi: 10.4172/2155-9872.1000e125.

Malekzad, Hedieh, Parham Sahandi Zangabad, Hamed Mirshekari, Mahdi Karimi, and Michael R Hamblin. 2017. "Noble Metal Nanoparticles in Biosensors: Recent Studies and Applications." *Nanotechnology Reviews* 6 (3): 301–29. doi: 10.1515/ntrev-2016-0014.

Malik, Parth, Varun Katyal, Vibhuti Malik, Archana Asatkar, Gajendra Inwati, and Tapan K Mukherjee. 2013. "Nanobiosensors: Concepts and Variations." *ISRN Nanomaterials* 2013: 1–9. doi: 10.1155/2013/327435.

Neethirajan, S, X Weng, A Tah, J O Cordero, and K V Ragavan. 2018. "Nano-Biosensor Platforms for Detecting Food Allergens – New Trends." *Sensing and Bio-Sensing Research* 18: 13–30. doi: 10.1016/j.sbsr.2018.02.005.

Noh, Hui-Bog, Pranjal Chandra, You-Jeong Kim, and Yoon-Bo Shim. 2012. "A Simple Separation Method with a Microfluidic Channel Based on Alternating Current Potential Modulation." *Analytical Chemistry* 84 (22). United States: 9738–44. doi: 10.1021/ac301351y.

Parveen, Suphiya, Ranjita Misra, and Sanjeeb K Sahoo. 2012. "Nanoparticles: A Boon to Drug Delivery, Therapeutics, Diagnostics and Imaging." *Nanomedicine: Nanotechnology, Biology, and Medicine* 8: 147–66. doi: 10.1016/j.nano.2011.05.016.

Pickup, John C, Faeiza Hussain, Nicholas D Evans, and Nabihah Sachedina. 2005. "In Vivo Glucose Monitoring: The Clinical Reality and the Promise." *Biosensors & Bioelectronics* 20 (10). England: 1897–1902. doi: 10.1016/j.bios.2004.08.016.

Prasad, Shalini. 2014. "Nanobiosensors: The Future for Diagnosis of Disease?" *Nanobiosensors in Disease Diagnosis* 3: 1–10. doi: 10.2147/ndd.s39421.

Rackus, Darius G, Michael D M Dryden, Julian Lamanna, Alexandre Zaragoza, Brian Lam, Shana O Kelley, and Aaron R Wheeler. 2015. "A Digital Microfluidic Device with Integrated Nanostructured Microelectrodes for Electrochemical Immunoassays." *Lab on a Chip* 15 (18). England: 3776–84. doi: 10.1039/c5lc00660k.

Roda, Aldo, Elisa Michelini, Martina Zangheri, Massimo Di Fusco, Donato Calabria, and Patrizia Simoni. 2016. "Smartphone-Based Biosensors: A Critical Review and Perspectives." *TrAC Trends in Analytical Chemistry* 79: 317–25. doi: 10.1016/j.trac.2015.10.019.

Salahandish, Razieh, Ali Ghaffarinejad, Seyed Morteza Naghib, Keivan Majidzadeh-A, Hossein Zargartalebi, and Amir Sanati-Nezhad. 2018. "Nano-Biosensor for Highly Sensitive Detection of HER2 Positive Breast Cancer." *Biosensors & Bioelectronics* 117 (October). England: 104–11. doi: 10.1016/j.bios.2018.05.043.

Sekretaryova, Alina N, Mats Eriksson, and Anthony P F Turner. 2016. "Bioelectrocatalytic Systems for Health Applications." *Biotechnology Advances* 34 (3): 177–97. doi: 10.1016/j.biotechadv.2015.12.005.

Sharifi, Majid, Mohammad Reza Avadi, Farnoosh Attar, Fariba Dashtestani, Hedayatollah Ghorchian, Seyed Mahdi Rezayat, Ali Akbar Saboury, and Mojtaba Falahati. 2019. "Cancer Diagnosis Using Nanomaterials Based Electrochemical Nanobiosensors." *Biosensors and Bioelectronics* 126: 773–84. doi: 10.1016/j.bios.2018.11.026.

Sireesha, Merum, Veluru Jagadeesh Babu, A Sandeep Kranthi Kiran, and Seeram Ramakrishna. 2018. "A Review on Carbon Nanotubes in Biosensor Devices and Their Applications in Medicine." *Nanocomposites* 4 (2). Taylor & Francis: 36–57. doi: 10.1080/20550324.2018.1478765.

Stern, Eric, James F Klemic, David A Routenberg, Pauline N Wyrembak, Daniel B Turner-Evans, Andrew D Hamilton, David A LaVan, Tarek M Fahmy, and Mark A Reed. 2007. "Label-Free Immunodetection with CMOS-Compatible Semiconducting Nanowires." *Nature* 445 (February). Nature Publishing Group: 519–22.

Turner, Anthony P F. 2013. "Biosensors: Sense and Sensibility." *Chemical Society Reviews* 42 (8). England: 3184–96. doi: 10.1039/c3cs35528d.

Walzik, Maria P, Verena Vollmar, Theresa Lachnit, Helmut Dietz, Susanne Haug, Holger Bachmann, Moritz Fath, et al. 2015. "A Portable Low-Cost Long-Term Live-Cell Imaging Platform for Biomedical Research and Education." *Biosensors & Bioelectronics* 64 (February). England: 639–49. doi: 10.1016/j.bios.2014.09.061.

Wang, Yanyun, Feng Zhou, Xiaoli Liu, Lin Yuan, Dan Li, Yanwei Wang, and Hong Chen. 2013. "Aptamer-Modified Micro/Nanostructured Surfaces: Efficient Capture of Ramos Cells in Serum Environment." *ACS Applied Materials and Interfaces* 5 (9): 3816–23. doi: 10.1021/am400469g.

Wen, Cong Ying, Ling Ling Wu, Zhi Ling Zhang, Yu Lin Liu, Shao Zhong Wei, Jiao Hu, Man Tang, et al. 2014. "Quick-Response Magnetic Nanospheres for Rapid, Efficient Capture and Sensitive Detection of Circulating Tumor Cells." *ACS Nano* 8 (1): 941–49. doi: 10.1021/nn405744f.

Wujcik, E K, H Wei, X Zhang, J Guo, X Yan, N Sutrave, S Wei, and Z Guo. 2014. "Antibody Nanosensors: A Detailed Review." *RSC Advances* 4 (82). The Royal Society of Chemistry 4: 43725–45. doi: 10.1039/C4RA07119K.

Xu, Xiaoxing, Songqin Liu, and Huangxian Ju. 2003. "A Novel Hydrogen Peroxide Sensor via the Direct Electrochemistry of Horseradish Peroxidase Immobilized on Colloidal Gold Modified Screen-Printed Electrode." *Sensors* 3: 350–60. doi: 10.3390/s30900350.

Yen, Swee Kuan, Parasuraman Padmanabhan, and Subramanian Tamil Selvan. 2013. "Multifunctional Iron Oxide Nanoparticles for Diagnostics, Therapy and Macromolecule Delivery." *Theranostics* 3: 986–1003. doi: 10.7150/thno.4827.

Zhang, Wenjun, Yunqing Du, and Ming L Wang. 2015. "Noninvasive Glucose Monitoring Using Saliva Nano-Biosensor." *Sensing and Bio-Sensing Research* 4: 23–29. doi: 10.1016/j.sbsr.2015.02.002.

Zhao, Junguo, J P O'Daly, R W Henkens, J Stonehuerner, and A L Crumbliss. 1996. "A Xanthine Oxidase/Colloidal Gold Enzyme Electrode for Amperometric Biosensor Applications." *Biosensors and Bioelectronics* 11 (5): 493–502. doi: 10.1016/0956-5663(96)86786-8.

Zhu, Ye, Pranjal Chandra, and Yoon-Bo Shim. 2013. "Ultrasensitive and Selective Electrochemical Diagnosis of Breast Cancer Based on a Hydrazine–Au Nanoparticle–Aptamer Bioconjugate." *Analytical Chemistry* 85 (2). American Chemical Society: 1058–64. doi: 10.1021/ac302923k.

[illegible]

15 Theranostic Advances of Graphene-Based Nanomaterial in Diabetes

Kunal Biswas, Siddharth Shaw, Sudhanshu Mishra, Debashis De, and Jaya Bandyopadhyay

CONTENTS

15.1 Introduction and Background 185
15.2 Diabetes: Causes, Symptoms, and Management 186
15.3 Conventional Glucose Biosensors and Therapeutic Strategies for Diabetes 187
15.4 Established Methods for the Synthesis of Graphene and Its Role as a Powerful Glucose-Sensing Agent 187
15.5 Graphene-Based Theranostics for Diabetes 188
15.6 Conclusion 189
Conflict of Interest 189
References 189

15.1 INTRODUCTION AND BACKGROUND

The nanotechnology industry has earned itself a good deal of fame by virtue of graphene, since its discovery in 2004 (Novoselov et al. 2004). Graphene, which is a universally renowned two-dimensional (2D) nanomaterial of carbon, has been one of the most coveted materials for diverse applications in biomedical engineering (Reina et al. 2017) sector. This has been made possible by closely interlinking discrete disciplines such as electronics, biology, and nanoscience in a bid to produce improved theranostic (Kumar et al. 2018; Zhao et al. 2017) approaches for the welfare of society. Owing to phenomenal thermal, mechanical, electrical, and optical properties and satisfactory biocompatible attributes of graphene (Girão et al. 2019), it has garnered huge demand in biomedical applications. The nanomaterial comprises a 2D flat single-layer sheet of carbon atoms (Girão et al. 2019) that possess sp^2 hybridization and forms a honeycomb (Girão et al. 2019) mesh-like structure. This structure acts as an essential building block for other carbon-related materials such as 3D graphite crystals (Geim and Novoselov 2007), carbon nanotubes (1D), and fullerenes (0D) on the basis of differential reconstitution of graphene sheets. In terms of enabling the attachment of biomolecules (Li et al. 2016) at different sites of the graphene sheet as well as promoting decent interactions between cells and material (Zhang et al. 2016), graphene provides abundant scope to do so by virtue of its excellent surface area.

As a result of the superior optical and electronic properties of graphene, it bears a unique electronic band that is an amalgamation of both semiconducting and metallic properties (Girão et al. 2019) of a material. Such an event causes the π electrons (Girão et al. 2019) in graphene to take on the role of ultrarelativistic particles (Girão et al. 2019) which can nearly match the speed of light while in motion. In the context of functional sensing devices, essential properties of graphene like transparency, zero band gap semiconductivity, and bioactivity (Girão et al. 2019) largely aid in the augmentation of sensitivity and selectivity of field-effect transistor (FET), electrochemical, and optical biosensors (Lawal 2018; Merkoçi 2016). Such potential devices could be useful for immunoglobulin-, DNA-, and enzyme-based biosensing strategies (Girão et al. 2019). The fabrication of an extremely sensitive monolayer graphene-based FET by Xu et al. (2017) for the execution of reproducible and reliable manifold assessment of DNA is an epitome of a graphene-based functional device.

The oxidized version of graphene, popularly known as graphene oxide (GO) (Girão et al. 2019), harbors multiple hydrophilic regions (Girão et al. 2019) that have the potential to facilitate good dissolution in water (Girão et al. 2019), owing to the presence of oxygen-based functional groups such as epoxy, hydroxyl, and carboxyl. These chemical groups are bonded to the carbon atoms in GO, which have a more reactive surface (Girão et al. 2019) along with hydrophobic sp^2 hybridized carbon centers (Girão et al. 2019), containing a host of sp^3 regions within it. The aforementioned factors are also responsible for giving rise to covalent and/or noncovalent bonding sites for polymers (Singh et al. 2018), biomolecules, and metals and also acquisition of a near-infrared (NIR) to visible fluorescence property. Diverse applications of GO, including sensing, imaging, and therapy (Kim et al. 2017; Muazim and Hussain 2017), have therefore been realized to be possible by biomedical engineers. A typical example proving GO's worth in simultaneous live cellular imaging and drug delivery strategies is seen by allowing the oxygen groups to diffuse and bind to anti-neoplastic drugs in a sufficiently large volume while inciting blue fluorescence (Cheng et al. 2018) at the same time, using a mild thermal annealing process. In another work, a multifaceted nanocomposite was developed as a cancer nano-theranostic (Zhang et al. 2017) tool by successfully mixing Bi_2Se_3 nanoparticles with GO nanosheets and polyvinylpyrrolidone; this nano-theranostic tool demonstrated a level of performance similar to that of a photoacoustic contrast agent and X-ray computer tomography. Such a diagnostic approach in collaboration with photothermal therapy can aid in finer visualization of tumors in vivo followed by effective clearance of cancerous cells permanently.

Glucose, as the primary source of carbon and energy (Li et al. 2019), has a very important role to play in our daily lives. Deficiency of glucose in living organisms hampers normal growth and division of cells (Li et al. 2019), along with a decrease in the yield of products (Saltiel 2001) in industrial food bioprocessing. On the other hand, abnormally high glucose can turn out to be harmful by triggering lactate formation via the glycolytic pathway (Li et al. 2019) and is a clinical biomarker (Kim et al. 2012; Patel et al. 2012; Bandodkar and Wang 2014) of diabetes (Schousboe et al. 2007; Heinemann and Schmelzeisen-Redeker 1998; Heller 1999; Rowinski et al. 2008). In an attempt to devise the perfect glucose sensor (Feldman et al. 2000) for diabetes monitoring, several have already been produced. However, certain limitations (Li et al. 2019) exist pertaining to these monitoring systems, such as poor reproducibility of sensor fabrication, lack of biocompatibility, and signal reversibility in conjunction with poor storage stability. Therefore, there is a requirement for reliable glucose sensors that can accurately measure the levels of physiological glucose (Li et al. 2019).

Target molecules and interactions between molecules (Song et al. 2013) are extensively detected using a technique known as fluorescence resonance energy transfer (Selvin 2000; Sapsford et al. 2006; Wu et al. 2017; Shi et al. 2015; Long et al. 2015). GO, as a novel quencher of fluorescence (Li et al. 2019) and an alternative to carbon nanotubes (Li et al. 2019), has attracted a lot of attention (Kasry et al. 2012; Liu 2012; Morales-Narváez and Merkoçi 2012). The presence of multiplex functional groups (Li et al. 2019) such as hydroxyl, carboxyl, and epoxy on the surface of GO enables it to easily dissolve in water (Li et al. 2019). In addition, attachment to other low-molecular-weight molecules like chitosan (CS) (Pang et al. 2012; Saghatforoush et al. 2014; Wang et al. 2012) also becomes possible by virtue of forming simple covalent bonds between compatible chemical groups. GO-based FRET biosensors (Zhu et al. 2015) have been developed on the basis of the maximum adsorption surface area (Li et al. 2019) in GO as well as its high possibility of accepting non-radioactive excitation energy from a fluorophore compound (Kim et al. 2009; Swathi and Sebastian 2009).

15.2 DIABETES: CAUSES, SYMPTOMS, AND MANAGEMENT

When a deviation in the glucose concentration of an individual takes place from the reference range of 80–120 mg/dL (4.4–6.6 mM), over an extended span of time (Guariguata et al. 2014), this condition is medically referred to diabetes or diabetes mellitus (Tehrani and Bavarian 2016; Tao et al. 2015).

Symptoms that are found to manifest in patients include excessive thirst, frequent urination and abnormal hunger, hyperosmolar coma, and diabetic ketoacidosis (World Health Organization, 2013; Kitabchi et al. 2009). Complications arising over a long-term period of time involve diabetic heart disease, kidney failure, foot ulcers, stroke, and partial or complete loss of vision. When there is insufficient insulin production (Gardner et al. 2007) by the cells of the pancreas or when the cells of the body in general become refractory to the naturally synthesized insulin, diabetes is caused.

Diabetes mellitus (World Health Organization, 2013) is of the following three types.

1. Type 1 diabetes is caused when there is inadequate insulin production in the individual's body. It is termed "juvenile diabetes" or "insulin-dependent diabetes mellitus" (IDDM), and its underlying cause is not yet defined.
2. Type 2 diabetes (T2D) is initiated with resistance to insulin, whereby insulin has reduced effect on the cells of the body. Chances are that insulin deficiency (Tao et al. 2015) may also arise with the development of the disease (Tao et al. 2015). This version is termed as "non-insulin-dependent diabetes mellitus" (NIDDM) or "adult-onset diabetes." Increased body weight and poor physical exercise are the main causes of this type of diabetes.
3. A third type of diabetes known as gestational diabetes (Tao et al. 2015) happens in pregnant women who acquire raised blood glucose levels without any past record of diabetes.

In order to avoid diabetes and also provide remedies against it, one should seek routine physical workout, a healthy diet, maintenance of normal body weight, and total tobacco abstention. People suffering from diabetes must also be vigilant about their dietary intake of sugar, alterations in blood pressure (BP), and maintenance of proper diet. Insulin shots (World Health Organization, 2014) are used to manage patients suffering from Type 1 diabetes. Medications for the treatment of T2D either involve the use of insulin or are independent of the hormone use (Tripathy 2012; Diabetes, 2014). A decrease in the blood sugar level (Irwin and Rippe 2012) can result from insulin usage and a few other oral medications. An essential step to curb T2D in obesity-affected patients would be to undergo weight loss surgery (Picot et al. 2009). The delivery of a baby (Cash 2014) also ensures the cure of gestational diabetes in the affected patient. Nearly 382 million people are suffering from diabetes all over the world, out of which 90% of the cases (Melmed et al. 2011; Shi and Hu 2014) are found to be T2D, as per a global survey report. This observation matches 8.3% of the worldwide population of adults (Shi and Hu 2014), and men and women have the same rates (Vos et al. 2012). As per a prediction, 592 million individuals suffering from diabetes may face death by 2035 (IDF 2013).

15.3 CONVENTIONAL GLUCOSE BIOSENSORS AND THERAPEUTIC STRATEGIES FOR DIABETES

Microvascular complications associated with Type 1 diabetes such as nephropathy, retinopathy, and neuropathy (DCCT Research Group, 1993) can be considerably lowered by providing intensive remedy for glucose. Unfortunately, this kind of treatment gives rise to increased cases of hypoglycemia in patients (Oliver et al. 2009). Therefore, it was necessary to titrate the glucose levels up to the treatment procedures for individual patients by doctors with the help of glucose-sensing systems (Oliver et al. 2009). This would help in the maintenance of euglycemic conditions in patients along with maximum isolation of hypo- and hyperglycemia. In order to regulate diabetes treatment, healthcare professionals still use point sample results for capillary blood glucose. Intensive research to establish novel, continuous, and non-invasive (Bandodkar et al. 2014; Talary et al. 2007) glucose sensing and monitoring platforms using the latest state-of-the-art technologies is underway (Oliver et al. 2009).

Interstitial fluid (ISF) can be regarded as a room having a little volume. Glucose from the blood diffuses into the ISF through a rapid and free exchange process (Oliver et al. 2009), while clearance of glucose from the ISF takes place by the cells (Oliver et al. 2009) present within the compartment. A host of factors such as metabolic rate, blood flow, and the rate of change of glucose concentration (Oliver et al. 2009) in the blood directly govern the concentrations of glucose in the ISF. As per earlier studies on glucose concentrations, the mean lag time of peak ISF glucose concentration with respect to the blood glucose concentration is ascertained to be 6.7 min (Boyne et al. 2003). It is seen that when blood glucose starts to drop, then the interstitial glucose concentration drops faster than the blood glucose concentration, indicating that glucose from the ISF may behave as a biomarker for hypoglycemia (Thomé-Duret et al. 1996).

An example of a non-invasive tool for detecting glucose, an optical transducer is a device that exploits the various properties of light to drive interactions between photons and glucose molecules. By varying the frequencies of light (Oliver et al. 2009), optical transducers aid in the detection of glucose in a concentration-based manner (Oliver et al. 2009). There is a better chance to detect glucose across the anterior chamber of the eye by spectroscopically measuring the transmitted or reflected light, owing to high signal-to-noise ratio from different parts of the skin and existence of lesser biochemical artifacts in comparison to ISF.

The wavelength of a mid-infrared (MIR) electromagnetic ray falls within the range of 2.5–50 μm. Absorption spectroscopy works on the principle of measuring the intensity of material-interacted and un-interacted beams of light at a particular wavelength. Underlying molecular stretching and bending result in the formation of absorption bands at a fixed wavelength in use. Both glucose concentrations in water and blood in vitro have been measured using MIR spectroscopy (Rossel et al. 2006; Farrah et al. 2007; Haas and Mizaikoff 2016; Ycas et al. 2018; Rodrigo et al. 2015). However, the considerable amount of background absorption by other molecules, especially water, gave rise to excessive noise and therefore glucose measurement was negatively affected. Besides this, on account of the short path length of MIR light through the tissues, the in-depth concentration range of the analyte could not be measured. In spite of such limitations, glucose concentrations in the whole blood are still being measured in vitro, considering the standard prediction error to be 0.95 mmol/L (Shen et al. 2003).

15.4 ESTABLISHED METHODS FOR THE SYNTHESIS OF GRAPHENE AND ITS ROLE AS A POWERFUL GLUCOSE-SENSING AGENT

Just like in other applied fields of science, there is a huge demand of graphene for biosensing work as well (Claussen et al. 2012; Meng et al. 2015; Nia et al. 2015; Shang et al. 2015; Shang et al. 2014; Yuan et al. 2014; Li et al. 2014; Li et al. 2015; Fang et al. 2015; Hwa and Subramani 2014; Zhang et al. 2015; Shahriary and Athawale 2015; Ye et al. 2014), which is attributed to its extraordinary electrocatalytic features and enhanced surface area. Major limitations on the mass-scale production of graphene in a cost-effective manner and on its flexibility of applications still exist despite so many years of research since its discovery in 2004 (Geim 2009; Geim and Novoselov 2007). There are a host of methods to synthesize graphene, including chemical vapor deposition (Mattevi et al. 2011), thermal reduction, and chemical and mechanical exfoliation techniques (Martinez et al. 2011; Chang et al. 2010). Normally, there are many advantages for producing graphene by thermal reduction method, viz., easy processing, cost-effectiveness, and convenience of custom-made patterning accompanied by reduction with laser (El-Kady et al. 2012; El-Kady and Kaner 2013). The process (El-Kady et al. 2012) whereby graphite oxide coated on a compact disc (CD) is lightly scribed into graphene-based electrodes for the fabrication of supercapacitors is considered as a significant application of graphene in terms of scalability and economy (Lin et al. 2014).

Tehrani and Bavarian (2016) previously manufactured electrodes from graphene by using the direct laser engraving method, which they further functionalized to detect the presence of glucose in biological samples. The graphene electrodes developed by this method are known as direct laser engraved graphene (DLEG). The merits associated with DLEG manufacture are the utilization of cheap raw materials in fabricating the electrodes, the ability to scale electrode production, and finally the favorable size and design of the sensor electrodes. In order to further augment the sensing potential of the DLEG electrodes, pulse electrodeposition technique was employed to deposit nanosized cubic copper nanoparticles (CuNCs) on the surface of DLEG (Jiang et al. 2014; Khalid et al. 2015; Luo et al. 2012; Mallesha et al. 2012) electrochemically. Such an approach boosted the sensitivity of the as-synthesized nanodevice, which could detect glucose present in saliva, human tears, and sweat in trace amounts (Figure 15.1).

15.5 GRAPHENE-BASED THERANOSTICS FOR DIABETES

Li et al. (2019) demonstrated the use of a graphene oxide–chitosan (GO–CS) complex and another complex of maltose-binding protein and recombinant phycocyanin (MBP–rPC) in the development of a nanoscale biosensing device capable of sensitive and speedy monitoring of glucose levels. When glucose is absent, the fluorescence by virtue of MBP–rPC in vitro is lowered upon interactive binding of MBP with CS (Medintz et al. 2003a; Medintz et al. 2003b). In contrast, stronger binding interaction between glucose and MBP takes place in comparison to the MBP and CS interaction, when glucose is present in the surrounding medium. This type of interaction mechanism gives rise to a fluorescence signal for MBP–rPC binding, indicating the presence of glucose in the sample being investigated.

Using the aforesaid biosensor, sensitive and selective detection of glucose could be made possible at a linear detection ambit of 0.1–1.0 mg glucose per ml (Li et al. 2019). The detection limit in the biological sample was determined to be ~0.05 mg/ml concentration of glucose (Li et al. 2019).

A graphene-based, soft electrochemical device was previously developed by Lee et al. (2016) for the purpose of measuring blood glucose levels in real time along with accurate and regulated delivery of anti-diabetic drugs for keeping sugar levels under control (Lee et al. 2016). The device comprises an array of components for collecting sweat from the skin, sensors for detecting humidity, pH, glucose, temperature, and mechanical strain in conjunction with a microneedle system that dictates drug delivery to the target site upon coming in contact with heat (Lee et al. 2016). On the surface of gold-doped graphene, uniform patterning of the sensor devices is carried out, and the sensor devices are subsequently interlinked with the help of a curvy gold mesh. Such an approach of suitable material selection and module assembly helps in enhancing the electrochemical performance of graphene while imparting substantial selectivity and sensitivity to the sensor for the detection of glucose (Lee et al. 2016) in biological samples. This system exhibits mechanical robustness, partial transparency, flexibility, and adaptability with the skin. Monitoring of glucose is initiated once relative humidity becomes 80% after 10–20 min of sufficient perspiration.

Proper stimulus from the pH and glucose sensors for hyperglycemia detection (Lee et al. 2016) must be received by the sensing device in order to activate the drug delivery mechanism using microneedles. A hydrophobic coating of tridecanoic acid is added to the polymeric microneedles holding metformin within, which is a potent drug against T2D. Therefore, the microneedles are protected against damage from moisture after encountering the skin and untimely release of the payload is also prevented. The mechanism of drug delivery (Lee et al. 2016) is such that as soon as the presence of increased glucose in the sample detected, the patch heater is sparked off, thereby warming the microneedles. Upon exceeding the transition temperature of tridecanoic acid, i.e., ~41°C, the embedded polymer gets dissolved in the ISF, resulting in the release of the drug. The microneedles can be well adjusted to regulate the flow of metformin at the desired quantity and to bear flexibility of titration.

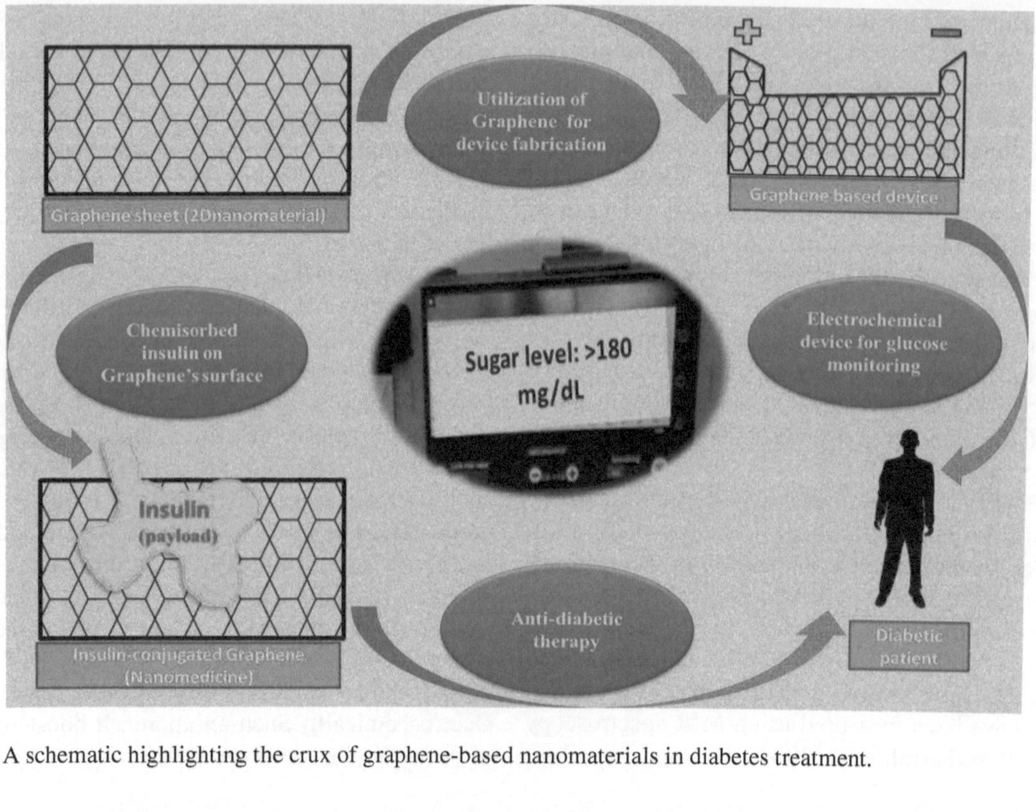

FIGURE 15.1 A schematic highlighting the crux of graphene-based nanomaterials in diabetes treatment.

Relevant characterizations of the device's modules and proper analysis of its bilateral operations have been carried out by Lee et al. (2016). In vivo experiments were carried out on diabetic mice for a span of several hours in order to prove the actual mode of operation of the theranostic device.

The presence of significant chemical groups such as epoxy, hydroxyl, and carboxyl on the top of graphene oxide (GO) sheets and the additional feature of greater surface area could make GO an ideal choice for stably conjugating it with insulin, a natural hormone that converts glucose to glycogen in the body for storage. Upon strong binding to GO, the bioavailability and activity of insulin in the physiological system could be raised, which otherwise gets disintegrated shortly after being injected into the patient. These features of GO make it a desirable vector for carrying and delivering the payload (insulin) to the target regions (receptor proteins). Such a strategy might help in maintaining proper integrity of the insulin molecules, which would result in a more pronounced effect on the control of blood sugar levels. A higher percentage of oxygen (~4.37%) in multi-layered graphene oxide (MLGO) sheets would ensure greater binding to insulin as compared to fewer oxygen groups (~1.63%) in single-layer graphene oxide (SLGO) sheets.

15.6 CONCLUSION

The biosensing and therapeutic capabilities of graphene against diabetes have been reviewed in the present article. High blood sugar in patients may lead to multiple complications that are difficult to treat without proper medication. High surface area and the presence of reactive oxygen species in graphene oxide nanosheets make it eligible for binding to anti-diabetic drugs and also strongly reacting with glucose. Based on these features, theranostic devices combining the functions of both sensor and drug delivery agents have been developed, demonstrating good sensitivity, specificity, and selectivity towards the target of interest.

CONFLICT OF INTEREST

The authors of the present article declare that no conflict of interest exists among the authors.

REFERENCES

Bandodkar, A.J. and Wang, J., 2014. Non-invasive wearable electrochemical sensors: a review. *Trends in Biotechnology*, *32*(7), pp. 363–371.

Bandodkar, A.J., Molinnus, D., Mirza, O., Guinovart, T., Windmiller, J.R., Valdes-Ramirez, G., Andrade, F.J., Schöning, M.J. and Wang, J., 2014. Epidermal tattoo potentiometric sodium sensors with wireless signal transduction for continuous non-invasive sweat monitoring. *Biosensors and Bioelectronics*, *54*, pp. 603–609.

Boyne, M.S., Silver, D.M., Kaplan, J. and Saudek, C.D., 2003. Timing of changes in interstitial and venous blood glucose measured with a continuous subcutaneous glucose sensor. *Diabetes*, *52*(11), pp. 2790–2794.

Cash, J., 2014. *Family practice guidelines* (3rd ed., p. 396). New York: Springer. ISBN 9780826168757.

Chang, Y.M., Kim, H., Lee, J.H. and Song, Y.W., 2010. Multilayered graphene efficiently formed by mechanical exfoliation for nonlinear saturable absorbers in fiber mode-locked lasers. *Applied Physics Letters*, *97*(21), p. 211102.

Cheng, S.J., Chiu, H.Y., Kumar, P.V., Hsieh, K.Y., Yang, J.W., Lin, Y.R., Shen, Y.C. and Chen, G.Y., 2018. Simultaneous drug delivery and cellular imaging using graphene oxide. *Biomaterials Science*, *6*(4), pp. 813–819.

Claussen, J.C., Kumar, A., Jaroch, D.B., Khawaja, M.H., Hibbard, A.B., Porterfield, D.M. and Fisher, T.S., 2012. Nanostructuring platinum nanoparticles on multilayered graphene petal nanosheets for electrochemical biosensing. *Advanced Functional Materials*, *22*(16), pp. 3399–3405.

Diabetes (2014) and "The top 10 causes of death Fact sheet N_310". World Health Organization (Oct 2013).

Diabetes Control and Complications Trial Research Group, 1993. The effect of intensive treatment of diabetes on the development and progression of long-term complications in insulin-dependent diabetes mellitus. *New England Journal of Medicine*, *329*(14), pp. 977–986.

El-Kady, M.F. and Kaner, R.B., 2013. Scalable fabrication of high-power graphene micro-supercapacitors for flexible and on-chip energy storage. *Nature Communications*, *4*, p. 1475.

El-Kady, M.F., Strong, V., Dubin, S. and Kaner, R.B., 2012. Laser scribing of high-performance and flexible graphene-based electrochemical capacitors. *Science*, *335*(6074), pp. 1326–1330.

Fang, Y., Zhang, D., Guo, Y., Guo, Y. and Chen, Q., 2015. Simple one-pot preparation of chitosan-reduced graphene oxide-Au nanoparticles hybrids for glucose sensing. *Sensors and Actuators B: Chemical*, *221*, pp. 265–272.

Farrah, D., Bernard-Salas, J., Spoon, H.W.W., Soifer, B.T., Armus, L., Brandl, B., Charmandaris, V., Desai, V., Higdon, S., Devost, D. and Houck, J., 2007. High-resolution mid-infrared spectroscopy of ultraluminous infrared galaxies. *The Astrophysical Journal*, *667*(1), p. 149.

Feldman, B., McGarraugh, G., Heller, A., Bohannon, N., Skyler, J., DeLeeuw, E. and Clarke, D., 2000. FreeStyle™: a small-volume electrochemical glucose sensor for home blood glucose testing. *Diabetes Technology & Therapeutics*, *2*(2), pp. 221–229.

Gardner, D.G., Shoback, D. and Greenspan, F.S., 2007. *Greenspan's Basic & Clinical Endocrinology*. McGraw-Hill Medical.

Geim, A.K., 2009. Graphene: status and prospects. *Science*, *324*(5934), pp. 1530–1534.

Geim, A.K. and Novoselov, K.S., 2007. The rise of graphene. *Nature Materials*, *6*, p. 183.

Girão, A.F., Serrano, M.C., Completo, A. and Marques, P.A., 2019. Do biomedical engineers dream of graphene sheets? *Biomaterials Science*, *7*, 1228–1239.

Guariguata, L., Whiting, D.R., Hambleton, I., Beagley, J., Linnenkamp, U. and Shaw, J.E., 2014. Global estimates of diabetes prevalence for 2013 and projections for 2035. *Diabetes Research and Clinical Practice*, *103*(2), pp. 137–149.

Haas, J. and Mizaikoff, B., 2016. Advances in mid-infrared spectroscopy for chemical analysis. *Annual Review of Analytical Chemistry*, *9*, pp. 45–68.

Heinemann, L. and Schmelzeisen-Redeker, G., 1998. Non-invasive continuous glucose monitoring in Type I diabetic patients with optical glucose sensors. *Diabetologia*, *41*(7), pp. 848–854.

Heller, A., 1999. Implanted electrochemical glucose sensors for the management of diabetes. *Annual Review of Biomedical Engineering*, *1*(1), pp. 153–175.

Hwa, K.Y. and Subramani, B., 2014. Synthesis of zinc oxide nanoparticles on graphene–carbon nanotube hybrid for glucose biosensor applications. *Biosensors and Bioelectronics*, *62*, pp. 127–133.

IDF, D., 2013. International Diabetes Federation Diabetes Atlas. Retrieved April 4, 2014. https://www.idf.org/e-library/epidemiology-research/diabetes-atlas/19-atlas-6th-edition.html

Irwin, R.S. and Rippe, J.M., 2012. *Manual of intensive care medicine*. Lippincott Williams & Wilkins.

Jiang, D., Liu, Q., Wang, K., Qian, J., Dong, X., Yang, Z., Du, X. and Qiu, B., 2014. Enhanced non-enzymatic glucose sensing based on copper nanoparticles decorated nitrogen-doped graphene. *Biosensors and Bioelectronics*, *54*, pp. 273–278.

Kasry, A., Ardakani, A.A., Tulevski, G.S., Menges, B., Copel, M. and Vyklicky, L., 2012. Highly efficient fluorescence quenching with graphene. *The Journal of Physical Chemistry C*, *116*(4), pp. 2858–2862.

Khalid, B., Meng, Q.H. and Cao, B., 2015. A non-enzymatic thermally reduced Cu nanoparticle based graphene-resorcinol benzaldehyde glucose sensor. *Materials Research Innovations*, *19*(2), pp. 91–96.

Kim, J., Cote, L.J., Kim, F. and Huang, J., 2009. Visualizing graphene based sheets by fluorescence quenching microscopy. *Journal of the American Chemical Society*, *132*(1), pp. 260–267.

Kim, D.H., Ghaffari, R., Lu, N. and Rogers, J.A., 2012. Flexible and stretchable electronics for biointegrated devices. *Annual Review of Biomedical Engineering*, *14*, pp. 113–128.

Kim, H.Y., Li, F., Park, J.Y., Kim, D., Park, J.H., Han, H.S., Byun, J.W., Lee, Y.S., Jeong, J.M., Char, K. and Lee, D.S., 2017. In vivo visualization of endogenous miR-21 using hyaluronic acid-coated graphene oxide for targeted cancer therapy. *Biomaterials*, *121*, pp. 144–154.

Kitabchi, A.E., Umpierrez, G.E., Miles, J.M. and Fisher, J.N., 2009. Hyperglycemic crises in adult patients with diabetes. *Diabetes Care*, *32*(7), pp. 1335–1343.

Kumar, R., Singh, R., Hui, D., Feo, L. and Fraternali, F., 2018. Graphene as biomedical sensing element: state of art review and potential engineering applications. *Composites Part B: Engineering*, *134*, pp. 193–206.

Lawal, A.T., 2018. Progress in utilisation of graphene for electrochemical biosensors. *Biosensors and Bioelectronics*, *106*, pp. 149–178.

Lee, H., Choi, T.K., Lee, Y.B., Cho, H.R., Ghaffari, R., Wang, L., Choi, H.J., Chung, T.D., Lu, N., Hyeon, T. and Choi, S.H., 2016. A graphene-based electrochemical device with thermoresponsive microneedles for diabetes monitoring and therapy. *Nature Nanotechnology*, *11*(6), p. 566.

Li, M., Bo, X., Mu, Z., Zhang, Y. and Guo, L., 2014. Electrodeposition of nickel oxide and platinum nanoparticles on electrochemically reduced graphene oxide film as a nonenzymatic glucose sensor. *Sensors and Actuators B: Chemical*, *192*, pp. 261–268.

Li, Z., Xie, C., Wang, J., Meng, A. and Zhang, F., 2015. Direct electrochemistry of cholesterol oxidase immobilized on chitosan–graphene and cholesterol sensing. *Sensors and Actuators B: Chemical*, *208*, pp. 505–511.

Li, D., Zhang, W., Yu, X., Wang, Z., Su, Z. and Wei, G., 2016. When biomolecules meet graphene: from molecular level interactions to material design and applications. *Nanoscale*, *8*(47), pp. 19491–19509.

Li, W., Jiang, T., Pu, Y., Jiao, X., Tan, W. and Qin, S., 2019. Glucose biosensor using fluorescence quenching with chitosan-modified graphene oxide. *Micro & Nano Letters*, *14*(3), pp. 344–348.

Lin, J., Peng, Z., Liu, Y., Ruiz-Zepeda, F., Ye, R., Samuel, E.L., Yacaman, M.J., Yakobson, B.I. and Tour, J.M., 2014. Laser-induced porous graphene films from commercial polymers. *Nature Communications*, *5*, p. 5714.

Liu, J., 2012. Adsorption of DNA onto gold nanoparticles and graphene oxide: surface science and applications. *Physical Chemistry Chemical Physics*, *14*(30), pp. 10485–10496.

Long, Q., Li, H., Zhang, Y. and Yao, S., 2015. Upconversion nanoparticle-based fluorescence resonance energy transfer assay for organophosphorus pesticides. *Biosensors and Bioelectronics*, *68*, pp. 168–174.

Luo, J., Jiang, S., Zhang, H., Jiang, J. and Liu, X., 2012. A novel non-enzymatic glucose sensor based on Cu nanoparticle modified graphene sheets electrode. *Analytica Chimica Acta*, *709*, pp. 47–53.

Mallesha, M., Manjunatha, R., Suresh, G.S., Melo, J.S., D'souza, S.F. and Venkatesha, T.V., 2012. Direct electrochemical non-enzymatic assay of glucose using functionalized graphene. *Journal of Solid State Electrochemistry*, *16*(8), pp. 2675–2681.

Martinez, A., Fuse, K. and Yamashita, S., 2011. Mechanical exfoliation of graphene for the passive mode-locking of fiber lasers. *Applied Physics Letters*, *99*(12), p. 121107.

Mattevi, C., Kim, H. and Chhowalla, M., 2011. A review of chemical vapour deposition of graphene on copper. *Journal of Materials Chemistry*, *21*(10), pp. 3324–3334.

Medintz, I.L., Clapp, A.R., Mattoussi, H., Goldman, E.R., Fisher, B. and Mauro, J.M., 2003a. Self-assembled nanoscale biosensors based on quantum dot FRET donors. *Nature Materials*, *2*(9), p. 630.

Medintz, I.L., Goldman, E.R., Lassman, M.E. and Mauro, J.M., 2003b. A fluorescence resonance energy transfer sensor based on maltose binding protein. *Bioconjugate Chemistry*, *14*(5), pp. 909–918.

Melmed, S., Polonsky, K., Larsen, P.R. and Kronenberg, H. *Williams text book of endocrinology*. (12th ed., pp. 1371–1435). Philadelphia, PA: Elsevier/Saunders Publ. (2011). ISBN 978-1-4377-0324-5.

Meng, L., Xia, Y., Liu, W., Zhang, L., Zou, P. and Zhang, Y., 2015. Hydrogen microexplosion synthesis of platinum nanoparticles/nitrogen doped graphene nanoscrolls as new amperometric glucose biosensor. *Electrochimica Acta*, *152*, pp. 330–337.

Merkoçi, A., 2016. Health and environment diagnostics using paper-based nanobiosensors. *Lab Chip*, *16*, pp. 3150–3176.

Morales-Narváez, E. and Merkoçi, A., 2012. Graphene oxide as an optical biosensing platform. *Advanced Materials*, *24*(25), pp. 3298–3308.

Muazim, K. and Hussain, Z., 2017. Graphene oxide—a platform towards theranostics. *Materials Science and Engineering C*, *76*, pp. 1274–1288.

Nia, P.M., Meng, W.P., Lorestani, F., Mahmoudian, M.R. and Alias, Y., 2015. Electrodeposition of copper oxide/polypyrrole/reduced graphene oxide as a nonenzymatic glucose biosensor. *Sensors and Actuators B: Chemical*, *209*, pp. 100–108.

Novoselov, K.S., Geim, A.K., Morozov, S.V., Jiang, D., Zhang, Y., Dubonos, S.V., Grigorieva, I.V. and Firsov, A.A., 2004. Electric field effect in atomically thin carbon films. *Science*, *306*(5696), pp. 666–669.

Oliver, N.S., Toumazou, C., Cass, A.E.G. and Johnston, D.G., 2009. Glucose sensors: a review of current and emerging technology. *Diabetic Medicine*, *26*(3), pp. 197–210.

Pang, Y., Cui, Y., Ma, Y., Qian, H. and Shen, X., 2012. Fluorescence quenching of cationic organic dye by graphene: interaction and its mechanism. *Micro & Nano Letters*, *7*(7), pp. 608–612.

Patel, S., Park, H., Bonato, P., Chan, L. and Rodgers, M., 2012. A review of wearable sensors and systems with application in rehabilitation. *Journal of NeuroEngineering and Rehabilitation*, *9*(1), p. 21.

Picot, J., Jones, J., Colquitt, J.L., Gospodarevskaya, E., Loveman, E., Baxter, L. and Clegg, A.J., 2009. The clinical effectiveness and cost-effectiveness of bariatric (weight loss) surgery for obesity: a systematic review and economic evaluation. *Health Technology Assessment*, *13*(41), pp. 1–214.

Reina, G., González-Domínguez, J.M., Criado, A., Vázquez, E., Bianco, A. and Prato, M., 2017. Promises, facts and challenges for graphene in biomedical applications. *Chemical Society Reviews*, *46*(15), pp. 4400–4416.

Rodrigo, D., Limaj, O., Janner, D., Etezadi, D., De Abajo, F.J.G., Pruneri, V. and Altug, H., 2015. Mid-infrared plasmonic biosensing with graphene. *Science*, *349*(6244), pp. 165–168.

Rossel, R.V., Walvoort, D.J.J., McBratney, A.B., Janik, L.J. and Skjemstad, J.O., 2006. Visible, near infrared, mid infrared or combined diffuse reflectance spectroscopy for simultaneous assessment of various soil properties. *Geoderma*, *131*(1–2), pp. 59–75.

Rowinski, P., Rowinska, M. and Heller, A., 2008. Liquid crystal membranes for serum-compatible diabetes management-assisting subcutaneously implanted amperometric glucose sensors. *Analytical Chemistry*, *80*(5), pp. 1746–1755.

Saghatforoush, L., Hasanzadeh, M. and Shadjou, N., 2014. Polystyrene–graphene oxide modified glassy carbon electrode as a new class of polymeric nanosensors for electrochemical determination of histamine. *Chinese Chemical Letters*, *25*(4), pp. 655–658.

Saltiel, A.R., 2001. New perspectives into the molecular pathogenesis and treatment of type 2 diabetes. *Cell*, *104*(4), pp. 517–529.

Sapsford, K.E., Berti, L. and Medintz, I.L., 2006. Materials for fluorescence resonance energy transfer analysis: beyond traditional donor–acceptor combinations. *Angewandte Chemie International Edition*, *45*(28), pp. 4562–4589.

Schousboe, A., Bak, L.K., Sickmann, H.M., Sonnewald, U. and Waagepetersen, H.S., 2007. Energy substrates to support glutamatergic and GABAergic synaptic function: role of glycogen, glucose and lactate. *Neurotoxicity Research*, *12*(4), pp. 263–268.

Selvin, P.R., 2000. The renaissance of fluorescence resonance energy transfer. *Nature Structural & Molecular Biology*, *7*(9), p. 730.

Shahriary, L. and Athawale, A.A., 2015. Electrochemical deposition of silver/silver oxide on reduced graphene oxide for glucose sensing. *Journal of Solid State Electrochemistry*, *19*(8), pp. 2255–2263.

Shang, L., Zhao, F. and Zeng, B., 2014. Sensitive voltammetric determination of vanillin with an AuPd nanoparticles– graphene composite modified electrode. *Food Chemistry*, *151*, pp. 53–57.

Shang, L., Zhao, F. and Zeng, B., 2015. Highly dispersive hollow PdAg alloy nanoparticles modified ionic liquid functionalized graphene nanoribbons for electrochemical sensing of nifedipine. *Electrochimica Acta*, *168*, pp. 330–336.

Shen, Y.C., Davies, A.G., Linfield, E.H., Elsey, T.S., Taday, P.F. and Arnone, D.D., 2003. The use of Fourier-transform infrared spectroscopy for the quantitative determination of glucose concentration in whole blood. *Physics in Medicine & Biology*, *48*(13), p. 2023.

Shi, Y. and Hu, F.B., 2014. The global implications of diabetes and cancer. *The Lancet*, *383*(9933), pp. 1947–1948.

Shi, J., Chan, C., Pang, Y., Ye, W., Tian, F., Lyu, J., Zhang, Y. and Yang, M., 2015. A fluorescence resonance energy transfer (FRET) biosensor based on graphene quantum dots (GQDs) and gold nanoparticles (AuNPs) for the detection of mecA gene sequence of Staphylococcus aureus. *Biosensors and Bioelectronics*, *67*, pp. 595–600.

Singh, D.P., Herrera, C.E., Singh, B., Singh, S., Singh, R.K. and Kumar, R., 2018. Graphene oxide: an efficient material and recent approach for biotechnological and biomedical applications. *Materials Science and Engineering C*, *86*, pp. 173–197.

Song, E., Cheng, D., Song, Y., Jiang, M., Yu, J. and Wang, Y., 2013. A graphene oxide-based FRET sensor for rapid and sensitive detection of matrix metalloproteinase 2 in human serum sample. *Biosensors and Bioelectronics*, *47*, pp. 445–450.

Swathi, R.S. and Sebastian, K.L., 2009. Long range resonance energy transfer from a dye molecule to graphene has (distance)– 4 dependence. *The Journal of Chemical Physics*, *130*(8), p. 086101.

Talary, M.S., Dewarrat, F., Huber, D. and Caduff, A., 2007. In vivo life sign application of dielectric spectroscopy and non-invasive glucose monitoring. *Journal of Non-Crystalline Solids*, *353*(47–51), pp. 4515–4517.

Tao, Z., Shi, A. and Zhao, J., 2015. Epidemiological perspectives of diabetes. *Cell Biochemistry and Biophysics*, *73*(1), pp. 181–185.

Tehrani, F. and Bavarian, B., 2016. Facile and scalable disposable sensor based on laser engraved graphene for electrochemical detection of glucose. *Scientific Reports*, *6*, p. 27975.

Thomé-Duret, V., Reach, G., Gangnerau, M.N., Lemonnier, F., Klein, J.C., Zhang, Y., Hu, Y. and Wilson, G.S., 1996. Use of a subcutaneous glucose sensor to detect decreases in glucose concentration prior to observation in blood. *Analytical Chemistry*, *68*(21), pp. 3822–3826.

Tripathy, B.B. (ed.) 2012.*RSSDI textbook of diabetes mellitus* (Rev. 2nd ed.). New Delhi: Jaypee Brothers Medical Publishers. (p. 235). ISBN 9789350254899.

Vos, T., Flaxman, A.D., Naghavi, M., Lozano, R., Michaud, C., Ezzati, M., Shibuya, K., Salomon, J.A., Abdalla, S., Aboyans, V. and Abraham, J., 2012. Years lived with disability (YLDs) for 1160 sequelae of 289 diseases and injuries 1990–2010: a systematic analysis for the Global Burden of Disease Study 2010. *The Lancet*, *380*(9859), pp. 2163–2196.

Wang, H., Chen, T., Wu, S., Chu, X. and Yu, R., 2012. A novel biosensing strategy for screening G-quadruplex ligands based on graphene oxide sheets. *Biosensors and Bioelectronics*, *34*(1), pp. 88–93.

WHO, Diabetes Fact sheet No. 312. WHO: October 2013. Retrieved March 25, 2014. https://www.who.int/news-room/fact-sheets/detail/diabetes

Wu, X., Song, Y., Yan, X., Zhu, C., Ma, Y., Du, D. and Lin, Y., 2017. Carbon quantum dots as fluorescence resonance energy transfer sensors for organophosphate pesticides determination. *Biosensors and Bioelectronics*, *94*, pp. 292–297.

Xu, S., Zhan, J., Man, B., Jiang, S., Yue, W., Gao, S., Guo, C., Liu, H., Li, Z., Wang, J. and Zhou, Y., 2017. Real-time reliable determination of binding kinetics of DNA hybridization using a multi-channel graphene biosensor. *Nature Communications*, *8*, p. 14902.

Ycas, G., Giorgetta, F.R., Baumann, E., Coddington, I., Herman, D., Diddams, S.A. and Newbury, N.R., 2018. High-coherence mid-infrared dual-comb spectroscopy spanning 2.6 to 5.2 μm. *Nature Photonics*, *12*(4), p. 202.

Ye, Y., Wang, P., Dai, E., Liu, J., Tian, Z., Liang, C. and Shao, G., 2014. A novel reduction approach to fabricate quantum-sized SnO 2-conjugated reduced graphene oxide nanocomposites as non-enzymatic glucose sensors. *Physical Chemistry Chemical Physics*, *16*(19), pp. 8801–8807.

Yuan, M., Liu, A., Zhao, M., Dong, W., Zhao, T., Wang, J. and Tang, W., 2014. Bimetallic PdCu nanoparticle decorated three-dimensional graphene hydrogel for non-enzymatic amperometric glucose sensor. *Sensors and Actuators B: Chemical*, *190*, pp. 707–714.

Zhang, B., He, Y., Liu, B. and Tang, D., 2015. Nickel-functionalized reduced graphene oxide with polyaniline for non-enzymatic glucose sensing. *Microchimica Acta*, *182*(3–4), pp. 625–631.

Zhang, B., Wei, P., Zhou, Z. and Wei, T., 2016. Interactions of graphene with mammalian cells: molecular mechanisms and biomedical insights. *Advanced Drug Delivery Reviews*, *105*, pp. 145–162.

Zhang, Y., Zhang, H., Wang, Y., Wu, H., Zeng, B., Zhang, Y., Tian, Q. and Yang, S., 2017. Hydrophilic graphene oxide/bismuth selenide nanocomposites for CT imaging, photoacoustic imaging, and photothermal therapy. *Journal of Materials Chemistry B*, *5*(9), pp. 1846–1855.

Zhao, H., Ding, R., Zhao, X., Li, Y., Qu, L., Pei, H., Yildirimer, L., Wu, Z. and Zhang, W., 2017. Graphene-based nanomaterials for drug and/or gene delivery, bioimaging, and tissue engineering. *Drug Discovery Today*, *22*(9), pp. 1302–1317.

Zhu, Y., Cai, Y., Xu, L., Zheng, L., Wang, L., Qi, B. and Xu, C., 2015. Building an aptamer/graphene oxide FRET biosensor for one-step detection of bisphenol A. *ACS Applied Materials & Interfaces*, *7*(14), pp. 7492–7496.

16 Role of Gold Nanoparticles (GNPs) in Cancer Diagnosis and Treatment

Kalpana Panati, Parasuraman Aiya Subramani, Dharaneeswara Reddy Devireddy, and Venkata Ramireddy Narala

CONTENTS

16.1 Introduction and Background of GNPs 193
16.2 Biomedical Applications of Gold Nanoparticles 194
16.2.1 Photothermal Therapy (PTT) 194
16.2.1.1 Gold Nanoshell-Based PTT 194
16.2.2 Cancer Imaging 195
16.2.2.1 In Vivo Imaging 196
16.2.2.2 Ex Vivo Imaging 196
16.3 Gold Nanoshell for Cancer Detection 197
16.3.1 Synthesis and Fabrication 197
16.3.2 Synthesis and Bioconjugation 198
16.3.3 Antibody Conjugation 198
16.3.4 In Vitro Nanoshell PTT 198
16.3.5 Mechanism of Gold Nanoshells on Cancer Cells 198
16.4 Conjugated GNPs for Hepatocellular Carcinoma (HCC) 198
16.4.1 Cell Culture 199
16.4.2 Animal Experiments 199
16.5 GNP-Based Delivery System 199
16.5.1 Cellular Uptake of the Drug 199
16.5.2 Cytotoxicity Assay 199
16.6 GNP-Mediated Cancer Immunotherapy 199
16.6.1 GNP-Mediated Adjuvant Delivery 200
16.6.2 GNP Cancer Vaccines 200
16.6.3 Gene Therapy 200
16.7 Assessment of Nanomaterial Risks 201
16.8 Conclusions 201
Acknowledgments 201
References 201

16.1 INTRODUCTION AND BACKGROUND OF GNPS

Cancer has been one of the most dreadful diseases of humankind since time immemorial. Cancer incidence and mortality (around 8.7 million) are expected to increase globally (Fitzmaurice, Allen et al. 2017). Conventional therapies against cancer, viz., chemotherapy, radiation therapy, etc., cause numerous side effects. One possible strategy to battle cancer is to use the cutting-edge approach called nanotechnology. Nanotechnology, the technology that deals with nanoscale materials, has become an interesting field in all areas of science and technology because of its various innovative applications. Nanomaterials, especially GNPs, are increasingly applied for cancer therapy (Li, Tan et al. 2017b). GNPs are very useful in theranostics (diagnosis and therapy) of cancer because they possess unique optical properties (Guo, Rahme et al. 2017). Light-induced heating of GNPs is a valuable property harnessed by scientists for photothermal therapy of cancer (Riley and Day 2017). The optical properties and subsequent photothermal property of GNPs are tunable and are functions of their shape and size (Chen, Ning et al. 2019). GNPs of various shapes and sizes can be synthesized with ease. Additionally, GNPs provide a nice platform upon which various other theranostic molecules can be functionalized (Chauhan, Chopra et al. 2017).

Libutti, Tamarkin et al. (2018) listed the limitations in conventional drugs and emphasized the advantages of GNPs in cancer therapy. The particle size of GNPs, which determines the light absorption and therapeutic properties of GNPs, also determines its toxicity (Lopez-Chaves, Soto-Alvaredo et al. 2018). Fortunately, therapeutic GNPs are large enough and were found to exert minimal toxicity in vitro and in vivo. Toxicity of GNPs is further reduced by encapsulation inside other nanomaterials (Capek 2017). Recently, GNPs also underwent clinical trials against various cancers (Libutti, Paciotti et al. 2010; Stern, Kibanov Solomonov et al. 2016). This chapter discusses biomedical applications of various GNPs in cancer detection and treatment.

16.2 BIOMEDICAL APPLICATIONS OF GOLD NANOPARTICLES

At present, a number of systems in the biomedical field have been incorporated with either inorganic or organic nanoparticles. More research activities are focused on the development of nanoparticles and nanostructures with suitable dimensions for different applications including diagnostics, drug delivery, colorimetric sensing, cancer immunotherapy, DNA hybridization detection, microarrays, and therapeutic purposes. The external functionalization of GNPs is an important requirement in order to target them to specific diseases, as it allows for selective interaction with the biomolecules or cells.

16.2.1 Photothermal Therapy (PTT)

GNPs are more advantageous than conventional PTT or radiation therapies that use high-energy electromagnetic waves for the treatment of cancer (Riley and Day 2017). Spherical GNPs are able to induce hyperthermia when excited by visible light (520 nm) due to its properties explained elsewhere. Hence, such GNPs are well suited for PTT of cancers. In one of the first studies, Pitsillides et al., 2003 used spherical 30-nm GNPs and short pulses of 520-nm laser to thermolyze GNP-labeled human lymphocytes (Pitsillides, Joe et al. 2003). Zharov, Galitovsky et al. (2003) was able to photolyze K562 cancer cells by incubating the cell lines with GNPs and then shining a 520-nm laser. Later, the same authors demonstrated the same phenomenon using other GNPs in breast (MDA-MB-231) cancer cell lines. When GNP-labeled cells undergo thermolysis in vitro, they form laser-induced bubbles and literally burst (Zharov, Galitovskaya et al. 2005a; Zharov, Kim et al. 2005b).

Pulse laser of a few nanoseconds is often used for GNP-induced PTT. Pulse laser is preferred for PTT of single metastatic cells or small cancers because of its accuracy and lesser damage to surrounding healthy tissues (Liu, Lloyd et al. 2008). However, pulse laser cannot be used for large tumor masses because much of the heat is lost during intervals between the laser pulses. As an alternative to the pulse laser, the continuous wave (CW) laser is preferred as it can produce more heat and kill tumors spread over a larger area. Examples of the use of CW lasers for PTT include continuous visible argon laser for the killing of oral squamous carcinoma (HSC 313 and HOC 3 Clone 8) cell lines by anti-EGFR-coated GNPs (El-Sayed, Huang et al. 2006) and targeted macrophage (RAW 264.7) destruction in a mixture of RAW 264.7 and CHO-K1 cells by CD-11b antibody–functionalized GNPs excited with 540-nm HeNe laser (Pissuwan, Cortie et al. 2007). In an earlier study, 40-nm GNPs were conjugated to anti-EGFR antibodies and targeted to human head and neck cancer cells. The nanoparticles induce cancer cell damage after the irradiation with an Ar^+ laser at 514 nm for 4 min, which is detected by dark-field light-scattering surface plasmon resonance (SPR) spectra on single cells, whereas healthy cells did not show the loss of cell viability under the same treatment (Huang, Jain et al. 2006; Pissuwan, Valenzuela et al. 2006; Huang, Qian et al. 2007b; Pissuwan, Cortie et al. 2007; Hleb, Hafner et al. 2008; Liu, Lloyd et al. 2008; Pustovalov, Smetannikov et al. 2008).

16.2.1.1 Gold Nanoshell-Based PTT

Though the spherical form of GNPs have better SPR absorption and PTT, various other forms of GNPs such as nanospheres, nanorods, nanoshells, and nanocages were tested for their ability to kill bacteria (Zharov, Mercer et al. 2006) and cancer cells (Loo, Lowery et al. 2005; Chen, Wang et al. 2007; Huang, Qian et al. 2007b; Tong, Zhao et al. 2007) by photothermal effect.

Though visible light seems promising in vitro, in vivo PTT of tumors requires deep penetrating light waves such as the near-infrared (NIR) light. It is found that NIR can penetrate deep into the human body, as it is not absorbed by body fluids. Fortunately, NIR-absorbing gold nanoshells were demonstrated to possess PTT in vitro and in vivo (Hirsch, Stafford et al. 2003) (Figure 16.1A, B). PEGylated nanoshells measuring 130 nm in diameter were injected into the bloodstream of mice having colon carcinoma via tail vein. Six hours after GNP administration, NIR light was shone over the tumors, which lead to their complete ablation (O'Neal, Hirsch et al. 2004) (Figure 16.1C, D). GNPs of 60–400 nm can be precisely targeted to tumors without labeling due to the established phenomenon called the enhanced permeability and retention (EPR) effect. The EPR effect is observed mainly because of the leaky nature of tumor vasculature. Hence, GNPs can passively enter into any actively dividing tumor. On the other hand, the lymphatic drainage system of tumors are impaired enough to not allow GNPs to be excreted as waste. As a result, GNPs that entered the tumor cells will be trapped inside them (Maeda 2001).

As mentioned above, small spherical GNPs with absorption at 500–600 nm are unsuitable for in vivo PTT. This drawback could be overcome by attaching many GNPs specifically to the cancer cells until they form a nanocluster of suitable size, and then irradiating it with NIR light. In one such study, Zharov et al. were successful in arraying numerous spherical 40-nm GNPs, which synergistically displaced (called red shift) the absorption peak of GNPs from ~500 nm (visible light) to ~1070 nm (NIR light) (Zharov, Galitovskaya et al. 2005a).

The stoichiometric number of GNPs required to kill a prostate cancer cell was found to be 5000 GNPs/cancer cell (Stern, Stanfield et al. 2007). The EPR effect may not hold good for solid-state tumors that have a necrotic center, since they remain far away from capillaries and blood flow. Such tumors are impossible to treat by either conventional or nanotechnology methods. However, it has been found that such solid tumor cells recruit higher numbers of macrophages.

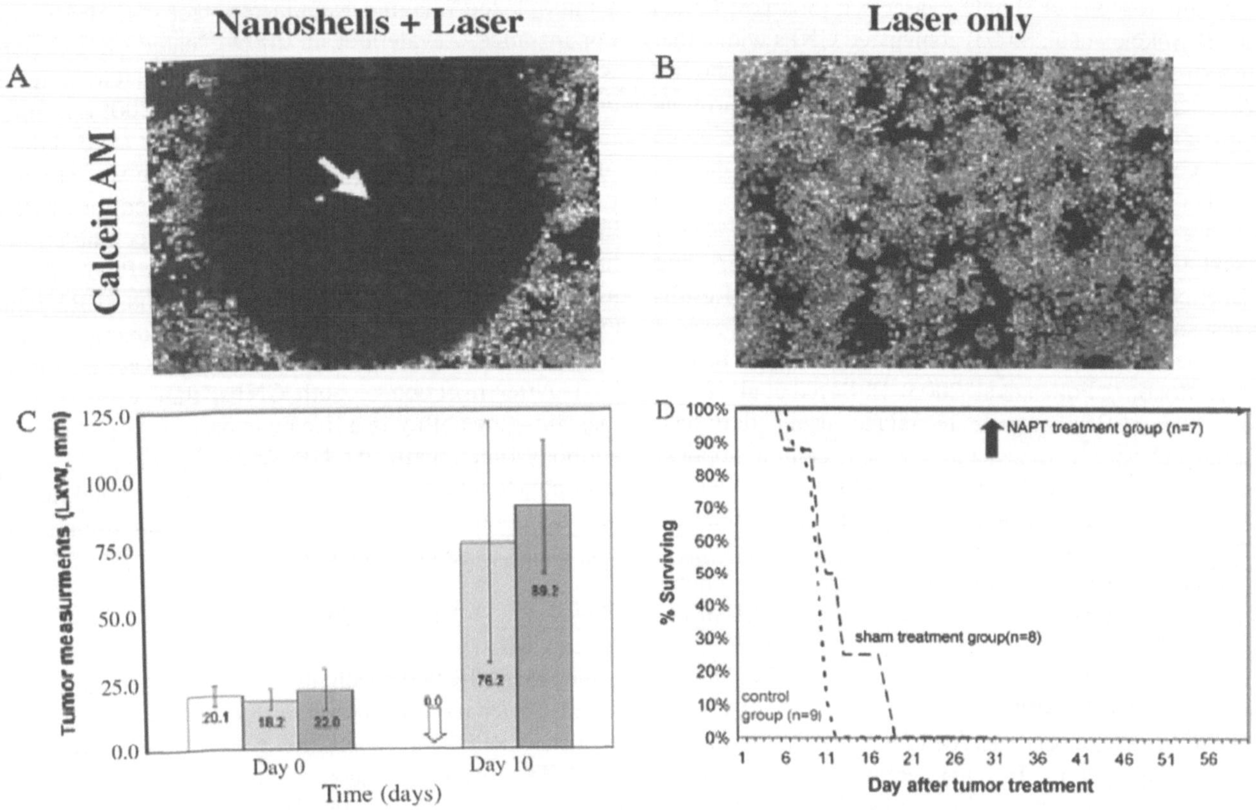

FIGURE 16.1 (A, B) Laser treatment (820 nm, 35 W/cm2, 4 min) of Sk-BR-3 cells with (A) and without (B) gold nanoshells. The loss of calcein AM staining in B indicates photothermal destruction induced by gold nanoshells. (C) Tumor size comparison before and after photothermal treatment for different groups. Blank: nanoshell + laser; gray: sham treatment; dark gray: no nanoshells and no laser. At day 10, the laser + nanoshell–treated tumors underwent complete necrosis, while two control groups did not show the same results, indicating the feasibility of gold nanoshells for photothermal tumor therapy. (D) Mice survival rate for different groups. Mice treated with both gold nanoshells and nanorods survive after 60 days of treatment. ((A) and (B) are reproduced with permission from Hirsch, Stafford et al. 2003. (C) and (D) are reproduced with permission from O'Neal, Hirsch et al. 2004.)

This recruitment is harnessed by nanotechnologists to design a cellular Trojan horse. Monocytes before their recruitment into solid tumors can be treated with NIR-absorbing nanoshells and can be used to kill solid tumors (Choi, Stanton-Maxey et al. 2007). GNPs were also tried as a vehicle for photodynamic therapy using phthalocyanine in HeLa cells (Wieder, Hone et al. 2006). Besides photothermal therapy, GNPs have also been investigated in other therapeutic studies.

16.2.2 Cancer Imaging

GNPs have a unique property of enhancing the Raman scattering signals of dye molecules adsorbed on their surface (Campion and Kambhampati 1998). This signal enhancement, which is referred to as surface-enhanced Raman scattering (SERS), has been reported to be as high as 1014- to 1015-fold (Qian, Peng et al. 2008). SERS-sensitive molecules adsorbed on the gold surface are called "Raman reporters" (Schutz, Muller et al. 2011) and are usually protected by a polymeric or silica coating. This ensures the prolonged adsorption of the Raman tags on the gold surface and the enhanced stability of the nanoparticle (Zavaleta, Smith et al. 2009; Zhang, Hong et al. 2011). GNP-mediated enhancement of Raman scattering signals not only aids picomolar amounts of target analyte to be detected but also aids a single target molecule to be identified. Since SERS offers picomolar sensitivity and has multiplexing capability, it may attain an important role in the molecular imaging field. SERS provides detailed spectroscopic information, which can be translated into imaging signals and adapted to an in vivo imaging system (Qian, Peng et al. 2008). GNPs are generally considered safe and have been safely administered to humans and used in clinical trials of cancer (Root, Andrews et al. 1954). High dosage of GNPs has been used safely as contrast agents for computed tomography (CT). Moreover, the fact that nanoparticle systems have been widely used for both imaging and image-guided therapy is because of the enhanced signal from a SERS nanoparticle (Reuveni, Motiei et al. 2011; Ock, Jeon et al. 2012). SERS of GNPs depends on the size, shape, and structure of the nanoparticles. For instance, the SERS signal of 60-nm GNPs is 105 times stronger than the emission of a fluorescein molecule (Schlucker 2009). Similarly, a highly sensitive 50-nm magnetic GNPs improved SERS and detection of microcystin-LR in orders of magnitude greater than 10^6 times (Hassanain, Izake et al. 2017). In the earlier attempts by Sokolov et al., the scattered light is collected in a reflection mode under single-laser wavelength excitation using a confocal microscope or simply a laser pen (Sokolov, Follen et al. 2003). Sokolov et al. (2003) conjugated GNPs with either anti-epidermal growth factor receptor (anti-EGFR) antibodies or bovine serum albumin (BSA) via nonspecific adsorption. When these conjugates were treated with cervical carcinoma cells and compared, anti-EGFR-conjugated GNPs showed much higher SERS than BSA-adsorbed nanoparticles due to the fact that it bound to the plasma membrane of the cancer cells. Next, the differences in SERS levels between BSA- and anti-EGFR-coated GNPs in cancer tissues were tested. Results of this test showed similar results in which anti-EGFR-coated GNPs showed stronger SERS in cancer tissues than in normal tissue (Sokolov, Follen et al. 2003). El-Sayed et al. (2005) improved the SERS imaging technique using dark-field microscopy (El-Sayed, Huang et al. 2005). A common optical microscope can be used to image cancer cells using GNPs. An anti-EGFR antibody conjugated to GNPs with polystyrenesulfonate linker was demonstrated to preferentially bind cancer cells and provide the required imaging contrast to be photographed using conventional optical microscope (El-Sayed, Huang et al. 2005).

16.2.2.1 In Vivo Imaging

Different Raman reporters adsorbed on same GNPs provide different Raman spectra. In other words, one can produce numerous SERS nanoparticles by simply changing the conjugated Raman reporters on the same GNPs. Alternatively, a multiplexed imaging Raman reporter can also be created. For instance, Zavaleta et al. 2009 created GNPs conjugated with 10 different Raman reporters and used it to image cells in deep tissue in vivo (Zavaleta, Smith et al. 2009) (Figure 16.2).

GNPs were also successfully used for subcellular imaging of cancer cells inside a live mouse. In this study, GNPs functionalized with Raman labels were used for SERS imaging in order to validate the nanoparticle accumulation in the tumor. Additional to that, these GNPs were PEGylated, which increased the stability and brightness of the SERS reporters. One set of PEGylated GNPs was then functionalized with anti-EGFR antibodies and the other set was not functionalized. After injection of both GNPs, they were assessed for their imaging ability in a live nude mouse. Mice injected with antibody-functionalized GNPs showed stronger SERS signals than non-functionalized ones. This is a pathbreaking finding in subcellular imaging of a live organism/mammalian model (Figure 16.3) (Qian, Peng et al. 2008).

16.2.2.2 Ex Vivo Imaging

Jokerst, Miao et al. (2011) prepared silicon-coated GNPs and conjugated NIR fluorescent dye (Alexa 647) to the silica coating of GNPs and a Raman reporter—trans-1,2-bis(4-pyridyl) ethylene (BPE)—between the gold surface and the silica shell. The resulting nanoparticle provided both NIR fluorescence and SERS signals. The purified nanoparticles were incubated with two different tumors ex vivo. They showed

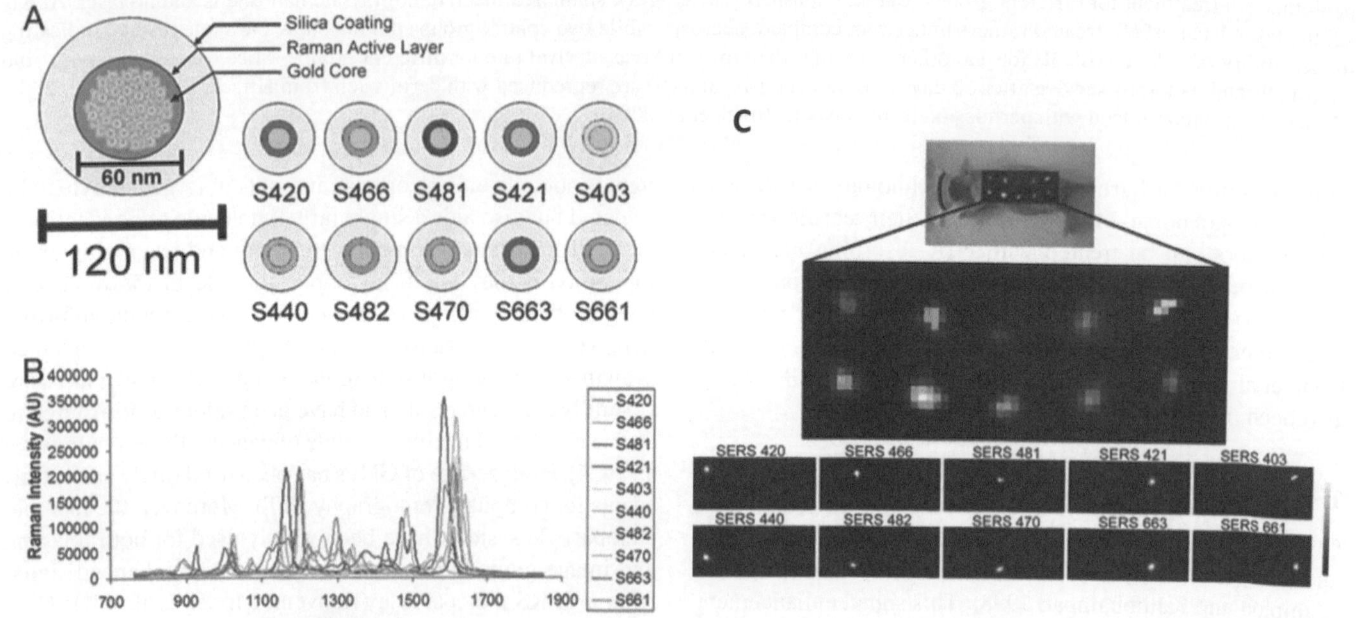

FIGURE 16.2 Schematic representation of a SERS Raman nanoparticle and graph depicting unique Raman spectra associated with each of the 10 SERS nanoparticles. (A) Schematic of a SERS Raman nanoparticle consisting of a 60-nm gold core with a unique Raman active layer coated with glass. The trade name of each SERS nanoparticle is depicted to the right, where a color has been assigned to the Raman active layer of each SERS nanoparticle. (B) Graph depicting Raman spectra of all 10 SERS nanoparticles; each spectrum has been assigned a color corresponding to its unique Raman active layer. (C) Evaluation of multiplexing 10 different SERS nanoparticles in vivo. Raman map of 10 different SERS particles injected subcutaneously in a nude mouse. Arbitrary colors have been assigned to each unique SERS nanoparticle batch injected. Panels below depict separate channels associated with each of the injected SERS nanoparticles (S420, S466, S481, S421, S403, S440, S482, S470, S663, and S661, respectively). (Reprinted with permission from Zavaleta, Smith et al. 2009.)

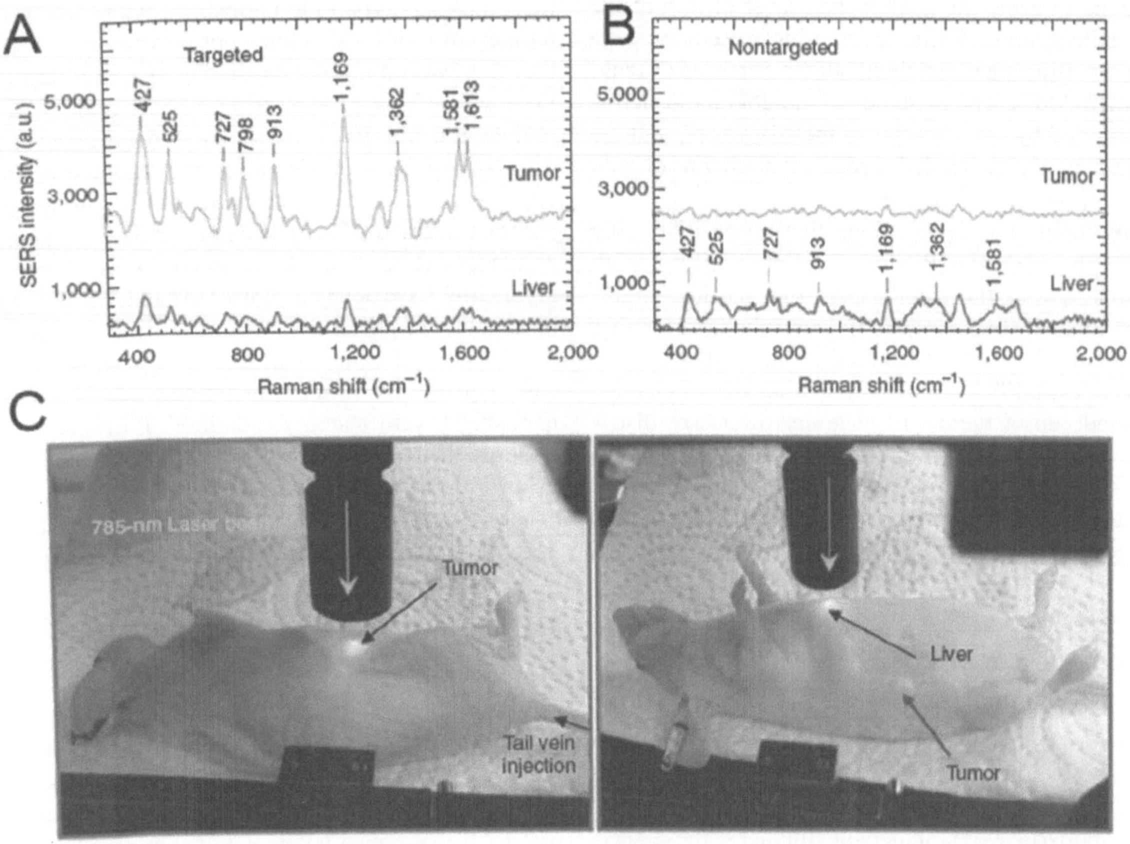

FIGURE 16.3 In vivo cancer targeting and surface-enhanced Raman detection by using ScFv antibody-conjugated gold nanoparticles that recognize the tumor biomarker EGFR. (A, B) SERS spectra obtained from the tumor and the liver locations by using targeted (A) and nontargeted (B) nanoparticles. Two nude mice bearing human head-and-neck squamous cell carcinoma (Tu686) xenograft tumor received ScFv EGFR-conjugated SERS tags or PEGylated SERS tags via tail vein injection. (C) Photographs showing a laser beam focusing on the tumor site or on the anatomical location of liver. In vivo SERS spectra were obtained from the tumor site (red) and the liver site (blue) with 2-s signal integration and at 785-nm excitation. The Raman reporter molecule is malachite green, with distinct spectral signatures as labeled in A and B. Laser power, 20 mW. (Reprinted with permission from Qian, Peng et al. 2008.)

that EGFR-positive A431 tumors produced a SERS signal that was 35 times higher than that of EGFR-negative MDA-435S tumors. The authors were also able to show that adjacent tissues with low-level EGFR expression had signal seven-fold lower than EGFR-positive tumors. In order to show that the signal was due to molecular targeting of EGFR, the authors also incubated the EGFR-positive tumor with a competitive probe. It did not generate any SERS signal. They observed that competitive inhibition resulted in SERS signal six-fold lower than the tumors incubated with the engineered SERS probe (Jokerst, Miao et al. 2011).

16.3 GOLD NANOSHELL FOR CANCER DETECTION

Nanotechnology-based technologies for prognosis are being developed for early therapy of cancers. Metal nanoshells are a new type of nanoparticles made up of a dielectric core and an ultrathin metallic outer layer (shell) (Loo, Lin et al. 2004). Typically, the inner core would be made up of silicon and the outer shell layer out of gold. Nanoparticles containing gold shells possess optical properties similar to regular GNPs. For instance, in regular GNPs, the loss of red coloration during aggregation of GNPs can be used in commercial applications such as home pregnancy test kits (Rojanathanes, Sereemaspun et al. 2008). However, regular spherical GNPs can absorb only wavelengths in the visible region of the UV-vis spectrophotometer. Unfortunately, a number of biological entities have their absorption maxima at this range and can interfere with the results. In contrast, the optical response of gold can be fine-tuned by varying the relative diameters of core and shell. By these manipulations, the absorption maxima of gold nanoshells can be varied from visible to NIR spectral regions (Loo, Lin et al. 2004).

16.3.1 Synthesis and Fabrication

The manipulation of absorption maxima (SPR peak) can be easily carried out by changing the ratio of the core size to its shell thickness. A procedure for preparing gold nanoshells of varying thickness is given elsewhere (Caruso, Spasova et al. 2001). First, silica cores are developed using the Stöber process, i.e., by reduction of tetraethyl orthosilicate in ethanol. A chemical seeding method using small GNPs of 2–4 nm as seeds was done to grow the golden shell layer on top of the synthesized silicon core. Amine-containing molecules were used for the crosslinking of gold and silicon particles (Suzuki and Kawaguchi 2005). The diameter of the gold nanoshell is largely determined by the diameter of the silica core, and the shell thickness can be controlled through the amount of gold deposited on the surface of the core.

Depending upon the application, thermosensitive gold nanoshells can be prepared (Radloff, Vaia et al. 2005). With the use of microgel as the core, factors like particle aggregation and gold shell thickness can be controlled. In another study, a gold nanoshell was assembled using a virus as core (Averitt, Sarkar et al. 1997). Due to the smaller size of viruses, nanoshells made up of them will have narrower size distributions and smaller diameters (80 nm) than those made up of silica. The practical risks involved in using silicon as the core material for nanoshell preparations and the parameters for safe production of gold nanoshells are reviewed elsewhere (Xu 2018).

16.3.2 Synthesis and Bioconjugation

The synthetic protocol developed for the fabrication of gold nanoshells is very simple in concept (Stöber, Fink et al. 1968). (1) Grow or obtain silica nanoparticles dispersed in solution. (2) Attach very small (1–2 nm) metal seed colloids to the surface of the nanoparticles via molecular linkages; these seed colloids cover the dielectric nanoparticle surfaces with a discontinuous metal colloid layer. (3) Grow additional metal onto the seed metal colloid adsorbates via chemical reduction in solution. This approach has been successfully used to grow both gold and silver metallic shells onto silica nanoparticles. Various stages in the growth of a gold metallic shell onto a functionalized silica nanoparticle are described elsewhere (Loo, Lin et al. 2004).

16.3.3 Antibody Conjugation

Ortho-pyridyl-disulfide-n-hydroxyl succinimide polyethylene glycol polymer (OPSS-PEG-NHS) is used to conjugate antibodies with gold nanoshells. Using sodium bicarbonate buffer, OPSS-PEG-NHS is resuspended in antibody solution at equal volumes. The reaction is allowed to proceed overnight on ice. After overnight incubation, unbound antibody is removed by centrifugation. Finally, the reaction was stopped by conjugating the nanoshells with inert agents like PEG (Barton et al. 2004).

16.3.4 In Vitro Nanoshell PTT

Nanoshells were widely tested for PTT of malignant cancer cells both in vitro and in vivo due to the fact that they can be excited by many different light sources. Some of the most recent and advanced studies that reported PTT of gold nanoshells are given in this section.

A nontoxic, polypeptide block copolymer was used as core instead of silicon and tested for PTT. The results showed that these novel polypeptide nanoshells were able to increase the temperature of local tissues upon exposure to NIR light, and that this was enough to cause damage to tumor cells (Mayle, Dern et al. 2017). Zein (protein from maize)-based biocompatible 100-nm gold nanoshells showed an SPR peak at 808 nm (NIR spectrum), which is highly promising for therapeutic applications. Zein nanoshells were also claimed to be a better contrast agent for X-ray imaging than iodine-based contrast enhancers (Chauhan, Arunkumar et al. 2018a). Computer algorithms have been developed to calculate the gold nanoshell parameters required for targeted killing of large tumor cells amidst normal cells (Paul 2018). A simulated 3D model of cancer using microfluidic technique was developed and gold nanoshells were tested for their PTT efficacy (Kalinowska, Grabowska-Jadach et al. 2019). A highly biocompatible, vitamin E–modified gold nanoshell was prepared which showed absorbance peak at 750 nm and promising tumor specific accumulation (Chauhan, Bukhari et al. 2018b).

16.3.5 Mechanism of Gold Nanoshells on Cancer Cells

Gold nanoshells follow a straightforward mechanism of killing action against cancer cells. First of all, the monoclonal antibody functionalized on the surface of the nanoshell binds specific receptors present in the surface of tumor. These bound nanoshells were then targeted by lasers of suitable wavelength. Nanoshells absorbing wavelengths at their absorption maximum convert light energy to heat energy and create bubbles of free radicals which in turn kill the cells (Carpin, Bickford et al. 2011).

PTT of gold nanoshells has been used for other applications as well. For instance, gold nanoshell–mediated conversion of light energy to heat energy can stimulate the contraction of myotube and activation of certain muscles (Li, Liu et al. 2017a; Marino, Arai et al. 2017).

16.4 CONJUGATED GNPS FOR HEPATOCELLULAR CARCINOMA (HCC)

Currently, hepatocellular carcinoma is the second leading cause of cancer deaths next to lung cancer due to poor prognosis, targeting, and therapy (Bruix, Gores et al. 2014). The multifunctional ability of GNPs is well studied in hepatocellular carcinoma in which GNPs were functionalized with diverse molecules and therapeutics to be used in different applications. Jha, Ramadori et al. (2017) functionalized GNPs with peptides derived from hepatitis-B virus and added them to HepG2 carcinoma cells. As expected, peptide-functionalized GNPs specifically targeted HepG2 carcinoma cells over normal hepatocytes and got internalized by the cancer cells. In another study, GNPs acted as a delivery agent for micro RNA, miR-375, into hepatocellular carcinoma cells in vitro and in vivo (Xue, Liu et al. 2016). The possible treatment of hepatocellular carcinoma by targeted delivery of therapeutic monoclonal antibodies (immunotherapy) is also possible with GNPs (Ma, Hui et al. 2016). Silva, Pinho et al. (2019) reviewed different nanotechnologies available for GNP-mediated treatment of hepatocarcinoma. In the following section, we will go through different success stories in GNP-mediated hepatocellular carcinoma therapy.

16.4.1 Cell Culture

The uptake of GNPs by hepatocytes, namely, HepG2 and HL-7702, is dependent upon the size of GNPs and the incubation time. The mechanism of killing hepatocellular cancer cells, C3A, was found to be dependent upon the size and inhibition of mitochondrial β-oxidation (Choi and Joo 2018; Wang, Wu et al. 2018). GNPs are also considered to be safe and compatible with normal cells (Cancino-Bernardi, Marangoni et al. 2018). Biological synthesis of GNPs using extracts from *Corchorus olitorius* showed excellent anti-proliferative activity against the HepG2 hepatocarcinoma cell line, colon carcinoma HCT-116, and breast adenocarcinoma MCF-7 (Ismail, Saqer et al. 2018). A pectin-capped GNP of 14 nm was able to load the anticancer drug doxorubicin with an efficiency of 78% and deliver it to HepG2 hepatocarcinoma cells via clathrin-dependent endocytosis (Borker and Pokharkar 2018). Chitosan-based GNPs containing the anticancer drug 5-fluorouracil was developed and administered to the HepG2 cell line. Results were compared with that of free 5-fluorouracil and GNPs with laser ablation. It was shown that the degree of cytotoxicity was in the order GNPs+5-fluorouracil+laser > GNPs+5-fluorouracil > free 5-fluorouracil (Salem, Sliem et al. 2018).

16.4.2 Animal Experiments

Small GNPs (2 nm) were found to improve the efficacy and tolerability of anticancer drug maytansine analogue-DM1 when it was administered to mice (Hale, Perrins et al. 2018). BSA-conjugated GNPs of 8 and 50 nm were able to increase the sensitivity of cancer cells towards X-ray irradiation in tumor-bearing mice (Liu, Piao et al. 2018). PEGylated GNPs containing 5-fluorouracil and encapsulated by growth factor–functionalized mesoporous silica nanoparticles showed very high cytotoxicity and improved image contrast to hepatocarcinoma (HepG2 xenograft) in nude mice (Babaei, Abnous et al. 2017). A similar and more sensitive experiment was carried out using doxorubicin containing hollow GNPs functionalized with various hepatocarcinoma-specific molecules in Sprague-Dawley rats. The results indicated that GNPs functionalized with tumor-specific molecules greatly increased the cytotoxicity and image clarity when compared to free doxorubicin or non-functionalized GNPs (Li, Zhou et al. 2016). Wang, Luo et al. (2016) synthesized GNPs that allow for the detection of hepatocellular carcinoma in vivo by CT and magnetic resonance (MR) scans.

16.5 GNP-BASED DELIVERY SYSTEM

GNPs are well suited for the delivery of many well-known anticancer drugs like methotrexate, doxorubicin, 5-fluorouracil, etc. At present, anticancer chemotherapeutics are heavily criticized for their side effects, including nonspecific killing of healthy cells, less bioavailability, and conferring resistance to tumors upon prolonged usage. In general, GNPs increase the cytotoxicity, bioavailability, specificity, etc., for these otherwise efficacious drugs. Some GNPs are designed in a way that makes normal cells tolerant towards cytotoxic anticancer drugs. In addition, PTT is a deadly option available that can increase the anticancer arsenal severalfold. The use of GNPs as a drug delivery system is the hot subject of many reviews (Kong, Zhang et al. 2017; Singh, Pandit et al. 2018). In the following sections, strategies and characterization methods for GNP-based delivery system are explained.

16.5.1 Cellular Uptake of the Drug

Bioavailability is a major issue that arises during the development of anticancer drugs (Ferreira, Nicoletti et al. 2016; Subramani, Panati et al. 2017). As discussed earlier, the properties of GNPs vary greatly with shape and size. This holds good for the cellular uptake of GNPs also. Different cell types uptake GNPs differentially. Hence, specialized techniques are required to target and force the uptake of GNPs in different cell types (Ahmad, Zamry et al. 2017). Unfortunately, there is no holdfast correlation between nanoparticle properties and cellular uptake. For instance, Kumar, Mutreja et al. (2017) synthesized GNPs with diameters of 18, 40, 60, and 80 nm and treated them with ovarian cancer cell lines OVCAR5, OVCAR8, and SKOV3. However, 18- and 80-nm GNPs showed higher cellular uptake. There are numerous studies reporting that different GNP functionalizations have improved their cellular uptake (He, Yang et al. 2017). Cellular uptake of GNPs can be visualized in a number of ways, including confocal and electron (SEM and TEM) microscopies.

16.5.2 Cytotoxicity Assay

In ideal settings, researchers expect the synthesized GNPs to be toxic only to cancer cells. However, there are various factors that affect the compatibility of GNPs to untargeted/healthy cells. Careful consideration of these factors can lead researchers to synthesize ideal GNPs with specific toxicity towards cancer cells. As with other properties, the size and shape of GNPs play a crucial role in determining their cytotoxicity. For instance, Hanan, Chiu et al. (2018) determined the cytotoxicity of biosynthesized GNPs to be inversely proportional to their size. The reviewers also stated that spherical and quasi-spherical GNPs are more potent cytotoxic materials than other shapes. The tetrazolium salts-based cytotoxicity method is a widely followed method for assessing the cytotoxicity of GNPs (Riss, Moravec et al. 2004), among other more sensitive methods including flow cytometry– and confocal microscopy–based methods.

16.6 GNP-MEDIATED CANCER IMMUNOTHERAPY

Cancer immunotherapy using GNPs is a promising approach in cancer treatment that aims to stimulate the cancer-impaired host's immune system. In general, cancer evades the host immune system by numerous mechanisms (Vinay, Ryan et al. 2015), which creates a huge challenge to researchers seeking to find a potent solution. With the advent of nanotechnology, especially GNPs, the face of immunotherapy has developed a lot. Already, there are numerous reports on the use of nanoparticles for delivery of antigens, vaccines, and adjuvants. Since immune cells are on the frontline to encounter any foreign particle, delivery of immunotherapeutics by nano/microparticles is considered to be preferable than conventional methods (Jewell, Lopez et al. 2011; Khlebtsov and Dykman 2011). Moon et al. reviewed the potentials of various nanoparticles in immunotherapy (Moon, Huang et al. 2012).

The use of GNPs for the delivery of vaccines and adjuvants for cancer therapy has been extensively reported elsewhere (Massich, Giljohann et al. 2010; Lee, Kwon et al. 2012; Wei, Chen et al. 2012). As mentioned earlier, GNPs have the intrinsic property of accumulation in the immune cells. This property can be harnessed to develop drug- and other ligand-functionalized GNPs for immunotherapy. Another important aspect is PTT, which can be modified into light-triggered drug delivery (Hirsch, Stafford et al. 2003; Huang, Jain et al. 2007a). Figure 16.4 illustrates some of the various possibilities in GNP-mediated immunotherapy.

In a preliminary study, Nguyen et al. 2012 were unsuccessful in inducing an inflammasome formation or macrophage activation after release of damage-associated molecular patterns (DAMPs), including ATP and uric acid, from gold nanoshells following PTT of tumors (Nguyen, Tran et al. 2012).

The negatively charged nucleic acids possess great affinity towards positively charged GNPs. This property is used to develop GNPs containing gene-silencing elements such as small interfering RNA (siRNA), micron RNA (miRNA), etc. In one study, Lu, Zhang et al. (2010) synthesized GNPs functionalized with siRNA against NF-κB subunit which could be released upon PTT. siRNA-mediated silencing of this subunit was found to be tumor specific. A number of other recent studies use GNPs functionalized with siRNAs against various targets. Conde, Oliva et al. (2016) were able to deliver a

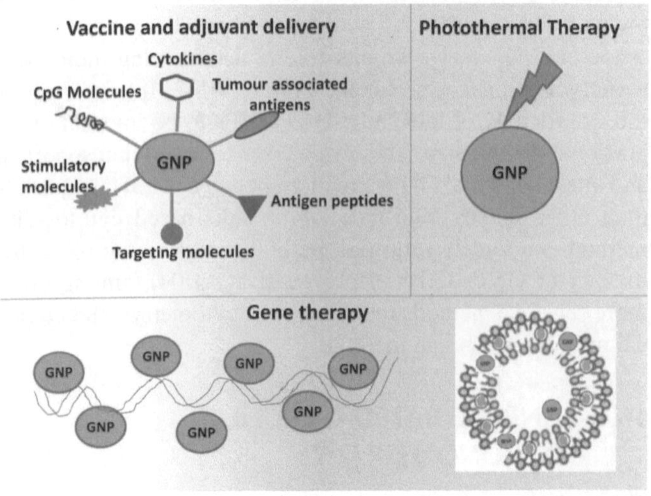

FIGURE 16.4 Gold nanoparticles in immuno-, photo-, and gene therapies of cancer.

triple-pronged approach to exploit various properties of GNPs such as in vivo imaging, delivery of siRNA, and a chemotherapy drug followed by PTT. A number of other immunotherapies using GNPs have been postulated using various immune molecules and combining different properties of GNPs, including optical and thermal properties.

16.6.1 GNP-Mediated Adjuvant Delivery

Cytokines like tumor necrosis factor (TNF) can be functionalized on GNPs and delivered to cancer cells. One such study was done by functionalizing TNF with 33-nm GNPs against prostate cancer in mice (Paciotti, Myer et al. 2004; Goel, Shah et al. 2009). Encouraged by successful results, Goel et al. took up this work and did clinical trials. GNPs conjugated with TNF were dubbed as Aurimune™ and administered to terminally ill patients with solid tumors (Shenoi, Iltis et al. 2013). Their results were overwhelming, and Aurimune™ is under phase II clinical trials now (Shao, Griffin et al. 2013).

16.6.2 GNP Cancer Vaccines

A cancer vaccine is envisaged to be a lifesaver for millions. However, this concept is still in its infancy. GNPs are promising vaccine carrier candidates, as they can be practically conjugated to molecules of different sizes (Chen, Hung et al. 2010). Cancer antigens are however difficult to identify and are poorly immunogenic, since many are self-antigens (Chesson and Zloza 2017). Nevertheless, GNPs with 80-nm diameter and carrying large doses of peptides were synthesized and tested in vitro for their ability to release interferon-gamma (IFN-γ) from cultured splenocytes. Results from the study showed that peptides in conjugation with GNPs were able to produce more IFN-γ response than individual peptides (Lin, Lunsford et al. 2013).

16.6.3 Gene Therapy

Certain genes that control cell cycle checkpoints have the ability to initiate either tumor death or immune activation. A wide range of such genes were tested successfully in clinical trials (Ginn, Alexander et al. 2013). As with the delivery of siRNA and miRNA, GNPs can be used in gene therapy (Ghosh, Kim et al. 2008; Ryou, Kim et al. 2010; Cebrian, Martin-Saavedra et al. 2011). For example, plasmids that encode tumor-associated antigens, cytokines, or other immune molecules can be used in conjugation with GNPs as cancer vaccines (Chlichlia, Schirrmacher et al. 2005).

Conventionally, DNA vaccines were administered by ballistic DNA delivery methods. The same approach was used by Mahvi et al. to transfect a plasmid encoding granulocyte-macrophage colony stimulating factor (GM-CSF) to tumor cells. GNP-mediated transfection yielded 58% protection to mice from tumor challenge. On the other hand, only 2% of non-transfected controls were protected (Mahvi, Burkholder et al. 1996). This work was further extended as a clinical trial by Cassaday et al., who improvised the DNA vaccine by adding gp100 to the plasmid construct. Results of the clinical trials showed stable expression of GM-CSF and gp100, and a high level of tolerance towards plasmid-encoded proteins with a moderate immune response (Cassaday, Sondel et al. 2007). DeLong et al. functionalized GNPs with genes encoding two hepatitis antigens complexed with protamine and then tested for their ability to deliver the plasmid DNA. They found that the GNP-DNA complexes, even though exposed to physical and chemical degradation agents, retained their structure and function in mice. The expression of hepatitis antigens, IFN-γ, and IL-4 were assessed by ELISA (DeLong, Akhtar et al. 2009). Unfortunately, a DNA-based vaccine cannot perpetuate by itself and requires several rounds of booster doses. Hung et al. 2001 prepared a papilloma virus vaccine using GNPs conjugated with plasmid. However, they fused papilloma virus antigen with that of VP22 protein from herpes simplex virus, which increases the intracellular transport of antigens. Booster doses of vaccine were also given two times. Upon challenge with tumor cells, they showed 50-fold increase in Ag-specific $CD8^+$ T cell precursors and reduction in tumor mass (Hung, Cheng et al. 2001).

Cheng et al. 2005 expanded on this work for the treatment of systemic tumors at multiple sites (Cheng, Hung et al. 2005). Wang et al. 2001 transfected tumor cells using GNPs functionalized with a plasmid encoding IL-12 and transferred it to mice. Later the same mice were vaccinated with oral melanoma vaccine. Results showed that the production of IL-12 by the tumor resulted in tumor suppression and mice survival (Wang, Murakami et al. 2001).

16.7 ASSESSMENT OF NANOMATERIAL RISKS

Even though GNPs are attractive anticancer therapeutic agents, particular attention should be given considering the toxicity of GNPs. Reports show acute toxicity of GNPs (10–50 nm in diameter) to be as low as 2000 mg nanoparticles/kg body weight of rats (Pokharkar, Dhar et al. 2009). However, chronic use of GNPs may cause discomfort to humans, including skin rashes (Russell, Langley et al. 1997). GNPs of different shapes and sizes were found to possess genotoxic and DNA-damaging effects in vitro (Hadrup, Sharma et al. 2015). GNPs were also shown to have caused mutations in zebrafish (Geffroy, Ladhar et al. 2012) and DNA damage in *Drosophila* (Sabella, Brunetti et al. 2011). Taking into consideration the above-mentioned effects, it is up to the discretion of patients to choose GNP-mediated therapy. Nevertheless, more research in this area is required in order to make GNPs more usable for cancer therapy.

16.8 CONCLUSIONS

GNPs are the promising nanotechnologies of the future for imaging tumors in vitro and in vivo, delivery and targeting of anticancer drugs, and improving tolerance and bioavailability of well-known anticancer drugs. There are certainly a few drawbacks of using GNPs in clinical settings, but considering the volume of research work carried out with GNP-mediated anticancer activities, we are convinced that GNP-mediated anticancer therapy will take over conventional cancer treatments in the near future.

ACKNOWLEDGMENTS

The financial support provided by University Grants Commission, New Delhi, India for Minor Research Project (No. F MRP-5520/15 (SERO/UGC) to KP is greatly acknowledged. VRN was supported by the University Grants Commission, Raman postdoctoral fellowship award in the United States (F.No. 5-90/2016 (IC)), and SERB grant (EMR/2017/000973) New Delhi.

REFERENCES

Ahmad, S., A. A. Zamry, et al. (2017). "Targeting dendritic cells through gold nanoparticles: A review on the cellular uptake and subsequent immunological properties." *Mol Immunol* **91**: 123–133.

Averitt, R. D., D. Sarkar, et al. (1997). "Plasmon resonance shifts of Au-coated Au2S nanoshells: Insight into multicomponent nanoparticle growth." *Phys Rev Lett* **78**(22): 4217–4220.

Babaei, M., K. Abnous, et al. (2017). "Synthesis of theranostic epithelial cell adhesion molecule targeted mesoporous silica nanoparticle with gold gatekeeper for hepatocellular carcinoma." *Nanomedicine (Lond)* **12**(11): 1261–1279.

Barton, J, Romanowski, M., Halas N., Drezek R. (2004). "Nanoshells as an optical coherence tomography contrast agent." SPIE proceedings 5316: 8 pages.

Borker, S., V. Pokharkar (2018). "Engineering of pectin-capped gold nanoparticles for delivery of doxorubicin to hepatocarcinoma cells: An insight into mechanism of cellular uptake." *Artif Cells Nanomed Biotechnol* **46**(sup 2): 826–835.

Bruix, J., G. J. Gores, et al. (2014). "Hepatocellular carcinoma: Clinical frontiers and perspectives." *Gut* **63**(5): 844–855.

Campion, A., P. Kambhampati (1998). "Surface-enhanced Raman scattering." *Chem Soc Rev* **27**(4): 241–250.

Cancino-Bernardi, J., V. S. Marangoni, et al. (2018). "Gold-based nanospheres and nanorods particles used as theranostic agents: An in vitro and in vivo toxicology studies." *Chemosphere* **213**: 41–52.

Capek, I. (2017). "Polymer decorated gold nanoparticles in nanomedicine conjugates." *Adv Colloid Interface Sci* **249**: 386–399.

Carpin, L. B., L. R. Bickford, et al. (2011). "Immunoconjugated gold nanoshell-mediated photothermal ablation of trastuzumab-resistant breast cancer cells." *Breast Cancer Res Treat* **125**(1): 27–34.

Caruso, F., M. Spasova, et al. (2001). "Multilayer assemblies of silica-encapsulated Gold nanoparticles on decomposable colloid templates." *Adv Mater* **13**(14): 1090–1094.

Cassaday, R. D., P. M. Sondel, et al. (2007). "A phase I study of immunization using particle-mediated epidermal delivery of genes for gp100 and GM-CSF into uninvolved skin of melanoma patients." *Clin Cancer Res* **13**(2 Pt 1): 540–549.

Cebrian, V., F. Martin-Saavedra, et al. (2011). "Size-dependent transfection efficiency of PEI-coated gold nanoparticles." *Acta Biomater* **7**(10): 3645–3655.

Chauhan, G., V. Chopra, et al. (2017). ""Gold nanoparticles composite-folic acid conjugated graphene oxide nanohybrids" for targeted chemo-thermal cancer ablation: In vitro screening and in vivo studies." *Eur J Pharm Sci* **96**: 351–361.

Chauhan, D. S., P. Arunkumar, et al. (2018a). "Facile synthesis of plasmonic zein nanoshells for imaging-guided photothermal cancer therapy." *Mater Sci Eng C Mater Biol Appl* **90**: 539–548.

Chauhan, D. S., A. B. Bukhari, et al. (2018b). "Enhanced EPR directed and imaging guided photothermal therapy using vitamin E modified Toco-Photoxil." *Sci Rep* **8**(1): 16673.

Chen, J., C. Ning, et al. (2019). "Nanomaterials as photothermal therapeutic agents." *Prog Mater Sci* **99**: 1–26.

Chen, J., D. Wang, et al. (2007). "Immuno gold nanocages with tailored optical properties for targeted photothermal destruction of cancer cells." *Nano Lett* **7**(5): 1318–1322.

Chen, Y. S., Y. C. Hung, et al. (2010). "Assessment of gold nanoparticles as a size-dependent vaccine carrier for enhancing the antibody response against synthetic foot-and-mouth disease virus peptide." *Nanotechnology* **21**(19): 195101.

Cheng, W. F., C. F. Hung, et al. (2005). "Characterization of DNA vaccines encoding the domains of calreticulin for their ability to elicit tumor-specific immunity and antiangiogenesis." *Vaccine* **23**(29): 3864–3874.

Chesson, C. B., A. Zloza (2017). "Nanoparticles: Augmenting tumor antigen presentation for vaccine and immunotherapy treatments of cancer." *Nanomedicine* **12**(23): 2693–2706.

Chlichlia, K., V. Schirrmacher, et al. (2005). "Cancer immunotherapy: Battling tumors with gene vaccines." Curr Med Chem – Anti-Inflamm Anti-Allergy Agents **4**(4): 353–365.

Choi, K., H. Joo (2018). "Assessment of gold nanoparticles-inhibited cytochrome P450 3A4 activity and molecular mechanisms underlying its cellular toxicity in human hepatocellular carcinoma cell line C3A." *Nanoscale Res Lett* **13**(1): 279.

Choi, M. R., K. J. Stanton-Maxey, et al. (2007). "A cellular Trojan Horse for delivery of therapeutic nanoparticles into tumors." *Nano Lett* **7**(12): 3759–3765.

Conde, J., N. Oliva, et al. (2016). "Local triple-combination therapy results in tumour regression and prevents recurrence in a colon cancer model." *Nat Mater* **15**(10): 1128–1138.

DeLong, R. K., U. Akhtar, et al. (2009). "Characterization and performance of nucleic acid nanoparticles combined with protamine and gold." *Biomaterials* **30**(32): 6451–6459.

El-Sayed, I. H., X. Huang, et al. (2005). "Surface plasmon resonance scattering and absorption of anti-EGFR antibody conjugated gold nanoparticles in cancer diagnostics: Applications in oral cancer." *Nano Lett* **5**(5): 829–834.

El-Sayed, I. H., X. Huang, et al. (2006). "Selective laser photothermal therapy of epithelial carcinoma using anti-EGFR antibody conjugated gold nanoparticles." *Cancer Lett* **239**(1): 129–135.

Ferreira, V. F., C. D. Nicoletti, et al. (2016). "Strategies for increasing the solubility and bioavailability of anticancer compounds: Beta-Lapachone and other naphthoquinones." *Curr Pharm Des* **22**(39): 5899–5914.

Fitzmaurice, C., C. Allen, et al. (2017). "Global, regional, and national cancer incidence, mortality, years of life lost, years lived with disability, and disability-adjusted life-years for 32 cancer groups, 1990 to 2015: A systematic analysis for the global burden of disease study." *JAMA Oncol* **3**(4): 524–548.

Geffroy, B., C. Ladhar, et al. (2012). "Impact of dietary gold nanoparticles in zebrafish at very low contamination pressure: The role of size, concentration and exposure time." *Nanotoxicology* **6**(2): 144–160.

Ghosh, P. S., C. K. Kim, et al. (2008). "Efficient gene delivery vectors by tuning the surface charge density of amino acid-functionalized gold nanoparticles." *ACS Nano* **2**(11): 2213–2218.

Ginn, S. L., I. E. Alexander, et al. (2013). "Gene therapy clinical trials worldwide to 2012 - An update." *J Gene Med* **15**(2): 65–77.

Goel, R., N. Shah, et al. (2009). "Biodistribution of TNF-alpha-coated gold nanoparticles in an in vivo model system." *Nanomedicine (Lond)* **4**(4): 401–410.

Guo, J., K. Rahme, et al. (2017). "Gold nanoparticles enlighten the future of cancer theranostics." *Int J Nanomed* **12**: 6131–6152.

Hadrup, N., A. K. Sharma, et al. (2015). "Toxicological risk assessment of elemental gold following oral exposure to sheets and nanoparticles – A review." *Regul Toxicol Pharmacol* **72**(2): 216–221.

Hale, S. J., R. D. Perrins, et al. (2018). "DM1 loaded ultra-small gold nanoparticles display significant efficacy and improved tolerability in murine models of hepatocellular carcinoma." *Bioconjug Chem* **30**(3): 703–713.

Hanan, N. A., H. I. Chiu, et al. (2018). "Cytotoxicity of plant-mediated synthesis of metallic nanoparticles: A systematic review." *Int J Mol Sci* **19**(6): 1725.

Hassanain, W. A., E. L. Izake, et al. (2017). "Gold nanomaterials for the selective capturing and SERS diagnosis of toxins in aqueous and biological fluids." *Biosens Bioelectron* **91**: 664–672.

He, B., D. Yang, et al. (2017). "Increased cellular uptake of peptide-modified PEGylated gold nanoparticles." *Biochem Biophys Res Commun* **494**(1–2): 339–345.

Hirsch, L. R., R. J. Stafford, et al. (2003). "Nanoshell-mediated near-infrared thermal therapy of tumors under magnetic resonance guidance." *Proc Natl Acad Sci U S A* **100**(23): 13549–13554.

Hleb, E. Y., J. H. Hafner, et al. (2008). "LANTCET: Elimination of solid tumor cells with photothermal bubbles generated around clusters of gold nanoparticles." *Nanomedicine (Lond)* **3**(5): 647–667.

Huang, X., P. K. Jain, et al. (2006). "Determination of the minimum temperature required for selective photothermal destruction of cancer cells with the use of immunotargeted gold nanoparticles." *Photochem Photobiol* **82**(2): 412–417.

Huang, X., P. K. Jain, et al. (2007a). "Gold nanoparticles: Interesting optical properties and recent applications in cancer diagnostics and therapy." *Nanomedicine (Lond)* **2**(5): 681–693.

Huang, X., W. Qian, et al. (2007b). "The potential use of the enhanced nonlinear properties of gold nanospheres in photothermal cancer therapy." *Lasers Surg Med* **39**(9): 747–753.

Hung, C. F., W. F. Cheng, et al. (2001). "Improving vaccine potency through intercellular spreading and enhanced MHC class I presentation of antigen." *J Immunol* **166**(9): 5733–5740.

Ismail, E. H., A. M. A. Saqer, et al. (2018). "Successful green synthesis of gold nanoparticles using a *Corchorus olitorius* extract and their antiproliferative effect in cancer cells." *Int J Mol Sci* **19**(9): E2612.

Jewell, C. M., S. C. Lopez, et al. (2011). "In situ engineering of the lymph node microenvironment via intranodal injection of adjuvant-releasing polymer particles." *Proc Natl Acad Sci U S A* **108**(38): 15745–15750.

Jha, S., F. Ramadori, et al. (2017). "Binding and uptake into human hepatocellular carcinoma cells of peptide-functionalized gold nanoparticles." *Bioconjug Chem* **28**(1): 222–229.

Jokerst, J. V., Z. Miao, et al. (2011). "Affibody-functionalized gold-silica nanoparticles for Raman molecular imaging of the epidermal growth factor receptor." *Small* **7**(5): 625–633.

Kalinowska, D., I. Grabowska-Jadach, et al. (2019). "Studies on effectiveness of PTT on 3D tumor model under microfluidic conditions using aptamer-modified nanoshells." *Biosens Bioelectron* **126**: 214–221.

Khlebtsov, N., L. Dykman (2011). "Biodistribution and toxicity of engineered gold nanoparticles: A review of in vitro and in vivo studies." *Chem Soc Rev* **40**(3): 1647–1671.

Kong, F. Y., J. W. Zhang, et al. (2017). "Unique roles of gold nanoparticles in drug delivery, targeting and imaging applications." *Molecules* **22**(9): E1445.

Kumar, D., I. Mutreja, et al. (2017). "Cytotoxicity and cellular uptake of different sized gold nanoparticles in ovarian cancer cells." *Nanotechnology* **28**(47): 475101.

Lee, I. H., H. K. Kwon, et al. (2012). "Imageable antigen-presenting gold nanoparticle vaccines for effective cancer immunotherapy in vivo." *Angew Chem Int Ed Engl* **51**(35): 8800–8805.

Li, J., J. Liu, et al. (2017a). "Remote control and modulation of cellular events by plasmonic gold nanoparticles: Implications and opportunities for biomedical applications." *ACS Nano* **11**(3): 2403–2409.

Li, J., M. Zhou, et al. (2016). "Hepatocellular carcinoma: Intra-arterial delivery of doxorubicin-loaded hollow gold nanospheres for photothermal ablation-chemoembolization therapy in rats." *Radiology* **281**(2): 427–435.

Li, Z., S. Tan, et al. (2017b). "Cancer drug delivery in the nano era: An overview and perspectives (Review)." *Oncol Rep* **38**(2): 611–624.

Libutti, S. K., G. F. Paciotti, et al. (2010). "Phase I and pharmacokinetic studies of CYT-6091, a novel PEGylated colloidal gold-rhTNF nanomedicine." *Clin Cancer Res* **16**(24): 6139–6149.

Libutti, S. K., L. Tamarkin, et al. (2018). "Targeting the invincible barrier for drug delivery in solid cancers: Interstitial fluid pressure." *Oncotarget* **9**(87): 35723–35725.

Lin, A. Y., J. Lunsford, et al. (2013). "High-density sub-100-nm peptide-gold nanoparticle complexes improve vaccine presentation by dendritic cells in vitro." *Nanoscale Res Lett* **8**(1): 72.

Liu, S., J. Piao, et al. (2018). "Radiosensitizing effects of different size bovine serum albumin-templated gold nanoparticles on H22 hepatoma-bearing mice." *Nanomedicine (Lond)* **13**(11): 1371–1383.

Liu, X., M. C. Lloyd, et al. (2008). "Enhanced imaging and accelerated photothermalysis of A549 human lung cancer cells by gold nanospheres." *Nanomedicine (Lond)* **3**(5): 617–626.

Loo, C., A. Lin, et al. (2004). "Nanoshell-enabled photonics-based imaging and therapy of cancer." *Technol Cancer Res Treat* **3**(1): 33–40.

Loo, C., A. Lowery, et al. (2005). "Immunotargeted nanoshells for integrated cancer imaging and therapy." *Nano Lett* **5**(4): 709–711.

Lopez-Chaves, C., J. Soto-Alvaredo, et al. (2018). "Gold nanoparticles: Distribution, bioaccumulation and toxicity. In vitro and in vivo studies." *Nanomedicine* **14**(1): 1–12.

Lu, W., G. Zhang, et al. (2010). "Tumor site-specific silencing of NF-kappaB p65 by targeted hollow gold nanosphere-mediated photothermal transfection." *Cancer Res* **70**(8): 3177–3188.

Ma, X., H. Hui, et al. (2016). "Enhanced immunotherapy of SM5-1 in hepatocellular carcinoma by conjugating with gold nanoparticles and its in vivo bioluminescence tomographic evaluation." *Biomaterials* **87**: 46–56.

Maeda, H. (2001). "The enhanced permeability and retention (EPR) effect in tumor vasculature: The key role of tumor-selective macromolecular drug targeting." *Adv Enzyme Regul* **41**: 189–207. Elsevier BV.

Mahvi, D. M., J. K. Burkholder, et al. (1996). "Particle-mediated gene transfer of granulocyte-macrophage colony-stimulating factor cDNA to tumor cells: Implications for a clinically relevant tumor vaccine." *Hum Gene Ther* **7**(13): 1535–1543.

Marino, A., S. Arai, et al. (2017). "Gold nanoshell-mediated remote myotube activation." *ACS Nano* **11**(3): 2494–2508.

Massich, M. D., D. A. Giljohann, et al. (2010). "Cellular response of polyvalent oligonucleotide-gold nanoparticle conjugates." *ACS Nano* **4**(10): 5641–5646.

Mayle, K. M., K. R. Dern, et al. (2017). "Polypeptide-based gold nanoshells for photothermal therapy." *SLAS Technol* **22**(1): 18–25.

Moon, J. J., B. Huang, et al. (2012). "Engineering nano- and microparticles to tune immunity." *Adv Mater* **24**(28): 3724–3746.

Nguyen, H. T., K. K. Tran, et al. (2012). "Activation of inflammasomes by tumor cell death mediated by gold nanoshells." *Biomaterials* **33**(7): 2197–2205.

O'Neal, D. P., L. R. Hirsch, et al. (2004). "Photo-thermal tumor ablation in mice using near infrared-absorbing nanoparticles." *Cancer Lett* **209**(2): 171–176.

Ock, K., W. I. Jeon, et al. (2012). "Real-time monitoring of glutathione-triggered thiopurine anticancer drug release in live cells investigated by surface-enhanced Raman scattering." *Anal Chem* **84**(5): 2172–2178.

Paciotti, G. F., L. Myer, et al. (2004). "Colloidal gold: A novel nanoparticle vector for tumor directed drug delivery." *Drug Deliv* **11**(3): 169–183.

Paul, A. (2018). "Computational study of photo-thermal ablation of large blood vessel embedded tumor using localized injection of gold nanoshells." *J Therm Biol* **78**: 329–342.

Pissuwan, D., C. H. Cortie, et al. (2007). "Gold nanosphere-antibody conjugates for hyperthermal therapeutic applications." *Gold Bull* **40**(2): 121–129.

Pissuwan, D., S. M. Valenzuela, et al. (2006). "Therapeutic possibilities of plasmonically heated gold nanoparticles." *Trends Biotechnol* **24**(2): 62–67.

Pitsillides, C. M., E. K. Joe, et al. (2003). "Selective cell targeting with light-absorbing microparticles and nanoparticles." *Biophys J* **84**(6): 4023–4032.

Pokharkar, V., S. Dhar, et al. (2009). "Acute and subacute toxicity studies of chitosan reduced gold nanoparticles: A novel carrier for therapeutic agents." *J Biomed Nanotechnol* **5**(3): 233–239.

Pustovalov, V. K., A. S. Smetannikov, et al. (2008). "Photothermal and accompanied phenomena of selective nanophotothermolysis with gold nanoparticles and laser pulses." *Laser Phys Lett* **5**(11): 775–792.

Qian, X., X. H. Peng, et al. (2008). "In vivo tumor targeting and spectroscopic detection with surface-enhanced Raman nanoparticle tags." *Nat Biotechnol* **26**(1): 83–90.

Radloff, C., R. A. Vaia, et al. (2005). "Metal nanoshell assembly on a virus bioscaffold." *Nano Lett* **5**(6): 1187–1191.

Reuveni, T., M. Motiei, et al. (2011). "Targeted gold nanoparticles enable molecular CT imaging of cancer: An in vivo study." *Int J Nanomed* **6**: 2859–2864.

Riley, R. S., E. S. Day (2017). "Gold nanoparticle-mediated photothermal therapy: Applications and opportunities for multimodal cancer treatment." *Wiley Interdiscip Rev Nanomed Nanobiotechnol* **9**(4): e1449.

Riss, T. L., R. A. Moravec, et al. (2004). Cell viability assays. Assay Guidance Manual. G. S. Sittampalam, N. P. Coussens, K. Brimacombe, et al. Bethesda, MD.

Rojanathanes, R., A. Sereemaspun, et al. (2008). "Gold nanoparticle as an alternative tool for a urine pregnancy test." *Taiwan J Obstet Gynecol* **47**(3): 296–299.

Root, S. W., G. A. Andrews, et al. (1954). "The distribution and radiation effects of intravenously administered colloidal Au198 in man." *Cancer* **7**(5): 856–866.

Russell, M. A., M. Langley, et al. (1997). "Lichenoid dermatitis after consumption of gold-containing liquor." *J Am Acad Dermatol* **36**(5): 841–844.

Ryou, S. M., S. Kim, et al. (2010). "Delivery of shRNA using gold nanoparticle-DNA oligonucleotide conjugates as a universal carrier." *Biochem Biophys Res Commun* **398**(3): 542–546.

Sabella, S., V. Brunetti, et al. (2011). "Toxicity of citrate-capped AuNPs: An in vitro and in vivo assessment." *J Nanopart Res* **13**(12): 6821–6835.

Salem, D. S., M. A. Sliem, et al. (2018). "Improved chemo-photothermal therapy of hepatocellular carcinoma using chitosan-coated gold nanoparticles." *J Photochem Photobiol B* **182**: 92–99.

Schlucker, S. (2009). "SERS microscopy: Nanoparticle probes and biomedical applications." *ChemPhysChem* **10**(9–10): 1344–1354.

Schutz M., C. I. Muller, et al. (2011). "Design and synthesis of Raman reporter molecules for tissue imaging by immuno-SERS microscopy." *J Biophotonics* **4**(6): 453–463.

Shao, J., R. J. Griffin, et al. (2013). "Photothermal nanodrugs: Potential of TNF-gold nanospheres for cancer theranostics." *Sci Rep* **3**: 1293.

Shenoi, M. M., I. Iltis, et al. (2013). "Nanoparticle delivered vascular disrupting agents (VDAs): Use of TNF-alpha conjugated gold nanoparticles for multimodal cancer therapy." *Mol Pharm* **10**(5): 1683–1694.

Silva, C. O., J. O. Pinho, et al. (2019). "Current trends in cancer nanotheranostics: Metallic, polymeric, and lipid-based systems." *Pharmaceutics* **11**(1): 22.

Singh, P., S. Pandit, et al. (2018). "Gold nanoparticles in diagnostics and therapeutics for human cancer." *Int J Mol Sci* **19**(7): E1979.

Sokolov, K., M. Follen, et al. (2003). "Real-time vital optical imaging of precancer using anti-epidermal growth factor receptor antibodies conjugated to gold nanoparticles." *Cancer Res* **63**(9): 1999–2004.

Stern, J. M., J. Stanfield, et al. (2007). "Efficacy of laser-activated gold nanoshells in ablating prostate cancer cells in vitro." *J Endourol* **21**(8): 939–943.

Stern, J. M., V. V. Kibanov Solomonov, et al. (2016). "Initial evaluation of the safety of nanoshell-directed photothermal therapy in the treatment of prostate disease." *Int J Toxicol* **35**(1): 38–46.

Stöber, W., A. Fink, et al. (1968). "Controlled growth of monodisperse silica spheres in the micron size range." *J Colloid Interface Sci* **26**(1): 62–69.

Subramani, P. A., K. Panati, et al. (2017). "Curcumin nanotechnologies and its anticancer activity." *Nutr Cancer* **69**(3): 381–393.

Suzuki, D., H. Kawaguchi (2005). "Gold nanoparticle localization at the core surface by using thermosensitive core-shell particles as a template." *Langmuir* **21**(25): 12016–12024.

Tong, L., Y. Zhao, et al. (2007). "Gold nanorods mediate tumor cell death by compromising membrane integrity." *Adv Mater* **19**: 3136–3141.

Vinay, D. S., E. P. Ryan, et al. (2015). "Immune evasion in cancer: Mechanistic basis and therapeutic strategies." *Semin Cancer Biol* **35** Suppl: S185–S198.

Wang, J., T. Murakami, et al. (2001). "Gene gun-mediated oral mucosal transfer of interleukin 12 cDNA coupled with an irradiated melanoma vaccine in a hamster model: Successful treatment of oral melanoma and distant skin lesion." *Cancer Gene Ther* **8**(10): 705–712.

Wang, P., Q. Wu, et al. (2018). "Evaluating cellular uptake of gold nanoparticles in HL-7702 and HepG2 cells for plasmonic photothermal therapy." *Nanomedicine (Lond)* **13**(18): 2245–2259.

Wang, R., Y. Luo, et al. (2016). "Hyaluronic acid-modified manganese-chelated dendrimer-entrapped gold nanoparticles for the targeted CT/MR dual-mode imaging of hepatocellular carcinoma." *Sci Rep* **6**: 33844.

Wei, M., N. Chen, et al. (2012). "Polyvalent immunostimulatory nanoagents with self-assembled CpG oligonucleotide-conjugated gold nanoparticles." *Angew Chem Int Ed Engl* **51**(5): 1202–1206.

Wieder, M. E., D. C. Hone, et al. (2006). "Intracellular photodynamic therapy with photosensitizer-nanoparticle conjugates: Cancer therapy using a 'Trojan horse'." *Photochem Photobiol Sci* **5**(8): 727–734.

Xu, X. (2018). "Potentials and pitfalls of gold-silica nanoshell as the exogenous contrast agent for optical diagnosis of cancers: A numerical parametric study." *Lasers Med Sci* **34**: 615–628.

Xue, H. Y., Y. Liu, et al. (2016). "Gold nanoparticles delivered miR-375 for treatment of hepatocellular carcinoma." *Oncotarget* **7**(52): 86675–86686.

Zavaleta, C. L., B. R. Smith, et al. (2009). "Multiplexed imaging of surface enhanced Raman scattering nanotags in living mice using noninvasive Raman spectroscopy." *Proc Natl Acad Sci U S A* **106**(32): 13511–13516.

Zhang, Y., H. Hong, et al. (2011). "Molecular imaging with SERS-active nanoparticles." *Small* **7**(23): 3261–3269.

Zharov, V. P., E. N. Galitovskaya, et al. (2005a). "Synergistic enhancement of selective nanophotothermolysis with gold nanoclusters: Potential for cancer therapy." *Lasers Surg Med* **37**(3): 219–226.

Zharov, V. P., V. Galitovsky, et al. (2003). "Photothermal detection of local thermal effects during selective nanophotothermolysis." *Appl Phys Lett* **83**(24): 4897–4899.

Zharov, V. P., J. W. Kim, et al. (2005b). "Self-assembling nanoclusters in living systems: Application for integrated photothermal nanodiagnostics and nanotherapy." *Nanomedicine* **1**(4): 326–345.

Zharov, V. P., K. E. Mercer, et al. (2006). "Photothermal nanotherapeutics and nanodiagnostics for selective killing of bacteria targeted with gold nanoparticles." *Biophys J* **90**(2): 619–627.

17 Graphene

A Revolutionary 2D Nanomaterial Heralding a New Horizon for Diagnostics and Therapeutic Applications

Kunal Biswas, Debashis De, and Jaya Bandyopadhyay

CONTENTS

17.1 Introduction and Background 205
17.2 Physico-Chemical Properties of Graphene 206
17.2.1 Carrying Capacity of Graphene Materials 206
17.2.2 Release Kinetics of Graphene Materials 206
17.2.3 Biocompatible Properties of Graphene-Based Material 206
17.2.4 Surface Area 206
17.2.5 Dimensions of the Material 206
17.2.6 Surface Chemistry 207
17.2.7 Purity of Graphene-Based Nanomaterials in Context of Biological Interaction 207
17.3 Comparison between Uncoated and Coated Drug Molecules 207
17.4 Graphenic Interaction with Active Biomolecular Molecules 208
17.5 Lipid and Protein Interactions 208
17.6 Oxidative Reactions 209
17.7 Biological Degradation 209
17.8 Potential for Human Exposure 209
17.8.1 Human Respiratory Tract Accumulation and Clearance 209
17.8.2 Observed Behavior of Plate-Like Materials 210
17.9 Toxicity and Biocompatibility 210
17.10 Drug Delivery 210
17.11 Gene Delivery 211
17.12 Bio-Imaging 211
17.13 Tissue Engineering 212
17.14 Biomedical Applications 212
17.15 Conclusion 212
References 212

17.1 INTRODUCTION AND BACKGROUND

The development of nano-based products requires well-coordinated efforts to understand their myriad toxicological effects and how their potential applications relate to their multiple toxicities (Bhattacharya and Mukherjee, 2008). Graphene, first discovered in 2004, has drawn much attention for its unique physico-chemical properties and ballistic electronic transport behavior (Neto et al., 2009). Applications of graphene involve transistors, conducting polymers, battery electrodes, supercapacitors, transport batteries, etc. (Reddy et al., 2010). The application of graphene nanosheets to biological systems requires the material characterization of the graphene nanomaterials, including an exploration of their different observable physico-chemical properties and biological implications, e.g., biomedical applications and environmental health and safety issues (Bianco, 2013). The implications of the graphene-based family for biological systems are not confined to biomedical applications but also include natural exposures to human health at large. Tissue structure and biological molecules are the common underlying target platform which is shared by both biomedical and environmental and safety organizations (Cellot et al., 2011). Understanding the variations in different graphene materials would be helpful in explaining the differential responses of biological systems to the graphene material. The high surface-to-volume ratio of the graphene sheets, bearing a theoretical value of 2600 m^2 per gram (Bhattacharya and Mukherjee, 2008), helps in the binding of different biological macromolecules, which results in various diagnostic and therapeutic applications in living systems. Due to the underlying physical activation and catalytic reaction, carbon black and associated materials abides the adsorptive interferences phenomenon, which harbours intrinsic hydrophobic surfaces.

The high effective surface area of graphene-based materials accelerates several crucial biochemical interactions such as Reactive Oxygen Species (ROS) quenching reactions, antioxidant deactivation reactions, etc.

In the present study, the therapeutic and diagnostic potentialities of graphene-based materials are discussed. The significance of basic underlying physico-chemical properties along with the fundamental electronic behavior of the graphene material have been analyzed, which could help in exploring the further applications of graphene-based materials to achieve different efficacious therapies.

17.2 PHYSICO-CHEMICAL PROPERTIES OF GRAPHENE

Graphene nanomaterial varies in terms of layer number, layer composition, surface chemistry, defect density or quality of the graphene nanosheets, and composition or purity (Pirkle et al., 2011). In this way, graphene materials are analogous to carbon nanotubes, which vary in wall number, diameter, length, surface chemistry, and the amount, composition, and physical form of the metal impurities. Graphene is a material that has unique electronic properties.

Graphene is an element consisting of exfoliated materials and with different unique surface electrons which help in the various electronic behaviors required for interaction with physical and biological systems (Hashimoto et al., 2004).

17.2.1 Carrying Capacity of Graphene Materials

The endurance capacity of the graphene vector is illustrated using the mechanical properties of the basic properties of carbon molecules (Novoselov et al., 2004). The load-bearing activity of graphene sheets is illustrated using the transport and stress-and-strain relations within the system. The fundamental relation between the load-bearing capacity of the graphene sheets and the surface attachments of different molecules over the vector enables the transport phenomenon of targeted drug delivery.

17.2.2 Release Kinetics of Graphene Materials

Graphene is used as a nanosized vehicle to transport cargo molecules. The transport behavior of the cargo molecules could also be related to the release kinetics of the entrapped molecule (Lazaridou et al., 2015). As the congestion of the cargo molecules entrapped inside the vector molecules prepared with graphene sheets increases, so does the frequency of release of the entrapped molecules from the vector, thus producing the targeted delivery mechanisms of the release systems. The entrapped cargo inside the vector could be designed by configuring the payload's intrinsic behavior. The nature of entrapment inside the vector dictates the release kinetics of the entrapped system.

17.2.3 Biocompatible Properties of Graphene-Based Material

Surface area, lateral dimensions (Wang et al., 2018), surface chemistry, and the intrinsic purity of the graphene materials dictate the overall biocompatible attributes of the graphene family nanomaterials (GFNs).

17.2.4 Surface Area

Biological and other biomedical interactions are basically governed by the available surface area, which typically bears a specific theoretical surface area of 2600 m^2/gram (Wang et al., 2018). Significant amounts of concentrated atoms are prevalent over nanomaterials of less than 100 nm. This huge concentration of surface atoms over the materials' surface enables an augmented chance of interaction between the nano-bio composites. Monolayer graphene bears more surface atoms than its non-monolayer counterparts, resulting in the formation of robust interaction between the graphene and the biological entity. As stated previously, the theoretical surface area of the graphene nanosheets seems to be ~2600 m^2/gram. This surface area seems to be higher than that of the other classes of nanomaterials studied in biological systems. Most of the surfaces of the graphene family nanomaterials are modulated by surface roughness and levels of undulations. The surface roughness of the graphene materials enables the more reactive nature of the graphene sheets towards biological systems. The surface area of the graphene sheets decreases with the increase in the layer number of graphene sheets. The stability of the nanomaterials is dictated by its inherent nano-dimension that is directly correlated with the available surface area of the nanomaterials. The primary particle surface area of the aggregated nanomaterials remains in the medium irrespective of the net aggregation of the accumulated particles bearing higher dimensions.

17.2.5 Dimensions of the Material

Several biological phenomena like cellular uptake and blood-brain barrier require the maximum dimensions of the materials and their particle sizes, which are dependent on the lateral dimension of the GFNs (Katsnelson and Novoselov, 2007) without affecting the specific surface area of the nanomaterials per se. Material uptake in biological systems depends on the lateral dimension, which represents the magnitude of endocytosis or phagocytosis in biological systems. Receptors existing among cells aid in the mediation of the uptake of the materials governed by the lateral size dimensions of the treated material. It is clearly understood that the lateral size of the graphene materials (Kim and Castro Neto, 2008) spans the size of the nanodimensional graphene oxide nanosheet, that lies within the range of approximate size of most of the proteins to that of most of the cells, i.e., 15–20 μm. There are potential biological threats if such GFNs possessing nano-dimensional particles get internalized by the cells upon contact. Layer numbering of the GFNs (Katsnelson and Novoselov, 2007) influences the material's deformability behavior, which determines the differential cellular uptake upon contact with the material's cellular context. There is a relatively negligible impact of deformation by the smaller flakes as compared to larger flakes bearing similar layer numbering.

17.2.6 Surface Chemistry

The surface of the graphene family material bears various material chemistry properties (Katsnelson and Novoselov, 2007) even before the bio-functionalization of the graphene sheets. The various chemical profiles of the pristine structures of the graphene sheets enable different applications towards different analytes. A different class of graphene nanomaterials, viz., graphene oxides, bear surfaces which exhibit partially hydrophobic behavior, with hydrophilic regions having contact angles of 40°–50°. This renders them capable of hydrogen bonding and complexing with metal ions that contain negative charges on the edge sites associated with carboxylate groups (Sun et al., 2008; Wang et al., 2010). Defect sites and the typical edge sites of the hydrophobic pristine graphene sheets are the suitable locations for different biochemical reactions to take place (Wang et al., 2011). Graphene sheets of other forms bear oxygen defects which are produced during the oxygen removal process. It has been previously reported that biomedical work has been carried out in the aqueous phase, because of its greater dispersibility (Wang et al., 2011). During the biological interaction of the graphene sheets, the long-term stability of the graphene nanosheets in culture and saline requires further functionalization (Cheng et al., 2017). In order to have biological interactions, surfactants and other stabilizing agents are required for a better interaction in the biological specimen.

Synthetic surfactants are used for the bio-functionalization of graphene nanosheets and for the necessary preparation of bio-compatible dispersion of graphene nanomaterials.

17.2.7 Purity of Graphene-Based Nanomaterials in Context of Biological Interaction

Residual metal catalysis is not present in the typical graphene sheets, as they are not produced catalytically (Pirkle et al., 2011). It is to be noted that residual intercalating molecules like chemical additives could be entrapped and unable to be washed away even after thorough washing (Pirkle et al., 2011). The purity of the graphene sheets is necessary for systematic biological interactions and elicitation of a proper biological response.

17.3 COMPARISON BETWEEN UNCOATED AND COATED DRUG MOLECULES

In order to visualize the effect of encapsulated drug molecules towards the target of interest, the efficiency of encapsulation is determined by the nature of coating material (Qiao et al., 2016). The release kinetics of the drug molecules is also determined by the unique physico-chemical properties of the vector molecules (Muhsin et al., 2016). The nature and kinetic release mode of drug molecules is associated with the vector capability. The complete material characterization of the vector molecules is required for understanding the toxicity of the vector molecules towards biological systems. The mechanistic toxicity of the vector molecules enables understanding of the possible effects of the encapsulating molecules towards the region of interest. Another region of interest for the carrier molecules is bio-dispersibility in lipids, resulting in direct and indirect toxicity. Long-term health effects must be given consideration while designing graphene nanomaterial drug delivery vectors. Such vectors enable the proper design of fluorescence-based bimolecular sensing along with other biological applications. The surface area of the graphene nanomaterials also plays a paramount role in their active relationship with biological systems. The materials' stiffness property also is very crucial for the design of drug delivery vectors. With an overall decrease in the number of layers present in graphene, the chances of adsorbing biological molecules greatly increase (Hou and Radadia, 2018). The mechanical rigidity of graphene matters significantly in terms of its pathological response to carbon nanotubes and fibers. Materials that are very thin like graphene oxide or monolayer graphene are vulnerable to damage by the weakest of forces, like the surface tension of water. On the contrary, materials bearing multiple layers may act as stiff bodies while interacting with cells in the living system. The material thickness or the layer number has a direct impact on the rise in the stiffness of the material. Thickness could be characterized by the deposition of thick films over the glass substrate using atomic force microscopy (AFM) or optical absorbance measurements (Lin et al., 2016). In terms of cellular uptake, renal clearance, and blood-brain barrier transport, another important aspect related to coated and uncoated drug delivery is the lateral dimensions of the graphene-based materials, which determine the optimum maximum dimension of the materials and other myriad biological phenomena that rely on the particle size and the different dimensions of the materials. The cellular uptake behavior of plate-like graphene nanomaterials is poorly understood, but the dependency on the lateral dimension seems to be exclusive to the processes of phagocytosis and endocytosis taking place inside the cells. The total number of receptors is also affected by the lateral dimension, which is required to facilitate uptake, as is the dimension of the lysosome inside, within which the payload should be packaged and internalized. In a nutshell, nano-graphene oxide sheets and their functionalized versions are effective sorbents for carrying hydrophobic compounds in biological systems, which are related to the already presented amphiphilicity of graphene oxide, with its patchwork of un-substituted and oxygenated graphenic surfaces.

17.4 GRAPHENIC INTERACTION WITH ACTIVE BIOMOLECULAR MOLECULES

Studies have determined that the most active area of research involves the DNA/RNA interaction with the graphene network. It has been reported that DNA/RNA shows a preferential interaction mechanism where single-stranded DNA (ssDNA) is intercalated over double-stranded DNA (dsDNA) molecules (Caridad et al., 2018). Additionally, attacks on the adsorbed nucleotides by nuclease enzymes could be avoided in this way by a steric mechanism and DNA intercalation by GFN bearing a small lateral dimension. Such modes of unique potential bio-interaction between two materials open up new opportunities for diverse applications. Lu et al. (2010) reported that a molecular beacon of an oligonucleotide sequence, when adsorbed onto the GO sheets, is liberated only when complementary deoxyribonucleic acid is present, indicating preferential binding of the single-stranded form in comparison to the double-stranded structure. Earlier study has shown that in the presence of complementary RNA, intracellular desorption takes place and GO helps DNA to be drastically protected against enzymatic cleavage. Analysis has also determined that nucleases are prevented from attacking DNA owing to net steric hindrance offered by DNA and also that the attack on the adsorbed-phase DNA can be avoided (Caridad et al., 2018). Other active molecules like dye-conjugated aptamers of ATP quench the fluorescence on being adsorbed onto graphene oxide until a duplex is formed between ATP and the aptameric sequence, the duplex structure disassociates to regain effective fluorescence behavior.

The selective capability of binding to GO by ssDNA and dsDNA strands along with the shielding capability of the strands from enzymes responsible for carrying out degradation make GO an attractive platform for delivery and sensing of DNA or RNA. The carrying capability and sensing performance of GO sheets get enhanced when the size dimension becomes <100 nm. Also enhanced are the potentialities of colloidal stability, dispersibility, minimal toxic value, and cellular uptake. It is further noticed that additional functionalization of GO enables its stable dispersion in biological media like saline, etc.; without this additional functionalization, GO is not easily soluble in biological media.

DNA adsorption on GO has been studied by Wu and coworkers. Adsorption seems to be enhanced in case of minute oligomers, along with high ionic strength and low pH. Also, ssDNA can be exchanged by free DNA in solution and can be comfortably desorbed by cDNA in solution but not increased by increasing temperature.

Now the question remains of what makes the preferential binding of ssDNA to GO more selective than that of dsDNA. The phenomenon was ascertained in an experimental fashion, but the role of base and GO interaction is also manifested. Various GO-based studies have been performed to understand relevant interactions taking place between GO and DNA strands. Several studies have reported that electrostatic repulsion takes place between the DNA bases and the graphenic surfaces. However, at low pH, the overall positive charge of the electrolytes can shield the negatively-charged groups of DNA and graphene, thereby causing the repulsive force to be weakened by effective protonation. This results in the intercalation of DNA strands in the graphenic structure.

Further evidence of GO attraction towards the GO surfaces is hydrogel synthesis, as illustrated by Ren et al. (2010). The researchers of this study mixed dsDNA with the GO surface and heated the mixture, which resulted in the uncoiling of the dsDNA to ssDNA strands and conversion to a hydrogel. The gelation is attributed to the non-covalent interaction of ssDNA with GO sheets, which is also evidenced by x-ray diffraction (XRD) studies. Similar study was pursued by Sheng et al. (2011), where ochratoxin A, a fungal contaminant, was detected by aptamer-based sensing. It is also reported that introduction of polyvinyl pyrrolidone (PVP) over the graphene surface increases the sensitivity of the technique.

It has been reported previously by Ren et al. (2010) that nano-sized Graphene oxide (nGO) sheets intercalate between the different base pairs of dsDNA, which are collectively known as the DNA intercalants. Ren et al. (2010) reported that in the presence of Cu^{2+} ions where it alone is not active, there is an occurrence of DNA cleavage. It could be argued that this phenomenon is taking place due to the planar architecture of the graphene sheet, which gets penetrated inside the base pairs, where the peripheral -COOH group provides the platform for Copper (Cu^{2+}) binding. The proximity of the copper ions to the DNA double helix enables the oxidative cleavage to the DNA sites (Ren et al., 2000). This breakthrough will bring the different angles of human toxicity, along with the possibilities of new discoveries of chemotherapeutic agents, to light in the near future.

17.5 LIPID AND PROTEIN INTERACTIONS

Graphene has been shown to interact with proteins and lipids in a hydrophobic manner (Demirel et al., 2018). The adsorption and quenching of two different macromolecules are aided by dye-labeled DNA on the topical surfaces of the graphene architecture. The desorption phenomenon is aided by the presence of cDNA upon the DNA exchange mode taking place in the solution of the DNA-graphene interface. The desorption phenomenon takes place at an elevated temperature of 90–95°C.

Due to the 2D structure of the graphene family nanomaterials (GFN), lipid molecules get a unique opportunity to interact in a strong way with the surfaces of these materials. The unique structural interactions between the lipid molecules and the graphene surfaces have been studied using coarse-grained molecular dynamic simulation (Wu and Yang, 2012). In the study, it was clearly noticed that the GFMs seem to penetrate inside the hydrophobic core of the lipid bilayer structure. The intra-localization effect in the lipid structure is negligible when the graphene structures have fewer layers. Such kinds of study enable understanding of the formation of stable and functional graphene–lipid hybrid structures. The fluid-gel-phase behavior of the typical lipid bilayer seems to be altered when interacted with other classes of nanomaterials. The positively charged nanomaterial seems to promote fluidity of the bilayer membrane upon electrostatic interaction with the head groups of the membranes, but in the case of negatively charged nanomaterials, this interaction seems to increase bilayer gelation. The lipid interaction with the GFMs seems to bear huge potential due to the intrinsically extended surface area of the graphene family nanomaterials. Similarly, the protein interaction phenomenon over the GFMs brings avenues for faster cellular uptake and toxic responses. The protein adsorption capacities over the GFNs are augmented due to the huge available surface area of the GFNs (Gu et al., 2015). Several interaction studies of protein and peptide with GFNs have been published, which have shown the mechanisms and different applications of the protein–GFN interaction interface. Peptide adsorption and sensing application of protease activity on GO sheets have been shown by several studies. The mechanism of fluorescently-tagged peptides were previously studied whereby the tagged peptides get eventually cleaved, leading to the detection of proteins such as thrombin and so on. There is a marked difference between the adsorption rates of the proteins and DNA molecules over the GO surfaces. Peptides are not protected from cleavages from proteases by the GO surfaces, unlike ssDNA, which is shielded by GO sheets from nuclease activity. The entire mechanism is anticipated to be related to the adsorbed states and the geometric confirmations of the enzymatically active sites. Several experimental and simulation studies were done by Zang et al., in which have shown with the single-amino-acid adsorption over GFN that positively charged amino acids and aromatic-group-bearing amino acids are easily adsorbed over the nGO sheets. Upon simulation results, using DFT studies Qing et al. have reported the adsorption of L-leucin amino acids over the graphene surfaces. Much more extensive study is required to explore the underlying phenomenon of protein–GFN interface (Qing et al., 2014).

17.6 OXIDATIVE REACTIONS

Antioxidant glutathione and its interaction with the carbon nanomaterial have been studied extensively. Different studies have shown the myriad catalytic activity of O_2-glutathione interactions that include GO sheets. The rate of glutathione activity over the GO sheets includes a few parameters like structural disorderliness in the material, the intensity of nitrogen doping over the GO structure, and the effective and available surface area of the GFN sheets. It was found that GO has a tremendous catalytic activity over living systems using the oxidative stress mechanism. Also, it was shown by Chang et al. (2011) that A549 cells were oxidatively damaged upon interaction with GO, which clearly provides an indication that GO induces oxidative stress upon confrontation with a living entity. The possible mechanism of the GO-induced cellular toxicity needs to be further explored.

17.7 BIOLOGICAL DEGRADATION

The capacity of biological durability or bio-persistence of carbon-based nanomaterials in living systems is determined by the nature and rate of resistance of carbonaceous nanomaterials to biological degradation, for example, in skin, lungs etc. Several researches have shown that enzymatic activity like that of horseradish peroxidase and hydrogen peroxidase could easily damage the intrinsic structure of GO, but that another form, reduced GO (rGO), is resistant towards such enzymatic attack (Zhang et al., 2011, 2015). Such studies help the understanding and design of GFN-based products which would be sustainable and biodegradable in the environment while bringing positive health impacts for human and other life forms.

17.8 POTENTIAL FOR HUMAN EXPOSURE

Nanomaterials are reported to be of potential risk when entering into the human body. There are various modes of entrance, including penetration through skin and lungs and through biomedical implants (Hurt et al., 2006). GFNs, however, are unique textured powdered material consisting of ultrathin particles bearing a high aspect ratio, lateral thinness, and high surface area, and substantial reports on adverse health effects are lacking for GFNs in human exposure.

17.8.1 Human Respiratory Tract Accumulation and Clearance

The mechanism of exhalation encompasses mucociliary movements in the respiratory system. Retention time and exposure depend upon the site of GFN deposition and how they interact with the airway surfaces (Su et al., 2016). When GFNs enter the human body, the pathway of deposition follows a few typical locations, like the mouth, throat, larynx, upper respiratory tract, etc. The deposition mechanism follows settling, inertia of deposition, and diffusivity in five regions of the human respiratory system (Su et al., 2016). The mechanism of exhalation involves the muco-ciliary movements. For example, deposited particles get faster clearance if they lie in the conducting airways of the tract, which enables an augmented rate of particle clearance by cough and ciliary movements. If the nanomaterial or GFN particles lie in the respiratory tract deeper in the lungs, then their retention time is increased, leading to difficulties in clearing the particles from the body (Su et al., 2016). It would be crucial to diagnose and monitor the pathways of particle deposition into the conducting pathways or other regions of the upper thoracic region for designing an effective release system for the ingested material. Goblet cells and serous cells, among others, are responsible for the generation of mucus, which protects the epithelial cells of the respiratory tract from incoming foreign bodies. Ingested material of aerosolic size dimensions is usually exhaled out from the body by the typical ciliary motions of the lining cells of the respiratory tract. The epithelium is responsible for the generation of mucus (Zhao et al., 2017), which is a structural and secretory product of the epithelium of the respiratory tract.

Due to the presence of glycan coating outside the mucin, the outer side of the mucin becomes the negatively charged structure. However, due to the presence of hydrophilic and glycosylated structures interspersing the mucin structure, there is a net formation of a positive–negative alternating structure (Zhao et al., 2017). The higher efficiency of mucus entrapment is illustrated by the underlying positive and negative domains in its native structure.

Low-affinity bonds can easily be formed on any surface of an interacting particle due to the presence of such alternating hydrophilic and hydrophobic domains in the native mucus structure (Zhao et al., 2017). The coordinated and rhythmic beating of the upper and lower respiratory cilia clears the mucus blanket and the external pathogenic debris from the human body (Su et al., 2016).

Due to their high aspect ratio and flattened structure, GFNs substantially differ from ingested spherical particles in terms of the basic phenomena of drag force and retention capacity. The exhalation of particles of different dimensions and sizes follows a typical dynamical shape factor, which enables the particle removal from the human body (Su et al., 2016). For GFNs, there has been scarce study of aerodynamic behavior and respiratory deposition patterns. Due to the anisotropic behavior in the intrinsic structure of the GFNs, there are ongoing studies of oblate structured material and the effects of shape on the movement of GFNs from the human respiratory system. GFNs from the transverse direction, owing to their ultrathin properties, resulted in minimized aerodynamic diameters compared to what was seen for their lateral dimensions (Su et al., 2016).

The effective size of the GFNs, i.e., their plate-like structure and crumpling nature, will determine their effective orientation and dynamics in the respiratory tract and the nature of their depositing pattern (Schinwald et al., 2012). Such size and shape dependency of the GFNs dictates the overall efficiency of the mucilary removal aided by ciliary movements from the respiratory tract upon ingestion. Upon ingestion of the GFNs into the respiratory tract, it is pretty clear that the ingested materials are required to undergo three clearance mechanisms in the respiratory tract, i.e., mucus, cilia, and epithelium, which must function in a coordinated manner for effective clearance of the ingested GFNs from the body. Effective clearance is required from the alveolar region of the lungs as well; the alveolar region is predominantly a gaseous functional region, so entrapment in the alveoli will lead to engulfment of the GFNs into the macrophagic clearance, which will take months for clearance. It is also reported that the mucus pore size can also be increased due to the adherence of hydrophobic materials like GFNs over the mucus layer. Microbial vulnerability would also be increased whenever ingested hydrophobic GFNs get enlarged in the respiratory systems, leading to toxicity insult in the human body (Ruiz et al., 2011).

17.8.2 Observed Behavior of Plate-Like Materials

There have been several reports on inhalation of particulates by industrial workers like workers working with the asbestos, who often suffer from asbestosis (Donaldson and Poland, 2012). Also, upon chronic exposure in the lungs of humans, silicate and talc among other industrial products lead to collagen deposition, causing difficulties in the normal breathing rate (Donaldson and Poland, 2012). Similarly, in graphite workers, chronic exposure to carbon-based materials like nanotubes leads to the formation of granulomas, which are aggregations of macrophage-bearing multinucleated cells surrounding foreign particles, resulting in difficulty in normal breathing in humans (Pacurari et al., 2010).

So, several precautionary steps must be taken for effective monitoring of the ingestion of GFN-based products into the human system.

17.9 TOXICITY AND BIOCOMPATIBILITY

Unlike for other carbon-based nanomaterials, like carbon nanotubes, carbon black nanomaterials, fullerenes, etc., GFN-based biological study has been scant. How GFNs would interact with the target cells of biological systems will be explored in this review in more detail, starting from the lower prokaryotes to mammalian cell systems along with the different clinical trial studies.

The toxicity of GFNs to bacterial systems has been studied previously by Biswas et al., (2018). It was noticed that functional groups generated in the as-prepared GFNs resulted in the formation of oxidative stress in the prokaryotic model. The detrimental effect is assumed to be minimized by the effective functionalization of the bare GFNs, resulting in reduced toxicity and an enhanced biocompatibility. Several studies have reported that GO has imparted toxicity to the mammalian cell line A549. Further study is required to explore the underlying phenomena of GFN and biological systems for effective design of GFN-based therapeutic drug and carriers.

17.10 DRUG DELIVERY

Drug delivery by GFNs has been studied by several researchers (Liu et al., 2008). It has been reported that for delivering water-insoluble and anticancer drugs like doxorubicin (DOX), several of the functionalizations approached have been adopted. Some studies have revealed the use of folic acid (FA) (Huang et al., 2011), some have reported functionalization of GO with PEG, and several have used chitosan; delivery becomes more effective and robust upon bio-functionalization of the GFNs, especially GO sheets (Liu et al., 2008).

GFNs remains stable and highly dispersible in biological suspensions and media like serum, media, phosphate-buffered saline (PBS), etc., where the magnitude of toxicity of bare GFNs becomes negligible as compared to un-functionalized GFNs. It is also been studied that the drug delivery mechanism gets enhanced by adhering and loading hydrophobic drugs into the GO sheets in a controlled fashion, by increasing the overall concentration of the loaded drug and GO sheet composites in physiological suspensions (Rana et al., 2011; Zhang et al., 2010). The mechanism of adherence over the GO sheets for the hydrophobic drug is achieved by the effective π–π interactions and the resulting hydrophobic forces prevalent between the two composites. It has been also reported that the carried drug molecules try to be released effectively in acidic lysosomes if they are adhered in a low-pH environment, which finally leads to their higher solubility and effective endocytosis as well as slower adsorption onto the GO sheets as a vector agent. GFN material, especially GO nanosheets, can hold the hydrophilic drug molecules more effectively than they can hold the hydrophobic drug molecules in normal physiological situations and under lower acidic conditions; based upon this, an effective drug delivery vehicle could be designed using GFNs (Zhang et al., 2010).

For the sparingly soluble drug molecules in the graphene-based vector materials, one previous study demonstrates the camptothecin (CPT)-loaded PEGylated graphene oxide carrier molecule (Liu et al., 2008). High solubility of the loaded drug along with higher magnitude of cytotoxicity has been exhibited in mammalian cell line HCT-116. The composite seems to be exponentially more potent than the CPT alone. In another report, PEGylated GO has been studied for the delivery of drug molecule rituxan using CD20 and antibody composites. Also, *in vitro* analyses of DOX-loaded PEGylated GO has been explored previously by authors (Sun et al., 2008).

Several studies have reported that gelatin is also a potential functionalizing compound which, when used with graphene, helps in the reduction of the bare graphene's toxic attributes in living systems. Such functionalized GFNs enable more drug dispersibility and higher release rate as compared to the bare graphene treated as the delivery vector for drugs like ibuprofen, etc., thereby simultaneously achieving a targeted drug delivery approach as well. Several cell lines like HepG2 have exhibited a higher response of cellular toxicity towards the chitosan-GO-CPT complex drug delivery vector (Liu et al., 2008). Besides chitosan as a composite material with GO sheets, iron- and cobalt-based compounds were also complexed with the GO sheets for achieving a controlled pH-triggered delivery mechanism (Yang et al., 2009). Also, controlled drug delivery is achieved with magnetically driven composites of iron-cobalt-GO complex, which have found applications ranging from phototherapy to near-infrared (NIR) therapy and in MRI studies (Sherlock et al., 2011). Simultaneous loading of doxorubicin and CPT was also examined in several studies (Zhang et al., 2010), which showed that the combinatorial activity of the multiple drug-loaded graphene-based carrier functionalized with covalently functionalized sulfonic acid exhibited a higher rate of cellular toxicity in MCF-7 cell lines compared to the CPT or doxorubicin alone. Such studies help us to understand that graphene-based drug delivery finds tremendous applications from magnetically driven to photothermal-based delivery and bio-imaging of biological systems.

17.11 GENE DELIVERY

For many biomedically significant approaches like gene delivery, co-delivery of gene and drug, and protein delivery, extensive research with graphene has been carried out, for which a substantial amount of literature exists. In this context, the use of a cationic polymer like polyethylenimine (PEI) to functionalize graphene is one of the many ways to mediate gene delivery (Zhang et al., 2011; Chen et al., 2011; Feng et al., 2011). Owing to high electrostatic interaction between PEI and negatively charged phosphate groups of DNA and RNA, the former has been widely explored for its use as non-viral gene vector. There are several advantages to chemically modify PEI, including cell selection, augmenting transfection efficiency, and minimizing cytotoxicity. However, the use of PEI for biomedical purposes is restricted on account of its high toxicity and low biocompatibility. Earlier work by Feng et al. (2011) and Chen et al. (2011) conveys how PEI with varying molecular weights can be harnessed for gene delivery purposes by functionalizing GO with it. The cytotoxic behavior of bare PEI has thus been shown to considerably reduce upon complexing it with GO, besides confirming graphene to be a novel nano-gene delivery vehicle bearing increased transfection efficiency. Upon employing PEI-functionalized GO as a nanovector to serially deliver Bcl-2-destined siRNA and the anticancer drug DOX to cells, features like enhanced transfection efficacy coupled with reduced cytotoxicity and increased anticancer efficacy are exhibited by the composite structure. There was absolutely no cytotoxic effect of the nanocarriers on the NIH/3T3 and PC3 cells. Disruption of the cellular endosomal membrane occurred when cells in the presence of rGO were irradiated by NIR rays. As a result of NIR irradiation, a localized heating effect was imparted to the cells, aiding in the rapid release of polyplexes. This led to improved transfection efficacy, which was screened with the help of luciferase gene reporter assay in the presence of NIR irradiation. In a similar manner, gene and drug delivery is also achieved using a complex of GO and chitosan (Bao et al., 2011). Such a complex assisted in the reduction of pDNA to stable nanoscale complexes. The ratio of phosphate and nitrogen in DNA greatly influenced the efficiency of transfection in HeLa cells. The presence of non-covalent forces between the surface of amine-terminated PEGylated GO and proteins was exploited for the delivery of payloads comprising huge amounts of proteins (Shen et al., 2012). A notable example of the PEG-GO-mediated delivery is the delivery of protein kinase A and ribonuclease A to the cell cytoplasm without hydrolysis or deprivation of any intracellular biological function. In separate research, a complex of titanium substrate having alternating layers of negatively (GO–COO^-) and positively (GO–NH_3^+) charged GO nanosheets was loaded with bone morphogenic protein-2 (BMP-2), exhibiting augmented loading efficiency and intact bioactivity (La et al., 2013). The formation of new bone material was initiated in mouse in vivo after Ti-GO-BMP2 conjugate was implanted into the mouse as compared to Ti or Ti-GO or Ti-BMP2 composites. The new conjugate was thereby shown to possess effective drug-carrying and drug delivery capacities.

17.12 BIO-IMAGING

The fact that GO-IONP-PEG could be used as the T2 contrast agent for MRI application in mice with tumors in vivo was shown by Ma et al. (2012) for the first time. The aforesaid complex also proved to be beneficial while performing magnetic field–guided localized photothermal ablation of cancer cells, as the complex displays good optical density in the range of NIR to visible electromagnetic spectrum. These findings highlight the significance of biocompatible multifunctional graphene-based nanocomposites in cancer theranostics. In a separate study, Shen et al. (2012) combined PEG-functionalized GO complex with gadolinium-diethylenetriamine-pentaacetic acid–poly(diallyl dimethylammonium) chloride (Gd-DTPA-PDDA) to give rise to a multifunctional nanocomposite as an MRI probe. In order to make the conjugate more specific towards a tumor, FA (Huang et al., 2011) was used to modify the same, which was subsequently charged with DOX. The as-prepared GO-based multifunctional conjugate demonstrated outstanding therapeutic potential and efficient imaging towards tumor and freely available gadolinium cations (Gd^{3+}) respectively. It was seen that a photosensitizer named chlorin e6 (Ce6), upon being loaded onto FA-functionalized GO, began to overaccumulate in tumor cells, causing better photodynamic efficacy when irradiated with electromagnetic rays (Huang et al., 2012). A multicomponent system comprising IONP (Fe_3O_4), fluorescent near-infrared probe cyanine 5.0 NHS (Cy5), and poly(amido amine) G4 dendrimer (PAMAM-64-NH2) was developed over the surface of GO by Wate et al. (2012) for imaging cells. The mechanism of cellular uptake by GO in gold–GO nanocomposite structures was investigated by Huang et al. via surface-enhanced Raman spectroscopy (SERS) (Huang et al., 2012).

17.13 TISSUE ENGINEERING

Owing to the supreme elastic and mechanical properties and good flexible behavior of graphene-based materials, they can potentially be utilized for reinforcement scaffolds in various applications ranging from electrospun fibers to biodegradable films (Liu et al., 2016). Several studies have shown that the compressive strength and tensile strength of GO-incorporated PVA hydrogels exhibited potentially good values of 132% and 36% as compared to the strength capacity of the materials by themselves (Zhang et al., 2011). The cyto-compatibility of the composite material seems to be minimized. Murine fibrosarcoma cell line L929 cell cultures did not exhibit any potential toxicity when the huge tensile property–bearing chitosan composite graphene films were tested (Fan et al., 2010). Several studies have also shown the potentialities of a graphene-reinforced nitrogen–GO composite scaffold tested for DNA origami patterning, which will open a new era for devising nanoscale tissue engineering scaffolds with higher precision and accuracy, with applications in biomedical to bio-nanodevice fabrication (Yun et al., 2012). Several crucial neurodegenerative disorders like Alzheimer's, which is mainly caused by the formation of amyloid beta plaques in the brain region, have been shown to be controlled using graphene-based therapeutic treatment (Mahmoudi et al., 2012). Due to the enhanced surface area of the graphene, there is a typical delay in the fibrillation procedure upon the adsorption process of amyloid monomers over GO sheets, leading to a comprehensive protective effect of GO sheets and protein-coated GO sheets in the progression and treatment of Alzheimer's.

17.14 BIOMEDICAL APPLICATIONS

The graphene nanomaterial family bearing augmented effective surface areas finds applications in various biomedical fields and disciplines. For instance, while carrying out biomolecular imaging using TEM, GFN has been used extensively in the substrate material (Nair et al., 2010). Also, due to its intrinsic conductive behavior, GFN has been widely employed in studies involving neural stimulation (Heo et al., 2011). Besides such studies, GFN, on account of its naturally corrugated surface structure, finds applications in single-nucleotide DNA sequencing technique. (Nelson et al., 2010; Heerema and Dekker, 2016).

17.15 CONCLUSION

The present chapter shows the tremendous potentialities of graphene family nanomaterials. In our chapter, several nomenclatures were used in order to abbreviate graphene family nanomaterials (e.g., GO for graphene oxide, rGO for reduced graphene oxide, etc.). In the current chapter, the significance of graphene-based drug delivery mechanisms has been shown, where huge amounts of payload could be loaded inside the graphene carrier molecule. Also, the targeted gene delivery and drug delivery avenues have been shown in the chapter, which has highlighted the significance of conventional and graphene-based unconventional routes of drug delivery. Such delivery mechanisms have produced targeted delivery mechanisms that advance upon previous delivery systems that have resulted in potential side effects to healthy cells while targeting a particular target cell or organ. Then, owing to graphene's intrinsic surface corrugation and the huge quantity of natural pores in its structure, the same has been employed for several biomedical applications like DNA sequencing, bio-imaging, etc. After that, due to the intrinsic high tensile strength, flexibility, and mechanical properties of the GFNs, the same has been explored in designing tissue engineering scaffolds and in several bio-nano medical implants. In a nutshell, because of the presence of several unique attributes of physico-chemical behavior ranging from high surface area to conductivity, GFNs will be the best choice in the near future for designing materials with applications ranging from novel drug delivery vectors to biomedical appliances bearing specific responses with minimized side effects.

REFERENCES

Bao, H., Y. Pan, and Y. Ping, et al. 2011. "Chitosan-functionalized graphene oxide as a nanocarrier for drug and gene delivery." *Small* 7, no. 11: 1569–78.

Bhattacharya, R., and P. Mukherjee. 2008. "Biological properties of "naked" metal nanoparticles." *Advanced Drug Delivery Reviews* 60, no. 11: 1289–306.

Bianco, A. 2013. "Graphene: Safe or toxic? The two faces of the medal." *Angewandte Chemie International Edition* 52, no. 19: 4986–97.

Biswas, K., D. De, and J. Bandyopadhyay, et al. 2018. "Differential antibacterial response exhibited by graphene nanosheets toward gram-positive bacterium *Staphylococcus aureus*." *IET Nanobiotechnology* 12, no. 6: 733–40.

Caridad, B., J.I. Paredes, and O. Pérez–Vidal, et al. 2018. "A biosupramolecular approach to graphene: Complementary nucleotide-nucleobase combinations as enhanced stabilizers towards aqueous-phase exfoliation and functional graphene-nucleotide hydrogels." *Carbon* 129: 321–34.

Cellot, G., F.M. Toma, and Z.K. Varley, et al. 2011. "Carbon nanotube scaffolds tune synaptic strength in cultured neural circuits: Novel frontiers in nanomaterial–tissue interactions." *Journal of Neuroscience* 31, no. 36: 12945–53.

Chang, Y., S. Yang, and J. Liu, et al. 2011. "In vitro toxicity evaluation of graphene oxide on A549 cells." *Toxicology Letters* 200, no. 3: 201–10.

Chen, B., M. Liu, and L. Zhang, et al. 2011. "Polyethylenimine-functionalized graphene oxide as an efficient gene delivery vector." *Journal of Materials Chemistry* 21, no. 21: 7736–41.

Cheng, C., S. Li, and A. Thomas, et al. 2017. "Functional graphene nanomaterials based architectures: Biointeractions, fabrications, and emerging biological applications." *Chemical Reviews* 117, no. 3: 1826–914.

Demirel, M.C., M. Vural, and M. Terrones. 2018. "Composites of proteins and 2D nanomaterials." *Advanced Functional Materials* 28, no. 27: 1704990.

Donaldson, K., and C.A. Poland. 2012. "Inhaled nanoparticles and lung cancer-what we can learn from conventional particle toxicology." *Swiss Medical Weekly* 142, no. 13547: 1–9.

Fan, H., L. Wang, and K. Zhao, et al. 2010. "Fabrication, mechanical properties, and biocompatibility of graphene-reinforced chitosan composites." *Biomacromolecules* 11, no. 9: 2345–51.

Feng, L., S. Zhang, and Z. Liu. 2011. "Graphene based gene transfection." *Nanoscale* 3, no. 3: 1252–57.

Gu, Z., Z. Yang, and L. Wang, et al. 2015. "The role of basic residues in the adsorption of blood proteins onto the graphene surface." *Scientific Reports* 5: 10873.

Hashimoto, A., K. Suenaga, A. Gloter, K. Urita, and S. Iijima. 2004. "Direct evidence for atomic defects in graphene layers." *Nature* 430, no. 7002: 870.

Heerema, S.J., and C. Dekker. 2016. "Graphene nanodevices for DNA sequencing." *Nature Nanotechnology* 11, no. 2: 127.

Heo, C., J. Yoo, and S. Lee, et al. 2011. "The control of neural cell-to-cell interactions through non-contact electrical field stimulation using graphene electrodes." *Biomaterials* 32, no. 1: 19–27.

Hou, B., and A.D. Radadia. 2018. "Differential stability of biosensing proteins on transferred mono/bilayer graphene." *ACS Biomaterials Science & Engineering* 4, no. 2: 675–81.

Huang, J., C. Zong, and H. Shen, et al. 2012. "Mechanism of cellular uptake of graphene oxide studied by surface-enhanced Raman spectroscopy." *Small* 8, no. 16: 2577–2584.

Huang, P., C. Xu, and J. Lin, et al. 2011. "Folic acid-conjugated graphene oxide loaded with photosensitizers for targeting photodynamic therapy." *Theranostics* 1: 240.

Hurt, R.H., M. Monthioux, and A. Kane. 2006. "Toxicology of carbon nanomaterials: Status, trends, and perspectives on the special issue." *Carbon* 44, no. 6: 1028–33.

Katsnelson, M.I., and K.S. Novoselov. 2007. "Graphene: New bridge between condensed matter physics and quantum electrodynamics." *Solid State Communications* 143, no. 1–2: 3–13.

Kim, Eun-Ah, and A.H. Castro Neto. 2008. "Graphene as an electronic membrane." *EPL (Europhysics Letters)* 84, no. 5: 57007.

La, W.G., S. Park, and H.H. Yoon, et al. 2013. "Delivery of a therapeutic protein for bone regeneration from a substrate coated with graphene oxide." *Small* 9, no. 23: 4051–60.

Lazaridou, A., K. Kritikopoulou, and C.G. Biliaderis. 2015. "Barley β-glucan cryogels as encapsulation carriers of proteins: Impact of molecular size on thermo-mechanical and release properties." *Bioactive Carbohydrates and Dietary Fibre* 6, no. 2: 99–108.

Lin, M.W., I.I. Kravchenko, and J. Fowlkes, et al. 2016. "Thickness-dependent charge transport in few-layer MoS2 field-effect transistors." *Nanotechnology* 27, no. 16: 165203.

Liu, C., H.M. Wong, and K.W.K. Yeung, et al. 2016. "Novel electrospun polylactic acid nanocomposite fiber mats with hybrid graphene oxide and nanohydroxyapatite reinforcements having enhanced biocompatibility." *Polymers* 8, no. 8: 287.

Liu, Z., J.T. Robinson, and X. Sun, et al. 2008. "PEGylated nanographene oxide for delivery of water-insoluble cancer drugs." *Journal of the American Chemical Society* 130, no. 33: 10876–77.

Lu, C.H., J. Li, and J.J. Liu, et al. 2010. "Increasing the sensitivity and single-base mismatch selectivity of the molecular beacon using graphene oxide as the "nanoquencher." *Chemistry–A European Journal* 16, no. 16: 4889–4894.

Ma, X., H. Tao, and K. Yang, et al. 2012. "A functionalized graphene oxide-iron oxide nanocomposite for magnetically targeted drug delivery, photothermal therapy, and magnetic resonance imaging." *Nano Research* 5, no. 3: 199–212.

Mahmoudi, M., O. Akhavan, and M. Ghavami, et al. 2012. "Graphene oxide strongly inhibits amyloid beta fibrillation." *Nanoscale* 4, no. 23: 7322–25.

Muhsin, M.D.A., G. George, and K. Beagley. 2016. "Effects of chemical conjugation of L-leucine to chitosan on dispersibility and controlled release of drug from a nanoparticulate dry powder inhaler formulation." *Molecular Pharmaceutics* 13, no. 5: 1455–66.

Nair, R.R., P. Blake, and J.R. Blake, et al. 2010. "Graphene as a transparent conductive support for studying biological molecules by transmission electron microscopy." *Applied Physics Letters* 97, no. 15: 153102.

Nelson, T., B. Zhang, and O.V. Prezhdo. 2010. "Detection of nucleic acids with graphene nanopores: Ab initio characterization of a novel sequencing device." *Nano Letters* 10, no. 9: 3237–42.

Neto, A.H.C., F. Guinea, N.M.R. Peres, K.S. Novoselov, and A.K. Geim. 2009. "The electronic properties of graphene." *Reviews of Modern Physics* 81, no. 1: 109.

Novoselov, K.S., A.K. Geim, and S.V. Morozov, et al. 2004. "Electric field effect in atomically thin carbon films." *Science* 306, no. 5696: 666–69.

Pacurari, M., V. Castranova, and V. Vallyathan. 2010. "Single-and multi-wall carbon nanotubes versus asbestos: Are the carbon nanotubes a new health risk to humans?" *Journal of Toxicology and Environmental Health, Part A* 73, no. 5: 378–95.

Pirkle, A., J. Chan, and A. Venugopal, et al. 2011. "The effect of chemical residues on the physical and electrical properties of chemical vapor deposited graphene transferred to SiO2." *Applied Physics Letters* 99, no. 12: 122108.

Qiao, D., H. Liu, and L. Yu, et al. 2016. "Preparation and characterization of slow-release fertilizer encapsulated by starch-based superabsorbent polymer." *Carbohydrate Polymers* 147: 146–54.

Qing, G., S. Zhao, and Y. Xiong, et al. 2014. "Chiral effect at protein/graphene interface: A bio inspired perspective to understand amyloid formation." *Journal of the American Chemical Society* 136, no. 30: 10736–42.

Rana, V.K., M.C. Choi, and J.Y. Kong, et al. 2011. "Synthesis and drug-delivery behavior of chitosan-functionalized graphene oxide hybrid nanosheets." *Macromolecular Materials and Engineering* 296, no. 2: 131–40.

Reddy, A.L.M., A. Srivastava, S.R. Gowda, H. Gullapalli, M. Dubey, and P.M. Ajayan. 2010. "Synthesis of nitrogen-doped graphene films for lithium battery application." *ACS Nano* 4, no. 11: 6337–42.

Ren, R., P. Yang, and W. Zheng, et al. 2000. "A simple copper (II)–l-histidine system for efficient hydrolytic cleavage of DNA." *Inorganic Chemistry* 39, no. 24: 5454–63.

Ren, H., C. Wang, and J. Zhang, et al. 2010. "DNA cleavage system of nanosized graphene oxide sheets and copper ions." *ACS nano* 4, no. 12: 7169–7174.

Ruiz, O.N., K.A.S. Fernando, and B. Wang, et al. 2011. "Graphene oxide: A nonspecific enhancer of cellular growth." *ACS Nano* 5, no. 10: 8100–7.

Schinwald, A., F.A. Murphy, and A. Jones, et al. 2012. "Graphene-based nanoplatelets: A new risk to the respiratory system as a consequence of their unusual aerodynamic properties." *ACS Nano* 6, no. 1: 736–46.

Shen, A.J., D.L. Li, and X.J. Cai, et al. 2012. "Multifunctional nanocomposite based on graphene oxide for in vitro hepatocarcinoma diagnosis and treatment." *Journal of Biomedical Materials Research Part A* 100, no. 9: 2499–506.

Shen, H., M. Liu, and H. He, et al. 2012. "PEGylated graphene oxide-mediated protein delivery for cell function regulation." *ACS Applied Materials & Interfaces* 4, no. 11: 6317–23.

Sheng, L., J. Ren, and Y. Miao, et al. 2011. "PVP-coated graphene oxide for selective determination of ochratoxin A via quenching fluorescence of free aptamer." *Biosensors and Bioelectronics* 26, no. 8: 3494–3499.

Sherlock, S.P., S.M. Tabakman, and L. Xie, et al. 2011. "Photothermally enhanced drug delivery by ultrasmall multifunctional FeCo/graphitic shell nanocrystals." *ACS Nano* 5, no. 2: 1505–12.

Su, W.C., B.K. Ku, and P. Kulkarni, et al. 2016. "Deposition of graphene nanomaterial aerosols in human upper airways." *Journal of Occupational and Environmental Hygiene* 13, no. 1: 48–59.

Sun, X., Z. Liu, and K. Welsher, et al. 2008. "Nano-graphene oxide for cellular imaging and drug delivery." *Nano Research* 1, no. 3: 203–12.

Wang, D., Kou, R., Choi, D., Yang, Z., Nie, Z., Li, J., Saraf, L.V., Hu, D., Zhang, J., Graff, G.L. and Liu, J. 2010. "Ternary self-assembly of ordered metal oxide– graphene nanocomposites for electrochemical energy storage." *ACS Nano*, 4, no. 3: 1587–1595.

Wang, T., M.D.J. Quinn, and S.M. Notley. 2018. "Enhanced electrical, mechanical and thermal properties by exfoliating graphene platelets of larger lateral dimensions." *Carbon* 129: 191–98.

Wang, Y., Li, Z., Wang, J., Li, J. and Lin, Y. 2011. "Graphene and graphene oxide: Biofunctionalization and applications in biotechnology." *Trends in Biotechnology*, 29, no. 5: 205–212.

Wate, P.S., S.S. Banerjee, and A.J. Badhwar, et al. 2012. "Cellular imaging using biocompatible dendrimer-functionalized graphene oxide-based fluorescent probe anchored with magnetic nanoparticles." *Nanotechnology* 23, no. 41: 415101.

Wu, D., and X. Yang. 2012. "Coarse-grained molecular simulation of self-assembly for nonionic surfactants on graphene nanostructures." *The Journal of Physical Chemistry B* 116, no. 39: 12048–56.

Yang, X., X. Zhang, and Y. Ma, et al. 2009. "Superparamagnetic graphene oxide–Fe 3 O 4 nanoparticles hybrid for controlled targeted drug carriers." *Journal of Materials Chemistry* 19, no. 18: 2710–4.

Yun, J.M., K.N. Kim, and J.Y. Kim, et al. 2012. "DNA origami Nano patterning on chemically modified graphene." *Angewandte Chemie International Edition* 51, no. 4: 912–15.

Zhang, C., S. Chen, and P.J.J. Alvarez, et al. 2015. "Reduced graphene oxide enhances horseradish peroxidase stability by serving as radical scavenger and redox mediator." *Carbon* 94: 531–38.

Zhang, L., J. Xia, and Q. Zhao, et al. 2010. "Functional graphene oxide as a nanocarrier for controlled loading and targeted delivery of mixed anticancer drugs." *Small* 6, no. 4: 537–44.

Zhang, L., Z. Lu, and Q. Zhao, et al. 2011. "Enhanced chemotherapy efficacy by sequential delivery of siRNA and anticancer drugs using PEI-grafted graphene oxide." *Small* 7, no. 4: 460–64.

Zhang, L., Z. Wang, and C. Xu, et al. 2011. "High strength graphene oxide/polyvinyl alcohol composite hydrogels." *Journal of Materials Chemistry* 21, no. 28: 10399–406.

Zhao, J., Y. Minami, and E. Etling, et al. 2017. "Preferential generation of 15-HETE-PE induced by IL-13 regulates goblet cell differentiation in human airway epithelial cells." *American Journal of Respiratory Cell and Molecular Biology* 57, no. 6: 692–701.

18 Applications of Nanosensor System in the Detection of Heavy Metals

Arun Kumar Pradhan and Soudamini Acharya

CONTENTS

18.1 Introduction 215
18.2 Types of Nanosensors 215
18.2.1 Chemical Nanosensors 215
18.2.2 Nano-Biosensors 216
18.2.3 Optical Sensors 216
18.2.4 Electromagnetic Nanosensors 216
18.2.5 Mechanical Nanosensors 216
18.3 Impact of Heavy Metals 216
18.4 Detection of Heavy Metals 216
18.4.1 Detection of Mercury (Hg) 217
18.4.1.1 Toxic Effects of Mercury 217
18.4.1.2 Sensors Used for Detection of Mercury Vapor 217
18.5 Detection of Arsenic (As) 218
18.6 Detection of Lead (Pb) 218
18.6.1 Sensors for Detection 219
18.6.2 Lead Detection 219
18.7 Detection of Chromium (Cr) 219
18.7.1 Sensors for Detection of Chromium 219
18.8 Conclusion 220
Acknowledgment 220
References 220

18.1 INTRODUCTION

Sensors are devices that can detect signals. The signals can be biomedical, optical, electrical, physical, mechanical, or electronic. Nanosensors are tiny sensors for the nanometer scale, i.e., about 1 to 100 nm (a nanometer is equivalent to 10^{-9} m). They can detect the presence of nanomaterials or molecules of that size or even smaller and give information about their behavior and characteristics at the macroscopic level. The environment surrounding us offers much information that can be monitored by sensors. The transduction of information to various forms of energy such as optical, mechanical, thermal, electrical, biochemical, etc., is the major task that is done by the sensors. In 1995, Weisendanger defined a nanosensor as a sensor that fulfills at least one of the following criteria: (1) the sensitivity of sensor is on the nanometer scale, (2) the size of the sensor is on the nanometer scale, and (3) the spatial confinement of the interaction of the sensor with the object is on the nanometer scale.

Many institutions have achieved great research advancements toward the development of nanosensors for many diverse applications within the medical, national security, aerospace, and integrated circuit fields. Engineered nanoscale equipment has many distinct and desirable characteristics, such as improved reactivity, catalytic efficiency, electrical conductivity, vigor, and magnetic properties over bulk material of similar composition. The major advantage of the nanoscale is its relatively massive surface area. Nanosensors are instruments that are used in places where it is difficult to reach, to monitor physical and chemical phenomena. They are also used in the organelles of cells to detect biochemical phenomena and in industries and environmental assessment to measure nanoscopic particles.

18.2 TYPES OF NANOSENSORS

18.2.1 Chemical Nanosensors

Chemical sensors are so sensitive that they can detect a single particular chemical or biological molecule. Usually, very small amounts of chemical vapors are detected with these nanosensors, which include, e.g., carbon nanotubes, zinc oxide nanowires, or palladium nanoparticles (NPs).

18.2.2 Nano-Biosensors

The main components of biosensors are a biological factor, a physiochemical transducer, and a detector. These use various techniques of recognition: the physical changes they detect can be piezoelectric (i.e., measuring the difference in mass), electrochemical (the difference between electrical deliveries is measured), optical (i.e., measuring on the basis of difference in force of light), or calorimetric (difference in heat is measured). This technology helps in early cancer detection and other diseases; even specific types of DNA can be detected.

18.2.3 Optical Sensors

For on-site recognition of heavy metals, optical nanosensors are used. The working principle of optical chemical sensors is based on the fluctuation of optical properties like emission, transmission, absorption, etc., due to the attachment of an immobilized marker (i.e., an organic dye) with the analyte (Gruber et al. 2017). The design of optical nanosensors is based on the choice of suitable solid matrix, suitable functioning, and the morphology and immobilization of nanoparticles (Ding et al. 2016).

18.2.4 Electromagnetic Nanosensors

Based on their physical mechanism, electromagnetic nanosensors are of two types:

- Those that measure electrical current
- Those that measure magnetism

The advantage of measuring the electrical current is use of a marker-free method rather than the use of dyes. Geng et al. (2005) studied the relation between hydrogen sulfide and gold nanoparticles. In this category, the vital component of the nanosensors is carbon nanotubes (CNTs). The inclusion of CNTs can be done vertically or embedded to form nanocomposite electrodes (Wang and Musameh 2003a; 2003b; Valentini et al. 2003; Santhanam and Ajayan 1996). CNTs are extensively chosen in nano-electronic applications such as flat-panel displays, ultrasensitive chemical and electromechanical sensors, and hydrogen storage devices (Di et al. 2004; Rakhi, Sethupathi, and Ramaprabhu 2008; Agnihotri and Rood 2005; Tan et al. 2008). Applications of CNTs for the removal of hazardous pollutants from water bodies have been extensively studied (Gaur and Shim 2008; Chen et al. 2009). Magnetic nanoparticles have unique properties which can be used in sensing technology for the rapid detection of targets without purification of samples.

18.2.5 Mechanical Nanosensors

Binh et al. (1994) proposed the earliest mechanical nanosensor for measuring the vibration and elastic nature of nanospheres attached to a tapered cantilever. This work is essential for applications in nano-device components.

18.3 IMPACT OF HEAVY METALS

Heavy metal pollution is one of the most serious environmental issues and is a major threat to global sustainability. Due to the increase in industrial activities, heavy metals are making their way into the environment. Heavy metals are highly toxic, carcinogenic, and hazardous to human health even at trace amounts (Aragay and Merkoc 2011). The combustion of high amounts of fossil fuels leads to the introduction of heavy metals into the atmosphere and the aquatic environment. Also, this release includes wastes that are generated from industry or domestic activities, along with pesticides and fertilizers that contain metals. Natural sources that contribute to heavy metal introduction into the environment include weathering or erosion of rocks (Aragay and Merkoc 2011).

Heavy metals may be either metals or metalloids having densities (gm/cm^3) four or five times greater than water and can be present in elemental, ionic, or complex forms (Ugulu 2015). These substances are retained in the ecosystem. To prevent the phenomenon of bioaccumulation and other related health problems, the monitoring of the heavy metals should be done.

Polluted water is a great threat to human life, welfare, and the mutual existence of society and the economy. The toxicity of heavy metals is due to their soluble nature in water and their non-biodegradability, extended half-life, and accumulation in different body parts (Arora et al. 2008). Some heavy metals like iron, copper, and zinc are biologically vital for organisms but become toxic in larger concentrations. However, others like lead, mercury, and cadmium are not biologically essential and can lead to very large amounts of free radical formation at very low concentrations (Hussein et al. 2005). Many natural and artificial processes constantly add chemical toxins, heavy metals, and other organic and inorganic pollutants to water, which needs to be adequately checked to provide a constant supply of clean drinking water to people and to control the adverse effects of pollution on ecosystems. Thus, a fast, accurate, and economical monitoring of heavy metals is very much desirable for the environment.

18.4 DETECTION OF HEAVY METALS

Broadly, the recognition of heavy metals is achieved by analyzing the optical or electronic properties of the nanomaterial. In recent years, extremely sensitive and selective sensors have been developed for the recognition of very low traces of heavy metal ions from biological and environmental samples (Ban, Pratihar, and Paul 2015). The oxidation state as well as other physical and chemical factors determines the toxicity of metal ions. Fluorescent chemosensors with fluorescence enhancement and quenching for the recognition of metal ions have been extensively studied by several groups (Makwana et al. 2014, 2016; Zhang, Liu, and Ye 2012; Yin et al. 2011; Kim et al. 2012; Wang et al. 2013). The quantification of heavy metals in the environment is very much required to locate a targeted region for remediation (Ali, Khan, and Anwar 2013; Melnick and Parkin 2014). The detection of heavy metal ions includes spectroscopic techniques such as inductively coupled plasma/atomic emission spectrometry (ICP-AES), ion chromatography ultraviolet-visible spectroscopy (IC-UV-vis), capillary electrophoresis (CE), X-ray fluorescence spectroscopy (XFS), microprobes (MP) (Kievit and Zhang 2011), atomic absorption spectrometry (AAS) (Fang, Ruzika, and Hansen 1984), Synchrotron X-ray fluorescence (SXRF) spectrometry (Arai et al. 2003), microwave plasma atomic emission spectroscopy (MP-AES) (Poirier et al. 2017; Ríos et al. 2017), inductively coupled plasma mass spectrometry (ICP-MS) (Yamakawa, Moriya, and Yoshinaga 2017; Shih et al. 2016), cold vapor atomic fluorescence spectrometry (CVASF) (Ebdon et al. 2002) and thermal decomposition mercury analysis (TDMA) (Sasamoto et al. 2016).

These techniques help to achieve perfect results in determining the level of heavy metals in the environment. Most of them cannot be used for in situ study, and they require very expensive machines as well as time-consuming processes. Therefore, to detect heavy metal on site, chemical sensors are extensively used. Nanomaterials are used to sense heavy metals based on their properties of high catalytic efficiency, high surface reactivity, great surface area, and strong adsorption ability. Nowadays, carbon-based, silicon-based, and metallic nanoparticles have been used to construct nanosensors to detect heavy metals. Another advantage of nano-scale sensors is the lab-on-a-chip (LOC) concept, which is only possible due to their tininess (Borah et al. 2016; Hoang et al. 2013).

18.4.1 Detection of Mercury (Hg)

Mercury, one of the top ten elements that pose major threats for public health, is a neurotoxicant. It is usually bioavailable, and when it enters the food web as methyl mercury (MeHg), it can be magnified as it passes from one trophic level to another. Additionally, the common forms of mercury that often come out of the industrial area are inorganic mercury, mercury (II) ions, and elemental mercury. Even a little exposure to mercury can lead to neurological and behavioral disorders (Wang and Shen 2000), heart disease (Zahir et al. 2005), and enduring injury to vital organs. Introduction to >1.6 ppb (or μg/kg body weight) and >4.0 ppb of mercury leads to neurocognitive effects and kidney damage, respectively (Chu and Porcella 1995; Keebaugh et al. 2006). A recent study revealed that neurotoxic MeHg was bioaccumulated in the pelagic food web, which later on will happen to be a food source for humans. When this infected food is taken up, food poisoning occurs among living beings on the earth. Taking up of infected food is a key channel of human exposure to heavy metals, more so than breathing and dermal contact.

Current statistics reveal that inorganic mercury comes out as waste from various industries. Most of the time, we knowingly or unknowingly come into contact with mercury as we are using mercury-based products like batteries, sphygmomanometers, fluorescent lightbulbs, thermometers, skin creams, and germicidal soap.

Natural emissions of mercury occur from land and water body surfaces, but the natural content of mercury varies because of unstable environmental conditions (Slemr and Langer 1992; Nriahu 1989). Mercury is emitted from volcanic eruptions, evaporation from land and water surfaces, burning of crop residues, and forest fires (Nriahu 1989; Cinnirella, Pirrone, and Allegrini 2008; Pirrone et al. 2010; Mason 2009). The main form of mercury is elemental mercury. Recently, studies have revealed that evasion of mercury vapor is caused due to differences in temperature and mercury concentrations at the air or water interface. Also, UV in sunlight contributes to mercury emission. Mercury emission from industries is the main source of release into the atmosphere, including coal-fired electric utilities and burning of municipal waste (Streets et al. 2005; Q. Wang and Shen 2000), along with manufacturing of non-ferrous metals and cement (Pirrone et al. 2010). The emission of mercury from coal combustion is the world's major source, with an estimated release of 810 tons per year (Q. Wang and Shen 2000; Keebaugh et al. 2006; Pirrone et al. 2010; Slemr and Langer 1992). Elemental mercury (Hg^0) that is released into the atmosphere remains for 0.5 to 2.0 years due to high vapor pressure and low water solubility (Slemr and Langer 1992; Slemr, Schuster, and Seiler 1985; Lindqvist and Rodhe 2017).

18.4.1.1 Toxic Effects of Mercury

Mercury is bioavailable when it enters the food web as methyl mercury (MeHg), which magnifies as it passes up the food chain. Mercury poisoning has been observed after exposure to Hg^0 vapors due to occupational labor (Satoh 2000; Sarikaya et al. 2010). Inhalation of Hg^0 vapors exceeding 500μg/m^3 may cause toxic effects to the central nervous system (CNS), including numerous symptoms like loss of memory, fever, malaise, cough, and neurocognitive disorders (Counter and Buchanan 2004; Satoh 2000). The most toxic form of mercury which is easily absorbed by body tissue is methyl mercury (Satoh 2000). Also, MeHg accumulates in aquatic living organisms, moreover at especially high concentrations in specific fish (Dabeka et al. 2004; Palheta and Taylor 1995). Consuming those fish leads to high risk of MeHg poisoning (Feng, Li, and Qiu 2006).

18.4.1.2 Sensors Used for Detection of Mercury Vapor

Nowadays, the sensors used for monitoring mercury vapor include surface acoustic wave-based sensors (SAW-based sensor) (Caron et al. 1997; Selid et al. 2009), piezoelectric sensors, nanosensors, and polymer carbon composite sensors (Severin and Lewis 2015).

18.4.1.2.1 Surface Acoustic Wave-Based and Piezoelectric Sensors

The surface acoustic wave-based sensor provides high sensitivity to mercury vapor (Caron et al. 1997; Selid et al. 2009). The difference in oscillation frequency determines the concentration of mercury vapor. This sensor provides high sensitivity for detecting vapor mercury at parts per million (ppm) and parts per billion (ppb) levels (Selid et al. 2009). The sensing mechanism of piezoelectric sensors is based upon the lower vibration frequency of a thin quartz plate which is coated with metallic gold film on both sides within the actual sensor. These sensors provide good response in the ppm range by measuring the mercury vapor levels. Limitations of these two sensors are that they show sensitivity to mercury vapor only over a limited temperature range and that regeneration can be completed only at high temperature, i.e., 200°C (Caron et al. 1997).

18.4.1.2.2 Gold Film/Nanowire Sensors

These sensors provide enhanced sensitivity towards the analyte, consume less energy, and could be developed as components of small portable devices for continuous on-site monitoring (Kauffman et al. 2010). It has been proven that gold film sensors have standard application use in the laboratory and as well as commercially. Gold films that are approximately 15 nm thick show prominent sensitivity to mercury vapors in the ppb range at room temperature. The main disadvantage is that they cannot be regenerated after use.

18.4.1.2.3 Polymer Carbon Composite Sensors

Sensors composed of polymer carbon composite sensors have been broadly responsive to low concentrations of volatile organic compounds (VOCs). They can detect up to 10-ppb levels of mercury vapor and can be regenerated at moderate temperature for use.

18.5 DETECTION OF ARSENIC (AS)

Arsenic in water used for drinking purposes is a matter of concern which causes a large impact on human health (Smedley and Kinniburgh 2002; Kirchner et al. 1998; Thornton 1996). More than 70 countries and around 130 million people across the globe are under threat for consumption of arsenic present in contaminated water and food. About 10 parts per billion is the accepted level of arsenic in drinking water (World Health Organization 1996). It requires the proper methodologies and instrumentation for quantifying and controlling these metals to detect the level of contamination. To measure the concentration of arsenic in water, complex preparation protocols and instruments of high quality and precision are currently required (Chatterjee et al. 2000). To meet this demand, surface plasmon sensors are used to overcome the problem of high price, short portability, and capacity to execute in situ measurements. The (*nanoSens*) autonomous sensors have been prepared to meet the demand of self-operating capabilities, affordable price, portability, usefulness in making in situ measurements, and possible facility for confirmation in a group of sensors committed to simultaneously mapping and monitoring arsenic concentrations at different places.

The *nanoSens* system comprises of four modules that accomplish diverse functions:

- Opto-electronic module
- Functionalization module
- Pretreatment module
- Communication module

All these may be built into a single transferable unit or used autonomously. The opto-electronic unit utilizes the well-known Kretschmann design, which is the most frequent approach for excitation of surface plasmon by prism pairing in a whole attenuation system (Kretschmann and Raether 1968). Many systems have previously used this type of equipment to give high-sensitivity measurement solutions.

According to the Kretschmann arrangement, p-polarized light is projected through a prism on a gold substrate functionalized with a monolayer which is able to combine with the analyte of interest. The reflected light from the gold substrate (changed intensity), which is the function of the incidence angle, is calculated by a photo detector. A minimum intensity is caused by the surface plasmon resonance for a given angle. This angle, termed the resonance angle, is dependent on the index of refraction at the crossing point where the analyte which is to be sensed is pointed.

Merits of developed system:

- The entire system can be incorporated into lightweight and transportable cases, or if required, each unit can be operated autonomously.
- It is economical (< 1800 USD). In the financial plan, the expenditure of the pretreatment part has not been integrated.
- The system is capable of detecting arsenic concentrations of less than 5 ppb. Measurements made from some types of sites that the sensor can reach can be made at concentrations of as little as 1 ppb.
- The sensor communicates by means of Ethernet.
- The substrates may be frequently re-functionalized.

The Kretschmann configuration selected shows high strength, devoid of noise thresholds changing the measurement in the range of 5–45°C. The wavelength chosen for the laser, i.e., 810 nm, proved to be accurate in terms of resolution meant for the sensor, which is reflected in the reduced distribution obtained in the reflectance curve.

18.6 DETECTION OF LEAD (PB)

Lead is a poisonous metal that tends to accumulate in the body. The complexes made up of lead are very stable inside the body and do not degrade easily, leading to lead toxicity. The distinguishing consequences of lead uptake can be seen in gastrointestinal, neuromuscular, and neuropathological symptoms. These are most severe after prolonged exposure, chiefly in children under six years of age, resulting in the decrease of physical and mental development. Lead poisoning can be fatal. Many consequences like anemia (Hsieh et al. 2017), hypertension (Toscano et al. 2017), immunotoxicity, and oxidative stress occurs. Considering the above-mentioned consequences, it is clear that a disposable, economical, and transportable sensor for Pb (II) detection is very useful for human health. So, it is essential to trace small amounts of lead in the environment to avoid lead poisoning in humans.

18.6.1 Sensors for Detection

Quantum dots (QDs) are semiconductor nanocrystals that have significant impact on research in lots of areas across the physical, chemical, and biological sciences. QD research has acquired thrust with the approval of QD bio-conjugates as fluorescent labels for biological imaging and discovery (Bruchez et al. 1998). Quantum dots have a number of merits over fluorescent organic dyes (Klarreich 2001), including superior brightness, photo-stability, symmetric and narrow emission peaks, and a size-tunable emission wavelength. They consist of wide excitation series, which is helpful when multiplexing. A single excitation source can be used with many quantum dots, each possessing different emission wavelengths (Clapp et al. 2005).

To construct nanostructures, normally DNA is the best biomolecule that joins with QDs. Tertiary structures of folded single-stranded DNA with the capacity to attach to a target molecule are called aptamers. Aptamers can be used as the molecular identification factor in a biosensor. For example, the thrombin-binding aptamer (TBA) is a 15-base oligonucleotide that combines with thrombin and is able to recognize Pb^{2+} ions. This recognition is based on the configuration of a G-quartet arrangement that TBA folds into with a Pb^{2+} ion at the core (W. Liu et al. 2011). Through a chemical selection method, aptamers can work together with a wide range of targets and can be chosen for new targets (Stoltenburg, Reinemann, and Strehlitz 2007). The relationship of an aptamer and its target usually causes a conformational alteration. This conformational alteration can be optically sensed with the utilization of fluorescent QDs and quenchers. A quencher absorbs energy from a donor molecule, such as a QD. This energy transfer event is known as fluorescence resonant energy transfer (FRET), or nanometal surface energy transfer (NSET) when metallic nanoparticles are used (Jennings et al. 2006).

18.6.2 Lead Detection

The NSET method follows 1/d4 distance dependence. Thus, for more quenching, the distance between the acceptor and the donor should be shorter. By contrast, FRET is based on a 1/d6 distance dependence, which means for quenching a shorter distance is essential (Yun et al. 2005). For recognition of Pb^{2+} ions, thrombin-binding aptamer acts as an extremely selective molecular recognition element. Lately Fu et al. have verified Pb^{2+} ion detection using the thrombin-binding aptamer as the molecular recognition element and resonant Rayleigh scattering for signal transduction. This method still uses a laser for excitation, though it is label free. Quantum dots can be energized with a simple and economical LED, which makes them appropriate for fieldwork (Fu et al. 2012). By means of organic fluorescent dyes and organic quenchers, TBA has also been exploited for the optical recognition of Pb^{2+} (Liu, Huang, and Chang 2009). DNA zyme, a type of oligonucleotide, has been used with QDs to Pb^{2+} ions. Wu et al. constructed a sensor using a Pb DNAzyme, QDs, and organic quenchers. Two dissimilar organic quenchers have also been used by Wu et al. for a multiplexed sensor.

Presently, the use of gold nanoparticles as quenchers is being studied. In this work, Pb^{2+} ion recognition is carried out with the TBA combined to a QD and gold nanoparticle quencher. Organic quenchers such as Iowa Black FQ absorb light between 420–620 nm, but silver NPs (AuNPs) take up a wider range of visible light than this. This simplifies the improvement of multiplexed assays by means of QDs having distinct emission wavelengths. QDs give a high fluorescence emission peak at an exact wavelength when it is excited by UV light. A G-quartet structure is formed when TBA (5'-GGT TGG TGT GGT TGG-3'), consisting of many deoxyguanosine (G) units, acts specifically with Pb^{2+} ions. The TBA remains in free coil state, but when G-quartet structure is formed it brings the QDs and AuNPs closer to each other. As a result, quenching of QD fluorescence takes place due to the NSET method.

18.7 DETECTION OF CHROMIUM (CR)

Chromium in the metallurgic, leather-tanning, electroplating, lumber, electricity generating, and other industries has delivered huge ecological blows to numerous sites that are contaminated with it (Miró and Estela 2004). Chromium (III) is a vital element that controls glucose lipid and protein metabolism in mammals. But chromium (VI) is unquestionably extremely toxic for biological systems, having mutagenic and probable carcinogenic properties (Zevin et al. 1997). Due to constant bioaccumulative consequences to the environment and living organisms, interest is rising in the determination of dangerous substances (chromium VI) at trace levels, like heavy metals.

18.7.1 Sensors for Detection of Chromium

To successfully determine and quantify Cr (VI) in solution, various methods have been reported (Dedkova, Shvoeva, and Savvin 2001). A few of these techniques depend on spectroscopic methods (Zevin et al. 1997), while others depend on mass-sensitive devices (Bergamini et al. 2007) or electrochemical detection (Svancara, Foret, and Vytras 2004). Of the spectroscopic techniques, most of them make use of diphenylcarbazide for the colorimetric determination of Cr (VI) and adsorption of Cr (VI) at low concentration, i.e., 10 ppm (Zevin et al. 1997). Recent efforts have been made to develop the visual detection of Cr (VI) using luminescence quenching or bacterial sensors or even chromophore (diphenylcarbazide, DPC) in both homogenous and heterogeneous systems.

Exploitation of matter at the nanoscale point has led to progress in opto-chemical sensors that show high power, as well as in the identification and signaling of a large variety of chemical group. The area of optical sensor technology demands a trouble-free yet common approach to the consistent design of optical chemical sensor-based nanostructure equipment for fast, responsive, careful, inexpensive, and exact identification of a wide variety of poisonous metal ions. An optical chemical nanosensor based on compactly patterned chromogenic receptor (DPC), as a selective binding site onto three-dimensional (3D) nanoscale structures, has been developed. The synthetic nanosensor can be used for an efficient recognition of Cr (VI) ions at low concentrations in the range of 0.07–200 ppb (Descalzo et al. 2005).

The efficient plan for optical nanosensors is based on compactly patterned chromogenic receptor (DPC), as a selective binding site onto 3D nanoscale structures. The capacity to exactly modify the nanoscale pore surfaces with DPC probe allows efficient sensing responses as a result of Cr (VI)-DPC binding events. A key element of this design policy is that the pore surface-enhanced polarity by dense dispersion of cationic surfactant such as dodecylethyldimethylammonium bromide (DDAB) led to increased accessibility and adsorptive characteristics of the DPC probe with high intrinsic mobility systems. Such flexibility in the fabricated nanosensors led to high flux of the metal analytes throughout the probe molecules without major kinetic hindrance. Therefore, these sensors can be used for a successful recognition of Cr (VI) ions at concentrations in the range of 0.07–200 ppb (Descalzo et al. 2005).

18.8 CONCLUSION

Different types of nanosensors are available to detect different heavy metals at various concentrations. This is a great achievement for us, which is only possible due to nanotechnology. Nowadays, we should focus on the increment of efficiency of nanosensors so that they will detect metals even if present in very low concentrations. In the future, the focus should be on the use of more stable biomolecules as sensing materials. The use of biomolecules may reduce the cost of nanosensors.

ACKNOWLEDGMENT

The authors show gratitude to Siksha 'O' Anusandhan (Deemed to be University), Bhubaneswar, for providing scientific support.

REFERENCES

Agnihotri, Sandeep, and Mark J Rood. 2005. "Adsorption Equilibrium of Organic Vapors on Single-Walled Carbon Nanotubes." *Carbon* 43 (2): 2379–88. doi: 10.1016/j.carbon.2005.04.020.

Ali, Hazrat, Ezzat Khan, and Muhammad Anwar. 2013. "Chemosphere Phytoremediation of Heavy Metals — Concepts and Applications." *Chemosphere* 91 (7): 869–81. doi: 10.1016/j.chemosphere.2013.01.075.

Aragay, Gemma, and Arben Merkoc. 2011. "Recent Trends in Macro-, Micro-, and Nanomaterial-Based Tools and Strategies for Heavy-Metal Detection." *Chemical Reviews* 111: 3433–58. doi: 10.1021/cr100383r.

Arai, Y, A Lanzirotti, S Sutton, J A Davis, and D L Sparks. 2003. "Arsenic Speciation and Reactivity in Poultry Litter." *Environmental Science and Technology* 37 (18): 4083–90.

Arora, Monu, Bala Kiran, Shweta Rani, Anchal Rani, Barinder Kaur, and Neeraj Mittal. 2008. "Heavy Metal Accumulation in Vegetables Irrigated with Water from Different Sources." *Food Chemistry* 111: 811–15. doi: 10.1016/j.foodchem.2008.04.049.

Ban, D K, S K Pratihar, and S Paul. 2015. "Controlled Modification of Starch in the Synthesis of Gold Nanoparticles with Tunable Optical Properties and Their Application in Heavy Metal Sensing." *RSC Advances* 5 (99): 81554–64. doi: 10.1039/C5RA16473G.

Bergamini, F, Daniela P Santos, Maria Valnice, and Boldrin Zanoni. 2007. "Development of a Voltammetric Sensor for Chromium (VI) Determination in Wastewater Sample." *Sensors and Actuators B: Chemicalmical* 123 (2): 902–8. doi: 10.1016/j.snb.2006.10.062.

Binh, V T, A L Levanuyk, and N Garcia. 1994. "A Mechanical Nanosensor in the Gigahertz Range: Where Mechanics Meets Electronics." *Surface Science Letters* 2584 (93): L224–28.

Borah, Sandhya B D, Tanujjal Bora, Sunandan Baruah, and Joydeep Dutta. 2016. "Groundwater for Sustainable Development Heavy Metal Ion Sensing in Water Using Surface Plasmon Resonance of Metallic Nanostructures." *Groundwater for Sustainable Development* 1 (1–2): 1–11. doi: 10.1016/j.gsd.2015.12.004.

Bruchez, M, M Moronne, P Gin, S Weiss, and A P Alivisatos. 1998. "Semiconductor Nanocrystals as Fluorescent Biological Labels." *Science* 281 (5385): 2013–16.

Caron, Joshua J, Reichl B Haskell, G Libby, Carl J Freeman, and John F Vetelino. 1997. "A Surface Acoustic Wave Mercury Vapor Sensor." In Proceedings of International Frequency Control Symposium, 133–9. IEEE. doi: 10.1109/FREQ.1997.638534.

Chatterjee, Amit, Yasuyuki Shibata, Jun Yoshinaga, and Masatoshi Morita. 2000. "Determination of Arsenic Compounds by High-Performance Liquid Nitrogen-Microwave-Induced Plasma Mass Spectrometry : An Accepted Coupling." *Analytical Chemistry* 72: 4402–12.

Chen, Yanling, Bingwen Lu, Qingkai Yang, Colleen Fearns, John R Yates, and Jiing-Dwan Lee. 2009. "Combined Integrin Phosphoproteomic Analyses and SiRNA-Based Functional Screening Identified Key Regulators for Cancer Cell Adhesion and Migration." *Cancer Research* 69 (8): 3713–20. doi: 10.1158/0008-5472.CAN-08-2515.

Chu, P, and D B Porcella. 1995. "Mercury Stack Emissions from U.S. Electric Utility Power Plants." *Water, Air, & Soil Pollution* 80: 135–44.

Cinnirella, Sergio, Nicola Pirrone, and Alessia Allegrini. 2008. "Modeling Mercury Emissions from Forest Fires in the Mediterranean Region." *Environmental Fluid Mechanics* 8: 129–45. doi: 10.1007/s10652-007-9048-1.

Clapp, Aaron R, Igor L Medintz, H Tetsuo Uyeda, Brent R Fisher, Ellen R Goldman, Moungi G Bawendi, and Hedi Mattoussi. 2005. "Quantum Dot-Based Multiplexed Fluorescence Resonance Energy Transfer." *Journal of the American Chemical Society* 127 (51): 18212–21. doi: 10.1021/ja054630i.

Counter, S Allen, and Leo H Buchanan. 2004. "Mercury Exposure in Children : A Review." *Toxicology and Applied Pharmacology* 198: 209–30. doi: 10.1016/j.taap.2003.11.032.

Dabeka, R, A D Mckenzie, D S Forsyth, and H B S Conacher. 2004. "Survey of Total Mercury in Some Edible Fish and Shellfish Species Collected in Canada in 2002." *Food Additives and Contaminants* 21 (5): 434–40. doi: 10.1080/02652030410001670184.

Dedkova, V P, O P Shvoeva, and S B Savvin. 2001. "Test Determination of Cu (II), Ni (II), and Cr (VI) in a Single Sample." *Journal of Analytical Chemistry* 56 (8): 758–62. doi: 10.1023/A:1016745929604.

Descalzo, Ana B, Knut Rurack, Hardy Weisshoff, M Dolores Marcos, Pedro Amoro, Katrin Hoffmann, Juan Soto, et al. 2005. "Rational Design of a Chromo- and Fluorogenic Hybrid Chemosensor Material for the Detection of Long-Chain Carboxylates." *Journal of the American Chemical Society* 127: 184–200. doi: 10.1021/ja045683n.

Di, Ze-chao, Yan-hui Li, Zhao-kun Luan, and Ji Liang. 2004. "Adsorption of Chromium (VI) Ions from Water by Carbon Nanotubes." *Adsorption Science & Technology* 22 (6): 467–74.

Ding, Yangjun, Shasha Wang, Jinhua Li, and Lingxin Chen. 2016. "Nanomaterial-Based Optical Sensors for Mercury Ions." *Trends in Analytical Chemistry* 82: 175–90. doi: 10.1016/j.trac.2016.05.015.

Ebdon, Les, Mike E Foulkes, Le Roux, and Riansares Muñoz-olivas. 2002. "Cold Vapour Atomic Fluorescence Spectrometry and Gas Chromatography-Pyrolysis-Atomic Fluorescence Spectrometry for Routine Determination of Total and Organometallic Mercury in Food Samples." *Analyst* 127 (8): 1108–14. doi: 10.1039/b202927h.

Fang, Z, J Ruzika, and E H Hansen. 1984. "An Efficient Flow-Injection System with On-Line Ion-Exchange Preconcentration for the Determination of Trace Amounts of Heavy Metals by Atomic Absorption Spectrometry." *Analytica Chimica Acta* 164: 23–39.

Feng, Xinbin, Guanghui Li, and Guangle Qiu. 2006. "A Preliminary Study on Mercury Contamination to the Environment from Artisanal Zinc Smelting Using Indigenous Methods in Hezhang County, Guizhou, China : Part 2. Mercury Contaminations to Soil and Crop." *Science of the Total Environment* 368: 47–55. doi: 10.1016/j.scitotenv.2005.09.036.

Fu, X B, F Qu, N B Li, and H Q Luo. 2012. "A Label-Free Thrombin Binding Aptamer as a Probe for Highly Sensitive and Selective Detection of Lead (II) Ions by a Resonance Rayleigh Scattering Method †." *Analyst* 137 (5): 1097–99. doi: 10.1039/c2an15980e.

Gaur, Anshu, and Moonsub Shim. 2008. "Substrate-Enhanced O2 Adsorption and Complexity in the Raman G -Band Spectra of Individual Metallic Carbon Nanotubes." *Physical Review B* 78: 125422. doi: 10.1103/PhysRevB.78.125422.

Geng, Junfeng, Michael D R Thomas, S Shephard, and Brian F G Johnson. 2005. "Suppressed Electron Hopping in a Au Nanoparticle/H 2 S System : Development towards a H 2 S Nanosensor." *Chemical Communications* 14: 1895–97. doi: 10.1039/b418559e.

Gruber, P, M P C Marques, N Szita, and T Mayr. 2017. "Integration and Application of Optical Chemical Sensors in Microbioreactors Pia." *Lab on a Chip* 17 (16): 2693–2712. doi: 10.1039/c7lc00538e.

Hoang, Chung V, Makiko Oyama, Osamu Saito, Masakazu Aono, and Tadaaki Nagao. 2013. "Monitoring the Presence of Ionic Mercury in Environmental Water by Plasmon-Enhanced Infrared Spectroscopy." *Scientific Reports* 3 (1175): 1–6. doi: 10.1038/srep01175.

Hsieh, Nan-hung, Shun-hui Chung, Szu-chieh Chen, Wei-yu Chen, Yi-hsien Cheng, and Yi-jun Lin. 2017. "Anemia Risk in Relation to Lead Exposure in Lead-Related Manufacturing." *BMC Public Health* 17 (389). doi: 10.1186/s12889-017-4315-7.

Hussein, H, S Farag, K Kandil, and H Moawad. 2005. "Tolerance and Uptake of Heavy Metals by Pseudomonads." *Process Biochemistry* 40: 955–61. doi: 10.1016/j.procbio.2004.04.001.

Jennings, T L, J C Schlatterer, M P Singh, N L Greenbaum, and G F Strouse. 2006. "NSET Molecular Beacon Analysis of Hammerhead RNA Substrate Binding and Catalysis." *Nano Letters* 6 (7): 1318–24.

Kauffman, Douglas R, Dan C Sorescu, Daniel P Schofield, Brett L Allen, Kenneth D Jordan, and Alexander Star. 2010. "Metal-Decorated Carbon Nanotubes." *Nano Letters* 10: 958–63. doi: 10.1021/nl903888c.

Keebaugh, Shawn, A Kaan Kalkan, Wook Jun Nam, and Stephen J Fonash. 2006. "Gold Nanowires for the Detection of Elemental and Ionic Mercury." *Electrochemical and Solid-State Letters* 9 (9): 88–91. doi: 10.1149/1.2217130.

Kievit, Forrest M, and Miqin Zhang. 2011. "Surface Engineering of Iron Oxide Nanoparticles for Targeted Cancer Therapy." *Accounts of Chemical Research* 44 (10): 853–62. doi: 10.1021/ar2000277.

Kim, Ha Na, Wen Xiu Ren, Jong Seung Kim, and Juyoung Yoon. 2012. "Fluorescent and Colorimetric Sensors for Detection of Lead, Cadmium, and Mercury Ion." *Chemical Society. Reviews* 41 (8): 3210–44. doi: 10.1039/c1cs15245a.

Kirchner, James W, Anne Weil, Ross Nickson, John Mcarthur, William Burgess, and Kazi Matin Ahmed. 1998. "Arsenic Poisoning of Bangladesh Groundwater." *Nature* 395: 338.

Klarreich, Erica. 2001. "Biologists Join the Dots." *Nature* 413 (6855): 450–52. doi: 10.1038/35097256.

Kretschmann, E, and H Raether. 1968. "Notizen: Radiative Decay of Non Radiative Surface Plasmons Excited by Light." *Zeitschrift Für Naturforschung A* 23 (12). doi: 10.1515/zna-1968-1247.

Lindqvist, Oliver, and Henning Rodhe. 2017. "Atmospheric Mercury-A Review." *Tellus B: Chemical and Physical Meteorology* 37 (3): 136–56. doi: 10.3402/tellusb.v37i3.15010.

Liu, Chi-wei, Chih-ching Huang, and Huan-tsung Chang. 2009. "Highly Selective DNA-Based Sensor for Lead (II) and Mercury (II) Ions." *Analytical Chemistry* 81 (6): 2383–87. doi: 10.1021/ac8022185.

Liu, Wei, Yan Fu, Bin Zheng, Sheng Cheng, Wei Li, Tai-chu Lau, and Haojun Liang. 2011. "Kinetics and Mechanism of Conformational Changes in a G-Quadruplex of Thrombin-Binding Aptamer Induced by Pb^{2+}." *Journal of Physical Chemistry B* 115 (44): 13051–6. doi: 10.1021/jp2074489.

Makwana, Bharat A, Disha J Vyas, Keyur D Bhatt, and Savan Darji. 2016. "Novel Fluorescent Silver Nanoparticles : Sensitive and Selective Turn off Sensor for Cadmium Ions." *Applied Nanoscience* 6 (4): 555–66. doi: 10.1007/s13204-015-0459-x.

Makwana, Bharat A, Disha J Vyas, Keyur D Bhatt, Vinod K Jain, and Yadvendra K Agrawal. 2014. "Highly Stable Antibacterial Silver Nanoparticles as Selective Fluorescent Sensor for Fe^{3+} Ions." *Spectrochimica Acta Part A: Molecular And Biomolecular Spectroscopy* 134: 73–80. doi: 10.1016/j.saa.2014.05.044.

Mason, Robert P. 2009. "Mercury Emissions from Natural Processes and Their Importance in the Global Mercury Cycle." In *Mercury Fate and Transport in the Global Atmosphere: Emissions, Measurements and Models*, edited by N. Pirrone, and R. Mason, 173–91. Springer Science + Business Media, LLC. doi: 10.1007/978-0-387-93958-2.

Melnick, Jonathan G, and Gerard Parkin. 2014. "Cleaving Mercury-Alkyl Bonds: A Functional Model for Mercury Detoxification by MerB." *Science* 317 (5835): 225–7. doi: 10.1126/science.1144314.

Miró, Manuel, and José Manuel Estela. 2004. "Application of Flowing Stream Techniques to Water Analysis Part III. Metal Ions : Alkaline and Alkaline-Earth Metals, Elemental and Harmful Transition Metals, and Multielemental Analysis." *Talanta* 63: 201–23. doi: 10.1016/j.talanta.2003.10.047.

Nriahu, J O. 1989. "A Global Assessment of Natural Sources of Atmospheric Trace Metals." *Nature* 338 (6210): 47–49. doi: 10.1038/338047a0.

Palheta, Dulcideia, and Andrew Taylor. 1995. "Mercury in Environmental and Biological Samples from a Gold Mining Area in the Amazon Region of Brazil." *Science of the Total Environment* 168: 63–69.

Pirrone, N, S Cinnirella, X Feng, R B Finkelman, H R Friedli, J Leaner, R Mason, andA B Mukherjee. 2010. "Global Mercury Emissions to the Atmosphere from Anthropogenic and Natural Sources." *Atmospheric Chemistry and Physics* 10 (13): 5951–64. doi: 10.5194/acp-10-5951-2010.

Poirier, Laura, Jenny Nelson, Greg Gilleland, Steve Wall, Lidia Berhane, and Francisco A Lopez-linares. 2017. "Comparison of Preparation Methods for the Determination of Metals in Petroleum Fractions (1000°F+) by Microwave Plasma-Atomic Emission Spectroscopy." *Energy and Fuels* 31 (8): 7809–15. doi: 10.1021/acs.energyfuels.7b00654.

Rakhi, R B, K Sethupathi, and S Ramaprabhu. 2008. "Synthesis and Hydrogen Storage Properties of Carbon Nanotubes." *International Journal of Hydrogen Energy* 33: 381–86. doi: 10.1016/j.ijhydene.2007.07.037.

Ríos, SE Gallego, Gustavo A Peñuela, Claudia María, and CM Ramírez Botero. 2017. "Method Validation for the Determination of Mercury, Cadmium, Lead, Arsenic, Copper, Iron, and Zinc in Fish Through Microwave-Induced Plasma Optical Emission Spectrometry (MIP OES)." *Food Analytical Methods* 10 (10): 3407–14. doi: 10.1007/s12161-017-0908-0.

Santhanam, K S V, and P M Ajayan. 1996. "Carbon Nanotube Electrode for Oxidation of Dopamine." *Bioelectrochemistry and Bioenergetics* 41: 121–25.

Sarikaya, Sezgin, Ozgur Karcioglu, Didem Ay, Asl Cetin, Can Aktas, and Mustafa Serinken. 2010. "Acute Mercury Poisoning : A Case Report." *BMC Emergency Medicine* 10 (1): 7.

Sasamoto, Takeo, Harunori Otani, Izumi Hirayama, Masaki Hayashi, Itoko Baba, Kenji Iida, Yasuhiro Tamura, et al. 2016. "Dietary Exposure to Dioxins, Polychlorinated Biphenyls, and Heavy Metals in the Tokyo Metropolitan Area from 1999 to 2014." In Persistent Organic Chemicals in the Environment: Status and Trends in the Pacific Basin Countries I Contamination Status, 85–106. ACS Symposium Series, Washington, DC.

Satoh, Hiroshi. 2000. "Occupational and Environmental Toxicology of Mercury and Its Compounds." *Industrial Health* 38: 153–64.

Selid, P D, H Xu, E M Collins, M S F Collins, and J X Zhao. 2009. "Sensing Mercury for Biomedical and Environmental Monitoring." *Sensors* 9: 5446–59. doi: 10.3390/s90705446.

Severin, Erik J, and Nathan S Lewis. 2015. "Relationships among Resonant Frequency Changes on a Coated Quartz Crystal Microbalance, Thickness Changes, and Resistance Responses of Polymer – Carbon Black Composite Chemiresistors." *Analytical Chemistry* 72: 2008–15.

Shih, Tsung-ting, Cheng-chuan Hsieh, Yu-ting Luo, Yi-an Su, and Ping-hung Chen. 2016. "A High-Throughput Solid-Phase Extraction Microchip Combined with Inductively Coupled Plasma-Mass Spectrometry for Rapid Determination of Trace Heavy Metals in Natural Water." *Analytica Chimica Acta* 916: 24–32. doi: 10.1016/j.aca.2016.02.027.

Slemr, F, and E Langer. 1992. "Increase in Global Atmospheric Concentrations of Mercury Inferred from Measurements over the Atlantic Ocean." *Nature* 355 (6359): 434–37. doi: 10.1038/355434a0.

Slemr, F, G Schuster, and W Seiler. 1985. "Distribution, Speciation, and Budget of Atmospheric Mercury." *Journal of Atmospheric Chemistry* 3: 407–34. doi: 10.3402/tellusb.v37i3.15010.

Smedley, P L, and D G Kinniburgh. 2002. "A Review of the Source, Behaviour and Distribution of Arsenic in Natural Waters." *Applied Geochemistry* 17: 517–68.

Stoltenburg, Regina, Christine Reinemann, and Beate Strehlitz. 2007. "SELEX — A (r) Evolutionary Method to Generate High-Affinity Nucleic Acid Ligands." *Biomolecular Engineering* 24: 381–403. doi: 10.1016/j.bioeng.2007.06.001.

Streets, David G, Jiming Hao, Ye Wu, Jingkun Jiang, and Melissa Chan. 2005. "Anthropogenic Mercury Emissions in China." *Atmospheric Environment* 39: 7789–7806. doi: 10.1016/j.atmosenv.2005.08.029.

Svancara, Ivan, Petr Foret, and Karel Vytras. 2004. "A Study on the Determination of Chromium as Chromate at a Carbon Paste Electrode Modified with Surfactants." *Talanta* 64: 844–52. doi: 10.1016/j.talanta.2004.03.062.

Tan, Xiaoli, Ming Fang, Changlun Chen, Shaoming Yu, and Xiangke Wang. 2008. "Counterion Effects of Nickel and Sodium Dodecylbenzene Sulfonate Adsorption to Multiwalled Carbon Nanotubes in Aqueous Solution." *Carbon* 6: 0–9. doi: 10.1016/j.carbon.2008.07.023.

Thornton, I. 1996. "Sources and Pathways of Arsenic in the Geochemical Environment : Health Implications." In *Environmental Geochemistry and Health*, edited by J. D. Appleton, R. Fuge, and G. J. H. McCall, 1996th ed., 153–61. doi: 10.1144/gsl.sp.1996.113.01.12.

Toscano, Cindy Medici, Maylla Ronacher Simões, Maria Jesus, Mercedes Salaices, and Dalton Valentim Vassallo. 2017. "Sub-Chronic Lead Exposure Produces B1-Adrenoceptor Downregulation Decreasing Arterial Pressure Reactivity in Rats." *Life Sciences* 180: 93–101. doi: 10.1016/j.lfs.2017.05.009.

Ugulu, Ilker. 2015. "Determination of Heavy Metal Accumulation in Plant Samples by Spectrometric Techniques in Turkey." *Applied Spectroscopy Reviews* 50: 113–51. doi: 10.1080/05704928.2014.935981.

Valentini, Federica, Aziz Amine, Silvia Orlanducci, Maria Letizia Terranova, Giuseppe Palleschi, Tecnologie Chimiche, Tor Vergata, and Ricerca Scientifica. 2003. "Carbon Nanotube Purification : Preparation and Characterization of Carbon Nanotube Paste Electrodes." *Analytical Chemistry* 75 (20): 5413–21. doi: 10.1021/ac0300237.

Wang, Ai-jun, Han Guo, Ming Zhang, and Dan-ling Zhou. 2013. "Sensitive and Selective Colorimetric Detection of Cadmium (II) Using Gold Nanoparticles Modified with 4-Amino-3-hydrazino-5-mercapto-1,2,4-triazole." *Microchimica Acta* 180: 1051–57. doi: 10.1007/s00604-013-1030-7.

Wang, Joseph, and Mustafa Musameh. 2003a. "Carbon Nanotube/Teflon Composite Electrochemical Sensors and Biosensors." *Analytical Chemistry* 75 (9): 2075–79. doi: 10.1021/ac030007+.

Wang, Joseph, and Mustafa Musameh. 2003b. "Enzyme-Dispersed Carbon-Nanotube Electrodes: A Needle Microsensor for Monitoring Glucose." *The Analyst* 128: 1382–85. doi: 10.1039/b309928h.

Wang, Qichao, and Wenguo Shen. 2000. "Estimation of Mercury Emission from Coal Combustion in China." *Environmental Science and Technology* 34 (13): 2711–13.

World Health Organization. 1996. "*Guidelines for Drinking Water Quality (Vol 2): Health Criteria and Other Supporting Information 2.*" World Health Organization.

Yamakawa, Akane, Katsutoshi Moriya, and Jun Yoshinaga. 2017. "Determination of Isotopic Composition of Atmospheric Mercury in Urban-Industrial and Coastal Regions of Chiba, Japan, Using Cold Vapor Multicollector Inductively Coupled Plasma Mass Spectrometry." *Chemical Geology* 448: 84–92. doi: 10.1016/j.chemgeo.2016.11.010.

Yin, Jun, Tao Wu, Jibin Song, Qian Zhang, Shiyong Liu, Rong Xu, and Hongwei Duan. 2011. "SERS-Active Nanoparticles for Sensitive and Selective Detection of Cadmium Ion (Cd^{2+})." *Chemistry of Materials* 23 (21): 4756–64.

Yun, C S, A Javier, T Jennings, M Fisher, S Hira, S Peterson, B Hopkins, N O Reich, and G F Strouse. 2005. "Nanometal Surface Energy Transfer in Optical Rulers, Breaking the FRET Barrier." *Journal of the American Chemical Society* 127 (9): 3115–19. doi: 10.1021/ja043940i.

Zahir, Farhana, Shamim J Rizwi, Soghra K Haq, and Rizwan H Khan. 2005. "Low Dose Mercury Toxicity and Human Health." *Environmental Toxicology and Pharmacology* 20: 351–60. doi: 10.1016/j.etap.2005.03.007.

Zevin, Michael, Renata Reisfeld, Ines Oehme, and Otto S Wolfbeis. 1997. "Sol-Gel-Derived Optical Coatings for Determination of Chromate." *Sensors and Actuators B: Chemical* 38–39 (1–3): 235–38. doi: 10.1016/S0925-4005(97)80210-9.

Zhang, M, YQ Liu, and BC Ye. 2012. "Colorimetric Assay for Parallel Detection of Cd^{2+}, Ni^{2+} and Co^{2+} Using Peptide-Modified Gold Nanoparticles." *Analyst* 137: 601–7. doi: 10.1039/c1an15909g.

Section 5

Nanobiotecnology: Health Safety Toxicity and Environmental Impact

19 Epigenetic Landscape–Modifying Nanoparticles

Dipta Sengupta and Debasis Nayak

CONTENTS

19.1 Introduction 227
19.2 Nanoparticles and Toxicity 228
19.2.1 DNA Methylation 228
19.2.2 DNA Methyltransferases 228
19.2.2.1 DNMT1 228
19.2.2.2 DNMT3A and DNMT3B 228
19.2.3 Histones 229
19.2.3.1 Histone Core Complex 229
19.2.3.2 Histone Modifications 229
19.3 MicroRNA 230
19.4 Classification of Nanoparticles 230
19.4.1 Organic Nanoparticles 230
19.4.2 Inorganic Nanoparticles 231
19.5 Synthesis of Nanoparticles 231
19.5.1 Physical Synthesis 231
19.5.2 Chemical Synthesis 231
19.5.3 Biological Synthesis 231
19.6 Effect of Nanoparticles on the Epigenetic Landscape 231
19.6.1 Effect of Nanoparticles on DNA Methylation 231
19.6.2 Effect of Nanoparticles on Histone Modifications 232
19.6.3 Effect of Nanoparticles in microRNA 232
19.7 Conclusion 232
References 232

19.1 INTRODUCTION

In the last era 'Epigenetics' have concerned many researchers and scientists to unravel the hidden truth of gene expression without changing the sequence in DNA. Epigenetics govern the molecular mechanism of gene expression which converts genetic information to phenotypic information. Each cells inherit the same genetic code but the expression of genes are regulated by its 'epigenetic landscape'; the term "Epigenetics" was coined by Conrad Hal Waddington (Waddington 1953) (Stern 2000). Epigenetic traits can pass through generations by mitosis or even meiosis with changing the primary DNA sequence. Recent discoveries have helped us to recognize the role of epigenetic landscape in the development of different diseases including cancer. The methylation in cytosine base of DNA, modifications of histones which changes the nucleosome dynamics and microRNA mediated genetic regulation forms the driving factors for epigenetic modifications. Nanoparticles and nanotechnology have now become synonymous with various therapeutic, diagnostic, fitness and cosmetic products. The emergence of nanotechnology in various fields may be due to their exceptional optical, magnetic and biological properties. However, the concept of nanotechnology was presented way back in 1957 by Prof. Richard Feynman, where in his famous lecture at Caltech University he proposed the idea of nanotechnology and said "there is plenty of room at the bottom". The term nanotechnology was coined in 1974 by Prof. Nario Taniguchi. Nanoparticles can be defined as very small objects that can act as a whole entity in terms of its physical, chemical and optical properties. Nanoparticles have a contrastingly different optical, mechanical, thermal and catalytic behaviour from their own bulk materials due to their increase in surface area to volume ratio. With increased application of nanoparticles starting from textiles, water and air purifiers, detergents, washing machines, paints, furniture, mobiles, television sets, ointments, sunscreens, creams, medicines, diagnostic instruments, bone cement, prosthetic materials, sutures and wound healing materials, they has invaded every household appliances and hence, the toxicity related with them cannot be ruled out (Nayak, Pradhan et al. 2015).

19.2 NANOPARTICLES AND TOXICITY

Reports by various research groups have exhibited the potential cytotoxicity and genotoxicity associated with nanoparticles both in vitro and in vivo (Satapathy, Siddharth et al. 2015, Nayak, Kumari et al. 2016). Almost 90% of toxicity studies by various research groups report the toxicity profiles of various metallic and organic nanoparticles produced via chemical, physical, or biological synthesis methods and their implication against cancerous cell lines. However, very limited research and review materials are available on the epigenetic modifications (DNA methylation, expression of histone proteins and microRNA [miRNA]) caused by various organic, inorganic, and polymeric nanoparticles.

19.2.1 DNA Methylation

In 1975 two separate papers autonomously suggested that methylation of cytosine in CpG dinucleotides may serve as an epigenetic mark in vertebrates. They proposed DNA sequence can get methylated de novo which could be inherited through cell divisions by enzymes that recognize hemi-methylated CpG palindromes and DNA binding protein recognize that and silences the gene expression. Many of these findings turned out to be correct, but the relationship between DNA methylation and gene silencing is still a Pandora's Box which is difficult to understand. DNA methylation prevents binding of transcription factors to the target sites, which eventually suppress gene expression. In contrast, unmethylated CpG promotes H3K4me3 enrichment, which facilitates gene expression. DNA methylation also plays an important factor in chromosomal integrity. A significant fraction of methylation is found in repetitive elements which prevent reactivation of endoparasitic sequences and cause chromosomal instability, translocation, and gene disruption (Esteller 2007). Non-CpG dinucleotide methylation also occurred in gene bodies of stem cells, which directly correlated with gene expression and depletion in protein binding sites and enhancers (Lister, Pelizzola et al. 2009). Non-CpG methylation decreases during differentiation but is again restored in induced pluripotent stem cells, which suggests that it plays a key role in origin and maintenance of the pluripotency state of cells (Lister, Pelizzola et al. 2009, Laurent, Wong et al. 2010). Newly identified DNA modifications are now being studied for their implications in healthy and diseased epigenetic regulation.

19.2.2 DNA Methyltransferases

DNA methylation in mammalian systems is done by catalytic activity of DNA methyltransferase (DNMT) enzymes. The DNMT family in mammalian system consists of five DNMT enzymes, which are DNMT1, DNMT2, DNMT3a, DNMT3b, and DNMT3L. Current studies show that among these DNMT1, DNMT3a, and DNMT3b have methyltransferase activity.

19.2.2.1 DNMT1

DNMT1, which works as a maintenance methyltransferase, is fundamentally essential to the DNA methylation machinery. DNMT1 accomplishes the majority of methylation copying to ensure that hemimethylated daughter strands in differentiated somatic cells faithfully maintain and propagate the proper DNA methylation pattern across successive cell generations (Kar, Deb et al. 2012). DNMT1 is the first mammalian DNA methyltransferase enzyme to be cloned and biochemically characterized (Jurkowska, Jurkowski et al. 2011). It exhibits optimal methyltransferase activity on hemimethylated DNA and localizes at the DNA replication foci during the S phase, properties that make it suitable for maintenance methylation.

The DNMT1 enzyme consists of two components; a smaller C-terminal catalytic domain and a large N-terminal domain with regulatory function. There are different motifs present in regulatory domain like a charge-rich domain that interacts with the Dmap1 transcriptional repressor, a nuclear localization signal, a PCNA (proliferating cell nuclear antigen) interacting domain, a replication foci targeting region, and a cysteine-rich Zn^{2+} binding domain of the CXXC type.

All the conserved motifs for cytosine-C5-MTases are present on the C-terminal domain of DNMT1. The C and N termini are connected via a lysine-glycine repeat hinge region (Patra and Bettuzzi 2009) (Dhe-Paganon, Syeda et al. 2011). DNMT1 is regulated by a number of intrinsic and extrinsic control points such as allosteric modulation, post-translational modifications, auto-inhibitory restraints, etc., which ensure the dynamic stability and functional competence of the enzyme (Qin, Leonhardt et al. 2011) (Kinney and Pradhan 2011). The enzyme is focused to specific genomic loci in a cell-dependent manner, and its enzymatic activity is tightly controlled to accomplish its job as a transcriptional repressor (Yokochi and Robertson 2002). DNMT1 plays a unique central role during embryogenesis for epigenetic reprogramming of germ line and zygotic lineages.

19.2.2.2 DNMT3A and DNMT3B

The DNMT3 family consists of DNMT3A and DNMT3B, which have similar amino acid sequences and other structural similarities. However, they are encoded by separate genes. Both proteins transfer methyl groups to unmethylated and hemimethylated substrates in similar fashion and without evidence of intrinsic sequence specificity beyond the CpG dinucleotide; DNMT3A has also been established to methylate CpA sites (Jurkowska, Jurkowski et al. 2011). DNMT3A and DNMT3B are highly expressed in embryonic tissues and undifferentiated embryonic stem (ES) cells and downregulated in differentiated cells. Similar to DNMT1, both DNMT3A and DNMT3B are indispensable for embryonic development in mice (Lan, Yu et al. 2010). The rare autosomal recessive human disorder known as ICF (centromere instability, immunodeficiency, and facial anomalies) syndrome is caused by point mutation in human DNMT3B.

The general architecture of both DNMT3 enzymes resembles DNMT1; each possesses an N-terminal regulatory part and a C-terminal catalytic part harboring all the conserved

C5 DNA MTase motifs. However, the N-terminal parts of DNMT1 and DNMT3A/3B are not related. In DNMT3A and DNMT3B, this section contains two defined domains: a cysteine-rich region called the ADD (ATRX-DNMT3-DNMT3L) domain, also known as the PHD (plant homeodomain) domain, and a PWWP domain. DNMT3A and DNMT3B catalytic domains share approximately 85% sequence similarity and are enzymatically active in their isolated forms, unlike DNMT1 (Hermann, Gowher et al. 2004). DNMT3A and DNMT3B mediate de novo methylation where new methyl marks are added to previously unmethylated cytosine around the time of implantation and are maintained throughout the lifespan. Another round of de novo methylation occurs during the development of the germ cells, and the methylation imprints are recognized in a gender-specific manner (Chiba, Hirasawa et al. 2008).

19.2.3 Histones

19.2.3.1 Histone Core Complex

In eukaryotic cells, DNA is packed in a regularly repeated structure which is commonly known as the nucleosomes. These nucleosomes are composed of repeated, 147-base pair (bp)-long units of DNA that are wrapped around histone octamers comprised of two copies of each histone, namely, H2A, H2B, H3, and H4 (Natoli 2010) (Deb, Kar et al. 2014). The hierarchical structure of nucleosome is a stable fundamental construction capable of expression and repression of genes by regulating the activities of enzymes that requires direct access to the DNA template, thereby regulating DNA replication, transcription, and translation that form the primary foundation of a cellular function. Depending on cell condition and stage along with DNA methylation of their cytosine bases within CpG repeats, histones are subject to numerous modifications in their random coil N-terminal tails, more than in their C-terminal tails and globular domains, which determine the access of different enzymes to the DNA template for multiple operations like replication, repair, and transcription.

19.2.3.2 Histone Modifications

The histone core itself is not enough to sustain the pattern of chromatin configuration. Chromatin modifier–arbitrated histone tail modifications are the main driver of the genetic regulation. Vincent Allfrey, in his studies in the 1960s, first showed that histones are post-translationally modified (Allfrey, Faulkner et al. 1964). In later years, it has been proven that there are a huge number of different histone post-translational modifications (PTMs). In 1997, the high-resolution X-ray structure of the nucleosome showed that PTMs could affect chromatin structure (Luger, Mader et al. 1997). The cell possesses various chromatin remodeling activities that can transform histones or move nucleosomes (Cole, Cui et al. 2016). From birth to death, a cell passes through a large genetic expression profile which is regulated by DNA and histone modifications. There are two types of chromatin in the genome, silent heterochromatin and active euchromatin. Each of these chromatin patterns associates with various sets of chromatin marks, tagged on DNA and histones where miRNAs also participate. In the early stages of embryo development in mammals, changes occur in genome-wide DNA methylation and histone modification patterns. Development of the embryo is controlled by Hox gene expression, which is involved in transcriptional regulation that maintains cell proliferation, and differentiation in the stem and progenitor cells is strictly regulated by Trithorax group (TrxG) and Polycomb group (PcG) proteins (Ringrose and Paro 2004).

It has been studied that different types of covalent, reversible PTMs occur in the histone tails of nucleosomes, which are mainly acetylation, methylation, phosphorylation, ubiquitylation, sumoylation, biotinylation, citrullination, ADP-ribosylation, and *cis–trans* isomerization (Arents and Moudrianakis 1995) (Latham and Dent 2007) (Patra, Patra et al. 2001). These PTMs take place on histone amino acid residues such as lysine (K), arginine (R), glutamic acid (E), serine (S), threonine (T), and proline (P), which are present in the tail domain or globular domain of the core histone regions (Deb, Kar et al. 2014) (Pradeepa, Grimes et al. 2016). Histone methylation mainly occurs at the K and R residues of H3 and H4 tails. Lysine residues experience mono- (K4me1), di- (K4me2), or tri- (K4me3) methylation at ε-nitrogen, whereas arginine shows either mono- or dimethylation with symmetric or asymmetric configurations (Vakoc, Sachdeva et al. 2006). According to the "histone code hypothesis," histone modifications can regulate different signaling pathways which can either activate or repress genes expression by *cis* (regulates downstream cellular activities by altering chromatin structure) or *trans* (regulates downstream cellular activities by creating a suitable binding site for effector proteins) mechanisms, or both (Hake and Allis 2006). Three histone H3 sites and one H4 methylation site are implicated in the activation of transcription marks: H3K4, H3K36, H3K79, and H4K16. H3K4me and H3K36me have been associated with transcriptional elongation. Not only lysine methylation but also H3S10 phosphorylations are related to transcriptional activation. However, the three other histone lysine methylation sites (H3K9, H3K27, and H4K20) and ubiquitylation of H2AK119 are accompanied by transcriptional repression (Kouzarides 2007) (Wang, Zhai et al. 2006). It has already been clearly established that anomalies with epigenetic events are a significant regulator of disease development and progression. The patterns of normal function are disrupted by the alterations in the function of histone-modifying complexes and subsequently disrupt the control mechanism of chromatin-based routes (Sawan and Herceg 2010).

In 2006, Bernstein et al. reported a novel chromatin modification pattern called "bivalent domains" (Bernstein, Mikkelsen et al. 2006). Bivalent domains consisting of the activating histone H3K4me3 and the repressive H3K27me3 mark is present in many promoters in ES cells. These bivalent domains are related to poise expression of developmental genes, such as transcription factors, and allow their activation when the time comes (Bernstein, Mikkelsen et al. 2006) (Voigt, Tee et al. 2013). In mammals, the germ line and global cellular development and functions are maintained and regulated by the histone modification pattern, where the silent heterochromatic structure is maintained by low levels of acetylation and high levels of H3K27, H3K9, and H4K20 methylation. Not only the modification mark but also different mark-associated protein group recruitment is associated with different structure. The recruitment of Polycomb group (PcG) proteins PC2 (a PRC1 protein complex) to H3K27me preserves the inactive state of the X chromosome (Aoto, Saitoh et al. 2008) (Mathiyalagan, Keating et al. 2014). Various studies have proven that histone PMTs play a key role in epigenetic machinery, deciding the nucleosomal architecture as well as defining chromatin function of cells.

19.3 MICRORNA

Lee and colleagues (1993) identified lin-4, the first miRNA, which participates in regulating the development timing in *Caenorhabditis elegans* (Lee, Feinbaum et al. 1993). The expression (negatively) of lin-14 mRNA is regulated by lin-4 via interaction with a complementary region in 3′UTR of lin-14. Many thousands of distinct miRNAs in the human genome have been explored with the advancement in technology (Bartel 2009). Histone modifications and DNA methylation play a substantial role in differential gene expression patterns (also regulated by miRNA). The gene expression for normal homeostasis is regulated by miRNA. When this regulation is disrupted, it causes several harmful manifestations and diseases. The miRNA genes are mainly transcribed by RNA pol-II into pri-miRNA (Rodriguez, Griffiths-Jones et al. 2004). However, the largest human miRNA cluster (C19MC) is transcribed by RNA Pol-III. Pri-mRNAs are of 1–3 Kb in length and further processed by Drosha (ribonucleases) and DiGeorge syndrome (22q11.2 deletion syndrome) critical region 8 (DGCR8) or Pasha in the nucleus, which forms 70- to 100-bp-long hairpin structures "pre-miRNA" (Lee, Ahn et al. 2003). From the nucleus, pre-miRNA is transported out of with help of exportin-5, and then processed by an RNase named Dicer. This enzyme possess two RNase III domains, dsRNA-binding domain, a helicase domain, a PAZ (Piwi– Argonaute–Zwille) domain, and a DUF283 domain (Bernstein, Caudy et al. 2001).

This PAZ domain permits the weak interaction with the 3′ end of single strand RNA (ssRNA) and 2-nucleotide at the 3′ overhang of double stranded DNA (dsDNA), which is managed by Drosha (Yan, Yan et al. 2003). The Dicer activity is dependent on this overhang. The PAZ domain of Dicer discovers the cleaved pre-miRNA by Drosha, helps the RNase III domain to chop out the stem and loop from the pre-miRNA, and further processes it into variable-length (18–25 nt) mature double-stranded miRNA (Hammond, Bernstein et al. 2000). Then, the double strand gets separated into the guide strand and the passenger strand. The guide (mature strand) gets incorporated into the RNA-induced silencing complex (RISC), while the passenger strand, usually signified as a star (miRNA*), gets degraded. The RISC complex (measure component of miRNA pathway) consists of miRNA, argonaute (Ago) proteins (Ago 1-4), and other protein factors. Ago proteins also help in miRNA biogenesis and maturation. The helicase causes the guide strand of miRNA to unwind for the target recognition and binding of miRNA into the RISC complex (Hutvagner and Zamore 2002). miRNAs principally bind to the 3′ untranslated region (3' UTR) of the target mRNAs, but current studies have revealed that they also bind to the 5' UTR of mRNA and the coding regions. It has been established that the "seed" region of 7–8 nt in the 5' end of miRNA is essential for its function (Lewis, Shih et al. 2003). Some research reports showed that the gene expression is regulated by miRNA, possibly by targeting the promoter-associated ncRNA (paRNA) and direct transcriptional silencing (Brennecke, Stark et al. 2005).

19.4 CLASSIFICATION OF NANOPARTICLES

Nanotechnology can be broadly divided into **wet nanotechnology** (deals with biological systems), **dry nanotechnology** (deals with surface chemistry and physical chemistry), and **computational nanotechnology** (deals with simulation and modeling) (Sudha, Sangeetha, Vijayalakshmi, & Barhoum, 2018). The primary approach for the synthesis of nanoparticles can be further divided into two types. The **top-down** method refers to a set of manufacturing technologies starting with a block of bulk material which is the same material as the base substrate. The regularly used top-down methods are drilling, milling, and grinding (Hoyt & Mason, 2008). The **bottom-up** method refers to the procedure of self-assembly of materials from cluster to cluster, molecule to molecule, or atom to atom on top of a base substrate. The adhesion of the surface layers to the base substrate is the most significant concern in the bottom-up approach. The most frequently used bottom-up methods are welding and riveting (Hoyt & Mason, 2008).

Based on their composition of base materials, nanoparticles are broadly divided into organic and inorganic nanoparticles (Table 19.1).

19.4.1 Organic Nanoparticles

Organic nanoparticles can be basically prepared using natural and synthetic materials. Most of the naturally organic nanomaterials, such as emulsions, lipid bodies, and proteinaceous coating in viruses, are made of organic materials. Organic nanoparticles such as liposomes, ethosomes, polymer–drug

TABLE 19.1
Types of Nanoparticles

Organic nanoparticles	Inorganic nanoparticles
Dendrimers	Metallic nanoparticles
Micelles	Quantum dots
Liposomes	Magnetic nanoparticles
Polymeric nanoparticles	

conjugates, and polymer–protein conjugates are most preferably used in the pharmaceutical sector. As most of the natural and synthetic organic nanoparticles are biodegradable in nature, they are favored as drug carriers for target-specific delivery. Further, the presence of various reactive and diverse functional groups aids in the encapsulation of various hydrophilic and hydrophobic drugs and thereby increases their efficacy by uniform and controlled release of the encapsulated drug (Jagannathan, Irvin, Blanton, & Jagannathan, 2006). Due to the formation of various non-covalent bonds such as hydrogen bonds, ionic bonds, and van der Waal's weak interactions, organic nanoparticles show diverse planes of geometry and sizes, thereby forming micelles, dendrimers, and vesicles (Lai, Daear, Löbenberg, & Prenner, 2014).

19.4.2 Inorganic Nanoparticles

In comparison to organic nanoparticles, inorganic nanoparticles are more uniform and can be modulated according to their application for biomedical usage. Most inorganic nanoparticles are formed from precipitation of their precursor inorganic salts and hence, based on the reaction conditions, they form various covalent or metallic bonds, ultimately forming a three-dimensional arrangement with fixed lattice (Rane, Kanny, Abitha, & Thomas, 2018). Organic nanoparticles are preferred for various drug delivery applications; however, inorganic nanoparticles, due to their unique physiochemical properties and diverse material and size-dependent modulation properties, are getting the attention of various pharmaceutical and biomedical branches as alternative therapeutics (Tsuzuki, 2009).

19.5 SYNTHESIS OF NANOPARTICLES

Based on the route employed, the synthesis of nanoparticles can be further divided into physical synthesis, chemical synthesis, and biological synthesis.

19.5.1 Physical Synthesis

In the physical route of synthesis, various electrical, laser, or evaporation–condensation methods are applied for the synthesis of nanoparticles. The main drawbacks of the physical route are the use of high temperature and pressure conditions, with huge consumption of electrical energy and space. Further, it also increases the temperature of the surrounding area and costs of production (Rane et al., 2018).

19.5.2 Chemical Synthesis

The chemical route of synthesis is very popular owing to its diverse range of reducing, capping, and stabilizing agents employed. Based on requirements, the size and surface modifications of the nanoparticles can be modulated to suit the specific application. The main drawbacks of the chemical route of synthesis are the use of toxic and harsh chemicals, along with the release of toxic and hazardous by-products that can increase the biohazard burden on the environment with additional purification costs before the product can be used (Ghorbani, Safekordi, Attar, & Sorkhabadi, 2011).

19.5.3 Biological Synthesis

This route is also popularly known as "green synthesis" of nanoparticle synthesis. This route is very diverse: the starting base material can be bacteria, algae, fungi, or higher plants. The microbial green synthesis route has certain drawbacks such as the maintenance of aseptic conditions, a low rate of synthesis, and additional purification steps to avoid the possibility of transferring the pathogenicity associated with the microorganisms to the nanoparticles. The use of plants as starting material has many advantages such as easy availability, diversity, large-scale production possibilities, and a spontaneous and rapid rate of synthesis. Moreover, the use of plants does not require any additional reducing, stabilizing, or capping agents and no additional hazardous by-products are formed, making the green synthesis route the most economical route for synthesizing nanoparticles (Gericke & Pinches, 2006).

19.6 EFFECT OF NANOPARTICLES ON THE EPIGENETIC LANDSCAPE

The term "epigenetics," which literally means "above genetics," describes an amalgamation of several phenomena such as DNA methylation, histone tail modifications, and expression of miRNA profiles. Epigenetic changes are basically heritable changes in terms of both phenotypes and genotypes without any modification or change in DNA sequences (Shyamasundar et al., 2015). With increases in emission of various environmentally hazardous substances due to the excessive use of nanoparticles in the textile industry, air and water purification systems, and fitness and health care items, cases of epigenetic modulation in cancers and cardiovascular and neurodegenerative disease have been reported (Kar et al., 2014).

19.6.1 Effect of Nanoparticles on DNA Methylation

DNA methylation mostly occurs in CpG dinucleotide–rich regions (at least 50% GC contained in a region of 200 bases), which are defined as CpG islands. Human gene promoters consist of 60% CpG islands, which remain unmethylated in normal cells but eventually promote differential methylation in a tissue-specific fashion during development and differentiation. DNA methylation is mainly related to gene silencing but it also plays a vital role in genomic imprint and X chromosome inactivation (Frohlich, 2012). Choi's research group for the first time reported epigenetic toxicity associated with engineered nanomaterials in MCF-7 breast cancer cells with CdTe quantum dots (QDs) (Choi, Brown, Szyf, & Maysinger, 2008). Their study exhibited expression of specific antioxidant genes and apoptosis genes, which affected DNA methylation levels at specific gene promoters as well as DNMT- and MBD-associated proteins. Another study by Gong et al. showed that on exposure of silicon dioxide on HaCaT cells, there is a reduced expression of DNMTs and MBD2, thereby inducing a global DNA hypomethylation (Gong et al., 2010). However, in an in vitro study, the exposure of copper oxide nanoparticles on murine neuronal cells did not show any changes in DNA methylation (Perreault et al., 2012). Further study on titanium dioxide (TiO_2) nanoparticles on exposure to HaCaT cells showed reduced methionine and glutathione levels, thereby causing DNA methylation deregulation (Tucci et al., 2013).

19.6.2 Effect of Nanoparticles on Histone Modifications

Research on cadmium telluride (CdTe) quantum dots on MCF-7 breast cancer cells revealed for the first time that nanoparticles can induce histone modifications by chromatin reorganization and dose-dependent decrease in acetylation of histone H3 (Choi et al., 2008). Similarly, a study on gold nanoparticle–treated fetal lung fibroblast cells showed decreased expression of histone H3 lysine27 trimethylation (H3K27me3), which is a repressor marker, when compared with a normal control group of cells (Ng et al., 2011). In another study, negatively charged thioglycolic acid–capped CdTe nanoparticles were observed to preferentially bind with positively charged histone proteins rather than to the nucleic acids, which confirms that charge modifications of nanoparticles can be utilized for targeted delivery to specific charge nuclear components (Conroy et al., 2008).

19.6.3 Effect of Nanoparticles in microRNA

During post-transcriptional changes of mRNA, miRNAs generally bind with them and control the expression profiles by supressing or degrading the target mRNAs (Sengupta et al., 2016). Among them, miR-155 is known to regulate several cellular pathways that are critical for carcinogenesis, inflammation, and cardiovascular diseases (Faraoni, Antonetti, Cardone, & Bonmassar, 2009). Various studies have stipulated the role of nanoparticles in inducing miRNA expression levels relating to cell processes including cell development and progressive diseases (Campani, De Rosa, Misso, Zarone, & Grimaldi, 2016). Gold nanoparticles, iron oxide nanoparticles, CdTe nanoparticles, and multi-walled carbon nanotubes exhibited altered expression of *PROS*-1 and miR-155, miR-183, Let-7a, miR-34s, miR-21, and miR-29a (Shyamasundar et al., 2015; Wong, Hu, & Baeg, 2017). Silver nanoparticles exhibited altered expression of various miRNAs that regulated bone mineralization in osteoblast cells, and they also targeted various genes involved in cell signaling, cellular stress, and overall degradation of various important cellular proteins (Mahmood et al., 2011).

19.7 CONCLUSION

In most of the studies, results upon treatment with nanoparticles have indicated their potentiality in inducing oxidative stress by damaging the genes responsible for maintaining the harmony between the permissible amount of oxidative stress and a harmful level, at which the reactive oxygen species (ROS) act on themselves. Although considerable research has been conducted to explore the toxicity profile of various nanoparticles such as Ag, Au, TiO_2, iron oxide, and zinc oxide nanoparticles for inducing epigenetic modulation though alteration of DNA methylation, histone acetylation, and expression profiles of miRNAs. However, much information is still missing on other types of histone modifications such as methylation, sumoylation, and phosphorylation. Hence, much research is required to explore the toxicity profiles of various nanoparticles which are important in textiles or fitness and medical industry but can be toxic with prolonged usage.

REFERENCES

Allfrey, V. G., R. Faulkner and A. E. Mirsky (1964). "Acetylation and methylation of histones and their possible role in the regulation of RNA synthesis." *Proc Natl Acad Sci U S A* 51: 786–794.

Aoto, T., N. Saitoh, Y. Sakamoto, S. Watanabe and M. Nakao (2008). "Polycomb group protein-associated chromatin is reproduced in post-mitotic G1 phase and is required for S phase progression." *J Biol Chem* 283(27): 18905–18915.

Arents, G. and E. N. Moudrianakis (1995). "The histone fold: a ubiquitous architectural motif utilized in DNA compaction and protein dimerization." *Proc Natl Acad Sci U S A* 92(24): 11170–11174.

Bartel, D. P. (2009). "MicroRNAs: target recognition and regulatory functions." *Cell* 136(2): 215–233.

Bernstein, B. E., T. S. Mikkelsen, X. Xie, M. Kamal, D. J. Huebert, J. Cuff, B. Fry, A. Meissner, M. Wernig, K. Plath, R. Jaenisch, A. Wagschal, R. Feil, S. L. Schreiber and E. S. Lander (2006). "A bivalent chromatin structure marks key developmental genes in embryonic stem cells." *Cell* 125(2): 315–326.

Bernstein, E., A. A. Caudy, S. M. Hammond and G. J. Hannon (2001). "Role for a bidentate ribonuclease in the initiation step of RNA interference." *Nature* 409(6818): 363–366.

Brennecke, J., A. Stark, R. B. Russell and S. M. Cohen (2005). "Principles of microRNA-target recognition." *PLoS Biol* 3(3): e85.

Campani, V., G. De Rosa, G. Misso, M. R. Zarone and A. Grimaldi (2016). "Lipid nanoparticles to deliver miRNA in cancer." *Curr Pharm Biotechnol* 17(8): 741–749.

Chiba, H., R. Hirasawa, M. Kaneda, Y. Amakawa, E. Li, T. Sado and H. Sasaki (2008). "De novo DNA methylation independent establishment of maternal imprint on X chromosome in mouse oocytes." *Genesis* 46(12): spc one.

Choi, A. O., S. E. Brown, M. Szyf and D. Maysinger (2008). "Quantum dot-induced epigenetic and genotoxic changes in human breast cancer cells." *J Mol Med (Berlin, Germany)*, 86(3): 291–302.

Cole, H. A., F. Cui, J. Ocampo, T. L. Burke, T. Nikitina, V. Nagarajavel, N. Kotomura, V. B. Zhurkin and D. J. Clark (2016). "Novel nucleosomal particles containing core histones and linker DNA but no histone H1." *Nucleic Acids Res* 44(2): 573–581.

Conroy, J., S. J. Byrne, Y. K. Gun'ko, Y. P. Rakovich, J. F. Donegan, A. Davies … Y. Volkov (2008). "CdTe nanoparticles display tropism to core histones and histone-rich cell organelles." *Small (Weinheim an Der Bergstrasse, Germany)*, 4(11): 2006–2015.

Deb, M., S. Kar, D. Sengupta, A. Shilpi, S. Parbin, S. K. Rath, V. A. Londhe and S. K. Patra (2014). "Chromatin dynamics: H3K4 methylation and H3 variant replacement during development and in cancer." *Cell Mol Life Sci* 71(18): 3439–3463.

Dhe-Paganon, S., F. Syeda and L. Park (2011). "DNA methyl transferase 1: regulatory mechanisms and implications in health and disease." *Int J Biochem Mol Biol* 2(1): 58–66.

Esteller, M. (2007). "Cancer epigenomics: DNA methylomes and histone-modification maps." *Nat Rev Genet* 8(4): 286–298.

Faraoni, I., F. R. Antonetti, J. Cardone and E. Bonmassar (2009). "miR-155 gene: a typical multifunctional microRNA." *Biochimica et Biophysica Acta*, 1792(6), 497–505.

Frohlich, E. (2012). "The role of surface charge in cellular uptake and cytotoxicity of medical nanoparticles." *Int J Nanomed* 7: 5577–5591.

Gericke, M. and A. Pinches (2006). "Biological synthesis of metal nanoparticles." *Hydrometallurgy* 83(1): 132–140.

Ghorbani, H. R., A. A. Safekordi, H. Attar and S. M. Sorkhabadi (2011). "Biological and non-biological methods for silver nanoparticles synthesis." *Chem Biochem Eng Q* 25(3): 317–326.

Gong, C., G. Tao, L. Yang, J. Liu, Q. Liu & Z. Zhuang (2010). "SiO(2) nanoparticles induce global genomic hypomethylation in HaCaT cells." *Biochem Biophys Res Commun* 397(3): 397–400.

Hake, S. B. and C. D. Allis (2006). "Histone H3 variants and their potential role in indexing mammalian genomes: the "H3 barcode hypothesis"." *Proc Natl Acad Sci U S A* 103(17): 6428–6435.

Hammond, S. M., E. Bernstein, D. Beach and G. J. Hannon (2000). "An RNA-directed nuclease mediates post-transcriptional gene silencing in Drosophila cells." *Nature* 404(6775): 293–296.

Hermann, A., H. Gowher and A. Jeltsch (2004). "Biochemistry and biology of mammalian DNA methyltransferases." *Cell Mol Life Sci* 61(19–20): 2571–2587.

Hoyt, V. W. and E. Mason (2008). "Nanotechnology." *J Chem Health Saf* 15(2): 10–15.

Hutvágner, G. and P. D. Zamore (2002). "A microRNA in a multiple-turnover RNAi enzyme complex." *Science* 297(5589): 2056–2060.

Jagannathan, R., G. Irvin, T. Blanton and S. Jagannathan (2006). "Organic nanoparticles: preparation, self-assembly, and properties." *Adv Funct Mater* 16(6): 747–753.

Jurkowska, R. Z., T. P. Jurkowski and A. Jeltsch (2011). "Structure and function of mammalian DNA methyltransferases." *Chembiochem* 12(2): 206–222.

Kar, S., D. Sengupta, M. Deb, A. Shilpi, S. Parbin, S. K. Rath and S. K. Patra (2014). "Expression profiling of DNA methylation-mediated epigenetic gene-silencing factors in breast cancer." *Clinical Epigenetics* 6(1): 1.

Kar, S., M. Deb, D. Sengupta, A. Shilpi, S. Parbin, J. Torrisani, S. Pradhan and S. Patra (2012). "An insight into the various regulatory mechanisms modulating human DNA methyltransferase 1 stability and function." *Epigenetics* 7(9): 994–1007.

Kinney, S. R. and S. Pradhan (2011). "Regulation of expression and activity of DNA (cytosine-5) methyltransferases in mammalian cells." *Prog Mol Biol Transl Sci* 101: 311–333.

Kouzarides, T. (2007). "Chromatin modifications and their function." *Cell* 128(4): 693–705.

Lai, P., W. Daear, R. Löbenberg and E. J. Prenner (2014). "Overview of the preparation of organic polymeric nanoparticles for drug delivery based on gelatine, chitosan, poly(d,l-lactide-coglycolic acid) and polyalkylcyanoacrylate." *Colloids Surf B: Biointerfaces* 118: 154–163.

Lan, S. Y., T. Yu, Z. S. Xia, Y. H. Yuan, L. Shi, Y. Lin, K. H. Huang and Q. K. Chen (2010). "Musashi 1-positive cells derived from mouse embryonic stem cells can differentiate into neural and intestinal epithelial-like cells in vivo." *Cell Biol Int* 34(12): 1171–1180.

Latham, J. A. and S. Y. Dent (2007). "Cross-regulation of histone modifications." *Nat Struct Mol Biol* 14(11): 1017–1024.

Laurent, L., E. Wong, G. Li, T. Huynh, A. Tsirigos, C. T. Ong, H. M. Low, K. W. Kin Sung, I. Rigoutsos, J. Loring and C. L. Wei (2010). "Dynamic changes in the human methylome during differentiation." *Genome Res* 20(3): 320–331.

Lee, R. C., R. L. Feinbaum and V. Ambros (1993). "The C. elegans heterochronic gene lin-4 encodes small RNAs with antisense complementarity to lin-14." *Cell* 75(5): 843–854.

Lee, Y., C. Ahn, J. Han, H. Choi, J. Kim, J. Yim, J. Lee, P. Provost, O. Radmark, S. Kim and V. N. Kim (2003). "The nuclear RNase III Drosha initiates microRNA processing." *Nature* 425(6956): 415–419.

Lewis, B. P., I. H. Shih, M. W. Jones-Rhoades, D. P. Bartel and C. B. Burge (2003). "Prediction of mammalian microRNA targets." *Cell* 115(7): 787–798.

Lister, R., M. Pelizzola, R. H. Dowen, R. D. Hawkins, G. Hon, J. Tonti-Filippini, J. R. Nery, L. Lee, Z. Ye, Q. M. Ngo, L. Edsall, J. Antosiewicz-Bourget, R. Stewart, V. Ruotti, A. H. Millar, J. A. Thomson, B. Ren and J. R. Ecker (2009). "Human DNA methylomes at base resolution show widespread epigenomic differences." *Nature* 462(7271): 315–322.

Luger, K., A. W. Mader, R. K. Richmond, D. F. Sargent and T. J. Richmond (1997). "Crystal structure of the nucleosome core particle at 2.8 A resolution." *Nature* 389(6648): 251–260.

Mahmood, M., Z. Li, D. Casciano, M. V. Khodakovskaya, T. Chen, A. Karmakar, A. S. Biris (2011). "Nanostructural materials increase mineralization in bone cells and affect gene expression through miRNA regulation." *J Cell Mol Med* 15(11): 2297–2306.

Mathiyalagan, P., S. T. Keating, X. J. Du and A. El-Osta (2014). "Chromatin modifications remodel cardiac gene expression." *Cardiovasc Res* 103(1): 7–16.

Natoli, G. (2010). "Maintaining cell identity through global control of genomic organization." *Immunity* 33(1): 12–24.

Nayak, D., M. Kumari, S. Rajachandar, S. Ashe, N. C. Thathapudi and B. Nayak (2016). "Biofilm impeding AgNPs target skin carcinoma by inducing mitochondrial membrane depolarization mediated through ROS production." *ACS Applied Materials & Interfaces* 8(42): 28538–28553.

Nayak, D., S. Pradhan, S. Ashe, P. R. Rauta and B. Nayak (2015). "Biologically synthesised silver nanoparticles from three diverse family of plant extracts and their anticancer activity against epidermoid A431 carcinoma." *J Colloid Interface Sci* 457: 329–338.

Ng, C.-T., S. T. Dheen, W.-C. G. Yip, C.-N. Ong, B.-H. Bay and L.-Y. Lanry Yung (2011). "The induction of epigenetic regulation of PROS1 gene in lung fibroblasts by gold nanoparticles and implications for potential lung injury." *Biomaterials* 32(30): 7609–7615.

Patra, S. K. and S. Bettuzzi (2009). "Epigenetic DNA-(cytosine-5-carbon) modifications: 5-aza-2-deoxycytidine and DNA-demethylation." *Biochemistry (Mosc)* 74(6): 613–619.

Patra, S. K., A. Patra and R. Dahiya (2001). "Histone deacetylase and DNA methyltransferase in human prostate cancer." *Biochem Biophys Res Commun* 287(3): 705–713.

Perreault, F., S. Pedroso Melegari, C. Henning da Costa, A. L. de Oliveira Franco Rossetto, R. Popovic and W. Gerson Matias (2012). "Genotoxic effects of copper oxide nanoparticles in Neuro 2A cell cultures." *The Science of the Total Environment* 441: 117–124.

Pradeepa, M. M., G. R. Grimes, Y. Kumar, G. Olley, G. C. Taylor, R. Schneider and W. A. Bickmore (2016). "Histone H3 globular domain acetylation identifies a new class of enhancers." *Nat Genet* 48(6): 681–686.

Qin, W., H. Leonhardt and F. Spada (2011). "Usp7 and Uhrf1 control ubiquitination and stability of the maintenance DNA methyltransferase Dnmt1." *J Cell Biochem* 112(2): 439–444.

Rane, A. V., K. Kanny, V. K. Abitha and S. Thomas (2018). "Chapter 5 – Methods for synthesis of nanoparticles and fabrication of nanocomposites." In S. Mohan Bhagyaraj, O. S. Oluwafemi, N. Kalarikkal and S. B. T.-S. of I. N. Thomas (Eds.), *Micro and Nano Technologies* (pp. 121–139). Woodhead Publishing.

Ringrose, L. and R. Paro (2004). "Epigenetic regulation of cellular memory by the Polycomb and trithorax group proteins." *Annu Rev Genet* 38: 413–443.

Rodriguez, A., S. Griffiths-Jones, J. L. Ashurst and A. Bradley (2004). "Identification of mammalian microRNA host genes and transcription units." *Genome Res* 14(10A): 1902–1910.

Satapathy, S. R., S. Siddharth, D. Das, A. Nayak and C. N. Kundu (2015). "Enhancement of cytotoxicity and inhibition of angiogenesis in oral cancer stem cells by a hybrid nanoparticle of bioactive quinacrine and silver: implication of base excision repair cascade." *Mol Pharm* 12(11): 4011–4025.

Sawan, C. and Z. Herceg (2010). "Histone modifications and cancer." *Adv Genet* 70: 57–85.

Sengupta, D., M. Deb, S. K. Rath, S. Kar, S. Parbin, N. Pradhan and S. K. Patra (2016). "DNA methylation and not H3K4 trimethylation dictates the expression status of miR-152 gene which inhibits migration of breast cancer cells via DNMT1/CDH1 loop." *Exp Cell Res* 346(2): 176–187.

Shyamasundar, S., C. T. Ng, L. Yue, L. Yung, T. Dheen and H. Bay (2015). "Epigenetic mechanisms in nanomaterial-induced toxicity." *Epigenomics* 7(3): 395–411.

Stern, C. D. (2000). "Conrad H. Waddington's contributions to avian and mammalian development, 1930–1940." *Int J Dev Biol* 44(1): 15–22.

Sudha, P. N., K. Sangeetha, K. Vijayalakshmi and A. Barhoum (2018). "Nanomaterials history, classification, unique properties, production and market." In *Emerging Applications of Nanoparticles and Architecture Nanostructures* (pp. 341–384). Elsevier.

Tsuzuki, T. (2009). "Commercial scale production of inorganic nanoparticles." *Int J Nanotechnol* 6(5–6): 567–578.

Tucci, P., G. Porta, M. Agostini, D. Dinsdale, I. Iavicoli, K. Cain and A. Willis (2013). "Metabolic effects of TiO2 nanoparticles, a common component of sunscreens and cosmetics, on human keratinocytes." *Cell Death Dis* 4: e549.

Vakoc, C. R., M. M. Sachdeva, H. Wang and G. A. Blobel (2006). "Profile of histone lysine methylation across transcribed mammalian chromatin." *Mol Cell Biol* 26(24): 9185–9195.

Voigt, P., W. W. Tee and D. Reinberg (2013). "A double take on bivalent promoters." *Genes Dev* 27(12): 1318–1338.

Waddington, C. H. (1953). "Epigenetics and evolution." *Symp Soc Exp Biol* 7: 186–199.

Wang, H., L. Zhai, J. Xu, H. Y. Joo, S. Jackson, H. Erdjument-Bromage, P. Tempst, Y. Xiong and Y. Zhang (2006). "Histone H3 and H4 ubiquitylation by the CUL4-DDB-ROC1 ubiquitin ligase facilitates cellular response to DNA damage." *Mol Cell* 22(3): 383–394.

Wong, B. S. E., Q. Hu and G. H. Baeg (2017). "Epigenetic modulations in nanoparticle-mediated toxicity. *Food Chem Toxicol* 109(Pt 1): 746–752.

Yan, K. S., S. Yan, A. Farooq, A. Han, L. Zeng and M. M. Zhou (2003). "Structure and conserved RNA binding of the PAZ domain." *Nature* 426(6965): 468–474.

Yokochi, T. and K. D. Robertson (2002). "Preferential methylation of unmethylated DNA by Mammalian de novo DNA methyltransferase Dnmt3a." *J Biol Chem* 277(14): 11735–11745.

20 Nanotoxicology
Evaluating Toxic Potential of Drug Nanoparticles

Pavani Gonnabathula, Nageswararao Mekala, and Praveen Kumar Relangi

CONTENTS

20.1 Introduction ... 235
20.2 Physicochemical Properties of NPs Related to Toxicity ... 236
20.2.1 Size ... 236
20.2.2 Surface Area ... 236
20.2.3 Shape ... 236
20.2.4 Chemistry of NPs ... 237
20.2.5 Surface Charge ... 237
20.3 Possible Toxicological Effects on Human Health ... 237
20.3.1 Hepatotoxicity ... 237
20.3.2 Nephrotoxicity ... 238
20.3.3 Cardiotoxicity ... 238
20.3.4 Inhalation Toxicity ... 239
20.3.5 Dermal Toxicity ... 239
20.3.6 Immunotoxicity ... 239
20.3.7 Genotoxicity ... 239
20.4 Importance of Nanoparticle Toxicity Evaluation ... 239
20.4.1 In Vitro Assays ... 240
20.4.1.1 LDH Assay ... 240
20.4.1.2 Protein Assay ... 240
20.4.1.3 MTT (3-(4,5-Dimethylthiazoyl-2)-2,5-Diphenyltetrazolium Bromide) Assay ... 240
20.4.1.4 Oxidative Stress Assay ... 240
20.4.1.5 Apoptosis ... 240
20.4.1.6 Necrosis ... 240
20.4.2 In Vivo Assays ... 240
20.4.2.1 Acute Toxicity ... 240
20.4.2.2 Multi-Dose Toxicity Studies ... 241
20.4.2.3 Developmental and Reproductive Toxicity ... 241
20.4.2.4 Genotoxicity Evaluation ... 241
20.5 Conclusion and Future Studies ... 241
References ... 241

20.1 INTRODUCTION

Nanotoxicology is the study of the toxicity of nanoparticles (NPs). Nanotoxicological studies are meant to define the extent to which nanomaterials pose threats to human health and negatively impact the environment. Nanoparticle toxicity, or nanotoxicity, mainly arises due to their specific characteristics like large surface-area-to-volume ratio and other physicochemical properties, which can lead to different biological activity compared with their bulk counterparts. In the past few decades, there has been huge advancement in utilizing nanotechnology in many fields, including biomedical sciences, due to NPs' novel and unique properties, which make them biologically more reactive. The most important biomedical applications of nanomaterials are in disease diagnosis, drug design, drug delivery, regenerative medicine, and treatment. The biomedical applications of nanomaterials in disease diagnosis, treatment, and monitoring to control biological functions have been referred to as "nanomedicine" (Moghimi et al., 2005). This magnifies the need for and importance of nanomaterials as drug molecules for many pathological conditions, but the potentially unpredictable and unwanted consequences after exposure to the nanomedicine have yet to be evaluated. Exposure of NPs adversely affects the normal physiology of humans as well as animals by disrupting biomolecules at the cellular level by invading the cells, possibly by interrupting the normal structure of tissues and organs; this is known as nanotoxicity. The toxicity depends upon physicochemical properties of NPs, especially the size of particles, shape, surface area and charge, chemical composition, and their stability. The underlying mechanism of nanotoxicity is under extensive research (Figure 20.1), but it is known that cytotoxicity and gene activation related to oxidative stress are a result of reactive oxygen species (ROS) generated upon exposure to NPs at the cellular and molecular level (Jebali, and Kazemi, 2013).

Insight into nanoparticle toxicity is important for human health and the environment. The route of administration, the dose, and the extent of tissue distribution of nanomedicine are vital parameters in nanotoxicity. Particles may go into the body via several routes, including ingestion, inhalation, and intravascular/intravenous and dermal penetration, which could possibly distribute NPs to any organ system and regulate cellular functions both physiologically and toxicologically. Physicochemical properties of nanoparticles influence nanotoxicity, and Figure 20.2 depicts the possible routes and factors affecting nanoparticle exposure and resulting toxicity.

20.2 PHYSICOCHEMICAL PROPERTIES OF NPs RELATED TO TOXICITY

20.2.1 Size

The size influences the surface area, which is a key factor in determining the toxicity of NPs. A reduction in size of nanoparticles has increased surface area; therefore, more chemical molecules can be attached to the surface, which increases the reactivity and thus the possibility of toxicity (Hyuk Suh et.al. 2009). Many studies have indicated that the smaller the NPs, the greater their cellular intake, distribution through tissues, and absorption into the bloodstream. For example, differential tissue distribution of gold nanoparticles has been observed based on size: 10-nm NPs were distributed widely in many organs, whereas larger particles (50–250 nm) were found only in spleen, liver, and bloodstream of intravenously injected rats (De Jong et.al., 2008).

20.2.2 Surface Area

In vitro studies showed that very small particles have more destructive and pathological power in the lungs than relatively larger particles because of their larger surface area, better affinity to conjugate, and energy sustainability (Oberdorster et al., 1994; Holsapple et al., 2005). Several studies using a variety of nanoparticles have shown that surface area is a crucial factor in rodent toxic manifestations (lung and other inflammatory responses caused by epithelial inflammation) (Holgate, 2010). With nanoparticles decreasing in size, surface area increases, resulting in a dose-dependent increase in oxidation and DNA-damaging capabilities of these nanomaterials (Risom et al., 2005) much greater than that seen for larger particles at doses of equal mass (Donaldson and Stone, 2003).

20.2.3 Shape

Nanomaterials come in a variety of forms including fibers, rings, tubes, rods, spheres, ovoids, etc. For many nanoparticles, including carbon nanotubes, allotropes, silica, nickel, titanium, and gold nanomaterials, shape-dependent toxicity has been reported (Ispas et al., 2009; Chithrani et al., 2006 and Hamilton et al., 2009). During endocytosis or phagocytosis, shape-dependent nanotoxicity influences the in vivo membrane wrapping processes (Verma and Stellacci, 2010). Endocytosis of spherical nanoparticles has been observed to be relatively less toxic than rod-shaped or fiber-like nanoparticles (Champion and Mitragotri, 2006) and, more importantly, that of spherical nanoparticles, regardless of whether they are homogeneous or heterogeneous (Lee et al., 2007). Non-sphere nanomaterials are more willing to flow through capillaries with other biological effects (Kim et al., 2012). It has been found that TiO_2 fibers are more cytotoxic than spherical entities (Hsiao and Huang, 2011).

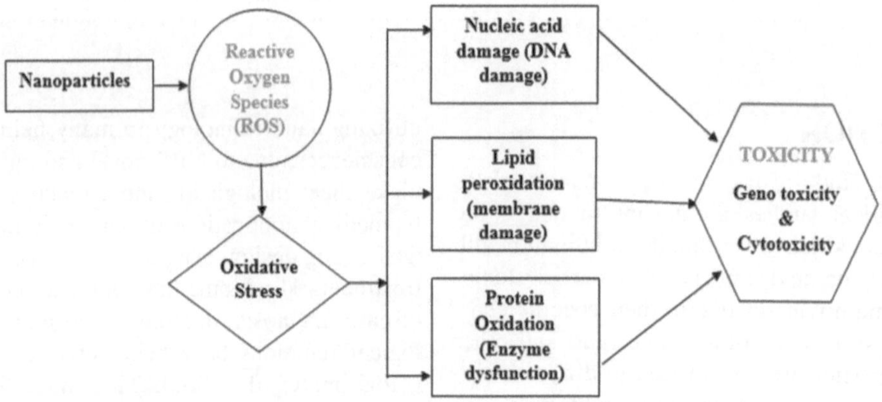

FIGURE 20.1 Possible mechanisms of toxicity due to NP exposure.

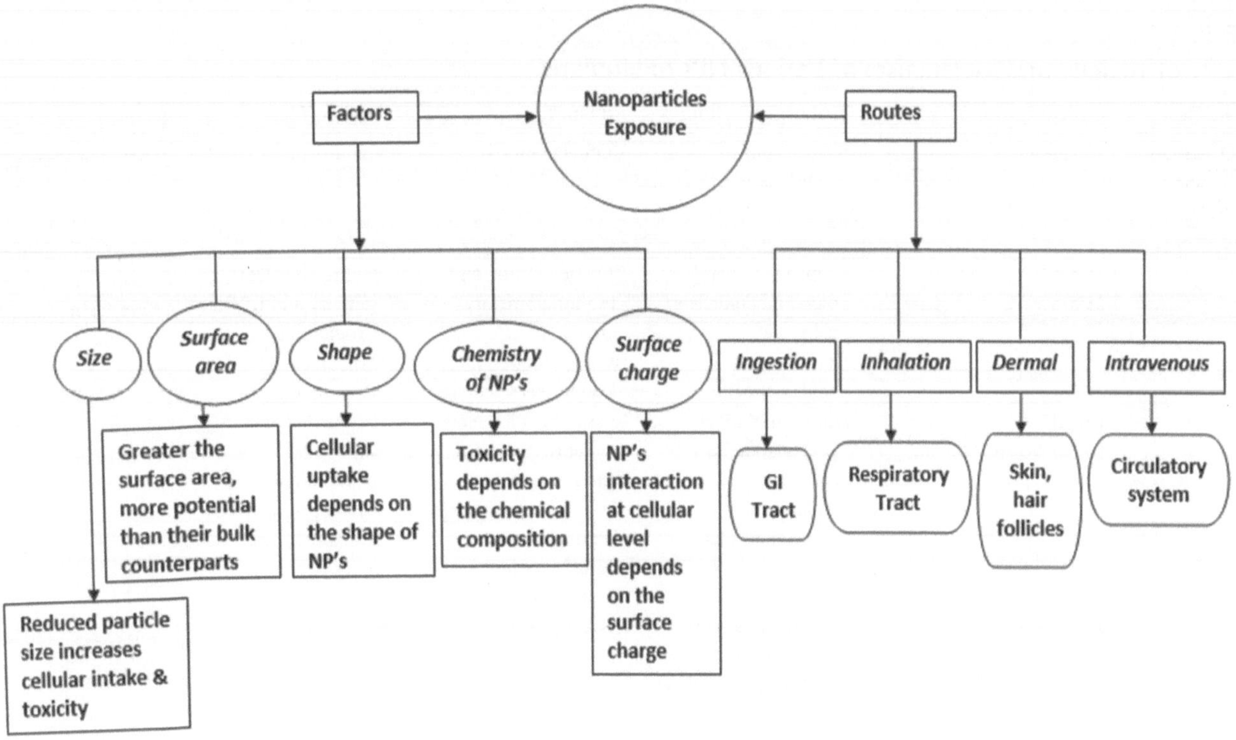

FIGURE 20.2 Factors affecting nanoparticle exposure and routes of administration.

20.2.4 Chemistry of NPs

However, it's also been emphasized that particle size plays an important role when considering the toxicity of nanoparticles, but the toxicity of identically shaped nanoparticles should be considered based on their chemistry. These studies highlight that the composition of nanoparticles additionally influences their toxicity problems. During a study using victimization of daphnids, algal species, and zebrafish as models of various trophic levels, it was identified that nanosilver and nanocopper with their soluble forms produced toxicity in all experimental organisms, whereas TiO_2 of the identical dimensions failed to cause any toxicity problems (Griffitt et al., 2008), which proves the role of composition in determining the toxicities of NPs.

20.2.5 Surface Charge

Surface charge plays a key role in nanoparticle toxicity, as it defines their interactions with biological systems to a large extent. Different aspects of nanomaterials, such as ability to penetrate blood-brain barriers, transmembrane permeability (Georgieva et al., 2011), colloidal behavior, plasma protein binding (Pietroiusti et al., 2011), and selective nanoparticle adsorption (Hoshino et al., 2004), are primarily controlled by the surface charge of nanoparticles. Cellular uptake of positively charged nanoparticles is considerable compared to that of neutral and negatively charged nanoparticles due to their enhanced plasma protein opsonization. In addition, hemolysis and platelet aggregation have also been shown to cause severe toxicity to the system.

20.3 POSSIBLE TOXICOLOGICAL EFFECTS ON HUMAN HEALTH

Metabolic pathways and absorption influence nanoparticles' mobility in human body and their accumulation in different organs; understanding this phenomenon requires a thorough examination of various organ systems. For instance, degraded nanoparticles exhibit peculiar molecular responses, tend to distribute across the body, and mostly accumulate in liver cells leading to cellular damage in the case of oral administration of the drug. Similarly, ingestion, inhalation, and dermal exposure may allow nanoparticles to enter the bloodstream, get distributed into major organs like lung, liver, and kidney, and dysregulate their physiology, which leads to toxic effects on human health. A summary of toxicity and mechanisms of various NPs to humans is given in Table 20.1.

20.3.1 Hepatotoxicity

The liver is the major organ for metabolism and detoxification of any xenobiotics, where blood is a carrier for these toxicants and sieved by the liver before distribution to other body parts. The rate of blood flow influences the concentration of toxicant accumulation to this organ. Likewise, liver is a targeted organ of nanoparticles, whether inhaled, absorbed via skin, or administered intravenously or through medical procedures for diagnostic purposes; the nanoparticles reach the blood circulation and finally are translocated to the liver. Based on the previous research, the entrapping of nanomaterials by the reticuloendothelial system makes the liver a principal target organ; therefore, nanoparticles might be potential

TABLE 20.1
Summary of Toxicity and Mechanism of Various NPs to Humans

Nanoparticles	Concentration/size	Test organ/cell lines	Mechanism	Toxic effects	References
Bare TiO, Mg, Ag, Au, and their triglyceride-coated forms	100 μg/mL	Suspensions of Balb/c skin cells	ROS generation	Cytotoxicity	Jebali and Kazemi (2013)
TiO	25 nm 25 nm	Human peripheral blood mononuclear cells. Human gastric epithelial cells	ROS generation and pro inflammatory effects. ROS generation may lead to DNA damage	Suppressed IDO activity and IFN-c production. Oxidative stress DNA damage	Becker et al. (2014), Botelho et al. (2014)
Ag	1 mg/L 10, 20, 40, 60, 100 nm	Human umbilical vein endothelial cells (HUVECS); human colon carcinoma cells	ROS generation; cell membrane damage; generation of inflammatory markers leading to mitochondrial dysfunction	Endothelial cell injury and dysfunction causing apoptosis; oxidative stress and cytotoxicity	Shi et al. (2014), Miethling-Graff et al. (2014)
Nickel oxide	20, 40, 60, 80, 100 μg/mL	Human pulmonary epithelial cell line BEAS 2B and A549	DNA damage	Inflammation and genotoxic effect in lung epithelial cells	Capasso et al. (2014)
ZnO	1–10 nm gradient concentrations of 0.01, 0.025, 0.5, 1.0, 1.5 μg/mL	Human polymorpho nuclear neutrophil (PMNS); human pulmonary adenocarcinoma cell line LTEP-a-2	Induced cell morphology and increased cell size; generation of ROS-inducing oxidative stress	Delay in human neutrophil apoptosis; cytotoxicity	Goncalves and Girard (2014), Wang et al. (2015)
Metal oxides CeO_2, TiO_2, Al_2O_3, ZnO	0.5–50 μg/mL, 5–20 nm	Human peripheral blood lymphocytes (PBLS)	Excess cell activation mechanism	Induced changes in expression level of adhesion molecules and c-x-c chemokine receptor type 4 (cxcr4) in these cells; T-cell proliferation upon cell exposure to TiO_2 and Al_2O_3	Lozano-Fernández et al. (2014)

hepatotoxins in certain situations based on the purpose of administration, dose, and other physicochemical properties (Hillyer and Albrecht, 2001). Researchers investigating whether certain nanoparticles may be hepatotoxic conducted in vitro studies using rat liver cell line BRL 3A to estimate the toxic potential of titanium oxide and silver nanoparticles. Tests for cytotoxicity, oxidative stress, and mitochondrial dysfunction established that silver was highly cytotoxic compared with titanium oxide particles (Hussain et al., 2005). Other studies indicated the cytotoxicity of nanosilver in human colon Caco2 and liver HepG2 cells (Sahu et al., 2014).

20.3.2 Nephrotoxicity

The kidney is the primary organ for excretion of many xenobiotics, and it has been stated that the kidney is a target organ for toxicity, which is applicable for nanomaterial toxicity as well (Wang et al., 2009; Yan et al., 2012). Studies indicated that rats exposed to ZnO nanoparticles induced cell membrane and mitochondrial damage in kidney, resulting in nephrotoxicity (Yan et al., 2012). Copper nanoparticle–induced nephrotoxicity was observed in exposed rats by renal proximal tubule necrosis in kidneys, which was confirmed by renal gene expression profiling (Liao and Liu, 2012). Therefore, it is evident that some nanomaterials are potential nephrotoxicants, depending on their physicochemical properties and route of administration and their toxicokinetics.

20.3.3 Cardiotoxicity

Studies on the cardiovascular toxicity of intratracheally instilled silica nanoparticles in rats concluded that the entrance of silica nanoparticles into systemic circulation through the alveolar capillary barrier is an indication of silica nanoparticle toxicity, depending on particle size and dosage. This was concluded after evaluating hematological parameters, oxidative stress, inflammatory reactions, dysfunction of the endothelium, and serum myocardial enzymes (Du et al., 2013). Both endothelial cell (in vitro) and zebrafish (in vivo) models were used to assess silica nanoparticle cardiovascular effects (Duan et al., 2013). Oxidative stress and apoptosis were reported as major factors for dysfunction of endothelial cells, which established that exposure to silica nanoparticles is a possible risk factor for failure of the cardiovascular system.

20.3.4 Inhalation Toxicity

Developing inhalable drug delivery systems via pulmonary administration has become a smart route to treat chronic diseases. Delivery of nanomaterials via pulmonary delivery system is a comparatively new concept, especially for lung cancer therapy (Lee et al., 2015). Nanoparticles are accumulated in the respiratory tract mainly by diffusion. Studies indicated that inhaled nanoparticles deposited within the alveolar areas of the rat lung (Oberdorster et al., 2004; Semmler et al., 2004) may cross biological membranes and reach cellular levels that would not be generally exposed to larger particles. For example, inhaled TiO_2 nanoparticles translocated into lung interstitial (Semmler et al., 2004; Ferin et al., 1992), caused in pulmonary overload in mice and rats with inflammation (Dobrovolskaia et al., 2009). For the study of nanomaterial toxicity, human lung cancer cell line A549 has been utilized as an in vitro cell culture model, and cytotoxic and genotoxic effects of nanosilver were investigated by measuring cellular uptake of nanomaterial using flow cytometry and atomic absorption spectroscopy, further concluding that silver nanomaterials induced ROS-mediated cytotoxic and genotoxic effects (Foldbjerg et al., 2011).

20.3.5 Dermal Toxicity

Skin, the body's largest organ, serves as the primary route for human exposure to chemicals. The use of nanomaterials by cosmetic industry is preferred due to NPs' ability to enhance cosmetic product solubility, transparency, and color. Nanomaterials with potential cosmetic uses include nanogold, nanosilver, nanocapsules, nanoemulsions, nano-crystals, dendrimers, fullerenes, hydrogels, liposomes, and nanoparticles of solid lipids (Raj et al., 2012). Nanomaterials are used in pharmaceuticals for diagnosis, treatment, and prevention of skin diseases due to their ability to increase water solubility of poorly soluble drugs, drug half-life, bioavailability, and sustained and targeted delivery of drugs (Dianzani et al., 2014), which allows drug delivery into the openings of hair follicles and diseased skin.

The toxicity of titanium dioxide from cosmetics is a controversial topic (Nohynek et al., 2007). There are concerns about titanium dioxide toxicity, since it is frequently used as a physical sunscreen as it reflects and scatters light rays of UVA (320–400 nm) and UVB (290–320 nm), the skin-damaging parts of the solar spectrum. However, TiO_2 also absorbs a significant amount of UV radiation, resulting in the generation of reactive oxygen species in aqueous media, including superoxide anion radicals, free hydroxyl radicals, singlet oxygen, and hydrogen peroxide. These reactive oxygen species can cause significant DNA damage (Gurr et al., 2005). Several other studies, however, reported that titanium dioxide caused chronic pulmonary inflammation in rats (again by instillation) (Oberdörster G et al., 1992) and had a proinflammatory effect in vitro using human endothelial cells (Peters et al., 2004).

20.3.6 Immunotoxicity

Information on the immunotoxicity of nanomaterials is limited. Their surface chemistry largely determines the compatibility of nanomaterials with the immune system (Dobrovolskaia and McNeil, 2007). A safety evaluation review of nanomaterial immunotoxicity tests shows that nanoparticles can both stimulate and/or suppress immune responses (Dobrovolskaia et al., 2009). Cytokine production can be modulated by nanomaterials (Di Gioacchino et al., 2011). They induce pro-inflammatory lung effects in investigational animals with increased expression of macrophage inflammatory protein (MIP)-1α, interleukin (IL)-1Ø, MIP-2, monocyte chemoattractant protein (MCP)-1, chemokine (C - C motif) ligand 17 (TARC), granulocyte macrophage colony-stimulating factor (GM-CSF), keratinocyte chemoattractant, and stress-activated mitogen-activating factor activation, which suggests that oxidative stress is the mechanism that leads to pro-inflammatory disease processes in the lung, especially allergy (Di Gioacchino et al., 2011; Petrarca et al., 2015). Cytotoxicity, immunotoxicity, and genotoxicity studies carried out for iron oxide and gold nanoparticles on human cells utilized a panel of cell-based assessments in various laboratories (Pfaller et al., 2010) to show that these nanoparticles induced very little or no cytotoxicity, genotoxicity, and immunotoxicity. Further investigation is required for the development and validation of procedures for reviewing the immunotoxicity of nanomaterials.

20.3.7 Genotoxicity

Several methods found in the literature were used to test nanomaterials for genotoxicity, including chromosomal aberrations, oxidative DNA damage, DNA strand breaks, mutations, DNA adducts, and micronucleus formation (Landsiedel et al., 2009; Xie et al., 2011). It has been testified that iron oxide nanoparticles induce genotoxicity as confirmed by the comet assay in vitro (Zuzana et al., 2011). The possible genotoxicity of nanosilver (20 nm) in human colon Caco2 cells and liver HepG2 in culture has been investigated using cytochalasin B-blocked micronucleus assay and in vitro micronucleus assay flow cytometric assay (Sahu et al., 2014).

20.4 IMPORTANCE OF NANOPARTICLE TOXICITY EVALUATION

Nanoparticles possess unique characteristics with respect to size, shape, surface area, surface charge, and drug composition and chemistry that will affect their functioning in the human body and ability to accumulate in different organ systems, which may lead to toxicity (Oberdorster et al., 2005) because of increased reactivity at the cellular level. This makes it a complex scenario to specifically evaluate their toxicity (Dhawan and Sharma, 2010; Pfaller et al., 2010). Existing screening methods for drug compounds do not apply to NPs; moreover, there are no currently accepted international methods and standard protocols for testing NP toxicity. Development of relevant testing methods to detect, analyze, and better understand NPs' impact on human health is of great importance. Some of the test methods are summarized in Table 20.2.

20.4.1 In Vitro Assays

In vitro assays define the characteristics of the nanoparticles based on their influence on mechanistic pathways. Assessment methods include cytotoxicity assays to identify the toxic impacts on the cell like lactate dehydrogenase (LDH) assay and protein assay, proliferation assays like methyl tetrazolium assay (MTT), apoptosis assays, necrosis assays, oxidative stress assays, etc.

20.4.1.1 LDH Assay

Lactate dehydrogenase is present in all of the body's cells. Any damage to the cell or tissue will result in the release of LDH into the blood, and the amount of LDH in blood serves as an indicator of cell damage. The cytotoxicity of drug nanoparticles can be assessed as it relates to cellular damage.

20.4.1.2 Protein Assay

Sulforhodamine B assay measures the cellular assay content, indicating the cell density. In this assay, total protein is stained followed by fixing the adherent protein; toxicity screening of the nanoparticles adherent to the protein measures cytotoxicity.

20.4.1.3 MTT (3-(4,5-Dimethylthiazoyl-2)-2,5-Diphenyltetrazolium Bromide) Assay

This is a colorimetric assay method to measure the activity of enzymes that reduce MTT, a yellow water-soluble tetrazolium dye to form purple formazan crystals. The spectrum of the samples dissolved in dimethyl sulfoxide (DMSO) at 550 nm gives a clear distinction of nanoparticle-treated and untreated samples, indicating the cytotoxicity.

TABLE 20.2
List of Test Methods to Detect and Analyze the Impact of NPs on Human Health

In vitro assays	• Cytotoxicity • Oxidative stress • Apoptosis • Necrosis
In vivo assays	• Acute toxicity • Multi-dose toxicity • Developmental and reproductive toxicity
Genotoxic evaluations	• Mutation assays • Cytogenetic assays

20.4.1.4 Oxidative Stress Assay

Nanoparticles inside the body stimulate oxidative, stress resulting in the generation of ROS and reactive nitrogen species (RNS), which can be detected using several methods. Examples include Amplex red assay, lipid peroxidation test by C11-BIODIPY assay, and TBA assay for malondialdehyde (Fantel, 1996).

20.4.1.5 Apoptosis

The accumulation of nanoparticles inside the organ system results in oxidative stress in cell culture systems. Abundant free radicals produced in this process cause DNA damage and apoptosis. This phenomenon can be better understood with the help of gel electrophoresis in differentiating cell death induced by apoptotic and necrotic modes, analyzed by the comet assay method. The treatment method involves lysing cells to remove cellular protein and allowing the damaged cells to move away from nucleus through electrophoresis, thus forming a tail of the comet; undamaged cells will remain in the nucleus, forming the head of the comet. The samples can be tagged with a customized fluorescent tag to the DNA.

20.4.1.6 Necrosis

Dyeing of the cells with suitable coloring agents like neutral red and trypan blue helps in differentiating living and dead cells. Neutral red forms hydrophobic bonds in lysosomal matrix, and nanoparticle accumulation weakens the surface of the cell, which leads to reducing uptake levels of the dye; thus, the dead cells can be differentiated. In a study, silver nanoparticle exposure lead to an instability in the endosome-lysosome system (Miranda et al., 2017). Trypan blue is an exclusion test: it enters the dead cells and is excluded from living cells. Integrity of the cell membrane can be assessed in the presence of zinc nanoparticles; significant cytotoxicity has been observed (Kononenko et al., 2017).

20.4.2 In Vivo Assays

This procedure includes assessment of toxic impacts of NPs in animal models like mice, rats, guinea pigs, and rabbits, which exhibit resemblance to metabolic and pharmacokinetic mechanisms of humans. The obtained data can be extrapolated to understand the effects on human physiology. Modifications in cell type, serum chemistry, and histopathology of organs upon exposure to nanoparticles prove their toxic impacts (Baker et al., 2008; Lei et al., 2008). The test results are acceptable by regulatory agencies (Keller, 2006).

20.4.2.1 Acute Toxicity

Short-term effects of the NPs in two different animal species and sexes are to be determined inside a 24-hr time frame. The route of drug administration should be chosen based on the route suggested for human intake like ingestion, inhalation, etc. In this case, a single drug is chosen to observe the effects; generally a maximum dosage of 2000 mg/kg is tested unless specified by a governing agency to test at higher limits, which cannot be more than 5000 mg/kg. The results from this study serve as precursors to design a suitable dosage for long-term studies (Rauckman and Soiefer, 2006).

20.4.2.2 Multi-Dose Toxicity Studies

Studies of the effect of smaller doses of nanoparticles over a period and their possible sub-chronic, chronic, and sub-acute impacts are collectively termed multi-dose studies. These are comprehensive studies of biochemical, functional, pathological, and/or physiological changes, including reversible effects in the planned recovery phase. The study duration ranges from a few weeks to a year. Long-term testing enables researchers to understand the effects over a larger fraction of the animal lifespan (e.g., nanoparticles encapsulating anti-tuberculosis or anti-cancer drugs). At the end of the dosing period, necropsy and histopathological examinations of different organs reveal the toxicity imposed by the NPs (Keller and Banks, 2006).

20.4.2.3 Developmental and Reproductive Toxicity

Developmental and reproductive toxicology (DART) study follows a set of harmonized global guidelines for testing to be carried out in three segments representing the entire reproductive cycle. These three test groups sequentially are from pre-mating and mating through implantation, the phase from gestation to major organ formation, and finally late organ formation (Keller, 2006).

1. Segment I – Reproduction and fertility studies
2. Segment II – Teratology or developmental toxicity study and
3. Segment III – Prenatal or postnatal testing

Scientific experiments are most frequently carried out in rats, rabbits, mice, and quite often monkeys and dogs to mimic metabolic and pharmacokinetic profiles in humans. Dosing is performed at approximately the same time period by the expected route of delivery in humans, with values usually directed by the maximum tolerated pharmacological dose. In addition to observations on general health, behavioral pattern, body weight, and food consumption, other factors scrutinized include checking vaginal smears, specific cohabitation, post-euthanization parameters like gestation uterine examination (day 13), term caesarean section, histopathological examination at the end of lactation, necropsy fetal examination, assessment of visceral organs and skeletal characteristics, and hardships during childbirth, along with postnatal examination related to the number of offspring, their weight, gender, or any indication of gross deformity, and their reflex, behavioral, and reproductive capabilities.

Thus, DART studies analyze the long-term effects of test molecules on complex and contentious interactions among biological, physiological, and endocrine systems by examining the animal offspring's female reproductive and epigenetic variables. Despite the limitations of the experimental models currently in use, significant investigation in this direction has played a critical role in summarizing developmental and reproductive toxicity mechanisms with the final objective of exactly predicting possible human risks.

20.4.2.4 Genotoxicity Evaluation

20.4.2.4.1 Mutation Assays

The bacterial mutation assays give a rapid and easy assessment of potentially mutagenic or carcinogenic impacts by identifying changes or reversals in normal organism phenotype. Examples include Ames test (Ames et al., 1975), mammalian cell mutation assay, and in vitro chromosome aberration test.

20.4.2.4.2 Cytogenetic Assays

The effects of metabolic activation, detoxification processes, and alteration of test materials by the intestinal bacteria are assessed in performing these assays. Such determinations allow for the assessment of both somatic and reproductive cell damage in the case of DNA damage. The tests most prevalently used include the bone marrow or peripheral blood cell micronucleus test, the bone marrow chromosome aberration test, and spermatological cell test, all of which are used to evaluate clastogenic activities.

20.5 CONCLUSION AND FUTURE STUDIES

Even though nanoparticle-encapsulated drugs offer a unique, precise efficiency in drug delivery, disease diagnosis, treatment, and monitoring, integrated research still warrants the formulation of well-defined and consistent experimental protocols to determine the toxicity associated with therapeutic nanoparticles in an accurate and predictive manner and to address the challenges posed by the toxicological issues of these therapeutic nanoparticles. A thorough understanding is also needed in the formulation of safe, biodegradable, and biocompatible nanoparticle systems. To help end users like drug formulators and surgeons in diagnosis and treatment to understand the biological implications of nanoparticle exposure, standardized protocols should be designed to help overcome cytotoxic, genotoxic, and related health risks implicated by nanoparticles in potential drug delivery systems.

REFERENCES

Ames BN, McCann J, Yamasaki E 1975. Methods for detecting carcinogens and mutagens with the Salmonella/mammalian-microsome mutagenicity test. *Mutat Res* 31: 347–364.

Baker GL, Gupta A, Clark ML, Valenzuela BR, Staska LM, Harbo SJ, Pierce JT, Dill JA 2008. Inhalation toxicity and lung toxicokinetic of C60 fullerene nanoparticles and microparticles. *Toxicol Sci* 101(1): 122–131.

Becker K, Schroecksnadel S, Geisler S, Carriere M, Gostner JM, Schennach H, Herlin N, Fuchs D 2014. TiO2 nanoparticles and bulk material stimulate human peripheral blood mononuclear cells. *Food Chem Toxicol* 65: 63.

Botelho MC, Costa C, Silva S, Costa S, Dhawan A, Oliveira PA, Teixeira JP 2014. Effects of titanium dioxide nanoparticles in human gastric epithelial cells *in vitro*. *Biomed Pharmacother* 68: 59.

Capasso L, Camatini M, Gualtieri M 2014. Nickel oxide nanoparticles induce inflammation and genotoxic effect in lung epithelial cells. *Toxicol Lett* 226: 28.

Champion JA, Mitragotri S 2006. Role of target geometry in phagocytosis. *Proc Natl Acad Sci U S A* 103(13): 4930–4934.

Chithrani BD, Ghazani AA, Chan WCW 2006. Determining the size and shape dependence of gold nanoparticle uptake into mammalian cells. *Nano Lett* 6(4): 662–668.

De Jong WH, Hagens WI, Krystek P, Burger MC, Sips AJ, Geertsma RE 2008. Particle size-dependent organ distribution of gold nanoparticles after intravenous administration. *Biomaterials* 29(12): 1912–1919.

Dhawan A, Sharma V 2010. Toxicity assessment of nanomaterials: methods and challenges. *Anal Bio Anal Chem* 398(2): 589–605.

Di Gioacchino M, Petrarca C, Lazzarin F, Di Giampaolo L, Sabbioni E, Boscolo P, Mariani-Costantini R, Bernardini G 2011. Immunotoxicity of nanoparticles. *Int J Immunopathol Pharmacol* 24(Suppl 1): 65S–71S.

Dianzani C, Zara GP, Maina G, Pettazzoni P, Pizzimenti S, Rossi F, Gigliotti CL, Ciamporcero ES, Daga M, Barrera G 2014. Drug delivery nanoparticles in skin cancers. *Biomed Res Int* 2014:895986.

Dobrovolskaia MA, Germolec DR, Weaver JL 2009. Evaluation of nanoparticle immunotoxicity. *Nat Nanotechnol* 4: 411–414.

Dobrovolskaia MA, McNeil SE 2007. Immunological properties of engineered nanomaterials (a review). *Nat Nanotechnol* 2: 469–478.

Donaldson K, Stone V 2003. Current hypotheses on the mechanisms of toxicity of ultrafine particles. *Ann Ist Super Sanita* 39(3): 405–410.

Du Z, Zhao D, Jing L, Cui G, Jin M, Li Y, Liu X, Liu Y, Du H, Guo C, Zhou X, Sun Z 2013. Cardiovascular toxicity of different sizes amorphous silica nanoparticles in rats after intratracheal instillation. *Cardiovasc Toxicol* 13(3): 194–207.

Duan J, Yu Y, Li Y, Yu Y, Sun Z 2013. Cardiovascular toxicity evaluation of silica nanoparticles in endothelial cells and zebrafish model. *Biomaterials* 34(23): 5853–5862.

Fantel AG 1996. Reactive oxygen species in developmental toxicity: review and hypothesis. *Teratology* 53(3): 196–217.

Ferin J, Oberdörster G, Penney DP 1992. Pulmonary retention of ultrafine and fine particles in rats. *Am J Respir Cell Mol Biol* 6: 535–542.

Foldbjerg R, Dang DA, Autrup H 2011. Cytotoxicity and genotoxicity of silver nanoparticles in the human lung cancer cell line A549. *Arch Toxicol* 85(7): 743–750.

Georgieva JV, Kalicharan D, Couraud PO, Romero IA, Weksler B, Hoekstra D, Zuhorn IS 2011. Surface characteristics of nanoparticles determine their intracellular fate in and processing by human blood-brain barrier endothelial cells in vitro. *Mol Ther* 19(2): 318–325.

Goncalves DM, Girard D 2014. Zinc oxide Nanoparticles Delay Human neutrophil Apoptosis by a de novo Protein Synthesis dependent and reactive oxygen species-Independent Mechanism. *Toxicol Vitro* 28: 926.

Griffitt RJ, Luo J, Gao J, Bonzongo JC, Barber DS 2008. Effects of particle composition and species on toxicity of metallic nanomaterials in aquatic organisms. *Environmental Toxicology and Chemistry: An International Journal* 27(9): 1972–1978.

Gurr JR, Wang ASS, Chen CH, Jan KY 2005. Ultrafine titanium dioxide particles in the absence of photoactivation can induce oxidative damage to human bronchial epithelial cells Toxicol 213: 66–73 and references therein.

Hamilton RF Jr, Wu N, Porter D, Buford M, Wolfarth M, Holian A 2009. Particle length-dependent titanium dioxide nanomaterials toxicity and bioactivity. *Part Fibre Toxicol* 6: 35.

Hillyer JF, Albrecht RM 2001. Gastrointestinal per sorption and tissue distribution of differently sized colloidal gold nanoparticles. *J Pharm Sci* 90: 1927–1936.

Holgate ST 2010. Exposure, uptake, distribution and toxicity of nanomaterials in humans. *J Biomed Nanotechnol* 6(1): 1–19.

Holsapple MP, Farland WH, Landry TD, Monteiro-Riviere NA, Carter JM, Walker NJ, Thomas KV 2005. Research strategies for safety evaluation of nanomaterials, Part II: toxicological and safety evaluation of nanomaterials, current challenges and data needs. *Toxicol Sci* 88: 12–17.

Hoshino A, Fujioka K, Oku T, Suga M, Sasaki YuF, Ohta T, Yasuhara M, Suzuki K, Yamamoto K 2004. Physicochemical properties and cellular toxicity of nanocrystal quantum dots depend on their surface modification. *Nano Lett* 4(11): 2163–2169.

Hsiao I, Huang Y 2011. Effects of various physicochemical characteristics on the toxicities of ZnO and TiOw2 nanoparticles toward human lung epithelial cells. *Sci Total Environ* 409(7): 1219–1228.

Hussain SM, Hess KL, Gearhart JM, Geiss KT, Schlager JJ 2005. In vitro toxicity of nanoparticles in BRL3A rat liver cells. *Toxicol Vitro* 19: 975–983.

Ispas C, Andreescu D, Patel A, Goia DV, Andreescu S, Wallace KN 2009. Toxicity and developmental defects of different sizes and shape nickel nanoparticles in zebrafish. *Environ Sci Technol* 43(16): 6349–6356.

Jebali A, Kazemi B 2013. Triglyceride-coated nanoparticles: skin toxicity and effect of UV/IR irradiation on them. *Toxicol Vitro* 27: 1847.

Keller KA 2006. Laboratory animals and in vitro test systems used in regulatory toxicology. In: Jacobson-Kram D., Keller KA (Eds.), *Toxicological Testing Handbook Principles, Applications, and Data Interpretation*, 2nd Ed. Informa Healthcare USA, Inc., New York, pp. 15–31.

Keller KA, Banks C 2006. Multidose general toxicology studies. In Jacobson-Kram D, Keller KA (Eds.), *Toxicological Testing Handbook Principles, Applications, and Data Interpretation*, 2nd Ed. Informa Healthcare USA, Inc., New York, pp. 149–184.

Kim ST, Chompoosor A, Yeh Y, Agasti SS, Solfiell DJ, Rotello VM 2012. Dendronized gold nanoparticles for siRNA delivery. *Small* 8(21): 3253–3256.

Kononenko V, Repar N, Marušič N, Drašler B, Romih T, Hočevar S, Drobne D 2017. Comparative in vitro genotoxicity study of ZnO nanoparticles, ZnO macroparticles and ZnCl2 to MDCK kidney cells size matters. *Toxicol In Vitro* 40: 256–263.

Landsiedel R, Kapp MD, Schulz M, Wiench K, Oesch F 2009. Genotoxicity investigations on nanomaterials: methods, preparation and characterization of test material, potential artifacts and limitations—many questions, some answers. *Mutat Res* 681(2–3): 241–258.

Lee M, Lim S, Kim C 2007. Preparation, characterization and in vitro cytotoxicity of paclitaxel-loaded sterically stabilized solid lipid nanoparticles. *Biomaterials* 28(12): 2137–2146.

Lee WH, Loo CY, Traini D, Young PM 2015. Inhalation of nanoparticle-based drug for lung cancer treatment: advantages and challenges. *Asian J Pharm Sci* 10(6): 481–489.

Lei R, Wu C, Yang B, Ma H, Shi C, Wang Q, Wang Q, Yuan Y, Liao M 2008. Integrated metabolomic analysis of the Nano-sized copper particle-induced hepatotoxicity and nephrotoxicity in rats: a rapid *in vivo* screening method for nanotoxicity. *Toxicol Appl Pharmacol* 232(2): 292–301.

Liao M, Liu H 2012. Gene expression profiling of nephrotoxicity from copper nanoparticles in rats after repeated oral administration. *Environ Toxicol Pharmacol* 34(1): 67–80.

Lozano-Fernández T, Ballester-Antxordoki L, Pérez-Temprano N, Rojas E, Sanz D, Iglesias-Gaspar M, Moya S, González-Fernández Á, Rey M 2014. Potential Impact of Metal oxide Nanoparticles on the Immune System: the Role of integrins, L-Selectin and the chemokine Receptor CXCR4. *Biol Med* 10: 1301.

Miethling-Graff R, Rumpker R, Richter M, Verano-Braga T, Kjeldsen F, Brewer J, Hoyland J, Rubahn HG, Erdmann H 2014. Exposure to silver nanoparticles induces size and dose-dependent oxidative stress and cytotoxicity in human colon carcinoma cells. *Toxicol Vitro* 28: 1280.

Miranda RR, Bezerra Jr AG, Oliveira Ribeiro CA, Randi MAF, Voigt CL, Skytte L, Rasmussen KL, Kjeldsen F, Filipak Neto F 2017. Toxicological interactions of silver nanoparticles and non-essential metals in human hepatocarcinoma cell line. *Toxicol Vitro* 40: 134–143.

Moghimi SM, Hunter AC, Murray JC 2005. Nanomedicine: status and prospects. *FASEB J* 19(3): 311–330.

Nohynek GJ, Lademann J, Ribaud C, Roberts MS 2007. Grey goo on the skin? *Nanotechnol Cosmet Sunscreen Saf Crit Rev Toxicol* 37: 251–277 and references therein.

Oberdörster G, Ferin J, Gelein R, Soderholm SC, Finkelstein J 1992. Role of the alveolar macrophage in lung injury: studies with ultrafine particles Environ. *Health Perspect* 97: 193–199.

Oberdorster G, Ferin J, Lehnert BE 1994. Correlation between particle-size, in-vivo particle persistence, and lung injury. *Environ Health Persp* 102: 173–179.

Oberdörster G, Oberdörster E, Oberdörster J 2005. Nanotoxicology: an emerging discipline evolving from studies of ultrafine particles. *Environ Health Perspect* 113: 823–839.

Oberdörster G, Sharp Z, Atudorei V, Elder A, Gelein R, Kreyling W, Cox C 2004. Translocation of inhaled ultrafine particles to the brain. *Inhal Toxicol* 16: 437–445.

Oberdörster G, Sharp Z, Atudorei V, Elder A, Gelein R, Lunts A, Kreyling W, Cox C 2002. Extrapulmonary translocation of ultrafine carbon particles following whole- body inhalation exposure of rats. *J Toxicol Environ Health A* 65: 1531–1543.

Peters K, Unger RE, Kirkpatrick CJ, Gatti AM, Monari E 2004. Effects of Nano-scaled particles on endothelial cell function in vitro: studies on viability, proliferation and inflammation. *Mater Sci Mater Med* 15: 321–325.

Petrarca C, Clemente E, Amato V, Pedata P, Sabbioni E, Bernardini G, Iavicoli I, Cortese S, Niu Q, Otsuki T, Paganelli R, Di Gioacchino M 2015. Engineered metal-based nanoparticles and innate immunity. *Clin Mol Allergy* 13(1): 13–18.

Pfaller T, Colognato R, Nelissen I, Favilli F, Casals E, Ooms D, Leppens H, Ponti J, Stritzinger R, Puntes V, Boraschi D, Duschl A, Oostingh GJ 2010. The suitability of different cellular in vitro immunotoxicity and genotoxicity methods for the analysis of nanoparticle-induced events. *Nanotoxicology* 4(1): 52–72.

Pietroiusti A, Massimiani M, Fenoglio I, Colonna M, Valentini F, Palleschi G, Camaioni A, Magrini A, Siracusa G, Bergamaschi A, Sgambato A, Campagnolo L 2011. Low doses of pristine and oxidized single-wall carbon nanotubes affect mammalian embryonic development. *ACS Nano* 5(6): 4624–4633.

Raj S, Jose S, Sumod US, Sabitha M 2012. Nanotechnology in cosmetics: opportunities and challenges. *J Pharm Bio allied Sci* 4: 186–193.

Rauckman EJ, Soiefer AI 2006. Acute toxicology. In Jacobson-Kram D Keller KA (Eds.), *Toxicological Testing Handbook Principles, Applications, and Data Interpretation*, 2nd Ed. Informa Healthcare USA, Inc., New York, pp. 135–147.

Risom L, Møller P, Loft S 2005. Oxidative stress-induced DNA damage by particulate air pollution. *Mutat Res* 592(1–2): 119–137.

Sahu SC, Roy S, Zheng J, Yourick JJ, Sprando RL 2014. Comparative genotoxicity of Nano silver in human liver HepG2 and colon CaCO2 cells evaluated by fluorescent microscopy of cytochalasin B-blocked micronucleus formation. *J Appl Toxicol* 34: 1200–1208.

Sahu SC, Zheng J, Graham L, Chen L, Ihrie J, Yourick JJ, Sprando RL 2014. Comparative cytotoxicity of Nano silver in human liver HepG2 and colon CaCO2 cells in culture. *J Appl Toxicol* 34: 1155–1166.

Semmler M, Seitz J, Erbe F, Mayer P, Heyder J, Oberdörster G, Kreyling WG 2004. Long-term clearance kinetics of inhaled ultrafine insoluble iridium particles from the rat lung, including transient translocation into secondary organs. *Inhal Toxicol* 16: 453–459.

Shi J, Sun X, Lin Y, Zou X, Li Z, Liao Y, Du M, Zhang H 2014. Endothelial Cell Injury and Dysfunction Induced by Silver Nanoparticles through Oxidative Stress via IKK/NF-κB Pathways. *Biomaterials* 35: 6657.

Suh WH, Suslick SK, Stucky GD, Suh YH 2009. Nano- technology, nanotoxicology, and neuroscience. *Pro Neurobiol* 87(3): 133–170.

Verma A, Stellacci F 2010. Effect of surface properties on nanoparticle-cell interactions. *Small* 6(1): 12–21.

Wang F, Gao F, Lan M, Yuan H, Huang Y, Liu J 2009. Oxidative stress contributes to silica nanoparticle-induced cytotoxicity in human embryonic kidney cells. *Toxicol Vitro* 23: 808–815.

Wang C, Wang H, Lin M, Hu X 2015. ZnO Nanoparticles Induced cytotoxicity on Human Pulmonary adenocarcinoma Cell Line LTEP-a-2. *Process Saf Environ Prot* 93: 265–273.

Xie H, Mason MM, Wise JP 2011. Genotoxicity of metal Nano- particles. *Rev Environ Health* 26(4): 251–268.

Yan G, Huang Y, Bu Q, Lv L, Deng P, Zhou J, Wang Y, Yang Y, Liu Q, Cen X, Zhao Y 2012. Zinc oxide nanoparticles cause nephrotoxicity and kidney metabolism alterations in rats. *J Environ Sci Health a Tox Hazard Subst Environ Eng* 47(4): 577–588.

Zuzana M, Alessandra R, Lise F, Maria D 2011. Safety assessment of nanoparticles cytotoxicity and genotoxicity of metal nanoparticles in vitro. *J Biomed Nanotechnol* 7(1): 20–21.

21 Nanobioremediation

A Novel and Sustainable Biological Advancement for Ecological Cleanup

Manoj Kumar Enamala, P. Divya Sruthi, Silpi Sarkar, Murthy Chavali, Induri Vasavi, and Chandrashekar Kuppam

CONTENTS

21.1 Introduction to Bioremediation 245
21.1.1 Aerobic versus Anaerobic 246
21.1.2 Nanobioremediation 248
21.1.3 Principle and Concepts 248
21.2 Role of Environment in Nanobioremediation 249
21.3 Role of Microorganisms in Balancing the Environment 249
21.3.1 Importance of Biological Sources (Bacteria and Fungi) 250
21.3.1.1 Bacteria 250
21.3.1.2 Fungi 250
21.3.2 Role of Soil and Soil Microorganisms 251
21.3.2.1 Microbial Activity in the Soil 251
21.4 Exploring the Role of Algae 252
21.5 Metabolic Pathways Involved in Treating Contaminated Sites 252
21.6 Applications of Nanobioremediation 252
21.6.1 Heavy Metal and Wastewater Remediation 253
21.6.2 Using Algae in Wastewater Remediation 254
21.6.3 Engineered Polymeric Nanoparticles for Soil Nanoparticles 254
21.6.4 Cleanup of Oil Spills 254
21.6.5 Degradation of PAHs 255
21.7 Conclusion 255
References 256

21.1 INTRODUCTION TO BIOREMEDIATION

In today's world, pollution is one of the major causes of the declining quality of the environment. There are various sources which help tackle environmental problems for the betterment of human health as well as the health of nature. There are various types of pollution that are classified based upon the kind of toxic wastes/heavy metals being released into the environment. "Remediate" is to solve a growing existing problem, and "bioremediate" means usage of biological organisms to solve environmental problems like polluted soil or contaminated groundwater. Bioremediation is a process used to treat contaminated media, including water, soil, and subsurface material, by altering environmental conditions to stimulate the growth of microorganisms and degrade specific pollutants. Bacteria, fungi, algae, protozoans, and other microorganisms are constantly at work breaking down organic matter in a non-polluted environment. Suppose a pollutant (organic) such as oil contaminates water or any environment: some of the microorganisms would die, while others capable of eating the organic pollution would survive. Bioremediation works by providing these pollution-eating organisms with fertilizer, oxygen, and other conditions that encourage their rapid growth. These microorganisms will break down the organic pollutant at a seemingly faster rate. In fact, bioremediation is often used to help clean up oil spills.

Bioremediation is a very important process of using organisms to neutralize or remove contamination from waste. This form of waste remediation uses no toxic chemicals, although it may use an organism that can be harmful under certain circumstances. A gross but simple explanation of bioremediation is the use of maggots in wound care control. Wounds that have contamination can have maggots introduced to them. The maggots then eat the contamination, allowing the wound to heal correctly. That is a form of medical bioremediation, but there are many other types that are used to control different types of waste contamination. Bioremediation provides a good cleanup strategy for some types of pollution, but as you might expect, it will not work for all. Nonetheless, bioremediation provides a technique for cleaning up pollution by enhancing the same biodegradation processes that occur in nature. For example, bioremediation may not provide a feasible strategy at sites with high concentrations of chemicals that are toxic to most microorganisms. These chemicals include metals such as cadmium or lead and salts such as sodium chloride. It also has the advantage of treating the contamination in place so that large quantities of soil, sediment, or water do not have to be dug up or pumped out of the ground for treatment. Bioremediation of a contaminated site typically works in one of two ways. In one case, ways are found to enhance the growth of whatever pollution-eating microbes might already be living at the contaminated site. In the second, which is a less common case, specialized microbes are added to degrade the contaminants, which can be in the presence or absence of oxygen. The concept of bioremediation is represented in Figure 21.1.

Most bioremediation processes involve oxidation-reduction (redox) reactions, where either an electron acceptor (commonly oxygen) is added to stimulate oxidation of a reduced pollutant (e.g., hydrocarbons) or an electron donor (commonly an organic substrate) is added to reduce oxidized pollutants (nitrate, perchlorate, oxidized metals, chlorinated solvents, explosives, and propellants). In both approaches, extra nutrients, minerals, vitamins, and pH buffers are added to optimize conditions for the microorganisms. In some cases, specialized microbial cultures are added (bioaugmentation) to further enhance biodegradation. Some examples of bioremediation-related technologies are bioventing, bioleaching, phytoremediation, mycoremediation, land farming, bioreactor, composting, bioaugmentation, rhizofiltration, and biostimulation (Figure 21.2).

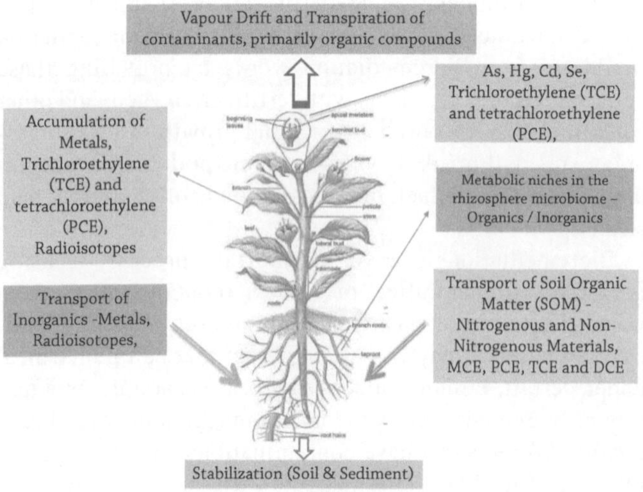

FIGURE 21.1 Concept of bioremediation.

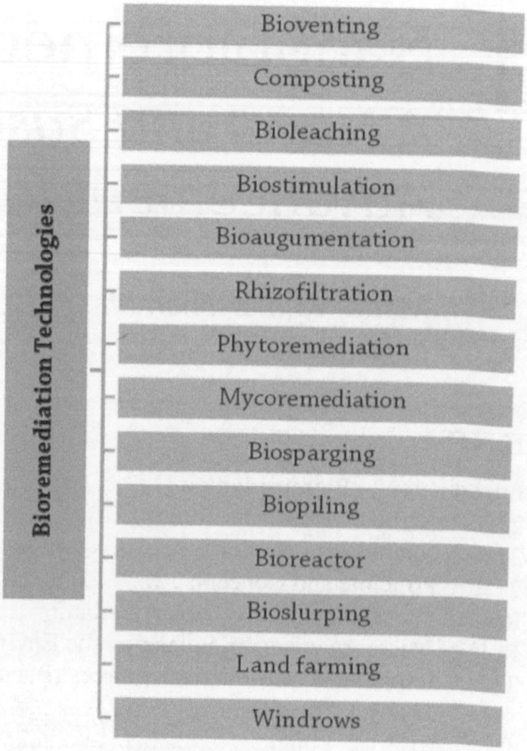

FIGURE 21.2 Bioremediation-related technologies.

21.1.1 Aerobic versus Anaerobic

Bioremediation processes take place both under oxic (with oxygen) or anoxic (without oxygen) conditions. When microorganisms like bacteria and fungi break down environmental contaminants for energy, the process is normally linked to the respiration of either oxygen or some inorganic molecule. The use of microorganisms with oxygen to break down environmental contaminants is referred to as aerobic bioremediation, and if it requires some other molecule besides oxygen, it is called anaerobic bioremediation. In aerobic bioremediation, microorganisms have oxygen and a contaminant available; they have the energy for growth and reproduction, which means the contaminant is broken down and removed from the environment. In anaerobic bioremediation, microorganisms have a contaminant and some molecule other than oxygen, which means there is energy available for the organism to grow and reproduce (contaminant serves as a food source). The main aspect in anaerobic bioremediation is that the organisms typically cannot carry out bioremediation in the presence of oxygen. In many cases, this is because oxygen interferes with the production or activity of the enzymes needed for anaerobic bioremediation.

The bioremediation process which involves an oxygen molecule as an electron acceptor, termed aerobic bioremediation, plays a crucial role in the oxidation of various hydrocarbons like petroleum, polyaromatic hydrocarbons (PAHs), phenols, and various other reduced pollutants. These acceptors can even be helpful in oxidizing other contaminants as well, like chlorinated ethenes (PCE, TCE, DCE, VC), chlorinated ethanes (TCA, DCA), chloromethanes (CT, CF), chlorinated cyclic hydrocarbons, various energetics (e.g., perchlorate, RDX, TNT), and nitrate.

Aerobic bioremediation is the most common oxidative bioremediation process, where oxygen is provided as the electron acceptor for oxidation of petroleum, polyaromatic hydrocarbons, phenols, and other reduced pollutants. Anaerobic bioremediation can be employed to treat a broad range of oxidized contaminants including chlorinated ethenes (PCE, TCE, DCE, VC), chlorinated ethanes (TCA, DCA), chloromethanes (CT, CF), chlorinated cyclic hydrocarbons, various energetics (e.g., perchlorate, RDX, TNT), and nitrate. A simple illustration to understand the types of bioremediation is given in Figure 21.3.

During the reaction process, the electron donor is oxidized whereas the electron acceptor is reduced. The most common and reliable electron acceptors and donors used in the bioremediation process are oxygen, nitrate, manganese (III and IV), iron (III), sulfate, carbon dioxide, and some pollutants like chlorinated solvents, explosives, oxidized metals, and radionuclides. The list of electron donors includes various sugars, fats, alcohols, natural organic material, fuel hydrocarbons, and a variety of reduced organic pollutants. The redox potentials for common biotransformation reactions are shown in Table 21.1.

A tabular column is portrayed to understand the various types of pollution which occur in our surroundings (Table 21.2). Among the various technologies available, bioremediation plays a major role in monitoring the health of the environment. Basically, bioremediation is defined as a process which involves the breakdown of organic and harmful compounds into simpler forms with the help of several microorganisms like bacteria, algae, and fungi. These microorganisms degrade the heavier compounds into simpler components like inorganic salts, carbon dioxide, and water. During the degradation process, the microorganisms release certain compounds which are nontoxic and non-harmful to the environment, and these metabolites are used in degrading the polluted/contaminated sites. In other words, it can also be defined as utilizing the microorganisms to neutralize the harmful pollutants from the polluted sites. This process occurs by natural means. The degradation of such compounds occurs by anaerobic as well as by aerobic digestion. Most of the bioremediation technologies involve oxidation and reduction reactions by exchanging/contributing electrons to reduce the various pollutants like nitrate, perchlorate, chlorinated solvents, etc. The various technologies available for bioremediation include phytoremediation, bioleaching, composting, mycoremediation, etc. (Alvarez and Illman, 2006).

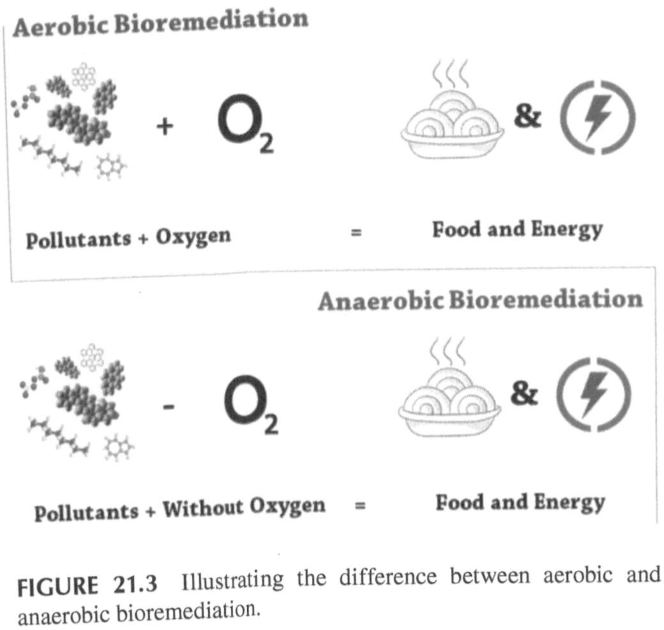

FIGURE 21.3 Illustrating the difference between aerobic and anaerobic bioremediation.

Basically, microorganisms are utilized in many ways in treating contaminated sites. Microorganisms selected according to the kind of contaminated area are supplied with essential nutrients which make them destroy the contaminants at the particular site. In the process of degrading pollutants by microorganisms, several environmental conditions play important roles based upon the situation present. The factors include variation in pH, temperature, growth of the microorganism, and even the percentage of moisture also depends on how far the process goes. The various microorganisms use their catabolic pathways for degrading these pollutants. The biodegradation of toxic compounds depends upon the condition of any microorganisms which are sensitive towards various environmental factors (Dzionek et al., 2016). The catabolic pathway is defined as a set of metabolic reactions which occur in the presence of oxygen and releases energy which is used in other reactions, like the anabolic reactions (Sharma et al. 2018).

Advantages:

Some of the major advantages of bioremediation are the following (Boopathy, 2000).

- Easy to match suitable microbes with the environmental conditions for degradation of the toxic substances.
- Completely a natural process.
- Almost no harmful side effects.
- The very high acceptance from regulatory authorities.
- Can be implemented in in situ as well as ex situ with conditions.
- Waste can be recycled.
- The turnaround is very fast to make the soil and water useful again for various other purposes.
- When the compounds/contaminants are degraded into various nontoxic compounds, this transfer does not involve the transfer of these contaminants into various other sites.
- Involves organic compounds/contaminants degraded into nontoxic compounds, and even these (organic compounds) are not transferred into any other site/area.

TABLE 21.1
Redox Potentials for Common Biotransformation Reactions

Process	Reaction	Redox potential (E_h in mV)
Aerobic	$O_2 + 4e^- + 4H^+ \rightarrow 2H_2O$	600 ~ 400
Anaerobic	Dechlorination of PCE	–600 ~ –400
Denitrification	$2NO_3^- + 10e^- + 12H^+ \rightarrow N_2 + 6H_2O$	500 ~ 200
Manganese(IV) reduction	$MnO_2 + 2e^- + 4H^+ \rightarrow Mn^{2+} + 2H_2O$	400 ~ 200
Iron(III) reduction	$Fe(OH)_3 + e^- + 3H^+ \rightarrow Fe^{2+} + 3H_2O$	300 ~ 100
Sulfate reduction	$SO_4^{2-} + 8e^- + 10H^+ \rightarrow H_2S + 4H_2O$	0 ~ –150
Fermentation	$2CH_2O \rightarrow CO_2 + CH_4$	–150 ~ –220

TABLE 21.2
Classification of Pollutants

Types of pollutants in nature	Description	Examples of pollutants
Quantitative pollutants	Already present in the environment and termed pollutants only when the concentration of carbon dioxide increases	Carbon Dioxide is the major one which exists as a quantitative pollutant
Qualitative pollutants	Any substance which is added into the environment by humans	Insecticides and pesticides
Primary pollutants	Emitted from exhaust fumes, ash, burning of fossil fuels, burning of plastics, etc.	Smoke, ash, nitric oxide, etc.
Secondary pollutants	Any substance formed by chemical reaction by burning of primary pollutants	Smog, ozone, sulfur trioxide
Biodegradable pollutants	Degraded by natural processes	Municipal sewage
Non-biodegradable pollutants	Degraded slowly (years) if at all	Aluminium, DDT, etc.

21.1.2 Nanobioremediation

Various environmental contaminants like heavy metals, organic, and inorganic pollutants being treated in combination with several nanoparticles/nanomaterials (NMs) and various plants and microbial sources like bacteria, fungi, etc., is termed "nanobioremediation." This technology is an emerging technology which is used in the removal of pollutants for cleaning the environment. Nanobioremediation technologies which are used currently include chemical and physical remediation, incineration, and bioremediation. With recent advances, bioremediation offers an environmentally friendly and economically feasible option to remove contaminants from the environment.

Nanobioremediation is a technique which helps in transforming the harmful pollutants into safer molecules using various microbes in combination with nanoparticles of less than 100 nm. Several nanoparticles like zinc, silver, gold, copper, etc., synthesized with numerous microorganisms as well as plants, are shown in the literature to be helpful in degrading the harmful pollutants. Researchers have researched widely by studying various combinations of nanoparticles synthesized and applied to remediation towards a cleaner and more sustainable environment. In terms of research, several researchers have used iron and zinc nanoparticles in cleaning up the environment. The iron nanoparticle technologies are currently used in cleaning up the groundwater as well as contaminated land areas by the process of absorptive/immobilization techniques and are even used in the reductive technologies, whereas zinc nanoparticles are used as a semiconductor photocatalyst having the capacity to degrade organic dyes. These zinc nanoparticles are found efficient in degrading a wide variety of compounds, dyes, organic acids, and even the pharmaceutical wastes which are left into the sewage system. Some of the various compounds being degraded using these metal nanoparticles (Davis et al., 2017) are given in Table 21.3.

When compared with various other metal nanoparticles, iron nanoparticles were found to possess/have more applications in degrading pollutants like pesticides, dyes, hydrocarbons, etc. (Davis et al. 2017).

21.1.3 Principle and Concepts

Nanobioremediation has the potential to reduce the overall costs in cleaning up large-scale contamination with reduced cleanup time. The basic principle of nanobioremediation is defined as degradation of organic wastes using nanocatalysts as a medium, which allows them to penetrate deep inside the contaminants and treat them safely without affecting the surroundings, with the help of several microorganisms (Yadav et al., 2017). These microorganisms are present everywhere in the world. However, these microorganisms have been found to have several advantages, like converting heavy metals into nontoxic forms with the help of mineralization of the organic contaminants to end products like CO_2 and H_2O or even to

TABLE 21.3
List of Organic Compounds Degraded into Simpler Compounds Using Nanoparticles

Nanoparticle	Degraded compounds
Iron	1. Chlorobenzene 2. Pyrene 3. RDX (hexahydro-1,3,5-trinitro-1,3,5-triazine) 4. Brominated methanes 5. Uranium 6. Pentachlorophenol 7. Several metal transition elements
Gold	1. Tertiary dye effluent 2. 4-Nitrophenol 3. Methylene blue
Zinc	1. Organic dyes 2. Brown CGG dye 3. Rhodamine B 4. Phenol 5. Congo red 6. Malachite green
Silver	– Organic dyes like methyl violet, methyl orange, methylene blue, 4-nitrophenol, Congo red and Coomassie brilliant blue
Copper	– Methylene blue, methylene orange, and dichloromethane

various metabolic intermediates which can be used as metabolites for the growth of these microorganisms. Moreover, the utilization of these toxic compounds leads to defense of their cell walls by production of degradative enzymes, which make them resistant to the various heavy metals which they consume. For a successful remediation process to take place, researchers should think in terms of a "designer microbes" approach where one can control the activity of microorganisms and understand their mechanism of growth activity in the contaminated sites and even monitor their response towards environmental changes. When these microorganisms absorb these contaminants, they develop special membranes around themselves which help them defend from the entry of foreign substances into the cells (Dixit et al., 2015).

The major concept which explains nanobioremediation is the size of the nanoparticles, because very tiny particles can to penetrate into the contaminated sites along with the various microorganisms like bacteria, fungus, etc. Many researchers have included this study as part of their research, because these nanoparticles can produce better results than microparticles, which are able to penetrate into the contaminated sites and engage in the cleanup process; however, this is much delayed when compared to the normal bioremediation technology. Some of the nano-based substances used for treating such contaminated sites are nanoclusters, nanomaterials, nanostructured materials, etc. These nanoparticles can be used further to immobilize the cells of microorganisms which can degrade or can be even used in the recovery of specific chemicals depending upon our interest.

21.2 ROLE OF ENVIRONMENT IN NANOBIOREMEDIATION

The organic compounds required to break down pollutants to clean up the environment are mostly microorganisms, fungi, and bacteria. Life is inextricably linked to the overall quality of the environment. Thus to preserve and sustain the environment, it is important to use nanobioremediation techniques, which is possible through the application of biotechnology. Environmental biotechnology applies genetic engineering, which improves the efficiency and cost by exploitation of microorganisms to reduce the environmental burden of toxic substances. Under circumstances of an environmental problem such as contaminated soil or groundwater, bioremediation is the use of biological organisms to solve such problems of contamination. Many major incidents in the past, such as the *Exxon Valdez* oil spill (Nixon and Michel, 2017), the Union-Carbide (Dow Chemicals) Bhopal disaster, large-scale contamination of Rhine River, progressive deterioration of aquatic habitats, release of radioactive material in the Chernobyl accident, etc., have revealed the necessity of preventing the escape of effluents into the surroundings.

21.3 ROLE OF MICROORGANISMS IN BALANCING THE ENVIRONMENT

There are several microorganisms in the environment growing at different rates which have been shown to possess a wide variety of functions and are considered to be ubiquitous. Many of the microorganisms are found in several habitats/environmental niches like the Dead Sea, the Himalayas, Yellowstone National Park, etc. These microorganisms play an essential role in recycling the gases present in the environment. The microorganisms are even responsible for driving the development of labile pools of nitrogen, hydrogen, sulfur, etc. A few examples of such organisms helping to balance the environment include *Halobacterium* (a salt marine/ocean bacterium) and *Chlamydomonas nivalis* (responsible for causing watermelon snow).

Various microorganisms have developed their cell wall structures to protect themselves from various damages caused by toxic pollutants by releasing wastes into freshwater lakes, etc., by various mechanisms like absorption, oxidation and reduction, uptake of nutrients, etc. These microorganisms, when used with various nanomaterials/particles, are very helpful in cleaning up the environment at a faster rate when compared with bioremediation. Apart from the normal bacteria which are found in water and air, the actinobacteria which are found in soil play an active role in the removal of xenobiotic heavy metals (Mahawar and Prasanna, 2018). When the process of nanobioremediation starts, the microorganisms which are present in the contaminated site release several extracellular enzymes which get mixed with xenobiotic compounds help them degrade the organic contaminants; this is the basic principle behind the degradation of organic compounds and can help in protecting the environment (Polti and Daniel, 2014).

21.3.1 Importance of Biological Sources (Bacteria and Fungi)

Bacteria and fungi are important potential sources. These microorganisms can even be used in combination to degrade various pollutants or discharge them into nature through various means. These sources are considered to be essential components in various ecological systems. Without the usage of nanomaterials, these sources will release various substances into the environment which interact with soils and organisms, whereas the nanomaterials present in the soil will certainly influence the nanoremediation of the contaminated soils. Some of the nanomaterials used in this particular field include carbon-related nanomaterials and metal nanomaterials. There are many carbon nanomaterials which are being investigated currently. Some studies reveal that several carbon nanomaterials, when introduced into the soil, enhance the activity of anammox bacteria when graphene oxide is used at a dose of 0.1 gl^{-1}. Bacteria grew 2–3 times better on the filters coated with 25 and 75 ml^{-1} graphene oxide than on the filters without graphene oxide.

21.3.1.1 Bacteria

Nowadays, bacteria are being used in various applications to protect the environment; among these, nanobioremediation processes are being considered for a variety of purposes. Various bacteria have adapted themselves in an efficient way for heavy metal pollution as well as degradation, mostly many of these same bacteria have shown up as part of the process of degrading the heavy metals and various other contaminants through the process of nanobioremediation. In addition to heavy metal remediation, there are several species recognized in recent times that have a role for even more intractable pollution, one example being *Deinococcus radiodurans*, a non-pathogenic solvent bacterium found in the soil which can be grown continuously. The major use of this bacterium is in reducing the radioactive wastes being discharged through nuclear power plants. There are four different clones available as well as several genes identified for this species which help in the degradation of metals present in the wastes. Some of the genes identified for metal degradation include the merA reductase gene, which codes for the mercuric ion reductase, which is highly toxic to humans (Brim et al., 2000). Some of the bacteria (Wan et al., 2018) used to degrade pollutants (Abatenh et al., 2017) are listed in Table 21.4.

TABLE 21.4
Various Bacteria Used in Bioremediation

Bacteria	Compounds degraded
Candida viswanathii	Phenanthrene, benzopyrene
Gloeophyllum striatum	Striatum pyrene, anthracene
Pseudomonas spp.	Aromatic hydrocarbons
Coprinellus radians	PAHs, methyl naphthalenes, dibenzofurans
Bacillus cereus	Diesel oil
Bacillus subtilis (Hay bacillus or Grass bacillus)	Phenols
Micro bacterium spp.	Aromatic hydrocarbons
B. brevis, B. licheniformis, B. sphaericus	Crude oil
Aneurinibacillus aneurinilyticus	Tolerates arsenic up to 4500 ppm (*helps in clearing arsenic-polluted water*)
Bacillus algicola (003-Phe1), *Rhodococcus soli* (102-Na5), *Isoptericola chiayiensis* (103-Na4), *Pseudoalteromonas agarivorans* (SDRB-Py1)	Degradation of hydrocarbons

Apart from the bacteria found in normal environments, bacteria found in marine environments are adapted to tolerate one of the most adverse environments available, i.e., the oceans, which can vary in temperature, pH, population regimes, and wind patterns. Since these bacteria are adapted to such harsh conditions, they can be even better used in degrading heavy metal compounds present in the environment. The major nutrients utilized by the marine bacteria for growth are sodium and potassium ions. Among them, the sodium is utilized in the production of indole from tryptophan, which helps in transport of substrates into the cell, which indirectly indicates how these marine bacteria if utilized will penetrate deep inside with the help of nanomaterials/nanoparticles and help in cleaning the polluted sites. Hence, marine bacteria are being utilized for various applications in the degradation of hazardous compounds (Dash et al., 2013). During bioremediation, process bacteria must adhere to surfaces through proper dispersion of cells.

Recently, a team of researchers developed a technique for attaching bacteria onto the surface of the inexpensive natural nanoparticles known as "Halloysites"; this technique roughens the surface and makes bacteria get easily attached to the surface and then help in the degradation of oil. The bacterium used in this research is *Alcanivorax borkumensis*, which is an alkanotrophic bacteria usually found in marine habitats (Panchal et al., 2018).

21.3.1.2 Fungi

Fungi are also considered to play important roles in the field of nanotechnology and in cleaning up toxic contaminants. Fungi are considered an important source because of the various extracellular syntheses they produce during growth, which result in the biosynthesis of nanoparticles. These nanoparticles are synthesized by the catalytic effects of the enzymes being secreted by fungi and have been found to be effective in the removal of pollutants and toxic effluents present in the environment. Some of the fungal sources used in the removal of pollutants are *Candida glabrata*, *Aspergillus niger*, *Aspergillus oryzae*, *Fusarium oxysporum*, *Coriolus versicolor*, *Fusarium semitectum*, and *Trichoderma viride*. The above sources have been utilized to synthesize various metal nanoparticles like silver, cadmium sulfide, gold, lead sulfide etc. (See Table 21.5).

TABLE 21.5
Various Fungi Used in Bioremediation

Fungi	Compounds degraded	From
Rhizophagus intraradices and *Robinia pseudoacacia*	Lead	Soil
Rhizophagus intraradices	Lead	Soil by inoculation into vetiver grass with *Glomus versiforme*
Arbuscular mycorrhizal fungi (AMF) and *Calendula officinalis*	Cadmium and lead	Soil
Phanerochaete	Diesel fuel	Diesel oil contamination
Crude extract from *Agaricus bisporus* (button mushrooms)	90% phenol	Polluted soils
AMF	Organic pollutants like benzene-, methyl tert-butyl ether, and ammonia	Wetlands and groundwater when inoculated into *Phragmites australis*
Aspergillus niger and *Phanerochaete chrysosporium*	Dyes such as malachite green, nigrosin, and basic fuchsin	Water
Aspergillus niger and *Phanerochaete chrysosporium*	Congo red, a carcinogenic dye	Water
Pleurotus ostreatus	Direct blue 14	Water
Fungal toxins	Petroleum fuels and phenols	Wastewater
Pleurotus ostreatus	Polychlorinated biphenyl (PCB) and 95% of PAHs	Contaminated soils
Aspergillus and other molds	Starches, hemicelluloses, celluloses, pectins, and other sugar polymers	–
Aspergillus spp.,	Intractable substrates as fats, oils, chitin, and keratin Paper and textiles (cotton, jute, and linen)	–
Ecuadorian fungus *Pestalotiopsis*	Polyurethane in aerobic and anaerobic conditions	Landfills
Edible mushrooms, *P. pulmonarius, L. squarrosulus* and *P. tuber-regium*	Different concentrations of crude oil, automotive gasoline oil (AGO), fresh engine oil, and spent engine oil	–
P. pulmonarius	Crude oil and palm kernel, Lead and zinc	Contaminated soils
Wood-decay fungi, especially white rot, secrete extracellular enzymes and acids	Lignin and cellulose	Woods, forests
Fungal extracellular ligninolytic enzymes like laccase and manganese peroxidase	Insecticide endosulfan, imazalil, thiophanate methyl, ortho-phenylphenol, diphenylamine, chlorpyrifos	Wastewater
Fungal extracellular ligninolytic enzymes like laccase and manganese peroxidase	Atrazine	Clay-loamy soils
Pleurotus, Aspergillus, Trichoderma	Lead, cadmium, nickel, chromium, mercury, arsenic, copper, boron, iron, and zinc	Marine environment, wastewater, and on land

21.3.2 Role of Soil and Soil Microorganisms

The contamination of soil is due to various activities being taken up by humans, which are leading to severe problems. The contaminants, when released into the environment, are causing major problems in agriculture, which researchers all over the world are trying to fix. The major contamination problem in agricultural soils is the release of various xenobiotics which are harming humans, livestock, agricultural crops, and wildlife and are even causing problems in balancing the environment and maintaining stable environmental conditions. This is all caused because of the soil being polluted, and this soil pollution is characterized by the stacking up of toxic compounds, chemicals, radioactive substances, etc., upon one another, which shows an adverse effect on the health of animals as well as the plants grown at the contaminated sites (Das and Adholeya, 2012). Apart from this, other causes of soil pollution are listed below:

- Discharge of industrial waste into the soil.
- Application of pesticides.
- Movement/discharge of contaminated water into the soil.

21.3.2.1 Microbial Activity in the Soil

Since microbes are present everywhere, they are unique in capturing/conquering the area they inhabit. Microbial activity has certain requirements like exact ambient temperature and pH to carry out chemical reactions, along with various other factors like nutrient availability, etc. Many organic pollutants do not degrade in the environment as soon as they are discharged. Some reasons for not getting immediate degradation of contaminants are as follows:

- High concentration of pollutants.
- Limited electron acceptors.
- Limited nutrients are available.
- Unfavorable environmental conditions like pH, temperature, redox status, etc.

Most of the organic matter in the environment is degraded only in the presence of the electron acceptor known as oxygen, and it takes place stepwise, which involves various microorganisms which participate in the mineralization process. Many researchers have shown that mixed communities of microbes have proven effective in mineralizing several pollutants like aromatic hydrocarbons, alkyl benzoyl sulfonates, etc. When these microbes are used in combination, they can significantly enhance the degradation of pollutants and prevent the accumulation of toxic intermediates. In oxygen-rich soils, electron acceptors are involved in microbial degradation. The microorganisms utilize various processes like denitrification; methanogenic and sulfate-reducing bacteria are used in this regard. If the soil is contaminated with chlorine, it takes a particularly long time to remove. Persistent chlorine chemicals include PCBs (polychlorinated biphenyls), PCE (perchloroethylene), etc. (Bollag et al., 1994).

21.4 EXPLORING THE ROLE OF ALGAE

Nowadays, apart from the normal flora of bacteria and fungi, there are several other species being researched for the betterment of nature as well as society. Among them, algae are well-known organisms which are currently used in the field of biofuel production and wastewater treatment, and also for removal of organic as well as heavy metals from the wastes being discharged by many industries. Among the various heavy metals, cadmium is considered to be the most toxic substance available currently; it has very ill effects on the health of humans. Hence, researchers have used algae known as *Distigmaproteus*, which is a Euglenophyta which was isolated from industrial waters and plays a possible role in cleaning up contaminated regions/places. The research (Farah and Shakoori, 2007) found that this algae (*Distigmaproteus*) reduced the cadmium contamination by 48% in 2 days, 68% in 4 days, 80% after 6 days, and 90% after 8 days, which shows that this species of algae is a potent source for metal detoxification and can be used in environmental cleanup (Farah and Shakoori, 2007). Another group of researchers discovered that *Trentipholia* (filamentous chlorophyte green algae which are found growing on the surfaces of buildings) can act as a bioindicator for metal pollution in the environment. These researchers have confirmed the presence of these particular algae by various analyses like SEM, X-ray fluorescence spectroscopy, and SEM-EDS. Algae accumulate such metals from the environment inside their cells for their proper metabolic functioning, thus making advantages for themselves as well as cleaning up the environment.

Cyanobacteria strains have also shown their potency in degrading hydrocarbon-polluted sites, and several strains like *Microcoleus chthonoplastes* and *Phormidium corium* have the ability to degrade alkanes. The reason for using the cyanobacterial strains is that they provide oxygen for efficient degradation of hydrocarbons when available with a group of other organisms. Researchers studying the various strains of cyanobacterial species available at a contaminated site found that *Leptolyngbya* spp., *Planktothrix* spp., *Tolypothrix* spp., *Anabaena* spp., *Oscillatoriales* spp., and *Arthronema* spp. were present and that they were responsible for the degradation of hydrocarbons at the site (Akoijam et al. 2015).

21.5 METABOLIC PATHWAYS INVOLVED IN TREATING CONTAMINATED SITES

Most contaminated sites are degraded under aerobic conditions (Das and Chandran, 2011). Initially, when the organic pollutants come in contact with the microorganisms, the intracellular attack is initiated and oxygen molecules are incorporated by the activation of several enzyme groups known as oxygenases and peroxidases. The degradation of these organic pollutants (for pathway, see Figure 21.4) is carried out in a step-by-step process in which several intermediates are formed through the intermediated central metabolism.

During this process, several enzymes participate in the degradation process, like cytochrome P450, which plays an important role in the microbial degradation of several oils and hydrocarbons depending upon the chain length and enzymes needed to initiate the degradation of such heavy molecules. Some of the enzymes used in the degradation of toxic compounds (Das and Chandran, 2011) in the environment are shown in Table 21.6.

21.6 APPLICATIONS OF NANOBIOREMEDIATION

There are numerous applications of nanomaterials in the field of environmental science like cleaning polluted water

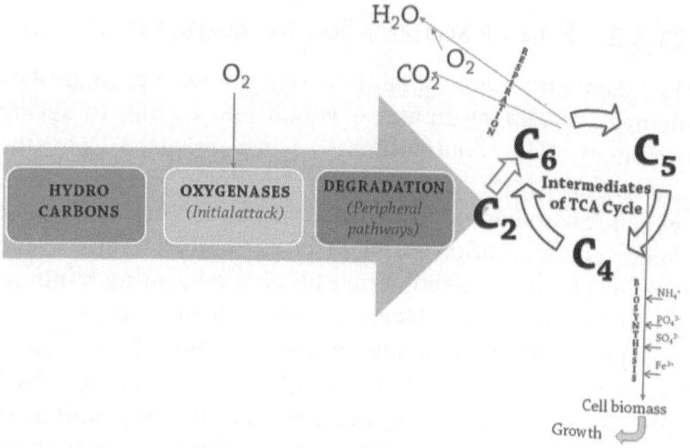

FIGURE 21.4 The metabolic pathway for degradation of hydrocarbons.

TABLE 21.6
Several Enzymes Used in Degrading Various Pollutants Available in the Environment

Enzyme	Substrates on which these enzymes act for degradation
Alkane hydroxylases	Fatty acids, alkylbenzenes, cycloalkanes
Eukaryotic P450	C_{10}-C_{16} alkanes, fatty acids
Bacterial P450	C_5-C_{16} alkanes and cycloalkanes
Soluble methane monooxygenases	C_1-C_8 alkanes, alkenes, and cycloalkanes
Dioxygenases	C_{10}-C_{30} alkanes

and detecting and cleaning other environmental contaminants (which include waste and toxic substances). The major advantages of using these technologies are that they have high competence, are selectable to selective metals, and do not have any specific nutritional requirements; there even exists a possibility of metal recovery. Various nanomaterials used currently include oxide nanomaterials, carbon-type nanomaterials, polymer nanomaterials, nanocomposites, and biological nanomaterials (Rizwan et al., 2014). Apart from these various applications, nanoparticles have been found as an effective way of treating heavy metals through the technology known as nanotechnology. The technology used for cleaning polluted sites with the help of nanotechnology is known as nanobioremediation; these technologies not only reduce costs of cleaning up contaminated sites on a large scale but also reduce the process time. In this chapter, we have discussed all the possible applications of nanoremediation with various biological sources being used in this particular field of technology (Dixit et al., 2015). Various categories of application of nanomaterials in bioremediation (nanobioremediation) are shown in Figure 21.5.

21.6.1 Heavy Metal and Wastewater Remediation

Nanoparticles with desirable shapes and sizes can be synthesized for the removal of heavy metals and other contaminants in the wastewater treatment. When nanoscience and nanoengineering are combined together, the end result is better opportunities for the applications being used currently in nanobioremediation. The combination of nanoscience and nanoengineering offers unimaginably better opportunities while restoring the original flavor of the treated wastewater; the wide variety of new properties they offer, when combined with conventional treatment methodologies, will make many improvements in the treatment methodologies of the future. Heavy metal pollution, when ingested, causes severe health effects for all living beings on this planet like hypertension, renal failure, emphysema, etc. Keeping this in mind, several researchers have developed their own technologies like graphene derivatives, carbon-based sorbents, activated charcoal, clay minerals, etc. Magnetite nanoparticles, when fixed on silica compound, are used for removal of a large number of toxic substances which usually exist in the environment.

Nanostructured silica is another way in which this method can be used in wastewater treatment to eliminate these heavy metal ions. Calcium, when used in combination with zinc nanoparticles, is being used effectively in removing lead ions from contaminated water. Metal oxidized nanomaterials in combination with nano-sized magnetite and titanium dioxide have also resulted in good removal of heavy metals from contaminated water. Huang et al. (2018) have synthesized titanate nanoflowers; their large surface area absorbs heavy metals with ease. These synthesized nanoflowers exhibit greater absorption capacity when compared to titanate nanotubes and nanowires.

Wastewater treatment is a complex process, taking into consideration various factors like quality standards, cost, and efficiency. In this regard, several researchers have found that iron nanoparticles (nano-sized zero-valency ions, otherwise known as NZVI) and polymer-coated nanoparticles play a significant role in the removal of heavy metals like Cr(VI), As(III), etc. The removal of heavy metals using this NZVI principle has been done using various media, showing that it has an outstanding ability to remove heavy metals from contaminated sources of water. Carbon tetrachloride, pentachlorobenzene, chloro- and dichlorobenzene, chloromethane, acid red, and acid orange are a few types of pollutants being released and also treated with the nano iron technology or NZVI technology.

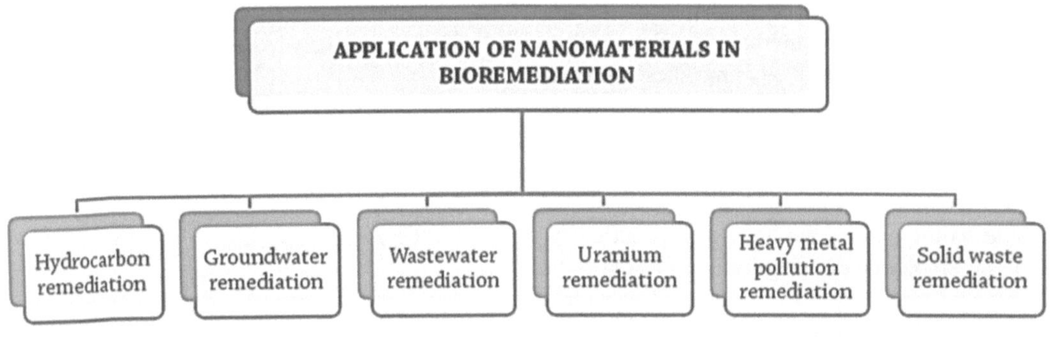

FIGURE 21.5 Applications of nanomaterials in bioremediation (nanobioremediation).

21.6.2 Using Algae in Wastewater Remediation

Currently, a huge amount of waste is being discharged into many freshwater lakes, which causes major pollution to most of the species which dwell in these freshwater bodies. The wastes which are discharged in the freshwater lakes are a major source of pollution. In current technologies, various microorganisms are used in the biological treatment of various wastewaters. Upon further exploration, microalgae have been found to possess better and more efficient activity when compared with various technologies available. These microalgae can be easily grown with some of the simple media constituents available. Algae can even grow with the nutrients available in the wastewater. The various dissolved salts like phosphates, nitrates, carbon sources, etc., are being absorbed by the algal cells (internally) and utilized in the growth of their biomass. Algae are also found helpful in the removal of heavy metals from wastewater. Various ions are absorbed by the cells of microalgae and are used in their metabolic pathways for the growth of cells. These algae can treat various kinds of wastes like sewage waste, agro-industrial wastes, hog farm wastes, effluent wastes, and wastes from various food industries. This process of wastewater treatment, in which photosynthetic algae species were combined with various nanomaterials, was found to be much more effective in the field of wastewater treatment. The various nano techniques used for cleaning up wastewater include nanoscale filtration techniques, absorption of various pollutants onto the surface of nanomaterials, and breakdown of contaminants by nanoparticle catalysts. The various membranes used in the treatment, including polymeric and ceramic membranes, have been found to have a great impact on wastewater treatment. Some of the absorbents being used for absorbing contaminants onto their surfaces include activated carbon, clay minerals, dendrimers, nano reactive membranes, metal oxides, etc.

21.6.3 Engineered Polymeric Nanoparticles for Soil Nanoparticles

There are several contaminants present in the groundwater like polynuclear aromatic hydrocarbons, which get attached to the soil and are therefore difficult to remove. Hence, researchers have synthesized a few particles known as amphiphilic polyurethanes (Rizwan et al., 2014) for soil contaminated with PAHs. Nanoparticles of this kind, when emulsified in water, lead to the synthesis of new molecules known as polyurethane acrylate anionomer (UAA), which are modified urethane acrylate precursor chains synthesized in the range of 17–97 nm. These particles have the ability to absorb the PAHs present in the groundwater. These particles are stable, and they are independent for concentration in the aqueous phase. The mobility of these particles is dependent on the charge density or the size of the water-soluble chains which reside on the particle surface. In a recent development, a solution of nanoparticles using iron was found to be nontoxic, highly reactive, and very effective at quickly removing soil contaminants. This engineered iron nanoparticle solution is effectively used in treating chlorinated hydrocarbons and various heavy metals present in the soil that are ultimately toxic to living beings. The process is as follows. The nanoparticles are first injected into the soil; these iron nanoparticles react with the contaminants, absorbing them. This mechanism can also be applied to treat industrial wastes containing chlorinated solvents. The products released after the reaction are harmless, and they release a by-product known as ethylene, which is used as a chemical product in the plastic industry.

21.6.4 Cleanup of Oil Spills

Oil spillage/petroleum spillage causes negative impacts on the health of living beings, which leads to the necessity of cleaning up soils present on the shoreline. There must be some effective oil remediation technologies for the cleanup process. There are several technological methods available like solvent extraction, phytoremediation, electrokinetic bioremediation, and thermal technologies, etc., which have made a certain impact in terms of cleaning up the oil spillage in the oceans (Lim et al., 2016). Hydrocarbon degradation rates in different environments are shown in Figure 21.6.

Currently, oil spills are causing major problems all around the world. Such events happen during wreckage of huge ships in the seas. During oil spills, huge quantities of hydrocarbons are released into the environment, thus causing great damage to the life in the oceans, and the hydrocarbons also slowly merge with local tributaries as well as other water bodies. In the year 2015, around 7,000 tons of oil was released into the marine environment through various pipelines, tankers, etc. (see Table 21.7). Therefore, for the treatment of such oil spill sites, a variety of microorganisms have played an important role through bioremediation because such microbes are better adjusted/adapted to the conditions requiring treatment by bioremediation. Among the various microbes identified, *Bacillus subtilis* has been shown to degrade pollutants like n-alkanes, which are considered to be the most toxic categories (Tao et al., 2016).

Alcanivorax cycloclasticus and *Rhodococcus* are considered to be the key organisms which play major roles in the

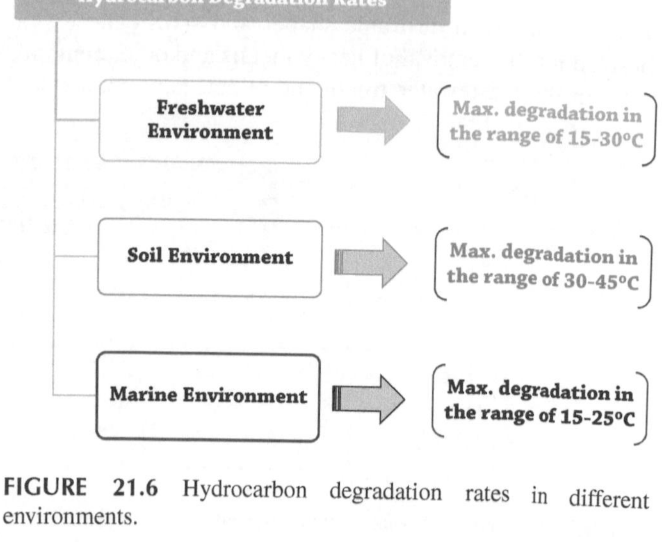

FIGURE 21.6 Hydrocarbon degradation rates in different environments.

TABLE 21.7
World's Largest Oil Spills Recorded in History

Location	Type of oil	Year of spill	Amount of oil spilled (millions of gallons)
Gulf of Mexico	Crude oil	2010	210
Caribbean Sea	Light crude oil	1979	9
Persian Gulf, Kuwait	Crude oil	1991	240–336
Saklanha Bay, South Africa	Light crude oil	1983	78.5
Mediterranean Sea, Italy	Crude oil	1991	45

degradation of petroleum hydrocarbons. Several other genera/strains which are efficient at degrading oil as well as abundant include *Acinetobacter, Bacillus, Pseudomonas, Micrococcus*, and *Alteromonas*. During the degradation of oil spills, researchers have identified several/numerous compounds like resins, asphaltenes, aromatics, etc., and have found that among the various classes of these hydrocarbons the polycyclic aromatic hydrocarbons are degraded much slower when compared with compounds of lower molecular weights.

Currently, nanotechnology is playing a crucial role in every field of science and engineering. Nanotechnology has also made its footprint in the field of oil spillage cleanup processes. Several new techniques have been introduced and are yet to be applied for clearing the oil spills using microbial communities. Some of the various oil-cleaning nanotechnologies which are available are described in Table 21.8.

The following are the mechanisms which are involved in the degradation of hydrocarbons. The hydrocarbon degradation is carried out by enzymatic degradation according to the steps listed.

1. The initial attack is by membrane-bound or intracellular oxygenases and peroxidases, where the hydrocarbon is activated and oxygen is incorporated.
2. The hydrocarbons are degraded in multiple steps in peripheral degradation pathways to substrates of important central intermediary metabolism pathways like the TCA cycle (6, 14). Consequently, the hydrocarbons are utilized by the microbes for respiration as well as for biosynthesis.

TABLE 21.8
Various Nanotechnologies/Materials Used in Oil-Cleaning Technology

Nanotechnology	Materials Used for Oil Spills
Magnetic nanocomposites	1. Magnetic carbon metal nanocomposites 2. Magnetic polymer nanocomposites 3. Iron oxide nanoparticles in organo clays 4. Hydrophobic core shell magnetic iron oxide nanoparticles
Filters/pads	1. Oil repellant/ hydrophilic cotton filter 2. Super-hydrophobic filter paper
Carbon nanostructures	1. Graphene worms 2. Carbon nanotube sponges 3. Vertically aligned CNT membranes
Aero gels	1. Hydrophobic aero gels 2. Rice husk-derived silica aero gels 3. Aero clays 4. Ambient pressure-dried silica aero gel

The bioremediation of hydrophobic compounds by microbial surfactants is done by two main mechanisms. The first one includes the emulsification process, and the second includes the formation of micelles by decreasing the interfacial tension and increasing the solubility.

21.6.5 Degradation of PAHs

Polycyclic aromatic hydrocarbons have originated from natural activities through various anthropogenic activities. These enter into the environment in various ways like chronic pollution/acute pollution through oil spills. In this regard, several researchers have been researching clearing the pollution caused by PAH degradation. Bacteria are considered as active organisms which play an important role in the degradation of organic pollutants. Benzopyrene is considered the most carcinogenic waste being released, and bacteria are capable of degrading even these carcinogens when grown in a liquid media. Several species of bacteria like *Pseudomonas, Agrobacterium, Bacillus, Burkholderia*, and *Sphingomonas* were found to degrade these PAHs effectively.

21.7 CONCLUSION

Nanotechnology has made our lives simpler, and the various unique characteristics it exhibits have made it a technology of choice by many researchers worldwide. Nanomaterials are being used in various applications like manufacturing building materials, treating various environmental pollutants, and even providing foods for microorganisms to feed on. Biosynthesis of these nanoparticles helps in minimizing the use of harmful chemicals and solvents. This process is simple as well as cost effective. Nanotechnology is revolutionizing the way we live. The unique characteristics of nanoparticles have made them the particle of choice in many fields, including the remediation of environmental pollutants. Eco-friendly synthesis of nanoparticles coupled with remediation can go a long way in promoting sustainability. Biosynthesis helps minimize the use of harmful chemicals and solvents and is simple, cost-effective, and time-saving. Nanomaterials exhibit unique physical and chemical properties; hence, they have received much attention from scientists and researchers in different areas of environmental sciences, specifically in bioremediation. Bioremediation provides a good cleanup strategy for some types of waste, but as expected, it will not be useful for all. For example, conventional bioremediation may not provide a feasible strategy at sites with high concentrations of chemicals that are toxic to most microorganisms. These include heavy metals and salt. Further, the advancement in science and technology has increased the standard of living, which directly or indirectly contributes to the increase in waste and toxic material. Therefore, the remediation of contaminants by the use of existing technology is not effective or efficient in cleaning up the environment. Hence, nanomaterials may be applied for bioremediation, which will not only have a less toxic effect on microorganisms but also improve the microbial activity of the specific waste and toxic material, which will reduce the overall time and cost. In this chapter, we have briefly summarized the major types of nanomaterials that have been used so far in the bioremediation of waste and toxic materials.

Nanomaterials have received appropriate attention by many researchers because they exhibit unique physical and chemical properties in the various fields of science and engineering and specifically in the field of bioremediation. The bioremediation process gives a good cleanup strategy if the polluted sites are heavily contaminated with pollutants that are toxic to most microorganisms dwelling there. Hence, the remediation of pollutants using the current technologies is not that effective and efficient in cleaning up the environment. The nanomaterials being used in these applications are seen to have a less toxic effect on the environment and also to allow progressive cleaning of contaminated sites. This chapter deals with various nanomaterials that are used in the field of bioremediation for treating the contaminated sites. Therefore, it can be definitely foreseen that nanotechnology will play an important role in the process of nanobioremediation, which can clean faster and safer than other processes.

It can be definitely concluded that nanoparticles and nanotechnological instruments play an efficient role in the process of nanobioremediation. Applying nanobioremediation to environmental hazards can clean them faster and safer than other methods and technology. It can be assumed that nanobioremediation maintains all three criteria, i.e., biological, social and economical. This chapter mainly focuses on the importance of nanomaterials in the degradation of waste and toxic substances. These nanomaterials not only increase the degradation of waste but also help to increase the efficiency of microorganisms in degrading waste and toxic materials. Therefore, nanotechnology has great applications in bioremediation. The most predominant heavy metals found in the environment include lead, mercury, copper, arsenic, cadmium, zinc, etc. Therefore, emerging technology in the field of nanomaterials will definitely treat the environment at a faster rate, which will enable us to specifically select the pollutants which need to be cleaned up and make the environment free from such contamination. NMs not only directly catalyze the degradation of waste and materials toxic to the microorganism, but also enhance the efficiency of microorganisms in the degradation of waste and toxic materials. This also shows that phytoremediation can be applied in the removal of toxic heavy metal from contaminated soil. Therefore, based on the above discussion, it can be said that, like in other scientific fields, nanotechnology has immense applications in bioremediation too.

Heavy metals that are predominantly considered as toxic materials that need immediate remediation are lead, mercury, copper, arsenic, cadmium, chromium, etc. Emerging applications of NMs will endeavor to find an effective remediation solution for removing these heavy metals. Furthermore, the specific control and design of NMs at the molecular level will give increased affinity, capacity, and selectivity for pollutants, which in consequence will lead to the reduced release of hazardous substances into the air and water, providing safe drinking water. This chapter finally concludes that the proposed nanobioremediation can be termed nanorenovogen, which is a future nanotechnology-based bioremediation process for contaminant removal. Due to its maturity and potentially effective nature, it is anticipated that nanobioremediation applications will increase with countless leaps towards boundless applications in the coming decades, and they will play a serious and key role in sustainable development.

REFERENCES

Abatenh, E., Gizaw, B., Tsegaye, Z., and Wassie, M. (2017). The role of microorganisms in bioremediation – A review. *Open J. Environ. Biol.* 2, 38–46.

Akoijam, C., Langpoklakpam, J. S., and Chettri, B. (2015). Cyanobacterial diversity in hydrocarbon-polluted sediments and their possible role in bioremediation. *Int. Biodeter. Biodegr.* 103, 97–104. doi: 10.1016/j.ibiod.2015.03.035.

Alvarez, P. J., and Illman, W. A. (2006). *Bioremediation and natural attenuation: Process fundamentals and mathematical models.* Wiley Subscription Services, Inc., A Wiley Company.

Bollag, J., Mertz, T., and Otjen, L. (1994). "Role of microorganisms in soil bioremediation." In *Bioremediation through Rhizosphere Technology* (ACS Symposium Series), 2–10. doi: 10.1021/bk-1994-0563.ch001, ACS Publications.

Boopathy, R. (2000). Factors limiting bioremediation technologies. *Bioresour. Technol.* 74, 63–67. doi: 10.1016/S0960-8524(99)00144-3.

Brim, H., Mcfarlan, S. C., Fredrickson, J. K., Minton, K. W., Zhai, M., Wackett, L. P., et al. (2000). Engineering Deinococcus radiodurans for metal remediation in radioactive mixed waste environments. *Nat. Biotechnol.* 18, 85–90.

Das, M., and Adholeya, A. (2012). "Role of microorganisms in remediation of contaminated soil." In *Microorganisms in Environmental Management:Microbes and Environment*, eds. T. Satyanarayana, B. Johri, and A. Prakash (Dordrecht/Heidelberg/London/New York: Springer), 81–112. doi: 10.1007/978-94-007-2229-3.

Das, N., and Chandran, P. (2011). Microbial degradation of petroleum hydrocarbon contaminants: An overview. *Biotechnol. Res. Int.* 2011, 941810. doi: 10.4061/2011/941810.

Dash, H., Mangwani, N., Chakraborty, J., Kumari, S., and Das, S. (2013). Marine bacteria: Potential candidates for enhanced bioremediation. *Appl. Microbiol. Biotechnol.* 97, 561–571. doi: 10.1007/s00253-012-4584-0.

Davis, A. S., Prakash, P., and Thamaraiselvi, K. (2017). "Nanobioremediation technologies for sustainable environment." In *Bioremediation and Sustainable Technologies for Cleaner Environment*, eds. M. Prashanthi, R. Sundaram, A. Jeyaseelan, and T. Kaliannan (Springer), 13–34. doi: 10.1007/978-3-319-48439-6.

Dixit, R., Malaviya, D., Pandiyan, K., Singh, U. B., Sahu, A., Shukla, R., et al. (2015). Bioremediation of heavy metals from soil and aquatic environment: An overview of principles and criteria of fundamental processes. *Sustainability* 7, 2189–2212. doi: 10.3390/su7022189.

Dzionek, A., Wojcieszy, D., and Guzik, U. (2016). Natural carriers in bioremediation: A review. *Electron. J. Biotechnol.* 23, 28–36. doi: 10.1016/j.ejbt.2016.07.003.

Farah, A. R., and Shakoori, R. S. A. R (2007). Heavy metal resistant Distigma proteus (Euglenophyta) isolated from industrial effluents and its possible role in bioremediation of contaminated wastewaters. *World J. Microbiol. Biotechnol.* 23, 753–758. doi: 10.1007/s11274-006-9291-5.

Huang, J-Q., Huang, Z., Wang, G., Wang, M-l., Cao, Y-G., and Hong, M-C. (2008). Facile synthesis of titanate nanoflowers by a hydrothermal route. *Crys. Grow. & Des.* 8, 2444–2446. doi: 10.1021/cg800030y.

Lim, M. W., Lau, E. Von, and Poh, P. E. (2016). A comprehensive guide of remediation technologies for oil contaminated soil — Present works and future directions. *Mar. Pollut. Bull.* 109, 14–45. doi: 10.1016/j.marpolbul.2016.04.023.

Mahawar, H., and Prasanna, R. (2018). Prospecting the interactions of nanoparticles with beneficial microorganisms for developing green technologies for agriculture. *Environ. Nanotechnol. Monit. Manag.* 10, 477–485. doi: 10.1016/j.enmm.2018.09.004.

Nixon, Z., and Michel, J. (2017). A review of distribution and quantity of lingering subsurface oil from the Exxon Valdez Oil Spill. *Deep Res. Part II* 147, 20–26. doi: 10.1016/j.dsr2.2017.07.009.

Panchal, A., Swientoniewski, L. T., Omarova, M., Yu, T., Zhang, D., Blake, D. A., et al. (2018). Bacterial proliferation on clay nanotube pickering emulsions for oil spill bioremediation. *Colloids Surf. B Biointerfaces* 164, 27–33. doi: 10.1016/j.colsurfb.2018.01.021.

Polti, M. A., and Daniel, J. (2014). "Role of actinobacteria in bioremediation." In *Microbial Bioegradation and Bioremediation*, ed. S. Das (Elsevier Inc), 269–286. doi: 10.1016/B978-0-12-800021-2.00011-X.

Rizwan, M., Singh, M., Mitra, C. K., and Morve, R. K. (2014). Ecofriendly application of nanomaterials: Nanobioremediation. *J. Nanoparticles* 2014, 1–7.

Sharma, B., Dangi, A. K., and Shukla, P. (2018). Contemporary enzyme based technologies for bioremediation: A review. *J. Environ. Manage.* 210, 10–22. doi: 10.1016/j.jenvman.2017.12.075.

Tao, K., Liu, X., Chen, X., Hu, X., Cao, L., and Yuan, X. (2016). Biodegradation of crude oil by a defined co-culture of indigenous bacterial consortium and exogenous Bacillus subtilis. *Bioresour. Technol.* 224, 327–332. doi: 10.1016/j.biortech.2016.10.073.

Wan, D., Lee, H., Kwon, B., Seong, J., Hyuk, U., Seok, B., et al. (2018). Biosurfactant-assisted bioremediation of crude oil by indigenous bacteria isolated from Taean beach sediment. *Environ. Pollut.* 241, 254–264. doi: 10.1016/j.envpol.2018.05.070.

Yadav, K. K., Singh, J. K., Gupta, N., and Kumar, V. (2017). A review of nanobioremediation technologies for environmental cleanup: A novel biological approach. *J. Mater. Environ. Sci.* 8, 740–757.

22 Nanotoxicity

Physicochemical and Biological (PK-PD) Perspectives

Mithun Rudrapal, Dipak Chetia, Khomendra Kumar Sarwa, and Soumya Bhattacharya

CONTENTS

22.1 Introduction 259
22.2 Nanotoxicity.. 260
  22.2.1 What Is Nanotoxicity? 260
  22.2.2 Environmental and Health Hazards 260
  22.2.3 Human Health and Safety Concerns.. 260
  22.2.4 Toxicity Risk Assessment 261
  22.2.5 Preventive Measures 261
22.3 Nanomedicine.. 262
  22.3.1 What Is Nanomedicine?... 262
  22.3.2 Why Nanomedicine?.. 262
  22.3.3 Nanobiopharmaceuticals 263
  22.3.4 Applications.. 263
22.4 Nanotoxicity: Physicochemical and Biological Aspects 264
  22.4.1 Physicochemical Aspects.. 264
  22.4.2 PK Aspects 264
  22.4.3 Mechanism of Toxicity 265
22.5 Conclusion 266
References.. 266

22.1 INTRODUCTION

Nanotechnology is defined as the intentional design, characterization, production, and application of materials, structures, devices, and systems by controlling their size and shape in the nanoscale range (1–100 nm). The term "nanotechnology" was coined by Norio Taniguchi in 1974 (www.nanowerk.com). Nanotechnology creates nanomaterials with modified particle properties and potential applications. Nanoparticles (NPs) play a central role in many natural processes and also exist abundantly in the environment, like ultrafine particles, coal and silica dust, asbestos, and mineral fibers. The commercial utility of nanomaterials include cosmetics, foods, drugs/medicine, fertilizers, clothing, computer/television screens, and biomedical devices/equipment. Nanomaterials are manufactured and released into the market without much safety testing or environmental impact assessment. For this reason, they impart certain adverse effects to humans, animals, and environmental systems. NPs have greater risk of producing biological toxicity than larger particles (Mohanty et al., 2009).

In recent years, the term "nanobiotechnology" has been introduced in the field of nanoscience and nanotechnology research. It refers to the combination of nanotechnology with biological systems. The application of nanobiotechnology in medicine and health care is termed "nanomedicine" (Fakruddin et al., 2012). It is the transformation of fundamental nanomaterials into nanoparticulate drug delivery systems intended for the diagnosis and treatment of human diseases (Salata, 2004). Thus, nanomedicine refers to the application of nanotechnology in the field of pharmaceutical and biomedical sciences. It encompasses the monitoring, repair, construction, and control of biological systems like the human body at the molecular level, using engineered nanodevices and/or nanodrugs. It is a large field ranging from in vitro and in vivo diagnostics to therapy, including targeted drug delivery and regenerative medicine (Sarwa et al., 2015a; Sarwa et al., 2015b).

Today, the growing concern of nanomedicine in healthcare and biomedical research is due to its higher therapeutic payload over conventional medicine. It is claimed that characteristic physicochemical properties, modified pharmacokinetic (PK) profile, and well-defined target specificity (at the molecular level in biological systems) make engineered nanostructures/nanodevices more effective as targeted delivery systems for diagnostics/drugs over conventional delivery systems. Some nanostructures/nanocarriers/nanodevices used in drug delivery systems/medicine/diagnostics include nanotubes, nanometals, nanowire, nanoparticles, nanocrystals, nanocapsules, etc. Despite the extensive use of nanomaterials in medicine, there is still little awareness regarding their biological toxic effects. The emergence of potential risk of nanotoxicity has become a serious health concern with regard to the efficacy and safety of nanomedicine (Bailey et al., 2004; Bekyarova et al., 2005; Cevc et al., 2004; Ekambaram et al., 2012; Mehnert et al., 2001; Pawar et al., 2016, Touitou et al., 2000).

Nanotoxicity occurs when a biological system induces toxic biological responses due to the interaction of cells/tissues with the particle properties (physical properties, chemical properties, or both) of nanostructures. The term "nanotoxicity" may be defined as the interaction between nanostructures and the biological system with the elucidation of the chemical relationship between physicochemical properties of nanostructure with biological properties like PK and consequent toxic responses. Nanotoxicology therefore includes the study of toxicity of nanomaterials to better understand and assess the health risks involved in the use of nanomaterials in medicine. The physicochemical properties of NPs, such as small size, large surface area, solubility, and flexible chemical composition/crystal structure that favor their use in nanomedicine with target-specific drug delivery and enhanced therapeutic efficacy, also contribute to their enhanced toxic effects. Experimental in vitro and in vivo studies showed that NPs cause cytotoxicity which is associated with oxidative stress (OS) induced by reactive oxygen species (ROS) (Shin et al., 2015). To circumvent adverse effects associated with nanomaterials/nanomedicine, nanotoxicity should be properly addressed considering the particle properties and/or surface chemistry of nanostructures/nanodrug delivery systems. Investigating and assessing the biological toxicity of nanomaterials therefore becomes an important topic of study in order to address nanotoxicity and associated health implications.

22.2 NANOTOXICITY

22.2.1 What Is Nanotoxicity?

According to OECD (Organization for Economic Co-operation and Development) guidelines, "toxicity" is the ability of a substance to cause poisonous effects resulting in severe biological harm or death after exposure to or contaminant with that substance (OECD, 2001). As described earlier, toxicity imparted by nanomaterials is termed nanotoxicity. More precisely, nanotoxicity is the science that deals with the study of adverse effects caused by nanomaterials/nanostructures in biological systems. The investigation and assessment of the toxic effects of nanomaterials in environmental or human health systems are the main domain of nanotoxicological research.

22.2.2 Environmental and Health Hazards

In the modern era, despite human civilization being benefitted by various aspects of engineered nanomaterials including all industries, biomedicine, and daily commodities, the intentional and unintentional exposure of human and environmental systems to nanomaterials like ultrafine particles might impart serious toxicity at both acute and chronic levels. NPs (natural and engineered) may contribute toxicity in different levels of the environment due to their different physicochemical properties (Walters et al., 2016). Engineered nanomaterials can enter into the environment through atmospheric emission or industrial transport and waste (Ray et al., 2009). Living components of the environment can be affected by toxic-level exposure to NPs. Due to the weak migration in soil, NPs may also accumulate in soil for longer durations, which renders toxic effects over the plant kingdom. NPs can retard or alter seed germination, plant elongation, or even biomass (Yang et al., 2017). Retardation of growth is observed in silica oxide (SiO_2) NP-exposed cotton. Rare earth metal NPs can retard the growth of vegetables and wheat. Exposure of carbon nanotubes (CNTs) reduces the production of growth hormone in rice seedlings (Qian et al., 2013). Accumulation of NPs in aquatic systems may cause severe adverse effects in aquatic diversity. The presence of NPs in the aquatic environment may lead to decreased fertility, behavioral abnormalities, and increased mortality rate in aquatic animals (Krysanov et al., 2010). Increased concentration of ionic silver is observed in tissues of *Cyprinus carpio*. Silver (Ag) NPs also show genotoxicity in mussel *Mytilus galloprovincialis* and dose-dependent toxicity in a zebrafish model. Titanium dioxide (TiO_2) in the presence of sunlight and humic acid brings about retardation of growth and DNA damage of *Danio rerio* (Jahan et al., 2017). In terrestrial animals, nanotoxicity is observed in pregnancy, fetus, and respiratory and excretory levels (Li et al., 2014). Owing to their small dimensions, NPs can behave like biomolecules, and their uncertain toxic effects in biological systems results in nanotoxicity, which has really become a potential risk factor for health and the environment.

22.2.3 Human Health and Safety Concerns

Nanomaterials are deliberately engineered and manufactured to have certain properties and characteristics with at least one primary dimension of less than 100 nm. Four classes of nanomaterials, viz., elemental carbon, carbon materials, metal and metal oxides, and ceramics, are the basic engineered nanomaterials. Humans can be exposed to nanomaterials via several routes such as inhalation, oral ingestion, injection, and the dermal route. Specifically, the respiratory system, the gastrointestinal tract, the circulatory system, the cardiovascular system, the reticuloendothelial system, and the central nervous system (CNS) are known to be primary targets that are adversely affected by NPs. In vivo experiments revealed that carbon-based NPs cause dose-dependent epithelioid granulomatous lesions in the lung and persistent interstitial inflammation on chronic exposure.

Carbon-based NPs include fullerene, single- or multi-walled CNTs, etc. (Bahadar et al., 2016). Due to chemical inertness, thermal stability, and flexible structure of CNTs, they are used in electronics, computers, supercapacitors, aircraft, textiles, implants, etc. (Khan et al., 2015). Because of increasing uses of different carbon nanomaterials, exposure of the population to nanomaterials is also increasing. Many studies proved the potential risk of nanomaterials in experimental animals. Inhalation and intra-tracheal instillation of single-walled and multi-walled CNTs (beyond permissible exposure limit) in animal models showed pulmonary inflammation (Kobayashi et al., 2017). Long nanoscale CNT fibers have been reported to develop serious risk of lung inflammation, which in turn induces pulmonary fibrosis through fibroblast proliferation and collagen deposition (Kobayashi et al., 2017). It has been reported that CNTs are more toxic than other ultrafine carbon black or silica dusts. Prolonged exposure may develop into cancer. Intra-tracheal instillation of multi-walled CNT may lead to genotoxicity through breaking of DNA, micronuclei, and mutation in lung. Graphene NPs are also responsible for toxic reactions in lungs and pleural space (Zhang et al., 2014).

Metallic NPs are also responsible for toxic effects in animals. Gold (Au) NPs have gained interest to be used as drug carriers, contrast agents, etc., because of their unique physicochemical properties and surface chemistry. Though gold is nontoxic and biocompatible, it may behave distinctly in the nano size range (Fratoddi et al., 2015; Jo et al., 2015). Accumulation of Au NPs in the intestinal epithelial cell may cause severe cytotoxicity due to cell proliferation, cell membrane damage, ROS generation, membrane depolarization of mitochondria, etc. (Yao et al., 2015). Because of having antimicrobial properties, silver (Ag) NPs are widely used in several consumer products such as detergents, cleansers, wound-dressing agents, etc. (Asharani et al., 2008). Bradycardia, axial curvature, and degeneration of body parts have been observed in embryos of zebrafish. Deposition of Ag NPs in CNS leading to CNS toxicity has been also observed. In experimental study, it has been observed that Ag NPs are more toxic than silver ions (Sambale et al., 2015). Zinc oxide (ZnO) NPs are used as skin care products, bottle coating, biosensors, photocatalytic agents, etc. ZnO NPs may cause cytotoxicity due to oxidative stress induced by ROS. Titanium oxide (TiO_2) NPs are the main ingredient in several cosmetics including sunscreen lotion, sports item, food colorant, packing items, etc. (Khan et al., 2015). In vivo studies showed that the inflammation may be induced on inhalation of TiO_2 NPs (Shi et al., 2013).

Moreover, cytotoxicity and genotoxicity have been observed with nanodrugs containing metallic ions like aluminum oxide, gold, copper oxide, silver, cerium oxide, zinc oxide, titanium oxide, and iron oxide. Although silica- and polymer-based nanomaterials have been deemed to be more biocompatible and relatively nontoxic, they may also cause oxidative stress and cytotoxicity. Furthermore, ceramic nanoparticles have been reported to exhibit cytotoxicity in the lungs, liver, heart, and brain, as well as teratogenic/carcinogenic effects (Bahadar et al., 2016).

22.2.4 Toxicity Risk Assessment

There is no standard methodology available for toxicity assessment of nanomaterials. However, the toxicity screening of NPs can be done by using the following three techniques: physicochemical characterization, in vitro assays (cellular and non-cellular), and in vivo (animal model) studies.

In physicochemical characterization, the parameters usually considered for the assessment of toxicity include the following: (1) size including surface area, size distribution, chemical composition (crystallinity, electronic properties, etc.); (2) solubility; (3) shape and aggregation; and (4) surface chemistry including surface structure, surface reactivity, surface groups, and coating. The magnitude of toxic (biological) responses depends on the physicochemical characteristics of NPs (Wani et al., 2011).

The toxicity of nanomaterials can also be studied in vitro on cultured cell lines. In in vitro assays, the effect of NPs on the morphology and metabolism of target cells (when exposed to NPs) is assessed. In this method, the portal-of-entry toxicity in lung, skin, and mucus membranes as well as systematic effects on target tissues/organ systems is assayed. In medicine, nanomaterials are usually employed as diagnostic reagents and therapeutic agents. The toxicity of nanomaterials in medicine is assayed by the measurement of protein interaction/protein binding studies. The surface properties of nanomaterials play an important role to identify specific toxicity induced by NPs either from the therapeutic effect of drugs produced or by measuring the interaction of diagnostic reagents with biomolecules.

The in vivo study is also used to study portal-of-entry toxicity and effects on target organs with disease-specific animal models. PK and pharmacodynamic (PD) parameters are considered in this method. The assessment of systematic responses such as oxidative stress, cytotoxicity, immune-inflammatory response, neurotoxicity, cellular disruption, membrane damage, and carcinogenesis is carried out in combination with in vitro and in vivo methods discussed above (Wani et al., 2011; Khan et al., 2017).

The physicochemical properties of nanomaterials are characterized and tested for their basal cytotoxicity, carcinogenic potential, genotoxicity, chronic toxicity, developmental toxicity, reproductive toxicity, and immune toxicity using the corresponding in vitro and also in vivo systems. In vitro tests assess uptake and transport characterization of NPs by conventional methods. Neutral red dye assays are used for cellular uptake studies. Lactate dehydrogenase, tetrazolium, and alamar blue dyes are usually adopted for cytotoxicity assays. Genotoxicity is assessed by Ames test and Comet assays (Demetzos, 2016; Bahadar et al., 2016). In recent days, novel biological tools and approaches such as genomics, proteomics, and metabolomics have been employed for the assessment of nanotoxicity. In conjunction with in vitro and in vivo assay methods, hyphenated analytical techniques like liquid chromatography-tandem mass spectrometry (LC-MS/MS), gas chromatography-tandem mass spectrometry (GC-MS/MS), liquid chromatography-nuclear magnetic resonance spectroscopy (LC-NMR), etc., play significant roles in the determination of nanoanalytes in the biological matrix.

22.2.5 Preventive Measures

During manufacturing of nanomaterials, unwanted exposure of nanomaterials may impart occupational health hazards to workers. Sometimes, accidental intake may also happen to healthcare professionals, patients, and attendants. In the case of nanodrugs/nanoformulations, unwanted exposure is expected to the general public due to the excretion of nanomaterials. Therefore, intensive monitoring systems based on physicochemical characterizations are required to be adopted for the control and prevention of nanotoxicity (Yadav et al., 2013).

Considering the potential risk of nanomaterials, it is essential to create certain measures for the prevention of nanotoxicity to reduce its effects on human health and the environment. Different organizations throughout the globe are currently working on this agenda. However, due to the lack of knowledge on nanotoxicity, it is very difficult to create specific measuring guidelines. In this section, different draft guidelines on handling nanomaterials are discussed as follows. On the basis of preliminary hazard analysis (PHA) in nanoengineered materials, the following stages are to be considered during production of engineered nanomaterials and their exposure to human health with a goal to minimize the risk of toxicity (Heng et al., 2010):

- Nature of chemicals/materials to be used
- All stages of production process
- Equipment and engineered control system
- Workers' job responsibilities
- Exposure chances/safety measurement

In the assessment of toxicity hazards in industry and research laboratories, one should thoroughly study the surface charge, surface chemistry, particle shape, particle size, solubility, toxicity hazard, fire hazard, and explosion hazard. Safety measures should be taken at all points of exposure. General safety measures can prevent the chances of exposure. Wearing gloves, lab coats, safety goggles, and face pieces can prevent dermal, ocular, and injection exposure of nanoparticles at the manufacturing level. Use of respiratory air filters like N100 can prevent respiratory exposure hazard (Heng et al., 2010).

Exposure control strategies, including prevention through design (PtD), are used to prevent nanotoxicity through exposure. Other strategies include engineering control and procedural control. One aspect of engineering control is the proper utilization and maintenance of air ventilation systems (HEPA, fume hood), air lock containers, distillation systems for evaporating solvent from a colloidal dispersion within an explosion-proof enclosure, and sensor controlling of production equipment. Procedural methods include proper training of personnel related to standard operating procedures (SOPs) for safe handling of nanomaterials, hazard analysis, personal protective equipment, waste disposal technique, etc. In addition, housekeeping, documentation, and medical surveillance of personnel are inevitable. Regular monitoring of medical conditions considering inflammatory responses, allergic conditions, and neurological, hematological, and gastrointestinal problems are essential to prevent long exposure to toxicity. Nanotoxicity to consumers caused by different medicine and healthcare products is not yet adequately established. Proper monitoring of product life periods, generation of premarketing safety data of engineered nanoparticles by manufacturers, and proper dissemination of safety guidelines among stakeholders are some regulatory approaches to be implemented for the prevention of nanotoxicity at the consumer level. However, research is ongoing to evade/circumvent unwanted interactions between nanomaterials and biological tissues. It has been investigated that surface coatings and other modifications may increase the safety of NPs in biological systems (Shi et al., 2013). However, protective surface coatings have got their own limitations, as they degrade with time soon after environmental interactions such as air exposure or ultraviolet irradiation. Apart from surface coating, the use of antioxidants such as quercetin, ascorbic acid, vitamin E, and other standard antioxidant medications may be beneficial in the prevention and/or treatment of nanotoxicity.

22.3 NANOMEDICINE

22.3.1 What Is Nanomedicine?

Nanomedicine refers to the use of nanomaterials such as nanodevices and nanoparticles in drug delivery and diagnostics. It ranges from the biomedical applications of nanomaterials to biosensors. It deals with the interactions of nanomaterials (surfaces, particles, etc.) or analytical nanodevices with human biological systems (cells, tissue, body fluids). Nanomedicine can be used to detect, treat, and prevent human diseases. It involves targeted drug delivery at the molecular level of certain biological systems. Some specific medical interventions include pulmonary delivery of drugs in respiratory diseases, curing diseases such as cancer, repairing damaged tissues, and also disease prevention (gene therapy) and diagnostic imaging of abnormal cells (Yadav et al., 2012; Singh et al., 2008).

22.3.2 Why Nanomedicine?

Nanomedicine (nanoscale drug delivery systems) or nanoscale devices in medicine are preferred (Agnieszka et al., 2012; Jain et al., 2010; Sarwa et al., 2013) over conventional drug delivery systems because of the following reasons:

- Smart nano tools (nanomaterials/nanodevices)
- Engineered materials (nanodrugs/nanocarriers/nano delivery systems)
- Imaging diagnostics
- Multifunctional therapeutics and better encapsulation efficiency
- Precise treatments of conditions such as cancer, and prevention and control
- Early detection or faster diagnosis of disease
- Repair of tissue deep within the body
- Site-specific targeted delivery of drugs
- Enhanced efficacy and reduced toxicity of drugs
- Enhanced solubility and oral bioavailability
- Improved stability of drugs or drug products

- Prolonged drug action (longer half-life) and controlled drug release
- Predictable drug release (able to cross biological barriers like blood brain barrier [BBB] and enter CNS)
- Biocompatibility (nontoxic, non-allergenic, and non-irritating) and biodegradable
- Less labor intensive and inexpensive
- Reduced dose and patient compliance (wider acceptability of routes of administration)

Nanodrug delivery systems provide control over the bio-distribution (BD) and PK of the encapsulated drug, and thus may lead to therapeutic efficacy superior to that of conventional drug delivery systems.

22.3.3 Nanobiopharmaceuticals

The biomaterials employed in the preparation of nanodrug delivery system are specially designed or engineered in order to improve their characteristics for drug delivery and targeting, imaging, and diagnosis (Pourmand & Abdollahi, 2012). Nanochips and nanoarrays are a class of electronic pharmaceuticals used to measure various biomarkers in biological samples to monitor disease formation and progression (Lammers et al., 2011; Abdelkader & Alany, 2012). Moreover, nanotheranostics contain both diagnostic and pharmacotherapeutic agents in the same formulation to achieve both diagnosis and treatment at a time. Nanodrug carries have attracted attention for their better drug delivery, release property, targeting, and therapeutic efficacy over conventional delivery systems (Nassiri & Abdollahi, 2017). There are many nanodrug carrier systems employed for the development of targeted nanodrug delivery systems/nanoformulations used in nanomedicine. They include carbon nanotubes, quantum dots, metallic nanoparticles, polymeric nanoparticles, solid lipid nanoparticles (SLNPs), dendrimers, ethosomes, liposomes, nanofibers, niosomes, polymer-based micelles, phytosomes, polymersomes, transferosomes, etc. These nanodrug carriers have special structural and functional properties. These properties may be altered and tailored by changing basic composition materials and percentage of their components, size, size distribution, solubility, and surface properties and reactivity by incorporation of specific chemical components/modification of surface functionalities (functional groups) in the biomolecules of nanostructures (Sharma et al., 2014). Nanodrug carriers are used to prepare nanobiopharmaceuticals/nanodrug delivery systems. Nanopharmaceuticals can be broadly classified into two categories (Bawa, 2008). The first category is that where the active drug acts both as a carrier system as well as the therapeutic entity having the intrinsic activity. Examples include multivalent dendrimers, cerium oxide, and platinum nanoparticles. In the second category, the active pharmaceutical agent is capsulated, entrapped, or coated to a distinct polymeric or other carrier system. Examples are liposomes, ethosomes, transferosomes, etc. In both cases, carriers must be able to transport the active agent to a specific biomolecular target site via various mechanisms like ligand adhesion or receptor binding and consequently impart maximum therapeutic activity with the least toxicity (Bawa 2010). In another way, nanopharmaceuticals can be classified on the basis of the polymeric nanomaterials (nanocarriers) used to prepare the delivery systems. Examples are liposomes, carbon nanotubes, dendrimers, colloidal gold, nanocrystals, fullerenes, etc. The therapeutic agent (e.g., drugs, proteins, nucleic acids, or genes) is packed/encapsulated inside the carrier system. The carrier systems are surface functionalized using ligands (NH_2, OCH_3, COOH, OH, etc.) or antibodies, so-called surface-functionalized nanopharmaceuticals. On the basis of chemical composition, nanopharmaceuticals can be polymeric-based (e.g., dendrimers), lipid-based (e.g., liposomes), or metal-based (ZnO NPs). However, the structural/functional properties of an ideal nanobiopharmaceutical/nanodrug delivery system are summarized in Table 22.1.

22.3.4 Applications

Nanomaterials have applications in biomedicine and health care. Nanobiopharmaceuticals have application as drug delivery systems/nanoformulations in nanoscale range (size ~50–100 nm). Different areas of applications are drug delivery, drugs, and therapy including in vivo imaging, in vitro diagnostics, biomaterials, theranostics, regenerative medicine, and active implants. Nanomaterials also have applications in tissue engineering, bioactive surfaces, biomarkers, biosensors, image enhancement devices, implant technology, etc. (Chavda 2016). Some common applications include quantum dots (NP contrast agents) for molecular imaging of malignant cells, nanoshells (gold nanoshells) for targeting tumor cells selectively, nanowire sensors (silicon nanowire), and CNTs for detection of cancer biomarkers.

Furthermore, nanodrug delivery systems are specially designed formulations for topical, oral, pulmonary, or parenteral

TABLE 22.1
Ideal Properties of Nanobiopharmaceuticals/Nanodrug Delivery Systems

Type	Properties/chemical composition
Physical properties	Size, size distribution, and surface area Shape (spheres, cubes, triangles, rods, quantum dots, nanoballs, etc.) Texture (mechanical, roughness, rigidity, porosity) Crystallinity and morphological features
Composition	Metallic (Ag, Au, Ce), iron oxide, silica, dendrimers, polymeric (natural/synthetic polymers/polymer conjugates), micelles, nanoshells, nanocapsules, nanogels, liposomes, ethosomes, etc.
Surface chemistry	Surface functionality (functional groups), surface charge (electrostatic charge), hydrophobicity (membrane solubility), etc.
Targeting property	Antibodies, proteins and peptides, polymers, nucleic acids (aptamers), small molecules, etc.

delivery of drugs and are meant for specific targeting and diagnosis of diseases. They are used for the treatment of diabetes, cancer, AIDS, neurodegenerative disorders like Alzheimer's disease and Parkinson's disease, and dermatological diseases. Nanocarrier-based protein/peptide (P/P) delivery systems are used for pulmonary/respiratory delivery and also for oral delivery of drugs. The targeted and controlled delivery of drugs into the CNS across the BBB has also been developed. Delivery of NPs that selectively target stem cells can be used for the treatment of musculoskeletal disorders (Sarwa et al., 2014). Some nanodrug delivery systems along with their therapeutic applications (Wilczewska et al., 2012) are shown in Table 22.2.

22.4 NANOTOXICITY: PHYSICOCHEMICAL AND BIOLOGICAL ASPECTS

22.4.1 Physicochemical Aspects

The chemical reactivity and biological activity of nano-sized materials are often greater than those of larger-sized particles. Their properties and applications are quite different from those of larger particles of the same substance. It is often seen that NPs pose greater risk of toxicity than larger particles. The factors primarily responsible for the biological toxicity of nanomaterials can be distinguished into certain basic physicochemical characteristics/particle properties, namely, size, surface area, surface charge, crystal structure, morphology/surface structure, chemical composition, solubility, and agglomeration/aggregation status. The potential of nanomedicine depends on the interaction of nanostructures with the biological systems (cells/tissue) at the molecular level. This phenomenon makes it more efficient as targeted drug delivery systems as well as more effective as diagnostic and/or therapeutic substances (Maravajhala et al., 2011; Khanna et al., 2015). To define the bio-efficacy of nanomaterials, understanding the concept of the drug targeting system is important. It involves biomolecular interaction mechanisms (between nanostructures and target biomolecules or receptors/protein molecules) usually termed the PD aspect of nanotherapeutics (Gatoo et al., 2014). In this mechanism, the coordinating behavior of three components, viz., the targeting moiety, the carrier, and the therapeutic drug, are required. The first one recognizes and binds the target/receptor molecule. The second one carries the drug substance. The third one provides a therapeutic action/biological effect at the desired site/organ system. Since such biological phenomena are highly target specific, the occurrence of adverse effects or toxicity is also certain. Nanotoxicity or cytotoxicity induced by nanomaterials is often target specific and/or target non-specific, which results from the interaction between the nanomaterial surface and cellular components (Shin et al., 2015).

TABLE 22.2
Some Nanodrug Delivery Systems and Their Therapeutic Applications

Drug	Type/material(s)	Therapy
CNTs		
Captopril	Multi-walled	Hypertension
Indomethacin	Multi-walled	Inflammation
Paclitaxel	Single-walled	Breast cancer
Polymeric nanocapsules		
Amphotericin-B	Polylactic acid (PLA)-b PEG	Neurodegenerative disorders
Zidovudine	Poly(isohexylcyanate)	AIDS
Stavudine	Polybutylcynoacrylate (PBCA)	HIV/AIDS
Liposomes		
Diclofenac	Small molecular liposome	Inflammation
Insulin	Protein liposome	Diabetes
Tobramycin	Small molecular liposome	Pulmonary infection
SLNPs		
Anticancer drugs	Phospholipids, triglycerides, alcohols, co-polymers, etc.	Cancer
Antiviral drugs		HIV/AIDS
NSAIDs		Inflammation

22.4.2 PK Aspects

As mentioned in the introduction section, physicochemical properties are primarily responsible for the toxicity of nanomaterials. Physicochemical factors influence PK behavior, which in turn ultimately determines PD property of nanomedicine (Yadav et al. 2013). The absorption, distribution, and cellular uptake processes of NPs are affected by physicochemical and surface properties (Figure 22.1). The role of physicochemical properties and their relation to pharmacokinetic properties in nanotoxicity (Figure 22.2) is discussed below:

1) **Size:** Size plays a vital role for PK such as absorption, distribution, and clearance of NPs. Moreover, it is also responsible for cellular uptake (by diffusion), binding, and molecular interaction followed by the induction of cytotoxicity. The absorption and distribution of specific particles within the body and the accumulation of particles in a particular part of the body are dependent on the particle's size and surface characteristics. Accumulation of NPs in the body without proper clearance may lead to severe toxicity. The main distribution sites of NPs are the metabolizing organs like the liver, kidney, and spleen. Deposition may also occur in other target organs like lungs, CNS, and growing fetus. Size distribution also affects PK of NPs. Variation in cellular uptake and distribution in vivo depends on variation in size of particles.
2) **Surface area:** In addition to size, particle surface area plays a critical role in cellular toxicity. The smaller a particle, the greater is its surface-area-to-volume ratio and the higher its chemical reactivity and biological activity.
3) **Surface charge:** This is nothing but the electrostatic charge of particles. It defines the surface reactivity of particles, which also plays an important role in toxicity. Thus, particles having the same composition can have different levels of toxicity depending on both the size and surface reactivity of nanoparticles. The extent of cytotoxicty is, therefore, primarily size dependant.

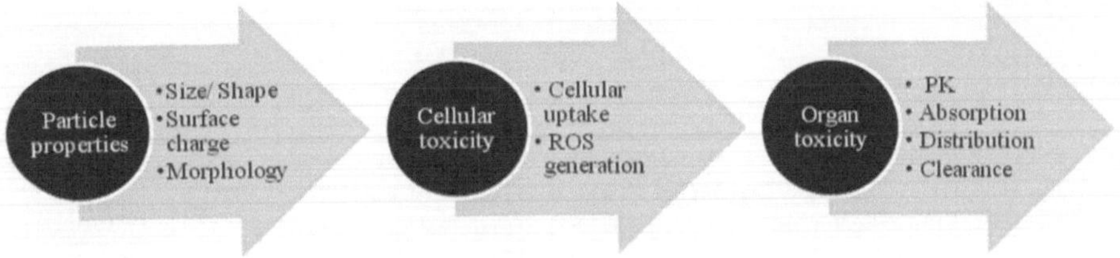

FIGURE 22.1 Relation between physicochemical properties, PK, and nanotoxicity.

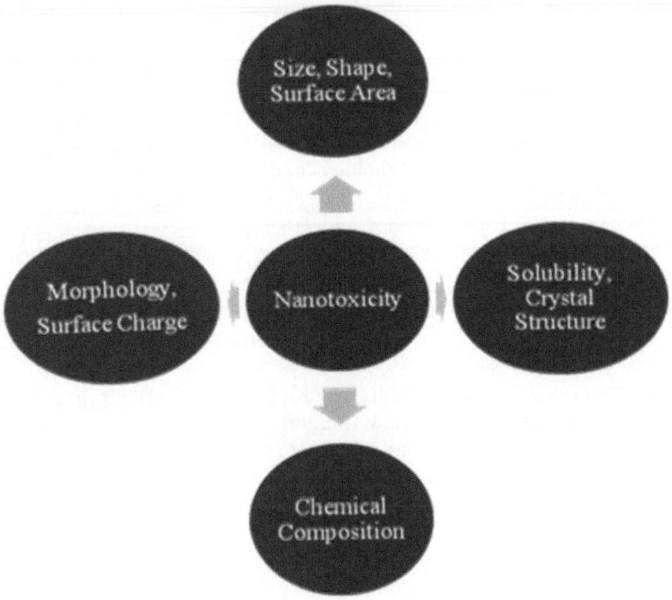

FIGURE 22.2 Physicochemical properties responsible for nanotoxicity.

Pharmacokinetic parameters are therefore considered to be critical for the investigation of biological toxicity of nanomedicine. The pharmacodynamics aspect is described below in the subsection on the mechanism of nanotoxicity (Liang et al., 2008, Ochekpe et al., 2009).

22.4.3 Mechanism of Toxicity

When NPs are ingested, inhaled, or absorbed through the skin, they can induce the formation of ROS. ROS are toxic by-products of cellular oxidative metabolism which occurs in the cell mitochondria. ROS include superoxide anion radicals, hydrogen peroxide, hydroxyl radicals, and singlet oxygen. Overproduction of ROS can cause oxidative stress, due to which cells fail to maintain normal physiological redox-regulated functions. Disturbance in the normal homeostasis mechanism leads to DNA damage, cytotoxicity, apoptosis, cancer initiation, unregulated cell signaling, and carcinogenesis. Increased reactivity of NPs results in increased production of ROS, which in turn induces OS, inflammation, and consequent damage to various biological/cellular components (such as membrane lipid, protein, enzymes, and even DNA). The production of ROS has been found in a diverse range of nanomaterials including carbon fullerenes, CNTs, and NP metal oxides. The production of ROS is considered the primary mechanism of nanotoxicity. Physicochemical factors mediate this biological process of nanotoxicity. Size is a key factor in determining the potential toxicity of a particle. NPs are much more mobile than larger-sized particles. They are more readily taken up by the human body and are able to cross membranes and gain access to cells, tissues, and organs that larger-sized particles normally cannot. As a result of their small size, NPs can even cross tissue junctions and cellular membranes and transport easily within cells. Inside cells, they are taken up by various cellular organelles. NPs may induce structural damage to the cell mitochondria or invade the cell nucleus and consequently cause serious DNA mutations leading to cell death or apoptosis (Figures 22.3 and 22.4). Major causes of nanotoxicity are presented in Figure 22.5. Furthermore, inflammation has been reported to directly cause toxicity and promote cell death through the induction of toxic by-products of inflammation such as ROS and complement proteins, as well as via receptor-induced apoptosis/necrosis (Gatoo et al., 2014).

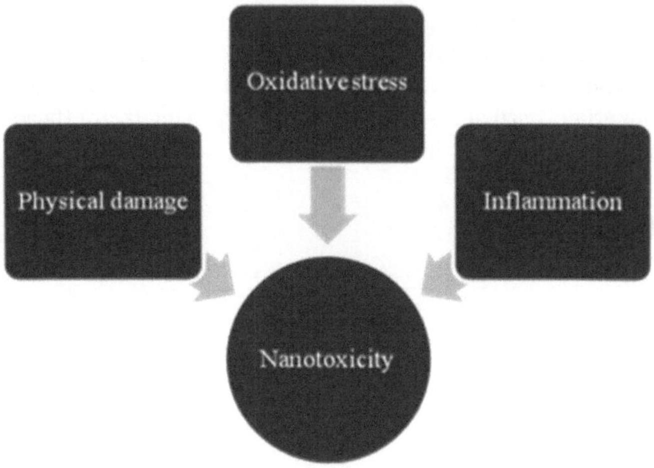

FIGURE 22.3 Main causes of nanotoxicity.

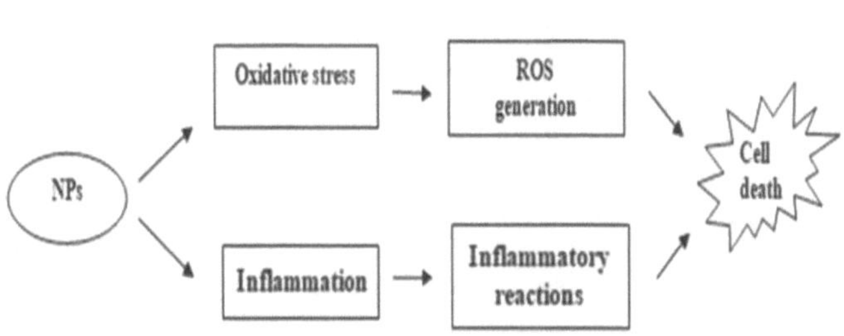

FIGURE 22.4 NP-induced toxicity: Major pathways involved.

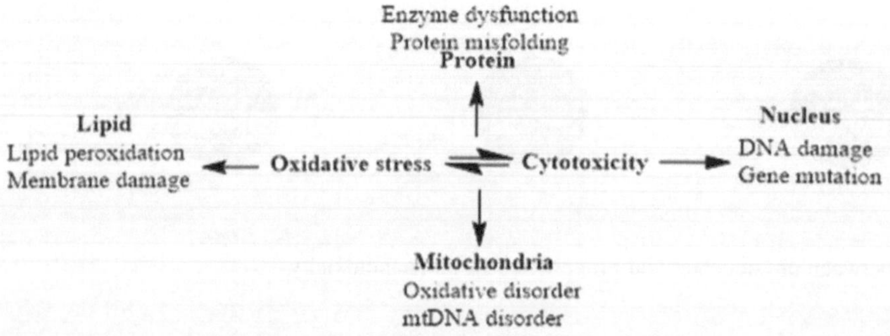

FIGURE 22.5 Different cellular levels of nanotoxicity.

22.5 CONCLUSION

Nanotechnology is a novel approach with wide applications, particularly in the field of biomedicine and health care. It is an effective tool for having potential applications in drug delivery and biopharmaceuticals. The physicochemical role and biological significance make nanomaterials effective as nanodevices and nanomedicine. However, the safe use of nanomaterials in biomedical devices, diagnostics, and medicine is a serious concern due to their toxicity in human health. Therapeutic implications or targeted delivery of nanomedicine is still lagging behind due to their safety issues. Nanotoxicity must therefore be addressed with certain resolutions for effective, sustainable, and safe nanotechnology. In view of limited awareness about fatal toxicities of nanomaterials, the further investigation of nanotoxicity is urgently required. Prior to the widespread introduction of nanomaterials in medicine and health care, their safety issues in biological systems should be thoroughly investigated, particularly considering the surface chemistry and other particle properties. Such a strategy should be exercised for the effective use of engineered nanomaterials, particularly in medicine, considering their efficacy and safety issues to human health.

REFERENCES

Abdelkader, H., and R. G. Alany. 2012. Controlled and continuous release ocular drug delivery systems: Pros and cons. *Current Drug Delivery* 9, no. 4: 421–430.

Agnieszka, A. K., E. Markiewicz, K. Niaz, and T. Jesionowski. 2012. Structural characterisation of ZnO particles obtained by the Emulsion Precipitation Method. *Journal of Nanomaterials*: 1–9.

Asharani, P. V., Y. L. Wu, Z. Gog, and S. Valiyavettil. 2008. Toxicity of silver nanoparticles in zebrafish models. *Nanotechnology* 19, no. 25: 255102. Available: https://www.nanowerk.com/nanotechnology/introduction/introduction_to_nanotechnology_1. php (accessed December 08 2018).

Bahadar, H., F. Maqbool, K. Niaz, and M. Abdollahi. 2016. Toxicity of nanoparticles and an overview of current experimental models. *Iranian Biomedical Journal* 20, no. 1: 1–11.

Bailey, R. E., A. M. Smith, and S. Nie. 2004. Quantum dots in biology and medicine. *Physica E* 25, no.1: 1–12.

Bawa, R. 2008. Nanoparticle-based therapeutics in humans: A survey. *Nanotechnology Law & Business* 5: 135–155.

Bawa, R. 2010. Nanopharmaceuticals for drug delivery – A review. *European Journal of Nanomedicine* 3, no. 1: 122–127.

Bekyarova, E., Y. Ni, E. B. Malarkey, V. Montana, J. L. McWilliams, R. C. Haddon, and V. Parpura. 2005. Applications of carbon nanotubes in biotechnology and biomedicine. *Journal of Biomedical and Nanotechnology* 1, no. 1: 3–17.

Cevc, G. 2004. Lipid vesicles and other colloids as drug carriers on the skin. *Advanced Drug Delivery Reviews* 56, no. 5: 675–711.

Chavda, H. 2016. Nanopharmaceuticals: The pharmaceuticals of the present era. *Elyns Journal Pharmaceutical Research* 1, no. 2: 106e.

Ekambaram, P., A. H. Sathali, and K. Priyanka. 2012. Solid lipid nanoparticles: A review. *Scientific Reviews and Chemical Communications* 2, no. 1: 80–102.

Fakruddin, M., Z. Hossain, and H. Afroz. 2012. Prospects and applications of nanobiotechnology: A medical perspective. *Journal of Nanobiotechnology* 10: 31.

Demetzos, C. 2016. *Pharmaceutical Nanotechnology: Fundamentals and Practical Applications*, 467–489. Springer.

Fratoddi, I., I. Venditti, C. Cametti, and M. Russo. 2015. How toxic are gold nanoparticles? the state-of-the-art. *Nano Research* 8, no. 6: 1771–1799.

Gatoo, M. A., S. Naseem, M. Y. D. Arfat, K. Qasim, and S. Zubair. 2014. Physicochemical properties of nanomaterials: Implication in associated toxic manifestations. *BioMed Research International* 2014: 1–8.

Heng, B. C., X. Zhao, S. Xiong, K. W. Ng, F. Y. Boey, and J. S. Loo. 2010. Toxicity of zinc oxide (ZnO) nanoparticles on human bronchial epithelial cells (BEAS-2B) is accentuated by oxidative stress. *Food and Chemical Toxicology* 48, no. 6: 1762–1766.

Jahan, S., I. B. Yusoff, Y. B. Alias, and A. F. B. A. Bakar. 2017. Reviews of the toxicity behavior of five potential engineered nanomaterials (ENMs) into the aquatic ecosystem. *Toxicological Reports* 4: 211–220.

Jain, N., R. Jain, N. Thakur, B. P. Gupta, D. K. Jain, J. Banveer, and S. Jain. 2010. Nanotechnology: A safe and effective drug delivery system. *Asian Journal of Pharmaceutical and Clinical Research* 3, no. 3: 159–165.

Jo, M., S. H. Bae, M. R. Go, H. J. Kim, Y. G. Hwang, and S. J. Choi. 2015. Toxicity and biokinetics of colloidal gold nanoparticles. *Nanomaterials (Basel)* 5, no. 2: 835–850.

Khan, I., K. Saeed, and I. Khan. 2017. Nanoparticles: Properties, applications and Toxicities. *Arabian Journal of Chemistry* (in press). doi: 10.1016/j.arabjc.2017.05.011.

Khan, S. H., M. H. Fulekar, and B. Pathak. 2015. Nanotoxicology – Health and environmental impacts: A review. *Journal of Environmental Nanotechnology* 4, no. 3: 55–73.

Khanna, P., C. Ong, B. H. Bay, and G. H. Baeg. 2015. Nanotoxicity: An interplay of oxidative stress, inflammation and cell death. *Nanomaterials (Basel)* 5, no. 3: 1163–1180.

Kobayashi, N., H. Izumi, and Y. Morimoto. 2017. Review of toxicity studies of carbon nanotubes. *Journal of Occupational Health* 59, no. 5: 394–407.

Krysanov, E. Y., S. Pavlov, T. B. Demidova, and S. Y. Dgebuadze. 2010. Effect of nanoparticles on aquatic organisms. *Biology Bulletin* 37, no. 4: 406–412.

Lammers, T., S. Aime, W. E. Hennink, G. Storm, and F. Kiessling. 2011. Theranostic nanomedicine. *Accounts of Chemical Research* 44, no. 10: 1029–1038.

Li, Y., Y. Zhang, and B. Yan. 2014. Nanotoxicity overview: Nano-threat to susceptible populations. *International Journal of Molecular Sciences* 15, no. 3: 3671–3697.

Liang, X. J., C. Chen, Y. L. J. Zhao, and P. C. Wang. 2008. Biopharmaceutics and therapeutic potential of engineered nanomaterials. *Current Drug Metabolism* 9, no. 8: 697–709.

Maravajhala, V., S. Papishetty, and S. Bandlapalli. 2011. Nanotechnology in development of drug delivery system. *International Journal of Pharmaceutical Sciences and Research* 3, no. 1: 84–96.

Mehnert, W., and K. Mader. 2001. Solid lipid nanoparticles-production, characterization and applications. *Advances in Drug Delivery Reviews* 47, no. 2–3: 165–196.

Mohanty, C., G. Arya, R. S. Verma, and S. K. Sahoo. 2009. Nanobiotechnology: Application of nanotechnology in therapeutics and diagnosis. *International Journal of Green Nanotechnology: Biomedicine* 1, no. 1: B24–B38.

Nassiri, K., and D. Abdollahi. 2017. Health risks associated with the pharmaceuticals in wastewater. *DARU: Journal of Pharmaceutical Sciences* 25: 9.

Ochekpe, N. A., P. O. Olorunfemi, and N. C. Ngwuluka. 2009. Nanotechnology and drug delivery Part 2: Nanostructures for drug delivery. *Tropical Journal of Pharmaceutical Research* 8, no. 3: 275–287.

OECD 2001. Glossary of statistical terms. Available: https://stats.oecd.org/glossary/detail.asp?ID=2734 (accessed December 22 2018).

Pawar, A. Y., R. J. Khanderao, and H. C. Laxmikant. 2016. Transfersome: A novel technique which improves transdermal permeability. *Asian Journal of Pharmaceutics* 10, no. 4: S425.

Pourmand, A., and M. Abdollahi. 2012. Current opinion on nanotoxicology. *DARU Journal of Pharmaceutical Sciences* 20: 95.

Qian, H., X. Peng, X. Han, J. Ren, L. Sun, and Z. Fu. 2013. Comparison of the toxicity of silver nanoparticles and silver ions on the growth of terrestrial plant model *Arabidopsis thaliana. Journal of Environmental Sciences* 25, no. 9: 1947–1956.

Ray, P. C., H. Yu, and P. P. Fu. 2009. Toxicity and environmental risks of nanomaterials: Challengesand future needs. *Journal of Environmental Science and Health C Environmental Carcinogenesis & Ecotoxicology Reviews* 27, no. 1: 1–35.

Salata, O. V. 2004. Applications of nanoparticles in biology and medicine. *Journal of Nanobiotechnology* 2: 3.

Sambale, F., S. Wagner, and R. Khaydarov. 2015. Investigations of the toxic effect of silver nanoparticles on mammalian cell lines. *Journal of Nanomaterials* 2015: 1–9.

Sarwa, K. K., B. Mazumder, and M. Rudrapal. 2015a. Topical ethosomal capsaicinoids attenuates edema and nociception in arthritic rats. *Drug Delivery* 22, no. 8: 1043–1052.

Sarwa, K. K., M.Rudrapal, B.Mazumder, and V. K.Verma. 2015b. Potential of capsaicin-loaded transferosomes in arthritic rats. *Drug Delivery* 22, no. 5: 638–646.

Sarwa, K. K., P. K. Suresh, M. Debnath, and M. Z. Ahmad. 2013. Tamoxifen citrate loaded ethosomes for transdermal drug delivery system: Preparation and characterization. *Current Drug Delivery* 10, no. 4: 466–476.

Sarwa, K. K., P. K. Suresh, M. Rudrapal, and V. K. Verma. 2014. Penetration of tamoxifen citrate loaded ethosomes and liposomes across human skin: A comparative study with confocal laser scanning microscopy. *Current Drug Delivery* 11, no. 3: 332–337.

Sharma, V. K., K. K. Sarwa, and B. Mazumder. 2014. Fluidity enhancement: A critical factor for performance of liposomal transdermal drug delivery system. *Journal of Liposome Research* 24, no. 2: 83–89.

Shi, H., R. Mayage, V. Castranova, and J. Zhao. 2013. Titanium dioxide nanoparticles: A review of current toxicological data. *Particle and Fibre Toxicology* 10, no. 15: 1–33.

Shin, S. W., I. H. Song, and S. H. Um. 2015. Role of physicochemical properties in nanoparticle toxicity. *Nanomaterials* 5, no. 3: 1351–1365.

Singh, O. P., and R. M. Nehru. 2008. Nanotechnology and cancer treatment. *Asian Journal of Experimental Sciences* 22, no. 2: 45–50.

Touitou, E., N. Dayan, L. Bergelson, B. Godin, and M. Eliaz. 2000. Ethosomes-novel vesicular carriers for enhanced delivery: Characterizationand skin penetration properties. *Journal of Controlled Release* 65, no. 3: 403–418.

Walters, C., E. Pool, and V. Somerset. 2016. Nanotoxicology: A review, 45–63, *Intech Open.*

Wani, M. Y., M. A. Hashim, F. Nabi, and M. A. Malik. 2011. Nanotoxicity: Dimensional andMorphological Concerns. *Advances in Physical Chemistry* 2011: 1–15.

Wilczewska, A. Z., K. Niemiorwicz, K. H. Markiewicz, and H. Car. 2012. Nanoparticles as drug delivery systems. *Pharmacological Reports* 64, no. 5: 1020–1037.

Yadav, S., M. Bansal, M. Goyal, and V. Singh. 2012. Nanomedicine: Current status and future implications. *Indian Journal of Clinical Practice* 23, no. 1: 6–9.

Yadav, N., S. Khatak, and U. V. S. Sara. 2013. Solid lipid nanoparticles- A review. *International Journal of Applied Pharmaceutics* 5, no. 2: 8–18.

Yang, J., W. Cao, and Y. Rui. 2017. Interactions between nanoparticles and plants: Phytotoxicity and defense mechanisms. *Journal of Plant Interactions* 12, no. 1: 158–169.

Yao, M., L. He, D. J. McClements, and H. Xiao. 2015. Uptake of gold nanoparticles by intestinal epithelial cells: Impact of particle size on their absorption, accumulation, and toxicity. *Journal of Agricultural and Food Chemistry* 63, no. 36: 8044–8049.

Zhang, Y., Y. Bai, J. Jia, N. Gao, Y. Li, R. Zhang, G. Jiang, and B. Yan. 2014. Perturbation of physiological systems by Nanoparticles. *Chemical Society Reviews* 43, no. 10: 3762–3809.

Index

1,2-dimyristyloxypropyl-3-dimethyl-hydroxylethyl ammonium bromide (DMRIE), 47
1,2-dioleoyl-3-dimethyl-hydroxyethyl ammonium bromide (DORIE), 47
1,2-dioleoyloxy-3-(trimethylammonio)-propane (DOTAP), 47
1,2-Distearoyl-sn-glycero-3-phosphocholine (DSPC), 47, 48
α-Lactalbumin, 66
β-Casein nanoparticle, 66
β-lactams, 72
β-Lactoglobulin, 66
ω-3 Poly unsaturated fatty acids, 66

Abrasions, 116
Abraxame, 66
Absorption, 236, 237
Acetazolamide, 48
Acetylated SNCs, 69
Active drug delivery, 47
Acute Toxicity, 241
Adenocarcinoma, 163, 199
Adipose-derived stromal cells (ADSCs), 126
Adjuvant, 193, 200, 202
Adsorption, 237
Aerobic, 246
Aerobic bioremediation, 246
Aero gels, 255
Agglomeration, 33
Agrobacterium sp., 124
Agro food industry, 82
AIDS, 115
Albumin based nanoparticles, 66
Alcanivorax borkumensis, 250
Algae, 29, 251
Algal Nutraceuticals, 101
Algal species, 30, 237
Alginate based nanoparticles, 66
Alkane hydroxylases, 253
Alkenyl succinic anhydride (ASA), 70
Alkoxide silica precursor, 12
Allotropes, 236
Aloe vera, 116
Alveolar capillary barrier, 239
Ames test, 261
3 aminopropyl triethoxysilane (APTES), 74
Amorphous Phase, 69
Amphiphilicity, 207
Amphotericin B, 57, 59
Amylopectin, 67, 69, 73
Amylose, 67, 69
Anabaena spp., 252
Anaerobic, 246
Analysis of Lipid, 56
Analyte, 216, 218, 220
Angiogenesis, 118, 120
Annamycin, 59
Anthracyclins, 56
Antibacterial, 30, 34
Anti-bacterial, 159
Antidiabetic, 100
Antifungal, 34
Antigen, 179–180
Antigen-presenting cell, 65
Anti-inflammatory, 34
Antimicrobial Therapy, 57
Antiobesity, 100
Antioxidant, 34, 209
Antioxidant defense, 123
Antiprotozoal, 34
Antiviral, 34
Apoptosis, 123, 179–181, 240
Applications in Food Industry, 58
Aptamer, 180, 208
Aquaculture, 31
Aromatics, 255
Arsenic, 218, 220–222
Arterial ulcers, 115
Arthronema spp., 252
ASA_{350}, 68
Asphaltenes, 255
Atomic absorption spectroscopy, 239
Atomic force microscopy, 6, 68, 207
Aurimune, 200

Bacillus algicola, 250
Bacillus cereus, 250
Bacillus subtilis, 250
Bacteria, 29, 249, 250
Bacterial pathogenicity, 30
Ballistic, 205
Band gap, 185
Barium nanoparticles, 91
Bedsores, 116
β-Carotene loaded starch nanocrystals (BSNCs), 68
Bifidobacterial, 101
Bifurcaria bifurcate, 38
Bilayered living cellular construct (BLCC), 117
Bilayer Fluidity, 49
Bile salts, 105
Bioaccumulation, 216
Bioactive lipids, 101
Bioaugmentation, 246
Bioavailability, 64, 73, 101, 105, 217
Biochip, 181
Biocompatible, 4, 5, 65–66, 74, 159, 172
Bioconjugation, 198
Biodegradability, 4, 64–67, 73
Biodegradation, 246
Bioenergy, 30
Bioengineered cell construct, 117
Biofuels, 30, 32
Bio-functionalization, 207, 210
Biogenic, 13, 16, 21, 24
Bioimaging, 169–171
Bioleaching, 246
Biological Parameters, 54, 56
Biological route, 81
Biological systems, 237
Biomarker, 186–187
Biomedical, 205–207, 209, 211–212, 215, 222
Biomedical applications, 194
Biomedicine, 260, 263, 266
Biomimetic, 16, 27
Bioreceptor, 177
Bioremediation, 87, 245, 246
Biosensing, 159
Biosensors, 11, 18, 163, 169, 171
Biostimulation, 246
Bioventing, 246
Bismuth nanoparticles, 91
Blood brain barrier (BBB), 207, 237
Blood circulation, 237
Bloodstream, 236
BM-MSCs, 126
Bottom up, 81
Bottom-up approach, 64–65
Brain Targeting, 58
Breast Cancer, 135
Brookfield viscosity, 69
Bruises, 116
Brunauer-Emmett-Teller analysis, 13
Buccal delivery, 147
Burkholderia pseudomallei, 21

Caco-2, 162
Cadexomer, 73
Cage proteins (CPs), 65
Calcium alginate, 163
Calorimetric, 216
Calotropis procera, 162
Cancer imaging, 195, 203
Cancer therapy, 56
Candida viswanathii, 250
Capping agents, 81
Caprolactone, 106
Carbohydrates, 31
Carbon, 123
Carbon nanotube, 215, 216, 220–222
Carbon nanotubes, 236, 260, 263
Carbonyl group, 68
Carcinogens, 180
Carcinoma, 135
Cardiotoxicity, 239
Cardiovascular, 34
Carotenoids, 104
Casein-based nanoparticles, 66, 163
Caseins, 104
Cassava starch nanocrystals (CSN), 67
Cationic liposomes, 47
Cell imaging, 159
Cell membrane damage, 238
Cell-penetrating peptides, 138
Cellular damage, 237
Cellular uptake, 237, 239
Ceramic nanoparticles, 3, 4
Characterization of liposomes, 54
Chemical nanosensors, 215
Chemical parameters, 54, 56
Chemical route, 81
Chemistry of NPs, 237
Chemokine, 239
Chemotherapeutics, 64
Chitosan, 4, 103, 159
Chitosan-based GNPs, 199
Chitosan based nanoparticles, 67
Chlamydomonas nivalis, 249
Chlorella, 101
Chlorella vulgaris, 32
Chlorinated ethanes, 246
Chlorinated ethenes, 246
Chloroform, 70
Chloromethane, 253

Chloromethanes, 246
Chlorophyta, 31
Cholesterol, 47
Cholesterol hemisuccinate (CHEMS), 47
Chordarilaes, 34
Chromium, 215, 219–222
Chromogenic, 21
Chromogenic receptor, DPC, 219, 220
Chromosomal aberrations, 239
Chronic pulmonary inflammation, 239
Chronic wounds, 115
Chrysophyta, 31
Ciprofloxacin, 72–73
Cisplatin, 59
Clean wound, 116
Clinical applications, 56
Close wounds, 116
Clostridium difficile, 163
Coarse-grained molecular dynamic simulation, 208
Coatings, 11, 12, 15, 21, 25
Cobalt nanoparticles, 136
Coccolith, 40
Coenzyme Q10, 105
Coenzymes, 100
Collagen, 118
Colloidal behaviour, 237
Colonized wound, 116
Comet assays, 261
Communicable disease, 160
Complementary DNA (cDNA), 208
Complementary metal oxide conductor, 180
Conducting polymer, 205
Contaminated wound, 116
Contamination, 245
Contraceptive, 34
Conventional liposomes, 47
Co-polymerization, 65
Copper, 30, 120
Copper nanoparticles, 135
Cosmetic applications, 58
Cosmetics, 100
Cost-effective, 32
Crush Injury, 116
Crystalline, 67–71
CTAB, 64
Cubic copper nanoparticles (CuNCs), 187
Curcumin, 105
Cutaneous, 116
Cyanobacteria, 252
Cyanobacterial, 33
Cyclometalated complexes, 137
Cytochalasin, 239
Cytochrome P450, 252
Cytogenetic assays, 241
Cytokine, 239
Cytotoxic, 34
Cytotoxicity, 108, 236, 238–239

D, L-lactic acid, 106
Dairy industry, 100
Daphnids, 237
Daunorubicin (DNR), 56, 59
Daunorubicin citrate, 59
Degradation, 64–66, 69–70
Deinococcus radiodurans, 250
Dendrimers, 81, 124, 138, 239
Denitrification, 248
Dermal penetration, 235, 237
Dermal toxicity, 239
Desmarestiales, 34
Detergent solubilization, 52
Detoxification, 237
Developmental and reproductive toxicology (DART), 7, 241
Diabetes, 115
Diabetes mellitus, 186
Diabetic foot ulcers, 115
Diadesmis gallica, 34
Dialysis, 5, 6, 52
Diatoms, 40
Dichlorobenzene, 253
Dictyotales, 34
Dimethyldioctadecyl ammonium bromide (DDAB), 47
Dioctadecyldimethyl ammonium chloride (DOGS), 47
Dioleoylphosphatidylethanolamine (DOPE), 47
Direct Laser Engraved Graphene (DLEG), 187
Distigmaproteus, 252
DNA adducts, 239
DNA damage, 124, 236, 239
DNA liposomes, 57
DNA methylation, 228
DNA methyltransferases, 228
DNA origami, 169, 171–172, 212
DNA repair, 123
DNA sensors, 21
DNA strand breaks, 239
Double emulsion method, 64
Doxorubicin, 65, 73
Doxorubicin (DXR), 59
DPPH assay, 32
Dressing materials, 117
Drug delivery, 3, 4, 6, 8, 80
Drug loading, 51
Drugs, 64–66
Dynamic light scattering, 6, 54
Dynamic light scattering (DLS), 120, 121

E. coli, 124
Electrical, 215, 216
Electrochemical, 216, 219, 221, 222
Electro-dropping system, 65
Electromagnetic, 215, 216
Electron acceptor, 246
Electron donor, 246
Electrospray method, 64
Elemental mercury, 217
Emulsification, 5, 64, 66, 102, 105
Emulsion stabilizer, 71
Encapsulation, 64, 66
Encapsulation efficiency, 4–6, 55
Endocytosis, 48, 236
Endothelial progenitor cell (EPC), 119
Energy-dispersive spectroscopy, 6, 7
Energy-dispersive x-ray analysis (EDAX), 120
Enhanced permeability and retention (EPR), 194, 202, 203
Enhanced permeation and retention effect (EPR), 48
Enhancement strategies, 144
Entrapping, 237
Enzymatic inhibitors, 144
Enzyme-linked immunosorbent assay, 178
Enzymes, 249
Epichlorohydrin, 73
Epigenetics, 227, 231
Epithelial inflammation, 236
Essential oils, 101
Ethanol Injection Method, 52
Etherification, 69
Ether Injection Method, 52
Ethyl carbamate, 21, 26
Euglenophyta, 31
Extracellular matrix, 123
Ex Vivo Imaging, 196

Fermentation, 248
Fertilizer, 245
FESEM, 120
Field–effect transistor (FET), 185
Field-flow fractionation (FFF), 54
Flow cytometric assay, 239
Flow cytometry, 181, 239
Fluorescein, 17, 19, 21, 22
Fluorescein isothiocyanate (FITC), 74
Fluorescence Resonance Energy Transfer (FRET), 186
Fluorescently labeled starch nanocrystal (FL-SNC), 74
Folate Receptors, 139
Foldscope, 169–170
Food and Drug Administration (FDA), 64–65, 115, 126
Food-based, 64–66
Food industry, 81
Food packaging, 64–65, 70–72
Fourier-transmission infrared spectroscopy, 6, 70
Free-flow electrophoresis (FFE), 54
Free radical polymerization (FRP), 70
Freeze-Thaw Extrusion Method, 52
French Pressure Cell Method, 52
FR-IR, 121
Fucoidan, 34
Fucus vesiculosus, 32, 33
Fullerene, 185, 239
Functional foods, 64
Fungi, 29, 84, 249, 250
Futuristic medicine, 64

Galactosylated -Chitosan -triptolide -nanoparticles (GC-TP-NPs), 67
Gastrointestinal, 34
Gel permeation chromatography, 52
Gene activation, 236
Gene delivery, 159
Gene expression, 82
Generally recognized as safe (GRAS), 65
Gene therapy, 56, 200, 202
Genetic engineering, 163
Genotoxic, 239
Genotoxicity, 239
Genotoxicity evaluation, 241
Gentamicin, 21
Germination, 82
Gloeophyllum striatum, 250
Glycine-arginine-glycine-aspartic acid-serine (GRGDS), 67
GM-CSF, 200, 201
Gold, 29, 159
Gold nanoparticles, 32, 33, 119, 134, 236
Gold nanoshell, 194, 195, 197, 198, 200
Graft-from, 70, 74
Graft-to, 70, 74
Granulocyte macrophage, 239
Graphene, 124, 168–169
Graphene chitosan, 124
Graphene Family Nanomaterials (GFNs), 206, 212
Green algae, 29
Green technology, 29

Halobacterium, 249
Heavy metals, 215–217, 219–223, 253
Hematomas, 116
Hemocompatibility, 4
Hemolysis, 237
Hepatocellular Carcinoma (HCC), 198, 199
Hepatotoxic, 238
Hepatotoxicity, 237
Hepatotoxins, 237
Hep-G2, 161, 198, 199
Histone, 229
Homeostasis, 117
Homeostasis phase, 117
Honey, 116
Hydrocarbons, 246
Hydrochloride drug–cyclodextrin complex, 53
Hydrogel, 163, 239
Hydrophilic, 5, 7, 8
Hydrophobic, 5, 7, 8, 64–65, 68, 70–71
Hydrophobicity, 66, 69
Hydroxy-ethyl starch, 73
Hyperbaric oxygen therapy, 117
Hyperosmolar coma, 186
Hypoglycemic, 34, 187

Ibuprofen, 53
ICP-OES, 120
Immobilization, 177–179, 181
Immune liposomes, 47, 48
Immunoassays, 181
Immunological adjuvants, 48
Immunosensors, 180
Immunostimulant, 171
Immunotherapy, 198–200
Immunotoxicity, 239
Incineration, 248
Incision, 116
Industrial production of liposomes, 52
Infected wound, 116
Inflammasome, 200
Inflammation, 117
Inflammation phase, 117
Ingestion, 236, 237
Inhalation, 236, 237
Inhalation toxicity, 239
Inorganic nanoparticles, 231
Insecticides, 81
Integrins, 139
Intercalation, 208
Interleukin (IL), 239
Interstitial fluid (ISF), 187
Intravascular/intravenous penetration, 236
Invasive injectable route, 146
In vitro assays, 240
In vitro drug release, 55
In vivo assays, 241
In vivo imaging, 195, 196, 200
Isoptericola chiayiensis, 250

Keratinocyte, 124
Keratinocyte chemoattractant, 239
Kirigami, 168, 171
Kretschman configuration, 218, 221

Label-based detection, 178
Label-free detection, 178, 180
Laceration, 116
Large unilamellar vesicles (LUV), 48
LDH assay, 240
LDL receptor, 139
Lead, 215–218, 220–222
Leakage analysis, 55
Lemanea fluviatilis, 32
Leptolyngbya spp., 252
Leukemia, 135
Ligand, 17, 239
Ligand coupling strategies, 53
Lipid nanoparticles, 138
Lipids, 31
LIPOPREP, 52
Liposomes, 46, 81, 239
Liquid chromatography, 6
Lithography, 80
Liver cancer, 135
Long-circulating liposomes (LCL), 47, 48
Low Density Lipoproteins (LDL), 139
Lung cancer, 135
Lung cancer cell line (A 549), 239
Lung interstitial, 239
Lymphocytes, 118
Lyngbya majuscule, 34
Lyophilization, 54

Macroalga, 34
Macrophage inflammatory protein (MIP), 239
Magnetism, 30
Magnetite nanoparticles, 91, 253
Maleimidophenyl butyrate-PE (MPB-PE), 53
Malnutrition, 82
Marine algae, 32
Matrix metalloproteinases (MMPs), 119
MCF-7, 135
MDA-MB-231, 135
Mechanical dispersion methods, 51
Mechanical nanosensor, 215, 216, 220
Medical bioremediation, 245
Melatonin-selenium nanoparticle, 162
Mercury, 215–218, 220–223
Mesoporous silica nanoparticles (MSNPs), 11, 13, 17–20, 22–27
Metabolic pathways, 252
Metabolic Pathways, 237
Metal arene complexes, 136
Metal-carbene complexes, 137
Metallic compounds, 31
Metallic nanoparticles, 134
Metallocenes, 136
Metalloids, 216
Metal nanoshells, 197
Metal toxicity, 40
Metal transport, 38
Methyl mercury, 217
Metronidazole, 53
Micelles, 80
Micro algae, 29, 30
Microarray, 181
Microcins, 72
Microelectro mechanical system (MEMS), 65
Microemulsion, 81
Microfluidic device, 169–170, 172
Microfluidics, 80
Microfluidic systems, 65
Microfluidization, 52
Microgrippers, 169–170
Micromachining, 178
Microneedle-based system, 65
Microorganisms, 249
Microparticles, 6, 249
Microscopy, 181
Microsurgery, 168–169
Microwave irradiation, 120
Mifepristone (MIF), 163
Mineralization, 248
Minerals, 246
Minicolumn centrifugation, 55
MiRNA, 230
Miscelles, 124
Mitochondrial damage, 238
Mitochondrial dysfunction, 238
Miura-ori, 169–170
Mizoroki–Heck cross-coupling reaction, 32
Molecular origami, 170
Monocyte chemoattractant Protein-1 (MCP), 239
MRNA, 82
Mucosal, 4
Multi-dose toxicity studies, 241
Multidrug- resistant (MDR), 119
Multilamellar vesicles (MLV), 48
Murine fibroblasts L929, 123
Murine fibrosarcoma cell line, 212
Mutation Assays, 241
Mutations, 239
Mycobacterium smegmatis, 18
Mycoremediation, 246
Myogenic, 88

NaCLO, Sodium hypochlorite, 70
Nanoagrochemicals, 82
Nanobelts, 180
Nanobiopharmaceuticals, 263
Nanobioremediation, 248, 249, 252
Nanobiotechnology, 3
Nano biotics, 124
Nanocapsules, 3, 4
Nano cargo, 17, 18
Nano-carriers, 4, 24, 27
Nano cellulose, 40
Nano coatings, 21, 26, 27
Nanocombs, 180
Nanocomposites, 15, 20, 21, 25, 27, 178
Nanocopper, 237
Nanocrystalline, 178–179
Nano-crystals, 239
Nanocubes, 33
Nanodevice, 29
Nanodots, 180
Nanodrug carriers, 263
Nanoelectromechanical systems, 178
Nanoemulsions, 80, 82, 239
Nanofertilizers, 82
Nanofibers, 117
Nanofluidics, 178
Nanogels, 124
Nanogold, 239
Nanomaterial, 80
Nanomaterials, 3, 8, 11, 17, 19, 25, 64–65, 69–71
Nanomaterial surface energy transfer (NSET), 219, 221
Nanomedicine, 235, 236
Nano-origami, 168, 172
Nanoparticles, 3–8, 11, 27, 29
Nanopesticides, 82
Nanopharmaceuticals, 263
Nanoplates, 33
Nanoprecipitation, 5, 64, 66
Nanorenovogen, 256
Nanoribbon, 180
Nanorobots, 168–169, 171–172
Nanorods, 33
Nano scaffolds, 126
Nanoscopic, 215
Nanosheets, 180
Nanosilica, 13, 18, 21–27

Nanosilver, 237–239
Nanospheres, 3, 4
Nanostructured lipid carriers, 150
Nanotechnology, 29, 259
Nanotoxicity, 235, 236, 260, 262, 264, 265
Nanotoxicology, 260
Nanotubes, 171, 178–181
Nanowires, 178, 180–181
Nanowire sensors, 218
Nasal, 4
Nasal delivery, 147
Navicula atomus, 34
Near-Infrared (NIR), 211, 212
Necrosis, 238, 240
Negative-pressure wound therapy, 117
Nephrotoxicants, 238
Nephrotoxicity, 238
Neurotoxic, 217
Nickel, 236
Nisin, 58
Non-cutaneous, 116
Non-invasive routes, 146
Norcantharidin, 67
N-pyridyldithiopropionyl-PE (PDP-PE), 53
Nuclear magnetic resonance, 6
Nuclear magnetic resonance (NMR), 54
Nuclease, 208–209, 211
Nutraceuticals, 30
Nystatin, 59

Obesity, 115
Ocular applications, 58
Ocular delivery, 149
OECD, 260
Oil-in-water, 5
Oil spills, 254
Open wounds, 116
Opsonization, 237
Optical fibers, 180
Optical transducer, 187
Opto-chemical sensors, 219
Opto-electronic module, 218
Oral, 4, 6
Oral delivery, 146
Organic nanoparticles, 230
Organometallic Complexes, 136
Origami, 167–172
Origamizer, 172
Oscillatoriales spp., 252
Osmudaria obtusiloba, 34
Oxidation, 236
Oxidation-reduction (redox), 246
Oxidative stress, 236, 238, 239
Oxidative stress Assay, 240

P53, 163
Paclitaxel, 65, 66
Padina gymnospora, 33
Papilloma virus, 201
Parenteral, 4, 6
Passive drug delivery, 48
Payload, 206–207, 211–212
Pediocin, 72
Pediocin AcH, 58
$PEGME_{550}$, 68
PEGylated, 194, 196, 197, 199
Peptides, 32
Perchloroethylene (PCE), 252
Pesticides, 248
Petroleum, 247
pH, 246
Phaeophyta, 31
Phagocytosis, 206–207
Pharmaceuticals, 69
Pharmacodynamic (PD), 261
Pharmacokinetic (PK), 264
Phenolic, 31
Phenols, 247
Phormidium valderianum, 34
Phosphatidyl choline, lecithin (PC), 47
Phosphatidyl ethanolamine, cephalin (PE), 47
Phosphatidyl glycerol (PG), 47
Phosphatidyl inositol (PI), 47
Phosphatidyl serine (PS), 47
Phospholipids, 46
Photodegradation, 64, 71
Photodynamic therapy, 18, 19, 27, 56
Photo-electrochemical systems, 179
Photolithographic, 169–170
Photosensitizer, 212
Photosynthetic activity, 82
Photothermal therapy, 18–20, 27, 194
PH-sensitive liposomes, 47
Physical parameters, 54
Physical route, 81
Physico–chemical, 205–207, 212
Physicochemical properties, 236
Phytoremediation, 246
Pickering emulsions, 71
Piezoelectric sensors, 217
Pigments, 32
Planktothrix spp., 252
Plant extracts, 116
Plant growth promotion, 21, 22
Plasma protein binding, 237
Platelet aggregation, 237
Platelet derived growth factor, 117
Platinum, 29
Platinum nanoparticles, 91
Point-of-care diagnostics, 178
Polarity, 69–70
Pollutant, 245
Poly (e-caprolactone) (PCL), 65–66, 68, 70
Poly (lactide co-glycolic acid), 4
Poly (tetrahydrofuran) (PTHF), 70
Polyaromatic hydrocarbons (PAHs), 246, 255
Poly-beta amino esters (PBAEs), 65
Polychlorinated biphenyls (PCBs), 252
Poly(ethylene glycol) methyl ether (PEGME), 68, 70
Polyglycolic acid, 4
Poly glycolic acid (PGA), 65
Polylactic acid (PLA), 4, 65, 124
Poly lactic-co-glycolic acid (PLGA), 65, 124
Polymeric conjugates, 81
Polymeric core, 64
Polymeric nanoparticles, 3–6, 8, 81
Polymerization, 64–65, 69–70
Poly (ethylene glycol) methyl ether, 70
Polymyxin, 72
(Poly propylene glycol) mono butyl ether (PPGBE), 70
Polysaccharide based nanoparticles, 66
Polysaccharides, 32, 65
Polystyrene, 70
Polyvinyl alcohol, 81
Polyvinyl pyrrolidone, 81
Poly(ε-caprolactone), 119
Porosity, 13
$PPGBE_{4000}$, 68
Pressure ulcers, 115
Proliferating cell nuclear antigen (PCNA), 228
Proliferation, 117
Proliferative phase, 117
Propolis, 116
Protein, 65–66, 68–69, 71
Protein assay, 240
Protein-based nanoparticles, 65
Proteins, 31
Pseudoalteromonas agarivorans, 250
Pseudomonas, 255
Pseudomonas fluorescens, 21, 23
Pseudomonas spp., 250
$PTHF_{1400}$, 68
Pulmonary applications, 58
Pulmonary delivery, 148
Pulse laser, 194
Pyrolysis, 12, 14, 24, 26
Pyrrophyta, 31

QAC-chitosan, 163
Quantum dot, 21, 80, 124, 219, 263
Quencher, 219

Raman, 120
Raman reporters, 195, 196
Raman scattering, 33
Reactive oxygen species (ROS), 123, 206, 236, 239, 260, 261, 265
Reducing, 29
Release kinetics, 206–207
Remodeling, 117
Remodeling Phase, 117
Resins, 255
Reticulo endothelial System, 237
Retinopathy, 187
Reverse-phase evaporation method, 52
Rheological properties, 69
Rhizocolonium fontinales, 34
Rhizofiltration, 246
Rhodamine, 15, 20–22, 26
Rhodococcus soli, 250
Rhodophyta, 31
Rifampin, 57
RNAi therapy, 124

Safety detectors, 64
Salidroside, 53
Salmonella gallinarum, 21
Salting out, 5, 64
Sargassum muticum, 34
Sargassum wightii, 32, 33
Scanning electron microscopy, 6
Sea weed, 33
Sediment, 246
Seed germination, 21, 26
Selenium, 29, 159, 161
Semiconductor, 179–180
Senescence, 163
Silane, 13, 14, 21
Silica, 236, 239
Silicon, 123
Silicon nanoparticles, 91
Siloxane intermediates, 12
Silver, 29, 159
Silver nanoparticles, 30, 260, 261
SiRNA therapy, 124
Size Analysis, 54
Size-exclusion chromatography (SEC), 54
Sizing of Liposomes, 52
Small unilamellar vesicles (SUV), 48
Smart drug delivery, 64
SNCs platelets, 68

Soil, 251
Soil nanoparticles, 254
Sol-gel, 12, 16, 25, 27, 81
Solid lipids, 239
Solvent dispersion methods, 52
Solvent evaporation, 5, 64
Sonication, 51
Sono chemical, 81
Sp^2 hybridization, 185
Spectroscopy, 187
Spirulina subsalsa, 34
Spray drying, 66
Squamous carcinoma, 194
Stability of Liposomes, 55
Staphylococcus aureus, 21
Starch nanocrystals (SNCs), 67–74
Starch nanocrystals loaded with streptomycin (SSNCs), 72–73
Stealth liposomes, 48
Stem cell, 119
Stent grafts, 169–170, 172
Steric hindrance, 208
Strain, 206
Strauroneis sp., 30
Streptomycin, 72
Strontianite nanoparticles, 92
Subcellular imaging, 196
((H-succinimidyl-4-)- p-maleimidophenyl) butyrate (SMPB), 53
Supercapacitor, 187, 205
Superconducting quantum interference devices, 179
Supercritical fluid, 6
Supercritical reverse-phase evaporation (SCRPE), 52
Super-paramagnetic, 179
Surface area, 236
Surface charge, 50, 237
Surface chemistry, 206
Surface-enhanced Raman scattering, 195
Surface hydration, 50
Surface plasmon resonance, 33
Surface-to-volume, 3, 6, 8
Surface to volume ratio, 64, 67
Surfactant, 207
Surfactant layer, 64
Surgical incision management system (SIMS), 117
Swelling study, 68
Synergistic antimicrobial activity, 72

Targeted drug delivery, 206, 211
Target-specific, 64
Terbium, 121
Tetra selmiskochinensts, 33
Theranostics, 11, 17, 188
Therapeutic applications, 58
Thermoplastic, 65
Thin-film hydration method, 52
Thrombin, 126
Tissue engineering, 72–74
Titanate nanoflowers, 253
Titanium, 120
Titanium nanoparticles, 91
Tolypothrix spp., 252
Top-down approach, 64
Toxicokinetics, 238
Traditional food, 31
Transdermal, 4, 6
Transducer, 177
Transesterification, 39
Transferrin, 138
Transferrin receptors, 138
Transforming growth factor-β, 117, 119
Transistor, 205
Transmembrane permeability, 237
Transmission electron microscopy, 6
Transmission Electron Microscopy (TEM), 212
Trentipholia, 252
Tridecanoic acid, 188
Tween, 64

Ultrarelativistic particles, 185
Ulva fasciata, 34, 38
Unilamellar vesicles (UV), 48
UV spectroscopy, 6

Vaccine adjuvant, 57
Vaccines, 64–66
Vascular surgery, 117
Vasoconstriction, 117
Venous leg ulcers (VLU), 117
Venous ulcers, 115
Verapamil, 53
Vesicle shape, 54
Victimization, 237
Vinca rosea, 116
Vincristine (VCR), 56, 59
Vincristine sulfate, 59
Vinorelbine, 59
Virus-like particles (VLPs), 65
Virus-mediated gene transfer, 57
Viscosity, 69, 73
Vitamins, 246

Waste degradation, 40
Whey protein nanoparticles, 66
World Health Organization (WHO), 82
Wound healing, 72–74, 115

Xanthomonas campestris, 21
Xenobiotics, 237, 238, 249, 251
X-ray diffraction, 6, 7, 67, 68, 120, 208
X-ray photoelectron spectroscopy, 6, 7
X-Ray Photoelectron Spectroscopy (XPS), 70

Yeast, 29

Zea mays, 21, 23, 26, 27
Zebrafish, 237, 239
Zeta potential, 6, 7, 55
Zinc, 123
Zirconium nanoparticles, 91
ZnO, 123
ZnONPs, 123, 136, 162

Taylor & Francis Group
an informa business
Taylor & Francis eBooks
www.taylorfrancis.com
A single destination for eBooks from Taylor & Francis with increased functionality and an improved user experience to meet the needs of our customers.
90,000+ eBooks of award-winning academic content in Humanities, Social Science, Science, Technology, Engineering, and Medical written by a global network of editors and authors.
TAYLOR & FRANCIS EBOOKS OFFERS:
A streamlined experience for our library customers
A single point of discovery for all of our eBook content
Improved search and discovery of content at both book and chapter level
REQUEST A FREE TRIAL
support@taylorfrancis.com
Routledge
Taylor & Francis Group
CRC Press
Taylor & Francis Group